Handbook of Research on Innovation, Differentiation, and New Technologies in Tourism, Hotels, and Food Service

Gonçalo Poeta Fernandes
CITUR, Polytechnic Institute of Guarda, Portugal & CICS, Universidade NOVA de Lisboa, Portugal

António Silva Melo
CiTUR, Polytechnic Institute of Porto, Portugal & CIDTFF, University of Aveiro, Portugal

A volume in the Advances in Hospitality, Tourism, and the Services Industry (AHTSI) Book Series

Published in the United States of America by
    IGI Global
    Business Science Reference (an imprint of IGI Global)
    701 E. Chocolate Avenue
    Hershey PA, USA 17033
    Tel: 717-533-8845
    Fax: 717-533-8661
    E-mail: cust@igi-global.com
    Web site: http://www.igi-global.com

Library of Congress Cataloging-in-Publication Data

Names: Fernandes, Gonçalo, 1969- editor. | Melo, António, 1967- editor.
Title: Handbook of research on innovation, differentiation, and new
  technologies in tourism, hotels, and food service / edited by Gonçalo
  Fernandes, António Melo.
Description: Hershey, PA : Business Science Reference, 2023. | Includes
  bibliographical references and index. | Summary: "Expand knowledge about
  tourism, hospitality and food service activities; Systematize knowledge
  about innovation in tourism, hospitality and food service; Know
  processes and digitalization media in tourism and hospitality;
  Disseminate products, services and differentiated tourist practices;
  Promote good practices and strategies in the operationalization of
  tourism and hospitality; Announce models of governance and innovation in
  tourism, hospitality and food service; Provide knowledge about
  management and planning models in tourism and hospitality sector;
  Encourage the transfer of knowledge between the research done and the
  professions in tourism sector; Stimulate entrepreneurship activities in
  tourism sector; Value heritage and cultural identities as assets for
  tourism and hospitality; Build scientific resources to evaluate Tourism,
  Hospitality sector and its trends; Promote applied research in tourism,
  hospitality and food service"-- Provided by publisher.
Identifiers: LCCN 2023029579 (print) | LCCN 2023029580 (ebook) | ISBN
  9781668469859 (hardcover) | ISBN 9781668469873 (ebook)
Subjects: LCSH: Tourism--Technological innovations. | Hospitality
  industry--Technological innovations. | Food service--Technological
  innovations.
Classification: LCC G156.5.I5 H355 2023 (print) | LCC G156.5.I5 (ebook) |
  DDC 338.4/791--dc23/eng/20230816
LC record available at https://lccn.loc.gov/2023029579
LC ebook record available at https://lccn.loc.gov/2023029580

This book is published in the IGI Global book series Advances in Hospitality, Tourism, and the Services Industry (AHTSI) (ISSN: 2475-6547; eISSN: 2475-6555)

British Cataloguing in Publication Data
A Cataloguing in Publication record for this book is available from the British Library.

All work contributed to this book is new, previously-unpublished material. The views expressed in this book are those of the authors, but not necessarily of the publisher.

For electronic access to this publication, please contact: eresources@igi-global.com.

Advances in Hospitality, Tourism, and the Services Industry (AHTSI) Book Series

Maximiliano Korstanje
University of Palermo, Argentina

ISSN:2475-6547
EISSN:2475-6555

MISSION

Globally, the hospitality, travel, tourism, and services industries generate a significant percentage of revenue and represent a large portion of the business world. Even in tough economic times, these industries thrive as individuals continue to spend on leisure and recreation activities as well as services.

The Advances in Hospitality, Tourism, and the Services Industry (AHTSI) book series offers diverse publications relating to the management, promotion, and profitability of the leisure, recreation, and services industries. Highlighting current research pertaining to various topics within the realm of hospitality, travel, tourism, and services management, the titles found within the AHTSI book series are pertinent to the research and professional needs of managers, business practitioners, researchers, and upper-level students studying in the field.

COVERAGE

- Hotel Management
- Customer Service Issues
- Cruise Marketing and Sales
- Food and Beverage Management
- Travel Agency Management
- Health and Wellness Tourism
- Service Management
- Sustainable Tourism
- Service Training
- International Tourism

IGI Global is currently accepting manuscripts for publication within this series. To submit a proposal for a volume in this series, please contact our Acquisition Editors at Acquisitions@igi-global.com or visit: http://www.igi-global.com/publish/.

Titles in this Series

For a list of additional titles in this series, please visit: www.igi-global.com/book-series

Prospects and Challenges of Global Pilgrimage Tourism and Hospitality

S.K. Gupta (Centre for Mountain Tourism & Hospitality Management, HNB Garhwal University, India) Lilibeth C. Aragon (College of International Tourism and Hospitality Management, Lyceum of the Philippines University, Philippines) Pankaj Kumar (Department of Tourism & Hospitality Management, Mizoram University, India) Madhurima (Department of Environmental Science, Mizoram University, India) and Rajesh Ramasamy (Department of Tourism & Hospitality Management, Mizoram University, ndia)
Business Science Reference • © 2023 • 310pp • H/C (ISBN: 9781668448175) • US $250.00

Cases on Traveler Preferences, Attitudes, and Behaviors Impact in the Hospitality Industry

Giuseppe Catenazzo (AUS American Institute of Applied Sciences, Switzerland)
Business Science Reference • © 2023 • 335pp • H/C (ISBN: 9781668469194) • US $215.00

Exploring Niche Tourism Business Models, Marketing, and Consumer Experience

Maria Antónia Rodrigues (CEOS.PP, ISCAP, Polytechnic of Porto, Portugal) and Maria Amélia Carvalho (ISCAP, Polytechnic of Porto, Portugal)
Business Science Reference • © 2023 • 300pp • H/C (ISBN: 9781668472422) • US $250.00

Women's Empowerment Within the Tourism Industry

Gül Erkol Bayram (Sinop University, Turkey) Syed Haider Ali Shah (Bahria University, Pakistan) and Muhammad Nawaz Tunio (Mohammad Ali Jinnah University, Karachi, Pakistan)
Business Science Reference • © 2023 • 345pp • H/C (ISBN: 9781668484173) • US $235.00

Global Perspectives on Human Rights and the Impact of Tourism Consumption in the 21st Century

Maximiliano E Korstanje (University of Palermo, Argentina) and Vanessa G.B. Gowreesunkar (Anant National University, India)
Business Science Reference • © 2023 • 263pp • H/C (ISBN: 9781668487266) • US $240.00

Dark Gastronomy in Times of Tribulation

Demet Genceli (Istanbul Kent University, Turkey)
Business Science Reference • © 2023 • 283pp • H/C (ISBN: 9781668465059) • US $205.00

Inclusive Community Development Through Tourism and Hospitality Practices

Vipin Nadda (University of Sunderland in London, UK) Faithfull Gonzo (University of Sunderland in London, UK) Ravinder Batta (Shimla University, India) and Amit Sharma (Shimla University, India)
Business Science Reference • © 2023 • 361pp • H/C (ISBN: 9781668467961) • US $215.00

IGI Global
PUBLISHER of TIMELY KNOWLEDGE

701 East Chocolate Avenue, Hershey, PA 17033, USA
Tel: 717-533-8845 x100 • Fax: 717-533-8661
E-Mail: cust@igi-global.com • www.igi-global.com

List of Contributors

Table of Contents

Detailed Table of Contents

Section 1
Robotization, Digitalization, and New Technologies in Tourism and Hospitality

Züleyhan Baran, Akcakoca Vocational School, Duzce University, Turkey
Şükran Karaca, Tourism Faculty, Sivas Cumhuriyet University, Turkey
Huseyin Baran, Art Design and Architecture Faculty, Duzce University, Turkey

New-gen technologies have profoundly impacted all aspects of life and various economic sectors. The tourism industry, known for its inclination towards innovation, has been quick to embrace technological advancements. In response to the global pandemic, tourism businesses such as hotels, food services, and transportation have increasingly utilized robotic systems to ensure social distancing, hygiene, and sanitation measures. However, digitization presents significant challenges for the tourism industry, requiring companies to adapt their operations to stay competitive. Automation has emerged as a highly beneficial trend, simplifying tasks and introducing innovative processes to tourism business models. This enables companies to provide personalized services tailored to the preferences of "digital tourists." Overall, new-gen technologies are reshaping the tourism industry and driving it toward enhanced efficiency and customer satisfaction.

Ahmed Chemseddine Bouarar, University of Medea, Algeria
Karolina Stojczew, Wrocław University of Economics and Business, Poland
Kamel Mouloudj, University of Medea, Algeria

COVID-19 has exerted a huge impact on the hospitality; the digital transformation was one of the most prompt and the most solutions that guarantee businesses' survival in the light of lockdowns and mobility restrictions imposed in a bid to limit the epidemic spread. Hence, this chapter sought to explore the impact of COVID-19 on the hospitality industry and hospitality industry's digital transformation in Poland, as well as to identify the obstacles that may stymie the shift towards digitalization in this industry. To this end, a literature review methodology was employed. The results revealed that COVID-19 induced a negative impact on the hospitality industry in Poland in 2020, but with the beginning of 2021 this industry began to recover gradually. Also, COVID-19 quickened digital transformation adoption. The results also indicated that a sound and effective digital transformation reduces costs and improves business' performance; however, digital transformation may encounter several hurdles such as lack of knowledge, poor digital infrastructures, and lack of digital experience.

Chapter 3

Sameera Khan, Vardhaman College of Engineering, Hyderabad, India
Dileep Kumar Singh, Narsee Monjee Institute of Management Studies, Hyderabad, India

A new wave of automation known as robotic process automation is revolutionizing company productivity and delivering excellent ROI. This book chapter examines the use of Robotic Process Automation (RPA) in the hospitality, travel, and food service industries. It focuses on how RPA handles routine activities, enhances operational effectiveness, and improves customer experiences. RPA streamlines operations and enables personalized client interactions in the travel and tourism sector. It automates front desk operations, inventory management, and back-office duties in hotels, optimizing resource allocation and enhancing visitor experiences. In the food service sector, RPA automates order processing, inventory management, and supply chain management, improving operational effectiveness and enabling personalized ordering experiences. The chapter covers popular RPA solutions, analyzes implementation difficulties, and emphasizes the future potential of RPA, including integration with cutting-edge technology. By automating mundane tasks and providing a basic framework for implementation, RPA offers significant advantages in industries like aviation, hotels, finance, tourism, and more. This chapter highlights the significance of RPA in enhancing productivity, reducing errors, and shaping the future of these industries.

Chapter 4

Hande Mutlu Ozturk, Pamukkale University, Turkey
Ozgur Guler, Pamukkale University, Turkey
Olcay Polat, Pamukkale University, Turkey

The main focus of this chapter is to examine the tourism income of Türkiye as a case country, taking into account the structure of the tourism industry and relevant economic and social indicators. Statistical methods are used to investigate the factors that influence tourism income and to demonstrate the impact of these variables. The chapter aims to identify the key factors that should be considered when planning tourism-related activities and to explore the suitability of different models for future predictions. In addition, the chapter explores the use of machine learning models, such as artificial neural networks (ANN) and gradient boosted regression trees (GBRT), to compare their performance with the established multiple linear regression model. Furthermore, the chapter adds to the existing literature on tourism economics and forecasting methods by examining the performance of different models in predicting tourism income and highlighting the importance of factors such as the country's image, safety, and transportation opportunities in shaping tourism income in Türkiye.

Chapter 5

Predrag Miroslav Vuković, Institute of Economic Sciences, Belgrade, Serbia
Marija Mosurović Ružičić, Institute of Economic Sciences, Belgrade, Serbia

A characteristic of rural tourism is the physical distance between supply, which is located in rural areas, and demand, which is located in cities. The role of tourist intermediaries is to connect supply and demand. Tourism distribution channels are seen as a "bridge" between supply and demand and the role of intermediaries is precisely to find ways to facilitate the exchange process. The lack of ICT application can lead to a decrease in demand, a decrease in economic activities, and thus can negatively affect the development of certain areas. The authors will specify the necessary steps which will enable the improvement of business activities and the promotion of rural tourist destinations, both at the micro level, and at the macro level. The expectation is that this approach will increase the income and stop the negative trends burdening the life in rural areas (depopulation, population migration to urban centres, the decline in macroeconomic indicators, etc.) that are present not only in Serbia but also in other countries.

Section 2
Innovation, Experiences, and Service Quality in the Hotel and Food Service Industry

Chapter 6

Nilgün Demirel III, Tourism Faculty, Igdir University, Turkey
Onur Çelen, Harmancık Vocational School, Bursa Uludag University, Turkey

The aim of this chapter is to reveal technology paradox, and technostress concepts' effects in tourism industry. In the research the impact of the technology on the tourism industry employees was determined through qualitative research steps. Descriptive results were obtained using phenomenological research design. 20 participants were interviewed face-to-face and videoconferenced, and the audio and video files obtained were deciphered. Three main themes and seven sub-themes were identified, and the data were classified by content analysis. Analyses performed with percentage values, total and frequency graphs, and descriptive results. It is emphasized that the negative effects of technology on employees are greater. It is a fact that technology creates stress. No matter how competent one is in the use of technology, the unhindered development of technology has a negative impact on employees in the tourism industry, as it is thought to lead to a decrease in the labor force and unemployment.

Chapter 7

Jeganathan Gomathi Sankar, BSSS Institute of Advanced Studies, India
Arokiaraj David, St. Francis Institute of Management and Research-PGDM, India

The tourism industry is rapidly adopting artificial intelligence (AI) to enhance customer experiences and improve service delivery. However, the adoption of AI has raised concerns about concierge chatbots, digital assistance, proactiveness, anthropomorphism, and security, and its impact on overall customer satisfaction. Therefore, the aim of this research is to measure the service quality of AI in the tourism industry, with primary data collected in Pondicherry. This study is a quantitative research study that utilized a survey method to collect primary data. A total of 350 respondents were targeted, with 307 valid responses obtained. The data were analyzed using confirmatory factor analysis and structural equation modeling. The study highlights that AI technology has a significant positive impact on the service quality of the tourism industry. This study contributes to the literature by providing empirical evidence to service quality in the context of AI technology and the importance of AI technology to enhance service quality and customer satisfaction.

Chapter 8

Mihály László Vörös, Edutus University, Hungary & HELIA Research Group, Hungary
Aleš Gačnik, Faculty of Tourism Studies, University of Primorska, Slovenia

The research study comprised by the book chapter investigates territorial identities and diversified features of the South-Danube region and presents a gastronomy tourism case study of an old culinary tradition of brewing and eating fish soup. This is not only a dish consumed with great frequency in the diet of local and regional residents but creates food offering and cooking demonstration of a summer gastronomy tourism festival organized every year in the town of Baja. In addition, this fish-dish is one of the most popular meals offered in almost every local and neighborhood restaurant's menu for tourist guests. The study covers short analysis on the healthiness of this dietary custom and highlights that this culinary feast and gastronomy tourism attraction also became a brand and cultural heritage which can contribute to enhance the image of the place and to promote sustainable development of gastronomy tourism.

Chapter 9

Alexandra Rodrigues Gonçalves, Universidade do Algarve, Portugal
Célia M. Q. Ramos, Universidade do Algarve, Portugal
Carina Viegas, Universidade do Algarve, Portugal

HoST Lab is an integrative approach that aims to innovate based on Mediterranean diet (MD) creating new products, services, and experiences, involving producers and distributors. The lab research also aims to know the emotions and sensations associated with eno-gastronomic experiences of the Mediterranean diet and their welcome among visitors-tourists, using a triangulation of traditional methods (interviews, focus groups, workshops), but also developing a digital solution for sharing results (webpage, digital survey, sentiment analysis). A set of reference indicators and a nutrition economic label will be developed and used to evaluate and monitor research results, both in a laboratory and in a real environment. The HoST Lab pretends to be a sharing and learning research space between the academy, the tourism, and the hospitality sector, in which proposals are tested, results transferred to the community, and well-being promoted among the local population and visitors, aspiring for a growing sustainable destination.

Chapter 10

Nesenur Altinigne, Istanbul Bilgi University, Turkey

This literature review aims to investigate the developments of virtual reality (VR) and augmented reality (AR) research in tourism marketing. This chapter also highlights fruitful directions for tourism marketing research regarding VR and AR applications. A total of 31 full-length articles published between 2010 to 2023 were retrieved from the Web of Science database and reviewed. The theoretical backgrounds of the articles were thoroughly examined, and a detailed report on the research progress of the theories and research methodologies are presented. Finally, future research directions for the improvement of the existing literature are explained.

Section 3
Development Dynamics, Governance, and Promotion of Tourism and Hospitality

Chapter 11

Albérico Travassos Rosário, GOVCOPP, IADE, Universidade Europeia, Portugal

The increasing wealth and economic development have dramatically driven rapid growth in the tourism industry, making it one of the fastest growing and developing industries worldwide. Tourism entrepreneurship should involve adopting a green business model innovation that is more efficient and leads to low carbon production to balance the needs of the tourists with those of their desired destinations. Despite sustainability being a core concept in current policies and trends in the last decade, most company managers in the tourism industry are yet to incorporate it into the agenda. Based on this research gap, a systematic review of the bibliometric literature was conducted, and data was synthesized from 80 documents identified through the Scopus indexation using. This chapter aims to evaluate the challenges and opportunities of innovation sustainability in tourism entrepreneurship, thus building a clear image of what should be done to overcome the obstacles and increase awareness of the need for sustainable tourism.

Chapter 12

Gorete Dinis, GOVCOPP, CITUR, Polytechnic Institute of Portalegre, Portugal
Zélia Breda, DEGEIT, GOVCOPP, University of Aveiro, Portugal

Few studies have analysed the accessibility level of information sources used by persons with disabilities when making tourism-related purchases. Consequently, the main objective of this chapter is to gain insights into whether Destination Management Organisations are actively developing inclusive destination promotion and advertising materials. To accomplish this, an exploratory study was conducted, with a specific focus on the tourism of Portugal. Portugal was chosen as the subject of the study due to its recognition by the World Tourism Organisation as the world's first accessible tourism destination. This recognition encompasses various aspects, including the official promotional tourism website (visitportugal. com), which features digital brochures. It is important to note that the exploratory nature of this study limited the ability to make direct comparisons with previous research. However, for future investigations, it is recommended that the framework employed in this study be applied to assess the accessibility of promotional materials from other DMOs and tourism stakeholders.

Chapter 13

Yassine El Bouchikhi, Al Akhawayn University, Ifrane, Morocco

With a focus on Morocco, this chapter reviews the literature on nation branding and its connection to social media in the context of tourism. The first part gives a foundation for understanding the many tactics used to distinguish one country from another by reviewing the literature on nation branding, social media, and tourism. The use of social media to market travel and to promote cultural and natural assets is examined in the second section. Then, an examination of the significant developments and trends in the Moroccan tourism industry are addressed. A Netnographic study is conducted on six influential media accounts to explore the strategies and tactics utilized to promote Morocco's image internationally.

Today, electronic word-of-mouth (eWOM) has a substantial impact on consumers' decisions to purchase tourism and travel services. Therefore, it is essential to have a comprehensive understanding of eWOM in tourism and hospitality research. The chapter aims to conduct a comprehensive examination of the existing body of knowledge pertaining to eWOM communication within the domain of tourism and hospitality. To this end, bibliometric data was obtained from the Web of Science database, utilizing the Bibliometrix R package. The PRISMA flowchart was employed to ensure a systematic examination, which included a total of 515 scholarly documents published between 2008 and 2021. The utilization of a word cloud facilitated the identification of the most prevalent terms within the field, while a thematic map was employed to reveal the themes that guide the existing body of knowledge. Furthermore, co-occurrence analysis was utilized to discern four distinct research themes that have emerged as particularly prominent within the field.

The study highlights the importance of accessible tourism, presents the actors of accessible tourism, including people with reduced mobility, and discusses their opportunities in tourism. The aim of the authors is to describe the ACT model and demonstrate its use in tourism. Creating an accessible environment provides comfort not only for people with reduced mobility but for almost everyone, so it has greater potential even from an economic point of view. The ACT model was born from the information obtained during the research in Hungary and summarizes and illustrates well what actors are involved in the topic of accessible tourism and what kind of relationship exists between them. Each actor and factor of the three levels of the theoretical model are important in order to ensure that accessibility is achieved, and that disabled people can participate in tourism in the same way as non-disabled people.

Preface

Welcome to the *Handbook of Research on Innovation, Differentiation, and New Technologies in Tourism, Hotels, and Food Service*. As editors, we are excited to present this comprehensive reference book, which delves into the intricate world of tourism and hospitality and explores the profound impact of innovation, differentiation, and new technologies within these industries.

Tourism and hospitality are assuming, in a growing and challenging way, complexity in the forms of their organization, in the means involved for their implementation, in the experiences required, in the associated technological support and in the links with the territories and their communities. Tourism, in its territorial relation and due to the transformations operated, promotes differentiated practices and consumption logics that require permanent monitoring, creativity and adaptation. On account of this an increasing number of services are associated and new functionalities are generated for its enjoyment.

The forms of organization in tourism, hospitality, and food service tend to be cherished more and more, and strategically planned, empowering organizations and territories with means and positioning that promote their competitiveness and qualification. The planning in tourism, hospitality and its sub-sectors requires a joint involvement of the different stakeholders, with emphasis on governance entities, able to create conditions for sustainability and qualification of destinations, providing the valuation of tourism assets, the creation of differentiated products or services, and their sustainability. The challenges and changes demand constant adaptation/readaptation, which promote instability and permanent updates in the means and positions, given the behavior of society, the economy demands, and social and environmental responsibility.

For the different tourism and hospitality players, it is crucial to be competent in their capacity to deal with permanent changes and reconfigurations, both in the markets and in the ways services are operated, promoting their competitiveness and forms of organization and cooperation that add value to their offer, their differentiation and, consequently, the valuation of their products and services.

Before changing contexts and the emergence of new demands, tourism players seek to broaden experiences and increase technological/digital supports that contribute to access a greater quantity and quality of data, supports, and resources that help make experiences safer, generate more accessible information, and enable a greater capacity to understand changes in the behavior of the demand and the experience promoted and desired. At the same time, tourism players per se and in corporate terms feel compelled to integrate, into their practices and strategies, greater concerns and investments concerning the triad of sustainability, namely, the environmental, economic and social aspects. It is crucial to promote the social responsibility of organizations in order to improve their processes continuously and obtain results that benefit all stakeholders, enhance human resources, generate greater concern for the communities and the environment where they are and with whom they interact. This involvement is decisive in the appreciation of the touristic destination, in the understanding of local ecocultural realities and in the development of the preservation process and creation of services with identity, socially responsible, contributing to create differentiation in a context of globalization and sustainability. In this context, differentiation, specialization, digitalization and dematerialization emerge as critical factors for the enhancement of the tourism offer of destinations, the resource mobilization, the effectiveness of organizational processes and the acceleration of entrepreneurial initiatives.

In light of these challenges and changes, this handbook aims to explore the current landscape of tourism and hospitality, sparking thoughtful reflection and fostering the sharing of innovative approaches to developing tourism products, destinations, and hospitality practices. It also emphasizes the significance of technology and competitiveness in the realms of tourism, hospitality, and food service.

The overarching objectives of this book are manifold: expand knowledge about tourism, hospitality, and food service activities; systematize knowledge on innovation within these sectors; understand the processes and mediums of digitalization in tourism and hospitality; promote products, services, and differentiated practices in tourism; showcase best practices and strategies for operationalizing tourism and hospitality; introduce models of governance and innovation specific to tourism, hospitality, and food service; provide insights into management and planning models within the tourism and hospitality sector; facilitate the transfer of knowledge between research and the professional realms of the tourism sector; stimulate entrepreneurship activities in tourism; value heritage and cultural identities as invaluable assets for tourism and hospitality; establish scientific resources for evaluating the tourism and hospitality sectors and their trends; and, lastly, promote applied research within the domains of tourism, hospitality, and food service.

The intended audience for this handbook is diverse and includes professors, academics, students, and researchers in higher education institutions; tourism management organizations at national and local levels; tour operators; hotel managers; food service managers; event organizers; administrative entities; destination management organizations; local development associations; tourism technicians; trainers specializing in tourism, hospitality, and food service management; professionals representing various subsectors of tourism; professional associations; entrepreneurs; and planners and managers of tourism destinations.

The book covers a broad spectrum of topics and binomials, encompassing areas such as tourism and local development, territorial branding and differentiated tourism products, tourism planning and local development policies, information technologies in tourism, hospitality and hotel management, food service management, local governance and public policies for tourism, territorial innovation in tourism and hospitality, entrepreneurship in tourism, tourism and hospitality entertainment, management and entertainment of tourist destinations, gastronomy and territorial identities, enogastronomy and tourism marketing, digitalization and technological training, niche tourism, innovative projects and good practices in tourism and hospitality, human resource management in tourism and hospitality, dematerialization of operations in tourism subsectors, management of operational processes and provision of tourism services, contexts of differentiation and specialization, professional and logistical challenges in tourism, and finally, robotization and new technologies in tourism, hotels, and food service.

The 15 chapters which this book is structured in three sections, which allow the framing and interpretation of dynamics and challenges regarding Section 1, "Robotization, Digitalization, and New Technologies in Tourism and Hospitality"; Section 2, "Innovation, Experiences, and Quality Service in the Hotel and Food Service Industry"; Section 3, "Development Dynamics, Governance, and Promotion of Tourism and Hospitality."

In Section 1, "Robotization, Digitalization, and New Technologies in Tourism and Hospitality," which includes five chapters, approaches are established on technological developments and the implications of digitalization in hospitality and tourism, with specificity on the themes of robotization and automation and management of hotel operations. This section highlights the contributions made to tourism and hospitality, its interactions and optimized development of operations and services are investigated.

Chapter 1, developed by Sameera Khan and Dileep Kumar Singh explores the significance of RPA in automating mundane tasks within the tourism, hotels, and food services industries. It discusses RPA use cases, the basic framework of RPA implementation, RPA tools, challenges in implementing RPA, and the future scope of RPA.

Ahmed Chemseddine Bouarar, Karolina Stojczew, and Kamel Mouloudj investigates within Chapter 2 the impact of COVID-19 on the hospitality industry in Poland and examines the industry's digital transformation as a response to the pandemic. It identifies obstacles that may hinder digitalization in the hospitality industry and highlights the benefits of a sound digital transformation.

In Chapter 3, Züleyhan Baran, Şükran Karaca, and Huseyin Baran focuses on how new-gen technologies have revolutionized the tourism industry. It highlights the adoption of robotic systems by tourism businesses, such as hotels, food services, and transportation, to ensure social distancing and enhance hygiene measures. The chapter explores the challenges and benefits of digitization in the tourism industry.

Hande Mutlu Ozturk, Ozgur Guler, Pamukkale, and Olcay Polat investigate the factors that influence tourism income in Türkiye on Chapter 4. It employs statistical methods to analyze the impact of these variables and compares the performance of different prediction models, including machine learning models, in forecasting tourism income. The chapter sheds light on the importance of factors such as the country's image, safety, and transportation opportunities in shaping tourism income.

Chapter 5 delves into the role of tourist intermediaries in connecting supply and demand in rural tourism. Predrag Miroslav Vuković and Marija Mosurović Ružičić emphasizes the importance of ICT application in improving business activities and promoting rural tourist destinations. This chapter aims to address the challenges faced in rural areas and proposes steps to enhance the development of these regions. The expectation is that this approach will increase the income and stop the negative trends burdening the life in rural areas (depopulation, population migration to urban centers, the decline in macroeconomic indicators, etc.) that are present not only in Serbia but also in other countries.

In 2, "Innovation, Experiences, and Quality Service in the Hotel and Food Service Industry," five chapters are included that address the issues of innovation, modernization and sustainability in hotel and restaurant services, with approaches applied in terms of different countries and analyses on behaviors and satisfaction levels. Ways of behaving in services and the incorporation of artificial intelligence and virtual reality are explained. Food traditions, the search for the preservation of gastronomic heritage and its incompliance with new tourist experiences are explored simultaneously. Processes of the impact of the evolution of technologies on hospitality and tourism are analyzed.

Nilgün Demirel İli and Onur Çelen examine in Chapter 6 the effects of technology paradox and technostress on employees in the tourism industry. Through qualitative research methods, the chapter identifies the negative impacts of technology on employees and highlights the potential decrease in the labor force. It emphasizes the need to mitigate the adverse effects of technology on the industry.

Chapter 7 investigates the service quality of Artificial Intelligence (AI) in the tourism industry, with a focus on Pondicherry. Jeganathan Gomathi Sankar and Arokiaraj David using a quantitative research approach, the chapter analyzes survey data to assess the impact of AI technology on service quality and customer satisfaction in tourism. The findings highlight the positive influence of AI on service quality. This study contributes to the literature by providing empirical evidence to service quality in the context of AI technology and the importance of AI technology to enhance service quality and customer satisfaction.

Mihály László Vörös and Aleš Gačnik explore the gastronomy tourism potential and cultural heritage of the South-Danube region in chapter 8. It examines a case study on the culinary tradition of brewing and eating fish soup, which has become a prominent gastronomy tourism attraction in the town of Baja. The chapter discusses the impact of this tradition on the local economy and sustainable development.

Chapter 9 introduces HoST Lab, an integrative approach that promotes innovation based on the Mediterranean Diet (MD) in the tourism sector. Alexandra Rodrigues Gonçalves, Célia Ramos, and Carina Viegas present in this chapter research on the emotional and sensory experiences associated with enogastronomic experiences of the MD. The chapter aims to create a sharing and learning research space between academia, tourism, and hospitality sectors, in which proposals are tested, results transferred to the community, and well-being promoted among the local population and visitors, aspiring for a growing sustainable destination.

Chapter 10 focuses on the developments of virtual reality (VR) and augmented reality (AR) in tourism marketing. Nesenur Altinigne provides an overview of the research progress in this field, highlights the theoretical backgrounds and methodologies employed, and suggests future research directions for improving the literature.

The last section, "Development Dynamics, Governance, and Promotion of Tourism and Hospitality," is composed of five chapters that bring studies on governance and marketing practices of tourism and hospitality, with emphasis on the issues of dissemination of destinations and accessibility. The chapters provide insight into different realities and problems associated with accessibility and the inclusive dimension of tourist destinations, as well as the exploration of communication models and supports for the dissemination of tourism in different territorial contexts.

Chapter 11 evaluates the challenges and opportunities of innovation sustainability in tourism entrepreneurship. It conducts a systematic review of bibliometric literature to identify the barriers to implementing sustainability practices in the tourism industry. The chapter, developed by Albérico Travassos Rosário aims to raise awareness and provide insights for promoting sustainable tourism practices. This research essay aims to evaluate the challenges and opportunities of innovation sustainability in tourism entrepreneurship, thus building a clear image of what should be done to overcome the obstacles and increase awareness of the need for sustainable tourism.

Gorete Dinis and Zélia Breda examine whether destination management organizations (DMOs) actively develop inclusive destination promotion and advertising materials in Chapter 12. It focuses on Tourism of Portugal as a case study and explores the accessibility level of promotional materials. The chapter highlights the importance of inclusive tourism and suggests future research directions.

Yassine El Bouchikhi in Chapter 13 reviews the literature on nation branding, social media, and tourism in the context of Morocco. It examines the use of social media in promoting travel, cultural assets, and the recent developments and trends in the Moroccan tourism industry. This chapter also includes a netnographic study of influential media accounts promoting Morocco's image.

Chapter 14 provides a comprehensive examination of eWOM communication in the tourism and hospitality industry. It analyzes existing scholarly literature on eWOM within this domain, including prevalent themes and research methodologies. The authors Mahmut Bakır and Ali Emre Sarılgan offer insights into the impact of eWOM on consumers' decisions and suggests future research directions.

The final chapter (15) focuses on accessibility in tourism, specifically addressing the ACT-Model. It emphasizes the importance of creating an accessible environment for all, including people with reduced mobility. The chapter developed by Borbála Gondos and Márta Nárai explores the roles of various actors involved in ensuring accessibility in tourism and highlights the economic potential of inclusive practices.

We hope that this handbook serves as an invaluable resource, providing profound insights and inspiration for both academics and professionals in the ever-evolving fields of tourism, hospitality, and food service. May it ignite new ideas, facilitate fruitful discussions, and contribute to the continued advancement and transformation of these industries.

Finally, we would like to thank all the authors for their contributions and commitment, allowing us to enrich this international publication, disseminating their research and projects and promoting a deeper and more up-to-date knowledge about technological advances in tourism and hospitality, the changes and challenges of these sectors. To highlight their concern to provide innovative studies that contribute to a better understanding of the current dynamics, problems and challenges of tourism, hospitality and food services.

Gonçalo Poeta Fernandes
CITUR, Polytechnic Institute of Guarda, Portugal & CICS, Universidade NOVA de Lisboa, Portugal

António Silva Melo
CiTUR, Polytechnic Institute of Porto, Portugal & CIDTFF, University of Aveiro, Portugal

Acknowledgment

We would like to acknowledge the assistance of all academics and researchers who contributed to the successful completion of this book, namely, the authors for their valuable contributions and commitment in the development of this publication and all the colleagues who took part in the review process and made it happens. Without their support and interest in promoting these territories of research, and disseminating its investigations, this publication would not have been possible to bring to the light.

Theoretical developments and achievements, the incorporation of new technologies and digital tools, territorial analysis of tourism products and services, case studies and good practices in the valorization of tourism, hotels, and restaurants, through new dynamics or forms of valorization are valuable contributions to this publication. The contribution of works from several countries in different continents, allows to analyze trends and different contributions and developments, of practices, technologies, and methodologies of analysis, for a better understanding of them and for the ways to be followed.

Section 1

Robotization, Digitalization, and New Technologies in Tourism and Hospitality

Chapter 1
Robotization and Smart Technologies in the Hospitality Industry

Züleyhan Baran
Akcakoca Vocational School, Duzce University, Turkey

Şükran Karaca
Tourism Faculty, Sivas Cumhuriyet University, Turkey

Huseyin Baran
Art Design and Architecture Faculty, Duzce University, Turkey

ABSTRACT

New-gen technologies have profoundly impacted all aspects of life and various economic sectors. The tourism industry, known for its inclination towards innovation, has been quick to embrace technological advancements. In response to the global pandemic, tourism businesses such as hotels, food services, and transportation have increasingly utilized robotic systems to ensure social distancing, hygiene, and sanitation measures. However, digitization presents significant challenges for the tourism industry, requiring companies to adapt their operations to stay competitive. Automation has emerged as a highly beneficial trend, simplifying tasks and introducing innovative processes to tourism business models. This enables companies to provide personalized services tailored to the preferences of "digital tourists." Overall, new-gen technologies are reshaping the tourism industry and driving it toward enhanced efficiency and customer satisfaction.

INTRODUCTION

The tourism industry is undergoing great transformation and unprecedented change. Digital processes and innovative solutions driven by new-generation technologies have led to the emergence of new players and models. The industry has gained a new dimension with smart technologies that offer unprecedented

DOI: 10.4018/978-1-6684-6985-9.ch001

application opportunities (Neuhofer et al., 2015). Hotels are one of the core structures of the tourism industry and new technologies in this field also encourage the development and innovation of the hotel industry. One important way to differentiate in the hotel industry is by offering added value through technology (Smartvel, 2020).

According to the International Federation of Robotics, a service robot is a type of autonomous robot that performs useful tasks for humans through sensing and adapting to different situations without human intervention (Paral, 2022). Service robots are defined as social intermediaries that can replace human service providers in service trials (Van-Doorn et al., 2017). Bowen and Morosan (2018) defined service robots as "physically embodied, artificial intelligence (AI) agents that can perform actions that have effects on the physical world." According to Ivanov, Webster, and Berezina (2017), service robots are "programmable, intelligent devices with a certain degree of autonomy, mobility, and sensory capabilities designed to perform a specific task" that are useful to humans. The term "social robot" is used to describe service robots that have the ability to interact and communicate with humans and follow social norms (Chi et al., 2020). Service robots are expected to play an increasingly important role in the hospitality and tourism industries, improving the service experience and quality (Mende et al., 2019). The use of robots in tourism and hospitality enterprises has the potential to enhance guest experiences and make them more efficient and enjoyable (Ivanov et al., 2017).

Robotic applications are widely used in manufacturing, military forces, medicine, and home care services. So, these applications are becoming increasingly common in hospitality and tourism (Murphy et al., 2017). The use of robots in the hospitality and tourism industry is one of the most modern, innovative, and advanced ever. The use of service bots ranges from basic AI chatbots to assist with the service process to sophisticated assistant bots that enhance the guest experience and satisfaction. As the number of companies using service bots increases, it is important to understand their impact on both business and customer satisfaction (Belanche et al., 2020). While some of these robots perform basic and routine tasks in hotels and restaurants, such as robotic floor cleaners (Murphy et al., 2017), the potential for their use in the industry is vast and varied.

The topics of AI and robotic technologies are rapidly spreading and widely used around the world, and are being studied by various disciplines in the literature. The field of tourism is also gaining attention as one of the disciplines in which research has been conducted in recent years. In this context, robots play a significant role in the application areas of the tourism sector (Kılıçhan & Yılmaz, 2020). Especially in light of the great developments in the field of information and communication technology, as well as the use of AI techniques in many areas, including tourism, smart technology has gained significant importance in the tourism industry today.

The objective of this chapter of the book is to provide an understanding of the concept of smart hotels and the application of new technologies in this field. It aims to create a discussion platform about the use of new technologies in smart hotels. To achieve this goal, the concept of smart hotels and the new-generation technology components that make up this concept will be explained based on the literature. Finally, a futuristic outlook will be presented in the conclusion chapter using the theoretical information obtained.

BACKGROUND

The hotel industry is putting more emphasis on smart and digital technology solutions and systems, such as AI, that can provide innovative solutions to meet the needs of tourists. As a result, the use of smart technologies is becoming more widespread every day (Kim & Han, 2020). China is a leading country in this field, having followed guidelines to build smart hotels for its tourism market since 2013. Many Chinese tech companies have been contributing to the digital transformation of traditional hotel business models to provide a better and more personalized tourist experience (The Economist, 2021). For instance, Fliggy (Alibaba Group's online travel platform) launched FlyZoo Hotel in Hangzhou in 2019, which offers a wide range of AI services. The hotel is considered a smart hotel because it uses many smart installations offered by Alibaba Group, according to Liang Bo, the hotel's Vice President. Similarly, Andy Wang, CEO of FlyZoo Hotel, notes that smart technologies are transforming the industry, and FlyZoo Hotel bridges the gap between hospitality and technology, inspiring and empowering tourists (Law et al., 2022).

Innovation is one of the key components of success in a competitive industry. However, in order to foster and implement it, it is crucial to be aware of the competition, potential risks, and challenges. Moreover, it can be challenging to predict whether tourism policymakers will support or obstruct investment in the development of innovative programs. For instance, some innovative strategies may clash with the traditional views of institutions, which could require collaboration with various stakeholders. Another factor to consider is crises within the tourism sector itself, as opposed to economic crises affecting a country. In such situations, requests for change from the private sector are often seen by politicians as pressing issues that require special attention and effort. (Rodríguez-Antón & Alonso-Almeida, 2020). Improving the tourist experience is a primary objective of innovative practices. Understanding tourists' perspectives on the use of new technologies in the hospitality industry is crucial in determining how well smart technology services can meet their expectations. The COVID-19 pandemic has accelerated the adoption of new technologies in the industry that help maintain, reduce, or eliminate social distancing (Davari et al., 2022). However, it is important to understand tourists' views on the use of these technologies to ensure that the services provided meet their expectations.

Smart Hotels

Smart hotel studies are derived from the field of global intelligent building studies and rely on advanced computer technologies that are constantly evolving (Frank et al., 2007; Doukas et al., 2007; Buckman et al., 2014; Leung, 2021). Smart hotels are prominent in countries such as the USA, China, Korea, and Singapore (Koo et al., 2013; Xu, 2018). Novotel Ambassador Seoul Dongdaemun in Korea stands out as a notable example of a smart hotel. The hotel is powered by AI from Hotels & Residences in Korea, and it offers AI room service with GiGA Genie, making it a new AI service platform (Gupta et al., 2022). In Singapore, the use of smart hotel technologies is supported by guidelines such as the "Smart Hotel Tech Guide 2018" and the "Technical Guide of The Smart Hotel 2019" published by the Singapore Hoteliers Association. These guides emphasize the importance of using technology in the tourist experience and are used in the hotel industry to provide better experiences. In China, smart hotels are being developed as an extension of smart tourism. The most prominent examples of smart hotels include the Penguin Hotel QQ chain and the FlyZoo Hotel opened by Alibaba Group (Luo & Pan, 2021).

Smart hotels are part of the broader trend in the hospitality industry toward the use of advanced technologies to enhance the customer experience. The concept of a "smart" hotel room involves a microprocessor-operated station that monitors essential parameters for the room's functioning, such as temperature, guest movement, and sensors. These stations are often connected to a central computer, allowing for centralized control of multiple rooms, floors, or even the entire hotel. In addition to room technology, smart hotels also offer guests self-check-in/check-out, mobile key access, remote room control, voice assistants, and digital guest services (Petrevska, 2016). However, technology plays an important role, smart hotels aim to provide a sustainable management model while still prioritizing guests' satisfaction. Ultimately, the hospitality industry is centered around people, and smart hotels aim to enhance the guest experience while maximizing efficiency. A timechart of smart hotel development worldwide can be seen in Table 1.

Table 1. Development of smart hotels

Year	Country	Example	Technology
2006	USA	Cobono Mountain Resort Pennsylvania	RFID system introduced
2009	USA	City Center Hotel Las Vegas	Implemented smart systems to identify personal preferences and improve accommodation experience
2009	China	Dragon Hotel Hangzhou	RFID and connected smart technologies, hotel management system
2013	China	National Tourism Administration	Official guide of "smart hotel construction and service"(LB/T 020-2013). This document provides guidance for hotel investors and operators in China and sets the quality and standards required for hotel construction and services.
2016	China	Smart Hotel Alliance	Celebrating China's Smart Tourism Year and establishing
2017	China	Penguin Hotel QQ	Development smart rooms
2018	Korea	Novotel Ambassador Seoul Dongdaemun	AI room service with GiGA Genie
2018	Singapore	The Singapore Hoteliers Association	"Smart Hotel Tech Guide 2018"
2018	China	FlyZoo Hotel Alibaba Group	It is the first hotel in the world to use a full-face recognition system. This hotel allows guests to check-in/check-out and check-in to their rooms using facial recognition technology. In addition, the hotel uses AI technology to understand guests' needs and provide better service
2019	Singapore	The Singapore Hoteliers Association	"Technical Guide of The Smart Hotel 2019"

The table highlights the development of smart hotels in different countries and the use of various technologies to enhance the customer experience. In China, the Dragon Hotel Hangzhou was the first to adopt smart technologies such as RFID and connected smart technologies, as well as a hotel management system developed by IBM. In Korea, the Novotel Ambassador Seoul Dongdaemun was the first to adopt AI room service through GiGA Genie. In Singapore, the development of smart hotels is supported by the publication of guidelines such as "Smart Hotel Tech Guide 2018" and "Technical Guide of The Smart Hotel 2019" by the Singapore Hoteliers Association. In China, the Penguin Hotel QQ chain and FlyZoo Hotel are examples of the latest smart hotel developments.

Internet of Things (IoT)

IoT, which stands for the "Internet of Things," is a global system consisting of interconnected computer networks that use standard internet protocols (Nunberg, 2012). In recent years, the IoT has enabled the emergence of elements that facilitate life by enabling communication of network-connected physical objects at any time and any place (Kosmatos et al., 2011). The IoT can be thought of as a global network system that provides a unique identity to each object, enabling communication from human to human, human to object, and object to object (Aggarwal & Lal Das, 2012). IoT defines a world in which almost everything can be connected and communicates intelligently, like never before. The term "connected" is often considered in terms of electronic devices such as servers, computers, tablets, and smartphones, but in the system called the Internet of Things, sensors and actuators embedded in physical objects are connected to each other through wired and wireless networks, and usually use the same Internet IP to connect to the Internet. These networks distribute huge amounts of data that flow to computers for analysis. Objects become tools to understand and respond quickly to complexity when they can both perceive the environment and communicate. The revolutionary aspect of all this is that these physical information systems can be coded and networked on the internet by intelligent technologies (Butler, 2020). This situation is considered an important detail that proves the availability of the IoT system by smart hotel systems (Han et al., 2021).

RFID (Radio Frequency Identification)

RFID (Radio Frequency Identification) technology is a crucial part of the IoT system (Liya et al., 2022) and can be applied in various fields such as agriculture, transportation, medical, and tourism where wireless network technology is used. RFID is also utilized in monitoring systems to track changes in the environment or specific geographic areas. It has the capability to reduce the labor required for inventory creation and security management effectively (Kaur et al., 2011). The first implementation of RFID technology in the hospitality industry was introduced at the Cobono Mountain Resort in Pennsylvania, USA in 2006. This technology enables guests to access their rooms and resort services using their keys or cards. USA in 2009, City Center Las Vegas Hotel implemented a smart system to determine their customers' personal preferences. This system allows for automatic check-in/check-out operations, controls heating-cooling systems, and simple functions such as room light curtains and restaurant reservations. The system also records past visits and the preferences of guests to provide more personalized services in future visits. In conclusion, RFID technology provides a convenient and personalized experience for guests in the hospitality industry by controlling simple functions such as room light curtains and making restaurant reservations. Additionally, the technology records past visits and preferences to enhance future customer service (Ren, 2014).

AI (Artificial Intelligence)

AI is a system that is based on large data processing capacities, and algorithms (Bulchand-Gidumal, 2022). John McCarthy organized the Dartmouth Conference in 1956, which was the first event focused on AI, and defined the term AI as "the science and engineering of intelligent machines, where intelligent machines are defined as those that can perform tasks that typically require human intelligence such as perception, reasoning, learning, and language understanding" (McCarthy, 2007). According to another definition, AI is a branch of computer science or the ability of a machine to imitate human behavior by simulating human intelligence (Webster, 2021). AI today offers services such as image recognition voice-activated search, and chatbots in mobile devices (Boden, 2017). The use of AI technology continues to grow with the advancement of algorithms development, access to new technologies becoming more affordable, and the participation of major technology companies in the tourism industry. Advanced technologies are required to enable smart hotel functions. AI technology is considered an important factor in the innovation of smart hotel services due to its technical advantages (Wang et al., 2020). On the other hand, AI is described as machine technology that understands, learns, and perceives like humans in the hospitality industry. From a practical implementation perspective, it is a smart machine system that has the ability to store and use information in the service process. In this context, it is described as a system that produces alternative solutions to human intelligence to help with the efficient use of all resources and to solve problems (Winston, 1993).

This table provides a high-level overview of the development of AI, including the key events and characteristics of each era. The first wave of AI, often referred to as "birth," refers to the early development of AI and the creation of basic computer programs and systems that could perform simple tasks. This period, which took place in the 1950s and 1960s, saw the creation of early AI technologies such as expert systems and decision trees. The second wave of AI, "development," saw the expansion of AI research and the creation of more advanced AI technologies. This period, which took place in the 1980s and 1990s, saw the creation of new AI technologies such as neural networks and genetic algorithms. The third wave of AI in the 1990s and present, "innovation," is often referred to as the current stage of AI development. This stage is characterized by the integration of AI technologies into a wide range of industries and applications, including healthcare, finance, transportation, and retail. Additionally, this wave is marked by the development of more advanced AI technologies, such as deep learning and reinforcement learning, or practical applications such as Siri, and Alexa, which are now being used to solve more complex problems. On the other hand, Perceptual Intelligence has emerged as a technology that aims to imitate human perception and intelligence in the field of AI. This technology grants machines the capability to perceive and understand sensory input through sound and vision. Perceptual intelligence refers to a type of AI that is designed to understand, interpret, and respond to sensory information from the physical world. It refers to the ability of AI systems to perceive, analyze, and understand data from a variety of sources, including images, videos, audio, and other forms of sensory data (Pentland, 2000). Perceptual intelligence is a key component of many AI applications, such as computer vision and speech recognition. For example, computer vision systems use perceptual intelligence to process and analyze images and videos, while speech recognition systems use perceptual intelligence to transcribe and interpret spoken language. In addition to its applications in specific domains, perceptual intelligence is also a critical component of more general AI systems that require a deep understanding of the sensory world. These systems often rely on machine learning algorithms, such as deep learning and reinforcement learning, to develop their perceptual intelligence over time (Pentland, 2001). However, the capability to comprehend, a crucial aspect of human intelligence, remains to be fully replicated. In a report released in 2020, it was noted that AI technologies in the field of Perceptual Intelligence have reached and even exceeded human standards, but the field of Cognitive Intelligence is still in its developmental stage (Li, 2021)s. In conclusion, the greatest advancements have been made in the field of Perceptual Intelligence, which is now considered the 3rd wave of AI technology development, and it is widely utilized across various industries.

AI Types Based on Approach

AI is usually divided into three approaches;

- Knowledge-based AI: At this level, the machine operates based on predefined knowledge. For example, a chatbot utilized provides pre-determined responses in customer service applications (Rodgers, 2020).
- Learning-based AI: At this level, the machine has learning capabilities in addition to its pre-defined knowledge. For example, a chatbot can learn from customer interactions, as well as from pre-programmed information (Nirala et al., 2022).
- Neural network-based AI: At this level, the machine acquires and performs through neural network algorithms. For example, designed to comprehend and respond to customer inquiries through the use of neural networks in customer service applications (Chen et al., 2022).

AI Types Based on Functionality

AI is usually divided into four basic functionalities. In Table 2, the types of AI based on their capacities are presented as a template:

Table 2. AI classification based on approaches

Category	Ability	Characteristics	Example
Mechanical	Automatically execute repetitive and routine tasks (Sternberg, 1997)	Mechanical AI is developed with restricted learning and application capabilities to ensure consistency. Not particularly smart.	Factory robots.
Analytical	Learning problems from the process in order to provide a solution using process information. (Sternberg, 1984-2005)	Analytical AI is considered "weak AI" as these AI applications can exhibit intelligent behavior, but cannot replicate human intuition.	Executing tasks by using the model created from learning the necessary knowledge and abilities
Intuitive	Thinking and adapting efficiently to new circumstances (Sternberg, 1984, 1999, 2005).	Intuitive AI is considered "strong AI" as it is designed to be more adaptable and function more like a human. Understanding is the most critical aspect.	Capable of producing original solutions to problems by utilizing prior knowledge and algorithms
Empathetic	Emotional (Empathetic) Intelligence (Goleman, 1996)	Empathic AI refers to a machine that can perceive, or at least simulate having emotions.	Robots that interact with humans using emotional intelligence features. Replica is utilized to comprehend human emotional states and provide appropriate responses. The Sophia Hanson robot is able to recognize and comprehend human emotional states by utilizing emotional intelligence technologies during human interactions.

Source: SHA, 2019

- Mechanical AI: AI systems can be used to automatically perform routine and repetitive tasks. For example, robots used in production lines in a factory can automatically carry out repetitive movements. In addition, AI systems can also be used in routine processes such as data entry or data processing (Huang & Rust, 2021). For example, in a call center application, AI system can automatically classify call records or, in a bill processing application, AI system can automatically verify invoice information (Vanneschi et al., 2018). AI systems are well-suited for performing routine and repetitive tasks, as the processes involved in these tasks typically have a fixed and standardized structure. These systems learn and execute operations based on the established model for such tasks (Fischer et al., 2020). This enables AI systems to perform routine and repetitive tasks instead of humans, allowing humans to focus on more valuable tasks.

- Analytical AI: Operational knowledge encompasses the specific knowledge and skillset required to carry out a particular task, such as operating a machine on a production line. The acquisition of operational knowledge and skills is an essential component of the learning process for individuals tasked with performing these types of duties (Friedlander & Zoellner, 2020). AI systems try to decode computing information using the learning process. For example, AI system gains an understanding of the knowledge and skills required to perform a task, and it then performs operations based on the learned model. In this way, AI systems can perform tasks that require processing knowledge instead of humans, so allowing humans to focus on more valuable tasks (Harris & Davenport, 2005). During the learning process, AI system identifies and learns from mistakes. In this manner, AI system continually improves its performance in processing information. This process of acquiring knowledge and skills for task completion can be considered learning.

- Intuitive AI: AI systems adapt effectively to new situations with the ability to think creatively. This process enables AI systems to generate unique solutions and find suitable solutions for new situations by using previously learned information and algorithms. According to Sternberg (1984), the ability to think creatively enables humans and AI systems to solve problems presented to them in ways that have not been solved before. In this process, AI system can produce unique solutions by using previously learned information and using the information learned during the learning process. Sternberg's (1999, 2005) creative thinking enables individuals and AI systems to solve problems in innovative ways, which were not used previously. In this process, AI system uses previously learned information and algorithms, can generate unique solutions, and find suitable solutions for new scenarios.

- Empathic AI: Emotional intelligence enables individuals and AI systems to identify and comprehend the emotions of others. This ability enables AI systems to perform in effective interactions with humans. It enables humans and AI systems to recognize and understand other people's emotions. In this process, AI system can give emotionally appropriate responses by recognizing and understanding other people's emotional states. AI system can influence other people's emotions and improve people's emotional state during these interactions. Empathic AI technologies and algorithms, as described by Goleman (1996), allow AI system to recognize and understand the emotions of others. AI is used in many various applications for example, smart systems can adjust room temperature, lighting, curtains or blinds automatically, leveraging AI technologies in hotels. Moreover, AI technologies can automate hotel booking processes and streamline check-in and check-out procedures.

AI Specific Systems Applies

According to AI Development Report (2011-2020), eight AI systems specifically applied in hospitality are highlighted:

- Machine Learning: It enables systems to learn and use from the data information.
- Robotics: It involves programming robots with AI technologies to automate tasks and improve efficiency. It is widely used for tasks such as transporting heavy materials in hospitality.
- Information Access: It enables systems to retrieve and analyze data to generate insights.
- NLP (Natural Language Processing): It allows systems to process, understand, and generate human language.
- Voice Recognition: It allows systems to recognize voice signals and convert them to text.
- Face Recognition: It allows systems to recognize face or image.
- Emotional Intelligence: It enables systems to detect and interpret human emotions.
- Social Intelligence: It enables systems to understand and engage social interactions.

In this context, AI technologies commonly used in hotels are shown in Table 3 presents a template of the commonly used AI technologies in hotels.

Table 3. Support Technologies and Specific Applications of AI

Technology	Explanation	Example
Robotics	Today's robots have the capability to move independently, execute repetitive and simple tasks, and provide information based on the data obtained from their actions	ServiceBots
Self-service software *	Self-service software is technology designed with user-friendly features that allow users to control their own service experience, providing electronic support without the need for interaction with a service representative. This technology offers limitless possibilities, from how they are used to how they present themselves. * *This technology, which is not technically considered AI, is still marketed and used as AI product in the industry due to its advanced capabilities*	Check-in/Check-out
Speech recognition	This technology is capable of recognizing and understanding spoken language. It listens to the speaker's voice to interpret the meaning and intention behind what is being said. To accomplish this, the audio signal is processed using machine-readable technology.	The participant is intelligent
Image Recognition	Video analytics refers to the use of computer algorithms to analyze and extract useful information from video footage. The goal is to use this information to support decision making and improve operations. It's often used to identify objects and detect patterns or behaviors in real-time. For instance, face recognition technology is a type of video analytics that uses unique features in captured images or videos to match them with stored templates for identification or authentication purposes.	Face recognition
Person-computer interaction	Virtual reality is a technology that creates a simulated environment that allows users to experience images, sounds, and sensations as if they were in a real-life setting. For example, virtual reality can be used to preview a hotel room from a distance before making a booking decision. Virtual reality glasses provide a fully immersive experience by putting users inside a 3D digital environment. In this artificial world, users can move around, interact with virtual objects, and experience the environment as if they were physically present.	Virtual reality (VR)

Source: SHA, 2018

Virtual Reality and Augmented Reality

Augmented Reality (AR) is considered a cutting-edge technology that operates through sophisticated algorithms and recognition, offering advanced services and is considered one of the world's leading technological innovations (Ara et al., 2021). AR technology enhances the functionality of mobile applications in industries such as health, tourism, education, and e-commerce with features such as motion tracking.

Virtual Reality (VR), an evolution of AR, is a technology that enables users to immerse themselves in a computer-generated virtual environment. VR technologies utilize multimedia devices and computer simulations to create a realistic experience for the user (Cao, 2016). These technologies typically include a head-mounted display and can display a room-sized virtual environment (Gold & Mahrer, 2018). With the visual experience provided by VR glasses such as Oculus, it is believed that future activities will increasingly take place in virtual environments (Huerta et al., 2019).

The main objective of virtual reality (VR) and augmented reality (AR) is to immerse users in a parallel digital environment that feels as real as possible. This technology has proven especially beneficial in digital marketing, as it allows marketers to bring the tourist experience closer to the consumer before they physically travel. For instance, a high-definition video that shows a picturesque beach with turquoise waters and blowing wind can evoke emotions and increase demand. Although VR and AR are mainly used in digital marketing, they also have practical applications in promoting lesser-known and far-off destinations. With the current travel restrictions due to the pandemic, the interest in virtual travel has been growing steadily.

Nowadays, VR applications in the hospitality industry are used as a support to make tourist activities dynamic and interesting. For example, by using these technologies in guided tours to the ruins, tourists can be immersed in historical events. For hotel chains and accommodation businesses, a virtual visit is offered before booking a room. By launching the Best Western Virtual Experience program in 2018, it aimed to provide immersive experiences to its guests. This allowed guests to better understand the property, its amenities, services and surroundings before booking. Thus, they managed to improve customer trust and communication and reduced the number of complaints by 71%.

Today, VR is utilized in the hospitality industry to enhance the dynamic and engaging nature of tourist activities. For instance, by incorporating VR technology into guided tours of historical sites, tourists can be fully immersed in the experience. In the case of hotels and other accommodation businesses, virtual visits are offered to potential guests before they book a room. In 2018, Best Western introduced its Virtual Experience program to provide an immersive experience for its guests. This program aimed to improve guests' understanding of the hotel's property, amenities, services, and surroundings, thus increasing customer trust and communication and reducing complaints by 71% (Camilleri & Camilleri, 2018).

VR technology is used in two main areas in the travel industry:

- First, it is used to increase the capacity of customers in the process of handling rooms and in the process of collecting information, to enable them to have a better understanding of the rooms and make quick decisions. For example; 360° VR photos can be given, which is a web application that does not require equipment.
- The second is used to provide a personalized and innovative experience during hotel stays. For example; can be given smart landscapes that offer an interactive experience.

Virtual reality technology has two primary applications in the travel industry:

- Enhancing the customer experience in room booking and information gathering, by providing a better understanding of the rooms and enabling faster decision-making. For instance, 360° VR photos can be made available through a web application that doesn't require any special equipment.
- Providing a personalized and innovative experience during hotel stays, such as interactive smart environments.

Examples of VR applications in hospitality are presented in Table 4 as a template.

Table 4. VR application in hospitality

Situation	Example
Before stay	Information and marketing regarding hotel rooms and facilities
Length of stay	To provide additional value to guests: - Offer the chance to fully immerse in local experiences from the comfort of their accommodations. - Utilize original content created specifically by the hotel, such as the history of the building and local stories, to enhance their stay.

Source: SHA, 2018

The benefits and considerations for the adoption of VR applications are presented in Table 5 as a template.

Table 5. Benefits and considerations VR adoption for hospitality

Benefits	Challenges
Elevating the overall brand experience. Boosting customer confidence and promoting quicker sales. Decreasing the time and effort required by the sales team for extended property inspections. Facilitating more streamlined cross-selling opportunities for travelers. A cutting-edge technology not accessible to all clients. Costly for virtual reality glasses with immersive experiences. Developing compelling advertising campaigns aimed at consumers.	A cutting-edge technology not accessible to all clients. Costly for virtual reality glasses with immersive experiences. Developing compelling advertising campaigns aimed at consumers. A cutting-edge technology not accessible to all clients. Costly for virtual reality glasses with immersive experiences. Developing compelling advertising campaigns aimed at consumers.

Source: SHA, 2018

As the hotel industry falls under the service sector, its offerings are intangible. The benefits and factors to consider when adopting VR in the hotel industry, as stated by Casaló et al. (2015), include:

- Taking experiential marketing to a new level
- Encouraging buyers to make quick decisions by boosting their confidence
- Reducing the time and effort required by the sales team for extended property inspections
- Offering easier cross-selling opportunities to foreign guests

- Limited accessibility for customers without VR viewing equipment
- High cost for VR glasses offering top-notch experiences
- The need for compelling content to captivate the consumer

Nowadays the widespread use of online booking, hotels now have a valuable opportunity to provide customers with panoramic views of their accommodations and food and beverage offerings on their website and through online travel agencies (OTA) platforms. Typically, tourists do not visit the property before making a reservation, as rooms are often reserved before finalizing travel plans. While not all customers have specialized equipment, such as VR glasses, to view VR content, even a limited experience viewed without such equipment is still more engaging than static photos. Additionally, augmented reality (AR) can also be used for entertainment purposes. However, providing a high-quality and private AR experience can be quite expensive, as the necessary equipment is costly.

The restaurant and catering industry is not immune to the use of VR technology. VR can be leveraged to create immersive culinary experiences, although it has yet to achieve realistic simulations of taste and smell. Restaurants can use VR technology to enhance their dining experience, for example, by adding a virtual show during meal service at a Caribbean restaurant. Some restaurants, like Sublimotion in Ibiza, are already utilizing this technology to offer more than just signature cuisine, but rather provide a multi-sensory experience for diners. While VR technology may not be appealing to the majority of the population, it is viewed as a niche with potential for growth and improvement in the future.

AR technology is a critical tool in enhancing the travel experience for tourists, making it easier, more enjoyable, and more empowering. Real-time camera translation systems, access to ratings and reviews of destinations, and software like "Google Lens" for Android phones are among the technologies that are currently in development and have the potential to make travel easier. It is important for the marketing and travel industry to pay close attention to these technologies and make efforts to optimize and improve them.

INNOVATIVE TECHNOLOGIES INTEGRATION INTO SMART HOTELS

Innovative Hotel Management Systems have become increasingly important in the hospitality industry due to the challenges posed by economic globalization and the growing demands of consumers for high-quality services (He, 2019). Traditional hotel service models are often characterized by regionalization and high degrees of commercialization, but they may not be effective in a fiercely competitive market where individual hotels resort to improper means to attract tourists (He, 2019; Xue et al, 2021). In this context, enterprises need to adopt a business attitude of excellence, constantly improving their hardware measures such as enterprise personnel, system, and facilities (Xue et al, 2021). To ensure the normal operation of the enterprise and meet the requirements of the new era, the original system needs to be improved and adapted to the changing information environment. The drawbacks of the old system may gradually appear, necessitating innovative management departments to take preventive measures to reduce the negative impact on enterprise development. In addition, the training of staff members is also essential to continuously upgrade their professional capabilities and knowledge reserves, enhancing their work efficiency and soft power (Feng, 2015). The new innovation mode has proven to be effective in addressing the conservative thinking of the old business model, which often leads to a lack of communication among different working layers of the enterprise and reduces work efficiency. The innovative approach promotes internal staff learning and exchange, bringing significant benefits to the enterprise. With the implementation of Innovative Hotel Management Systems, hotels can meet the demands of modern life and consumer preferences, which lays a good foundation for the further improvement of enterprise interests (Xue et al, 2021). In summary, Innovative Hotel Management Systems have become crucial for enterprises to succeed in a fiercely competitive market. The integration of hardware and software measures, the adaptation to the changing information environment, and the continuous improvement of staff members' capabilities and knowledge reserves are critical to the effective implementation of these systems. The innovative approach fosters communication, learning, and exchange among different working layers, bringing substantial benefits to the enterprise.

The Smart World and Smart Cities plan (Abdoullaev, 2011) was created by China as part of the Five-Year Tourism Plan developed by IBM, designed to modernize and enhance the tourism industry (Tu & Liu, 2014; Zhang, 2016). For this purpose, various technological solutions such as data analytics, AI, IoT, and blockchain were offered to tourism organizations and businesses. The aim of the plan was to improve customer experiences and increase the effectiveness of the tourism sector by making the country an attractive tourist destination. In this context, the "Smart Hotel" model was first implemented in 2009 through a partnership between Dragon Hotel Hangzhou and IBM. Under the agreement between the two businesses, the hotel will be expanded and reconstructed, and the RFID and connected smart technologies developed for the Smart Hotel model will be used. This partnership will be carried out within the "Smart World" strategy proposed by IBM and the hotel industry and will be accepted as a guide for the construction and service of smart hotels. This development has increased the use of technology in the hotel industry and provided personalization of hotel services (Zhang et al., 2012).

Automated Services

Generally, service bots are preferred in areas such as customer service, production, and cleaning, as they attempt to enhance human-computer emotional interaction and understand customer emotions through technologies such as self-check-in/check-out, smart assistant, face recognition, voice recognition, and email recognition (Frank et al, 2017).

Self-Check-In/Check-Out

In a traditional hotel, check-in is performed at the reception, while a smart hotel offers two alternative check-in options: through a mobile phone app or kiosks. Face recognition technology is used for identity and visa verification, and automated service software records personal and payment information. Upon completion of all transactions, the guest can unlock their room using either the electronic key in the mobile app or a physical room card from the kiosk. Check-out can be performed in the same way. The specific applications of this technology are detailed in Table 6 based on a guide developed by the Singapore Hotel Association and presented as a template.

Technically, robotic systems connect three essential components for the hospitality industry:

Table 6. Specific application of self-check-in/check-out system

Situation	Example
Before stay	Personal data and accommodation preferences, passport information Credit card information Personalized marketing promotion
Length of stay	Identity and visa verification with optical character and biometric recognition Card activation with electronic access from a mobile phone or automatic distribution of room cards via kiosks (Automatic check-in machine) Remote room control (air conditioning, lights and TV) with the app Direct communication with the application for questions and requests (food-beverage, cleaning and reservation, etc.) Personalized marketing promotion Automatic check-out
Post stay	Lost property, invoice and contact information Personalized marketing promotion Reminder to share experience on social networks

Source: SHA, 2018

- PSB (Police Station Bureau) - a system that facilitates the transmission of guest information to the security office by scanning the guest's ID before their stay in hotels in China.
- PMS (Property Management System) - a system that automates hotel operations such as guest reservations, guest information, and online bookings.
- OTA (Online Travel Agency) - a platform that allows for the booking and payment of rooms through a mobile device.

The specific applications of this technology are outlined in Table 7, based on a guide developed by the Singapore Hotel Association.

Table 7. Benefits of adopting self-check-in system

Benefits	Adoption elements
Reducing the waiting time **Offering more comfort** **Best guest experience** **Personalized experiences** **Encourage consumption** **Ensuring security in the pandemic**	Mobile apps have limited download rates, especially for unconventional customers. There are risks that could lead to a breach of user privacy. User experience interfaces should be appropriately designed to encourage usage. Physical personnel can complement the use of automated service applications.

Source: SHA, 2018

Smart assistant

The smart assistant with voice recognition in hotel rooms is similar to a smart speaker in a home. The smart speaker is a voice-controlled device equipped with a personal assistant that offers a range of services such as information search, music playback, and conversational capabilities (Nakanishi et al., 2020). Table 8 presents examples of smart assistant applications in hotels as a template.

Table 8. Implementation of smart assistant

Topics	Categories
Reception requests	Comments and complaints Cleaning service Sign out Facility reservation Care Transport Wake-up call
Smart room	Temperature Lights Curtains Media devices
Emergency alerts **The weather forecast** **Guest guide** **Calls**	
Linking personal accounts	Calendar notes Shopping list

Source: Buhalis & Moldavska (2021)

Besides all the functions of a smart assistant at home, it also has special functions when used in a hotel room. Rooms are the centerpiece of hotel service and are where guests will spend the majority of their time during their stay. In this sense, the smart speaker enables guests to easily and comfortably contact reception to request services or control all the devices in their room, enhancing their overall experience. Upon returning to their room, they can lie on their beds, close the curtains, turn on the TV, and start to unwind, just as they would at home. Table 9 presents the smart assistant benefits and considerations in hotels as a template.

Table 9. Smart assistant benefits and considerations

Benefits	Adoption elements
The ability to free up human resources and reduce operating costs thanks to the perfect interconnection of workflows **Combining self-updating operating systems** **Better experience for guests**	Guest resistance; ● Age/demographic characteristics ● Importance of human service ● Current habits Complex integrations Staff training requirement

Source: Buhalis & Moldavska (2021)

Its benefits for the hotel also make the hotel's operating system more efficient by freeing up staff with voice recognition technology and providing a personalized experience for guests to enhance their stay. However, incorporating smart assistants into the hotel management system can be a challenging process. The assistant needs to be compatible not only with the hotel's operating system but also with all the smart devices in the room, such as curtains, lights, audio, TV, etc. Additionally, staff must be trained to handle guest requests made through the assistant and to assist customers, especially elderly or technology-resistant guests, in using the assistant effectively.

Face Recognition

Face recognition is widely used in technologies that provide an intelligent experience at the hotel. For example, it plays an important role in features such as the previously mentioned service robots and self-check-in system. The functional aspects of the facial recognition application in the hotel, the benefits it provides and the need to be adopted are presented in Table 10 as a template.

Table 10. Face recognition application

Function	Example
Reception	Fast registration and room lock processes of the guests, Automatic detection of guest arrivals, fast forwarding of guest profiles to reception and personal selling suggestions. Reducing waiting time by directing more staff through video detection of crowds in the lobby
Arrangement	To determine the food and beverage rights of the guests
Security	Reducing the need for intense patrols for human resources and monitoring of CCTV images with a smart security video system. To follow and identify unauthorized or suspicious people. Tracking and managing people more effectively.
Sales & Marketing	To detect guest emotional states, expressions and profiles and increase additional sales opportunities. Tracking and analyzing guests' routes and identifying sales areas

Source: SHA, 2019

These applications can be basically divided into three categories:

- Identification: Facial recognition can be used for identity verification instead of manual checks of identity documents and personal information, such as reservation details.
- Demand Assessment: Facial recognition can be used to assess customer demand and reduce wait times, allowing for actions to be taken to enhance the customer experience. For example, at the FlyZoo Hotel, the system can pre-program elevators when customers leave their rooms and walk towards them, eliminating the need to wait.
- Emotional Perception: Facial recognition can be used to gain a better understanding of a customer's satisfaction and needs, though its technology at this stage is not advanced enough to accurately detect real emotions and satisfaction levels. However, this is a future direction of technology development.

The benefits and considerations of adopting facial recognition technology are presented in hospitality in Table 11 as a template.Formun ÜstüFormun Altı

Face recognition technology, similar to other AI technologies, increases efficiency by performing tasks more rapidly, lowering operational expenses, and improving the customer experience. Additionally, this technology offers a distinct security advantage compared to other AI technologies, as it can prevent fraud. However, it is important to ethical and transparency concerns that arise from the fact that many current AI algorithms are "black box" (Li, 2021), and the process by which data is collected and processed are not transparent. Face recognition technology could infringe on individuals' privacy security without proper adherence to privacy protection laws.

Table 11. Benefits and considerations of adopting face recognition technology

Benefits	Adoption elements
The ability to free up human resources and reduce operating Increasing operational efficiency by automating manual and labor-intensive work. Providing clearer information to better make planning decisions Improving tourist safety and experience. Reducing operating costs and increasing revenue generation opportunities. To reduce losses and theft and increase security.	The risks of user privacy violations; inform about applicable privacy regulations. High investment budget for hardware such as smart cameras and system components. System reliability, risk of system failure and idle time. It may require high-end hardware and high video storage capacity that improves video analysis and resolution.

Source: Buhalis & Moldavska, 2021

Robotization

Robot refers to autonomous machine systems that perform the task for which it is programmed (Decker, 2008). AI robots aim to create systems capable of human-like thinking and learning through technologies such as machine learning, classification, prediction, and NLP. Robots are the most typical application of AI and often use machine intelligence for routine and repetitive tasks (Frank et al, 2017). Table 12 presents the usage areas of robot technologies in hotels as a template.

Table 12. Application of robots in hotel services

Area	Example
Welcoming, greeting, and transporting customers.	*Cheetah Greetbot*: The Cheetah Greetbot is a robot developed by Xiaomi that is used to greet and serve guests in hotel rooms. It facilitates check-in/check-out procedures, room availability checks, and access to hotel services.
Delivering guest services and food orders to rooms	*Robot Run*: The Robot Run at Henn-na Hotel Nagasaki in Japan employs roboserve robots to fulfill guests' food orders and service needs.
Presenting treats to customers in the restaurant	*Siyanchaoren*: The Siyanchaoren restaurant robot used in China is capable of performing tasks such as food delivery and cooking through the use of sensors, cameras, and robotic arms.
Preparing food, ice cream and drinks	*Purple honor robot*: The Purple Honor robots used in China are capable of performing tasks such as preparing, cooking, and serving food.
Delivering and picking up luggage to rooms.	*Bellhop*: he Bellhop robot used at the Los Angeles San Gabriel Sheraton Hotel uses walking technology within the hotel and takes precautions to avoid obstacles and pedestrians. It delivers guests' luggage to their rooms, making check-in more efficient and comfortable.

Source: SHA, 2019

However, the implementation of robotic systems used is a crucial issue in hotels. Understanding the benefits such as providing uninterrupted customer support, providing fast and accurate answers, reducing the workload, increasing customer satisfaction as a concept, providing cost savings, and collecting and analyzing statistical data will make the use of this technology widespread. Table 13 presents the main benefits and important features of robots as a template in hospitality.

Table 13. Benefits and considerations for adopting service robots in hospitality

Benefits	Adoption Considerations
Using an innovative approach in hotel marketing strategies to increase brand awareness.	Renting robots instead of purchasing them can reduce investment costs
Optimizing business processes to increase efficiency by reducing repetitive manual tasks and freeing staff to focus on more valuable customer interactions and essential business services.	The existing building infrastructure can pose mobility challenges for the adoption of robots, such as uneven floors and narrow aisles.
Improving guest satisfaction through the reduction of wait times and an increase in the factor of innovation.	Systems such as Wi-Fi should be seamlessly integrated with the autonomous robots
Performing tasks with increased accuracy and consistency.	It is recommended that hotel staff receive training in resolving basic problems.
The use of robots for deliveries instead of in-room service by male staff may increase comfort levels for female guests at the hotel.	It is recommended to employ technical personnel as they can quickly repair or recover broken robots without having to wait for suppliers, reducing downtime.

Source: SHA, 2018

Robotic Technology in Hotel Kitchens

As robotic technology continues to advance, it is becoming more common for machines to replace human workers in various industries. These robots are capable of performing tasks such as creating chain learning algorithms and using 3D pointer trajectories to carry out production and service tasks. In order to accomplish these tasks, the robots are programmed with information about the objects and properties that they will be working with. This programming is typically done through the use of targeted training images. One area where robots have become particularly useful is in the food industry. Robots are equipped with autonomous systems that provide cognitive support, allowing them to perform complex tasks with ease. Overall, while the increasing use of robots in the workforce may have some drawbacks, it also presents many exciting opportunities for innovation and efficiency in various industries (Pfau et al., 2019). Examples of the application of robotic technologies in the F&B department in the hotel industry are as follows (Feller, 2021):

Robot in the kitchen: There are many innovations in robotics used in the food industry, including salad robots, automatic pizza robots, fast food machines, bread-making robots, and virtual dark food processors (Feller, 2021). Robot chefs are able to prepare noodles, hamburgers, coffee, sushi, grills, and drinks (Ivanov et al., 2017). One notable example can be found at the Henn-na Hotel in Japan, where a robot chef prepares "ekonomiyaki" pancakes. A visitor who witnessed the robot in action reported that it was able to efficiently mix the dough, cook the pancakes with the use of two spatulas, and even wrap the finished product with mayonnaise and dried green algae without dropping a single pancake (Grey, 2016)

Robot Waiter (Server)/Robot Busser: Keenon Robotics (2022), a leading company in intelligent robotics, has introduced a range of reliable and effective robots in the hospitality industry due to the ongoing shortage of employees and high labor costs caused by the pandemic. One of the applications provided by the company is the server robot, which is specifically designed to serve customers and transport used plates and glasses for a more efficient guest service experience. It is equipped with the latest AI technology, including GPS technology. The use of robots as waiters is becoming increasingly common in the hotel food and beverage industry. Restaurant operators have been known to turn to robotic waiters when staff is unable to keep up with orders or when the number of waiters is limited (Cheong et al., 2016). Automated waiters and robots can assist restaurant staff during busy times, but excessive use of robots can result in layoffs for some employees (Ivanov & Webster, 2020). The Henn-na Hotel in Japan is the first hotel in the world to use human-like robots to serve its guests (Alexis, 2017). Pizza Hut has also hired the humanoid robot Pepper to take customer orders through voice recognition and AI-based technologies. Pepper not only takes orders and delivers them to the kitchen, but also accepts payments (Ivanov et al., 2017).

Robot Host/Stewardess: As robots are being used to drive sales, the Tanuki Restaurant in Dubai employs a host robot to greet guests upon entering the restaurant (Prideaux, 2019). The robot host can communicate with guests, offer discount coupons, and encourage repeat visits (Ivanov & Webster, 2020). Robot hosts can be seen as an alternative to human hosts for tech-savvy restaurants or those targeting younger customers. Interacting with these robots could be a unique experience for tech-savvy customers, adding an element of fun to their dining experience (Berezina et al., 2019)

Delivery Robot/Robotic Butler: In 2014, the Starwood Group introduced two robotic butlers named ALO at the Aloft Hotel. These butler robots allowed hotel staff to deliver necessary items, such as toothbrushes, towels, and water, directly to guest rooms (Crook, 2014). Instead of receiving cash tips, ALO asks guests to provide feedback and rewards high votes with a dance performance (Trejos, 2014). At the Flyzoo Future Hotel, guest's check-in using passport scans at kiosks and access their rooms with face recognition technology. The hotel's robot butlers also provide in-room services, such as turning on lights and closing curtains (Saiidi, 2019).

Robot Bartender: The Robot Bartender can come in both robotic arm and human form (Tussyadiah et al., 2020). Typically, the robot bartender is equipped with the ability to interact with guests, take and serve beverage orders, and perform its functions at the hotel bar (Giuliani et al., 2013). For example, Swiss bartender "Barney," created by F&P Robotics AG, is a fully automated machine capable of ***preparing*** dozens of cocktails to exact specifications, self-sterilizing, and even cracking jokes while serving food and drinks to customers (Smith, 2021). The bartender typically consists of two robotic arms positioned beneath the bottles at the bar (Berezina et al., 2019)

3D robotic system: 3D printing technology has progressed quickly and has enabled digitization of the entire manufacturing ***process***. It has gained popularity in the food industry due to its digital model that facilitates automation. One of the most widespread applications of 3D printing is food modeling. With the advent of new printing techniques, 3D printing technology is not only used for various food shaping purposes but also for micro-level food shaping (Chunhua & Guangqing, 2020). The primary objective of using the 3D Robotic system in the kitchen is to offer customizable products, optimize food parameters, and ensure precise preparation through 3D printing. Human limitations in the cooking process prevent the food from being prepared under the optimal taste and texture conditions. The implementation of the 3D Robotic system in the food industry addresses all crucial aspects such as proper data input, accurate parameter determination, process control, cooking degree and timing. To this end, Moley Robotics provides robotic tools for use in the kitchen. Established in 2015 with the goal of developing innovative food robot systems and a global taste and unlimited food variety, Moley Robotics stands out with its cutting-edge technology and unique designs in the kitchen. Two of the 3D robotic tools it offers are (Moley, 2015);

- Shadow *Robot* Hand; it all started with the realization that effectors with a three-fingered grip were stabilized at the level. Later, with the advancement of research towards creating a fully functional hand, the design of the robotic hand adopted the biological properties of human muscles. In this direction, the Shadow robot hand can mimic the function of muscles and can execute many movements in a timely manner (Tuffield & Elias, 2003). Replicating the wrist, which is the most complex structure of the human body, Shadow comprises 20 motors, 24 joints, and 26 microcontroller mechanisms. Shadow is considered one of the closest robotics kitchen tools to human hand sensitivity in countries such as the USA, China, and Japan (Barakazi, 2022).

- Moley Robotic Kitchen; It is a home kitchen-based robotic system designed to assist humans in meal preparation. This system comprises sensors, actuators, and other robotic components and is controlled by software that predicts the user's next action and provides personalized assistance. All these components are interconnected over the network and compare historical data in the database with the current sensor data, thus monitoring the cooking process and significantly simplifying preparation, especially with the use of 3D Robotics systems (Mizrahi & Zoran, 2023)

CONCLUSION

Digitalization and advancements in new-generation technology have had a significant impact on the tourism industry. In terms of tourism demand, it has made it necessary to adopt new technologies that allow for the provision of personalized and interactive services for tech-savvy tourists. Furthermore, success in an increasingly competitive environment is achievable only through the use of smart technologies, by adopting innovative methods and increasing competitiveness. AI, robotics, and new-generation virtual reality technologies have started to be integrated into tourism, leading to the emergence of "smart tourism" and "smart hotels." In the hospitality industry, which is a crucial component of the tourism enterprise, the use of robots can provide a competitive advantage for companies in the future as consumer markets and technology continue to evolve (Ivanov et al., 2017). In service-based industries, the interactions and activities of robots differ greatly and these differences are critical. Robots can perform a range of complex tasks and provide specialized services, completing tasks that take longer for humans to perform.

It is crucial for service organizations to understand and acknowledge the role that robots will play in their businesses and how it will affect their customers, to ensure that everyone is satisfied during this emerging trend (Lukanova & Ilieva, 2019). It is widely believed that tourists are not opposed to new technologies and that any dissatisfaction that may arise will not be due to the acceptance of new technologies, but because the expected smart experiences are not yet available (Murphy et al., 2020). On the other hand, these new technologies are seen as highly intriguing and it is believed that they can bring added value to hotels. The most intriguing and valuable technologies are considered to be robotics, virtual reality, and voice recognition applications, which are among the latest advancements in technology.

Digitalization, robotization, and new technological advancements are developments that can significantly impact the tourism industry's supply chain. In this context, devising strategies to address the following issues will aid in attracting more customers to the tourism supply.

- Digital marketing: The use of digital technologies can enhance the marketing of tourism products and services, thereby reaching a wider customer base.
- Improved service quality: The implementation of robotization and digital technologies can improve the efficiency and service quality of the tourism industry's supply chain.
- Digital reservations: Utilizing digital technologies can simplify the reservation process, making it easier for customers to book tourism services.

- Multi-channel sales: The adoption of digital technologies can enable the tourism industry to sell its products and services through multiple channels, thereby reaching a broader customer base.
- Digital destination management: The use of digital technologies can facilitate the management and planning of tourism services, helping the industry attract more customers.

In the context of smart hotels, new-generation technologies, and robotization can bring about several advancements in the areas of tourism, hotel management, and food services

- Smart rooms: Digital technologies can help make rooms smart and configure them according to customers' wishes.
- Smart energy management: Digital technologies can help reduce costs and reduce environmental impacts by increasing energy efficiency.
- Smart food and beverage service: Digital technologies can help make food and beverage services more effective and efficient.
- Digital check-in/check-out: Digital technologies and robots can help make check-in/check-out faster and more efficient and increase customer satisfaction.
- Smart tourism management: Digital technologies can help make tourism management more effective and efficient.
- Robotic service attendants: Robots can assist customers in food and beverage services, tasks such as cleaning and maintenance, and check-in/check-out.
- Augmented reality and virtual reality technologies: Augmented reality and virtual reality technologies can help increase experiences and increase customer satisfaction in tourism and hospitality

Advancements in next-generation technologies and robotization can enhance the speed, efficiency, and customer-centricity of services in the tourism, hospitality, and food services sectors. The implementation of the smart hotel concept can bring several benefits, including increased customer satisfaction, reduced costs, and a reduced environmental impact through the integration of digital technologies in the fields of tourism, hospitality, and food services. However, it is important to note that these technologies must be effectively managed and integrated with human interaction for optimal results.

The integration of next-generation technologies in the tourism, hotel, and food service industries may have some implications on the operations of these industries and the overall customer experience:

- Improved service quality: Digital technologies and robotization can help businesses across industries improve efficiency and service quality.
- Transformation for workers: Robotization can help workers reduce their workload and focus more on quality and speed, but may also involve the risk of some workers being replaced by robots.
- More opportunities: Digital technologies and robotization can create more opportunities for entrepreneurs looking to invest in the tourism, hotel, and food service industries.
- Digitalization: Digital technologies can help businesses in the tourism, hotel, and food service industries digitize and reach more customers.
- Greater security and privacy: Digital technologies can help keep customers' data more secure and protect their privacy.

These technological advancements can aid in formulating strategies for the future of the tourism, hotel, and food service industries, thereby ensuring their long-term success. The speedy growth of digital technologies offers more efficient and effective service opportunities in hotels and food and beverage services in the tourism industry. In particular, the implementation of smart robots can speed up the check-in and check-out procedures in hotels and provide quicker and more convenient service in food and beverage services, enhancing the competitiveness of the tourism sector and boosting customer satisfaction and loyalty. As a result, there is a need for a better comprehension and progression of the relationship between tourism and digital technologies.

Finally, the contributions of quantum robot technology, another new generation technology, to the tourism sector are also considered among the important changes that will occur in the future. Quantum computers are a technology that combines sensors, the internet, and other tools, which are programmed to process information faster and more efficiently than conventional computers and robots. This technology has the potential to bring many benefits to the tourism industry. For example, quantum computers can help travel and tourism businesses better analyze customer preferences and demands using advanced technologies such as AI and machine learning. This can result in improved service quality and efficiency.

Quantum sensors and the internet have the potential to improve safety and quality in the tourism sector. Furthermore, the use of quantum robots can help businesses to be more efficient and effectively manage their resources in the industry. Although quantum robots are not yet widely adopted in the tourism sector, businesses are starting to recognize the potential benefits of quantum technologies. As these technologies become more advanced and widely used, they are likely to bring even more benefits to the tourism industry. However, it is important to note that quantum technologies are still in their early stages of development and need to be properly regulated. The future development of quantum technologies is expected to bring even more advancements to the tourism sector, leading to more advanced and intelligent tourism systems. The use of quantum robots can also help businesses to optimize the use of time and resources

Quantum robots are still in the early stages of adoption in the tourism industry, but tourism businesses are starting to recognize their potential benefits. The wider use of quantum technologies in the industry may bring additional benefits over time. However, it is important to keep in mind that these technologies are still developing and their implementation needs to be properly regulated. In the future, the continued development of quantum technologies will have a significant impact on the tourism sector, leading to even more advanced and smart tourism systems. The use of quantum robots can help businesses to be more efficient and optimize the use of time and resources. Despite their potential, quantum robots are not yet widely used in the tourism industry, but the recognition of their benefits is increasing

Tourism businesses are beginning to recognize the potential benefits of quantum technologies, and their wider use in the industry may bring additional benefits over time. However, it is important to remember that quantum technologies are still in their early stages of development and need to be properly regulated. The future development of these technologies will greatly enhance the tourism sector, leading to more advanced and intelligent tourism systems. Despite their potential, quantum technologies are not yet widely used in the tourism sector. Nevertheless, tourism businesses are starting to evaluate the potential benefits of these technologies. It should be noted that their implementation needs to be properly regulated to ensure their proper use and development. The development of quantum technologies will continue to contribute to the advancement of the tourism sector, leading to even more sophisticated and smart tourism systems.

SUGGESTIONS

The following suggestions can be made regarding the effects of robotization and innovative technologies on the tourism, hotel, and food service sectors:

- Industry leaders should carefully evaluate the impacts of robotization and digital technologies and strive to understand how these technologies can benefit their businesses.
- Businesses should prioritize investments in up-to-date and effective technologies that meet the needs and expectations of customers
- Employee training should be given priority, and employees should be educated about robotization and the use of digital technologies.
- In addition to robotization and digital technologies, businesses should also invest in sustainable and environmentally friendly solutions.
- The security and protection of customers' privacy should be a top priority for businesses when using robotics and digital technologies

By following these recommendations, businesses in the tourism, hotel, and food service sectors can enhance customer satisfaction by maximizing the advantages of robotization and innovative technologies.

In the future, it may be advisable to conduct the following academic studies on robotization and digital technologies in the tourism, hotel, and food service sectors:

- A comprehensive analysis of the effects of robotization and digital technologies on the tourism, hotel, and food service industries.
- An investigation of how robotization and digital technologies enhance service quality in accordance with customer expectations and needs.
- An examination of the impacts of robotization and digital technologies on employees, particularly focusing on employee training solutions.
- An investigation of how a sustainable tourism and environmentally friendly approach can be integrated with robotization and digital technologies.
- An exploration of how to ensure the security and privacy of customers in the context of digital technologies and robotization processes

Such studies can provide valuable insights and guidance for businesses in the tourism, hotel, and food service industries and can uncover crucial strategies for the future of these industries.

REFERENCES

Abdoullaev, A. (2011). Keynote: A smart world: A development model for intelligent cities. In *Proceedings of the 11th IEEE International Conference on Computer and Information Technology (CIT)*. IEEE.

Aggarwal, R., & Lal Das, M. (2012) RFID Security in the Context of Internet of Things. *First International Conference on Security of Internet of Things*, (pp. 51-56). ACM. 10.1145/2490428.2490435

Alexis, P. (2017). R-Tourism: Introducing the Potential Impact of Robotics and Service Automation in Tourism. *Ovidius University Annals, Series Economic Sciences, 17*(1)

Ara, J., Bhuiyan, H., Bhuiyan, Y. A., Bhyan, S. B., & Bhuiyan, M. I. (2021). Comprehensive analysis of augmented reality technology in modern healthcare system. *International Journal of Advanced Computer Science and Applications, 12*(6), 840–849. doi:10.14569/IJACSA.2021.0120698

Barakazi, M. (2022). The use of Robotics in the Kitchens of the Future: The example of'Moley Robotics'. *Journal of Tourism and Gastronomy Studies, 10*(2), 895–905. doi:10.21325/jotags.2022.1021

Belanche, D., Casaló, L. V., Flavián, C., & Schepers, J. (2020). Service robot implementation: A theoretical framework and research agenda. *Service Industries Journal, 40*(3-4), 203–225. doi:10.1080/02642069.2019.1672666

Berezina, K., Ciftci, O., & Cobanoglu, C. (2019). Robots, artificial intelligence, and service automation in restaurants. In *Robots, artificial intelligence, and service automation in travel, tourism and hospitality*. Emerald Publishing Limited. doi:10.1108/978-1-78756-687-320191010

Boden, M. A. (2017). *Inteligencia artificial*. Turner.

Bowen, J., & Morosan, C. (2018). Beware hospitality industry: The robots are coming. *Worldwide Hospitality and Tourism Themes, 10*(6), 726–733. doi:10.1108/WHATT-07-2018-0045

Buckman, A. H., Mayfield, M., & Beck, S. B. (2014). What is a smart building? *Smart and Sustainable Built Environment, 3*(2), 92–109. doi:10.1108/SASBE-01-2014-0003

Buhalis, D., & Moldavska, I. (2021). In-room voice-based AI digital assistants transforming on-site hotel services and guests' experiences. In *Information and Communication Technologies in Tourism 2021: Proceedings of the ENTER 2021 eTourism Conference, January 19–22, 2021* (30-44). Springer International Publishing.

Bulchand-Gidumal, J. (2022). Impact of artificial intelligence in travel, tourism, and hospitality. In *Handbook of e-Tourism*. Springer International Publishing. doi:10.1007/978-3-030-48652-5_110

Butler, D. (2020). Computing: Everything, Everywhere. *Nature, 28*(3), 402–440. PMID:16554773

Camilleri, M. A., & Camilleri, M. A. (2018). *Understanding customer needs and wants*. Springer International Publishing. doi:10.1007/978-3-319-49849-2_2

Cao, S. (2016). Virtual reality applications in rehabilitation, human-computer interaction. *Theory, Design. Development in Practice*, 3–10.

Casaló, L. V., Flavián, C., Guinalíu, M., & Ekinci, Y. (2015). Avoiding the dark side of positive online consumer reviews: Enhancing reviews' usefulness for high risk-averse travelers. *Journal of Business Research, 68*(9), 1829–1835. doi:10.1016/j.jbusres.2015.01.010

Chen, S. S., Choubey, B., & Singh, V. (2021). A neural network based price sensitive recommender model to predict customer choices based on price effect. *Journal of Retailing and Consumer Services, 61*, 102573. doi:10.1016/j.jretconser.2021.102573

Cheong, A., Lau, M. W. S., Foo, E., Hedley, J., & Bo, J. W. (2016). Development of a robotic waiter system. *IFAC-PapersOnLine*, *49*(21), 681–686. doi:10.1016/j.ifacol.2016.10.679

Chi, O. H., Denton, G., & Gursoy, D. (2020). Artificially intelligent device use in service delivery: A systematic review, synthesis, and research agenda. *Journal of Hospitality Marketing & Management*, *29*(7), 757–786. doi:10.1080/19368623.2020.1721394

Chunhua, S., & Guangqing, S. (2020). Application and development of 3D printing in medical field. *Modern Mechanical Engineering*, *10*(3), 25–33. doi:10.4236/mme.2020.103003

Crook, J. (2014). *Starwood introduces robotic butlers at aloft hotel in Cupertino.* Tech Crunch. https://techcrunch.com/2014/08/13/starwood-introduces-robotic-butlers-at-aloft-hotel-in-palo-alto/#:%7E:text=Starwood%2C%20one%20of%20the%20world's,around%20guests%20and%20use%20elevators

Davari, D., Vayghan, S., Jang, S., & Erdem, M. (2022). Hotel experiences during the COVID-19 pandemic: High-touch versus high-tech. *International Journal of Contemporary Hospitality Management*, *34*(4), 1312–1330. doi:10.1108/IJCHM-07-2021-0919

Decker, M. (2008). Caregiving robots and ethical reflection: The perspective of interdisciplinary technology assessment. *AI & Society*, *22*(3), 315–330. doi:10.100700146-007-0151-0

Doukas, H., Patlitzianas, K. D., Iatropoulos, K., & Psarras, J. (2007). Intelligent building energy management system using rule sets. *Building and Environment*, *42*(10), 3562–3569. doi:10.1016/j.buildenv.2006.10.024

Feller, C. (2021). 10 robots automating the restaurant industry. *Fast Casual.* https://www.fastcasual.com/blogs/10-robotsautomating-the-restaurant-industry/

Feng, J. T. (2015). Innovation and integration of hotel management under the new situation. *China Business*, *3*(1), 32–34.

Fischer, M., Imgrund, F., Janiesch, C., & Winkelmann, A. (2020). Strategy archetypes for digital transformation: Defining meta objectives using business process management. *Information & Management*, *57*(5), 103262. doi:10.1016/j.im.2019.103262

Frank, M., Roehrig, P., & Pring, B. (2017). *What To Do When Machines Do Everything: How to Get Ahead in a world of AI, algorithms, bots, and big data.* John Wiley & Sons, Inc.

Frank, O. L. (2007). Intelligent building concept: The challenges for building practitioners in the 21st century. [AARCHES J]. *J. Assoc. Archit. Educ. Niger.*, *6*(3), 107–113.

Friedlander, A., & Zoellner, C. (2020). Artificial intelligence opportunities to improve food safety at retail. *Food Protection Trends*, *40*(4), 272–278.

Giuliani, M., Petrick, R. P. A., Foster, M. E., Gaschler, A., Isard, A., Pateraki, M., & Sigalas, M. (2013). Comparing task-based and socially intelligent behavior in a robot bartender. Paper presented at the *ICMI 2013 – 2013 ACM International Conference on Multimodal Interaction*, (pp. 263–270). ACM.

Gold, J., & Mahrer, N. E. (2018). Is virtual reality ready for prime time in the medical space? A randomized control trial of pediatric virtual reality for acute procedural pain management. *Journal of Pediatric Psychology*, *43*(3), 266–275. doi:10.1093/jpepsy/jsx129 PMID:29053848

Goleman, D. (1996). Emotional intelligence. Why it can matter more than IQ. *Learning*, *24*(6), 49–50.

Grey, A. (2016). *This robot chef wants to know how you like your pancakes.* We Forum. https://www.weforum.org/agenda/2016/10/robot-chef-makes-pancakes-japan-hennna/

Gupta, S., Modgil, S., Lee, C. K., Cho, M., & Park, Y. (2022). Artificial intelligence enabled robots for stay experience in the hospitality industry in a smart city. *Industrial Management & Data Systems*, *122*(10), 2331–2350. doi:10.1108/IMDS-10-2021-0621

Han, Y., Niyato, D., Leung, C., Miao, C., & Kim, D. I. (2021). A Dynamic Resource Allocation Framework for Synchronizing Metaverse with IoT Service and Data. *Computer Science and Game Theory*, *21*(1), 43–58.

Harris, J. G., & Davenport, T. H. (2005). Automated decision making comes of age. *MIT Sloan Management Review*, *46*(4), 2–10.

He, Z. (2019). Integration and innovation of hotel management under the new trend. *Modern Marketing*, *4*, 112–119.

Huang, M. H., & Rust, R. T. (2021). A strategic framework for artificial intelligence in marketing. *Journal of the Academy of Marketing Science*, *49*(1), 30–50. doi:10.100711747-020-00749-9

Huerta, E.M.L., García, A.E., & Nava, M.R.Z. (2019). Cordodes: Realidad Aumentada, el futuro del Turismo.

Ivanov, S., & Webster, C. (2020). Robots in tourism: A research agenda for tourism economics. *Tourism Economics*, *26*(7), 1065–1085. doi:10.1177/1354816619879583

Ivanov, S., Webster, C., & Berezina, K. (2017). Adoption of robots and service automation by tourism and hospitality companies. *Revista Turismo & Desenvolvimento (Aveiro)*, *27*(28), 1501–1517.

Kaur, M. Sandhu, N. Mohan, & P. S. Sandhu. (2011). RFID Technology Principles Advantages Limitations Its Applications, *International Journal of Computer and Electrical Engineering*, *3*(1), 151–157.

Kılıçhan, R., & Yılmaz, M. (2020). Artificial Intelligence and Robotic Technologies In Tourism And Hospitality Industry. *Erciyes Üniversitesi Sosyal Bilimler Enstitüsü Dergisi*, *3*(1), 353–380. doi:10.48070/erusosbilder.838193

Kim, J. J., & Han, H. (2020). Hotel of the future: Exploring the attributes of a smart hotel adopting a mixed-methods approach. *Journal of Travel & Tourism Marketing*, *37*(7), 804–822. doi:10.1080/10548408.2020.1835788

Koo, C., Shin, S., Kim, K., Kim, C., & Chung, N. (2013). *Smart tourism of the Korea: A case study.* CORE.

Kosmatos, E. A., Tselikas, N. D., & Boucouvalas, A. C. (2011). Integrating RFIDs and Smart Objects into a Unified Internet of Things Architecture. *Advances in Internet of Things: Scientific Research*, *1*(1), 5–12. doi:10.4236/ait.2011.11002

Law, R., Ye, H., & Chan, I. C. C. (2022). A critical review of smart hospitality and tourism research. *International Journal of Contemporary Hospitality Management, 34*(2), 623–641. doi:10.1108/IJCHM-08-2021-0986

Leung, R. (2021). Hospitality technology progress towards intelligent buildings: A perspective article. *Tourism Review, 76*(1), 69–73. doi:10.1108/TR-05-2019-0173

Li, Y. (2021). *Hoteles inteligentes y nuevas tecnologías aplicadas en la industria hotelera. Estudio del caso de FlyZoo Hotel.* Zaragoza: Facultad de Empresa y Gestión Pública de la Universidad de Zaragoza. https://zaguan.unizar.es/record/106337/files/TAZ-TFM-2021-324.pdf

Liya, M. L., Aswathy, M., & Jayakrishnan, V. M. (2022, June). An Overview of Radio Frequency Identification systems. In *2022 7th International Conference on Communication and Electronics Systems (ICCES)* (530-535). IEEE. 10.1109/ICCES54183.2022.9835782

Lukanova, G., & Ilieva, G. (2019). Robots, artificial intelligence, and service automation in hotels, In: Ivanov, S., Webster, C., (eds) Robots, artificial intelligence, and service automation in travel, tourism and hospitality. Emerald Publishing Limited.

Luo, X., & Pan, Y. (2021). A Study on the customer experience design through analyzing smart hotels in China. *Journal of the Korea Convergence Society, 12*(3), 115–124.

McCarthy, J. (2007). *What is Artificial Intelligence?* Stanford University, Computer Science Department.

Mende, M., Scott, M. L., van Doorn, J., Grewal, D., & Shanks, I. (2019). Service robots rising: How humanoid robots influence service experiences and elicit compensatory consumer responses. *JMR, Journal of Marketing Research, 56*(4), 535–556. doi:10.1177/0022243718822827

Mizrahi, A. B., & Zoran, A. (2023). Digital gastronomy testcase: A complete pipeline of robotic induced dish variations. *International Journal of Gastronomy and Food Science,* 31.

Moley. (2023). *The future is served at work in the kitchen.* Moley. https://moley.com/

Murphy, J., Hofacker, C., & Gretzel, U. (2017). Dawning of the age of robots in hospitality and tourism: Challenges for teaching and research. *European Journal of Tourism Research, 15,* 104–111. doi:10.54055/ejtr.v15i.265

Murphy, R. R., Adams, J., & Gandudi, V. B. M. (2020). How robots are on the frontlines in the battle against COVID-19. *Smithsonian Magazine.* https://www.smithsonianmag.com/innovation/how-robots-are-on-front-linesbattle against-covid-19-180974720/

Nakanishi, J., Baba, J., Kuramoto, I., Ogawa, K., Yoshikawa, Y., & Ishiguro, H. (2020, October). Smart speaker vs. social robot in a case of hotel room. In *2020 IEEE/RSJ International Conference on Intelligent Robots and Systems (IROS)* (11391-11396). IEEE. 10.1109/IROS45743.2020.9341537

Neuhofer, B., Buhalis, D., & Ladkin, A. (2015). Technology as a catalyst of change: Enablers andbarriers of the tourist experience and their consequences. In I. Tussyadiah & A. Inversini (Eds.), *Information and Communication Technologies in Tourism 2015* (pp. 789–802). Springer. doi:10.1007/978-3-319-14343-9_57

Nirala, K. K., Singh, N. K., & Purani, V. S. (2022). A survey on providing customer and public admin-istration based services using AI: Chatbot. *Multimedia Tools and Applications, 81*(16), 22215–22246. doi:10.100711042-021-11458-y PMID:35002470

Nunberg, G. (2012). *The Advent of the Internet.* Citation Times.

Paral, T. (2022). Robotics. *Smart Manufacturing: The Lean Six Sigma Way,* 311-329.

Pentland, A. (2000). Perceptual user interfaces: Perceptual intelligence. *Communications of the ACM, 43*(3), 35–44. doi:10.1145/330534.330536

Pentland, A. (2001). Perceptual intelligence. In *Handheld and Ubiquitous Computing: First International Symposium, HUC'99 Karlsruhe,* (pp. 74-88). Springer Berlin Heidelberg.

Petrevska, B., Cingoski, V., & Gelev, S. (2016). From smart rooms to smart hotels. *Zbornik radova sa XXI međunarodnog naučno-stručnog skupa Informacione tehnologije-sadašnjost i budućnost, Žabljak.*

Pfau, J., Porzel, R., Pomarlan, M., Cangalovic, V. S., Grudpan, S., Höffner, S., & Malaka, R. (2019). Give MEANinGS to robots with kitchen clash: a VR human computation serious game for world knowledge accumulation. In *Entertainment Computing and Serious Games: First IFIP TC 14 Joint International Conference.* Springer International Publishing.

Prideaux, S. (2019). Robot host welcomes guests at new Dubai Mall restaurant. *The National News.* https://www.thenationalnews.com/lifestyle/food/robot-host-welcomes-guests-at-new-dubai-mall-restaurant-1.813177

Ren, X. X. (2014). *Research on the Construction and Application of the Smart Hotel Construction Evalu-ation Index System [D]* [Doctoral dissertation, Thesis for Master Degree in Hebei Normal University]

Rodgers, W. (2020). *Artificial intelligence in a throughput model: Some major algorithms.* CRC Press.

Rodríguez-Antón, J. M., & Alonso-Almeida, M. D. M. (2020). COVID-19 impacts and recovery strategies: The case of the hospitality industry in Spain. *Sustainability (Basel), 12*(20), 8599. doi:10.3390u12208599

Saiidi, U. (2019). *Facial recognition is coming to hotels.* CNBC. https://www.cnbc.com/2019/10/04/alibaba-china-flyzoo-hotel-uses-facial-recognition-tech-and-robots.html

SHA, Singapore Hotel Association. (2018). Smart Hotel Technology Guide. Using Technology to Navi-gate the Guest Experience Journey.

SHA, Singapore Hotel Association (2019). Smart Hotel Technology Guide. *Using Technology to Trans-form the "Heart of House".* SHA.

SHTG (Smart Hotel Technology Guide). (2018). *Using technology to navigate the guest experience journey.* SHA. https://sha.org.sg/userfiles/ckeditor/Files/Smart%20Hotel%20Technology%20Guide%20 2018.pdf

SHTG (Smart Hotel Technology Guide). (2019). *Using technology to navigate the guest experience journey.* SHA. https://sha.org.sg/userfiles/ckeditor/Files/Smart%20Hotel%20Technology%20Guide%20 2019.pdf

Smartvel. (2020). *Smart Hotels: what to expect from the future of hospitality.* Smartvel. https://blog. smartvel.com/blog/smart-hotels-what-to-expect-from-the-future-of-hospitality

Smith, C. (2021). Barney' the Swiss robot bartender can mix dozens of cocktails, tell jokes and sanitise itself. *The Drinks Business.* https://www.thedrinksbusiness.com/2021/04/barney-the-swiss-robot-bartender-can-mix-dozens-of-cocktails-tell-jokes-and-sanitise-itself/

Sternberg, S. R. (1981). Parallel architectures for image processing. *Real-Time Parallel Computing: Imaging Analysis*, 347-359. Book Authority.

Sternberg, S. R. (1984). Parallel processing in machine vision. *Robotica*, 2(1), 33–40. doi:10.1017/S026357470000881X

Sternberg, S. R. (2013, March). Environmental Research Institute of Michigan. In *Biomedical Images and Computers: Selected Papers Presented at the United States-France Seminar on Biomedical Image Processing, St. Pierre de Chartreuse, France, May 27–31, 1980 (Vol. 17*, p. 294). Springer Science & Business Media.

The Economist (2021). Thanks to the pandemic, diplomats have a bigger, better toolkit. *The Economist.*

Trejos, N. (2014). *Ready for the Hotel Industry's First Robotic Butler?* https://www.usatoday.com/story/travel/hotels/2014/08/12/aloft-hotels-starwood-robotic-bultler/13954231/

Tsinghua-Chinese Academy of Engineering Joint Research Center for Knowledge and Intelligence, Centre for Intelligent Research, Institute of Artificial Intelligence, Tsinghua University, Chinese Association for Artificial Intelligence. (2020). *Report on Artificial Intelligence Development 2011-2020.* Aminer. https://static.aminer.cn/misc/pdf/zpAIreport2020.pdf

Tu, Q., & Liu, A. (2014). Framework of smart tourism research and related progress in China. In *International conference on management and engineering (CME 2014)* (140-146). DEStech Publications, Inc.

Tuffield, P., & Elias, H. (2003). The shadow robot mimics human actions. *The Industrial Robot, 30*(1), 56–60. doi:10.1108/01439910310457715

Tussyadiah, I. P., Zach, F. J., & Wang, J. (2020). Do travelers trust intelligent service robots? *Annals of Tourism Research, 81*, 102886. doi:10.1016/j.annals.2020.102886

Van Doorn, J., Mende, M., Noble, S. M., Hulland, J., Ostrom, A. L., Grewal, D., & Petersen, J. A. (2017). Domo arigato Mr. Roboto: Emergence of automated social presence in organizational frontlines and customers' service experiences. *Journal of Service Research, 20*(1), 43–58. doi:10.1177/1094670516679272

Vanneschi, L., Horn, D. M., Castelli, M., & Popovič, A. (2018). An artificial intelligence system for predicting customer default in e-commerce. *Expert Systems with Applications, 104*, 1–21. doi:10.1016/j.eswa.2018.03.025

Wang, W., Kumar, N., Chen, J., Gong, Z., Kong, X., Wei, W., & Gao, H. (2020). Realizing the potential of the internet of things for smart tourism with 5G and AI. *IEEE Network, 34*(6), 295–301. doi:10.1109/MNET.011.2000250

Webster, C. (2021). Demography as a driver of robotics. *ROBONOMICS: The Journal of the Automated Economy*, *1*, 12–12.

Winston, P. H. (1993). *Artificial Intelligence* (III. Edition). Addison-Wesley Publishing Company the computer intelligence. USA: Massachusetts.

Xu, X. (2018). Research on the construction and development of smart hotels from the perspective of serving customers. In *2018 2nd International Conference on Education Science and Economic Management (ICESEM 2018)* (975-978). Atlantis Press. 10.2991/icesem-18.2018.228

Xue, Y., Fang, C., & Dong, Y. (2021). The impact of new relationship learning on artificial intelligence technology innovation. *International Journal of Innovation Studies*, *5*(1), 2–8. doi:10.1016/j.ijis.2020.11.001

Zhang, J. (2016). Weighing and realizing the environmental, economic and social goals of tourism development using an analytic network process-goal programming approach. *Journal of Cleaner Production*, *127*, 262–273. doi:10.1016/j.jclepro.2016.03.131

Zhang, L., Li, N., & Liu, M. (2012). On the Basic Concept of Smarter Tourism and Its Theoretical System. *Luyou Xuekan*, *27*(5), 66–73.

ADDITIONAL READING

McClements, D. J., & Grossmann, L. (2022). *Next-generation plant-based foods: design, production, and properties*. Springer Nature. doi:10.1007/978-3-030-96764-2

Richards, G. (2021). Evolving research perspectives on food and gastronomic experiences in tourism. *International Journal of Contemporary Hospitality Management*, *33*(3), 1037–1058. doi:10.1108/IJCHM-10-2020-1217

Tuomi, A., & Ascenção, M. P. (2023). Intelligent automation in hospitality: Exploring the relative automatability of frontline food service tasks. *Journal of Hospitality and Tourism Insights*, *6*(1), 151–173. doi:10.1108/JHTI-07-2021-0175

Yulia, I., Wulan, S., & Bilqis, L. D. R. (2022). History of traditional culinary and gastronomy and preservation efforts through culinary tourism (case study: Taoge Tauco (Geco) in Sukabumi City, West Java, Indonesia. In *Current Issues in Tourism, Gastronomy, and Tourist Destination Research* (pp. 373–377). Routledge. doi:10.1201/9781003248002-49

Zaragoza-Sáez, P., Marco-Lajara, B., & Ubeda-Garcia, M. (2022). Digital skills in tourism. A study from the Next Tourism Generation (NTG) Alliance. *Measuring Business Excellence*, *26*(1), 106–121. doi:10.1108/MBE-11-2020-0151

KEY TERMS AND DEFINITIONS

AI: Artificial Intelligence involves creating computer systems with human-like intelligence capabilities.

AR: Augmented Reality involves overlaying digital information on the real-world environment viewed through a device such as a smartphone or a computer.

Chatbot: A computer program that simulates a conversation with human users using text or voice-based interactions

Next-Gen Technology (NGT): NGT refers to cutting-edge advancements and innovations in various fields that aim to improve efficiency and provide new solutions. It includes technologies such as AI, 5G, IoT, quantum computing, robotics, and others.

SeviceBots: Robots designed to support and serve people through physical and social interactions.

Smart Technologies: Certain products and services that add value to the tourist experience by promoting higher interaction, **co**-creation, and personalization, using technology that enhances the experience.

VR: Virtual Reality is a computer-generated environment that can be interacted with using special equipment such as stereo-imaging goggles.

Chapter 2
An Analytical Study on Digital Transformation in the Poland Hospitality Industry

Ahmed Chemseddine Bouarar
https://orcid.org/0000-0001-8300-9833
University of Medea, Algeria

Karolina Stojczew
Wrocław University of Economics and Business, Poland

Kamel Mouloudj
https://orcid.org/0000-0001-7617-8313
University of Medea, Algeria

ABSTRACT

COVID-19 has exerted a huge impact on the hospitality; the digital transformation was one of the most prompt and the most solutions that guarantee businesses' survival in the light of lockdowns and mobility restrictions imposed in a bid to limit the epidemic spread. Hence, this chapter sought to explore the impact of COVID-19 on the hospitality industry and hospitality industry's digital transformation in Poland, as well as to identify the obstacles that may stymie the shift towards digitalization in this industry. To this end, a literature review methodology was employed. The results revealed that COVID-19 induced a negative impact on the hospitality industry in Poland in 2020, but with the beginning of 2021 this industry began to recover gradually. Also, COVID-19 quickened digital transformation adoption. The results also indicated that a sound and effective digital transformation reduces costs and improves business' performance; however, digital transformation may encounter several hurdles such as lack of knowledge, poor digital infrastructures, and lack of digital experience.

DOI: 10.4018/978-1-6684-6985-9.ch002

INTRODUCTION

COVID-19 pandemic has affected the majority sectors in almost all countries (Bouarar et al., 2020). However, the level of influence was different by sector and country (Mouloudj et al., 2020). Accordingly, the negative impact differed in the stages of the COVID-19 crisis, place, and from one industry to another. From the first moments of the COVID-19 spread, researchers were quick to study its impacts on the hospitality industry (Davahli et al., 2020; Gursoy & Chi, 2020; Temelkov, 2022; Thams et al., 2020) and the tourism industry (Bouarar et al., 2020; Korinth & Ranasinghe, 2020; Madani et al., 2020; Manczak & Gruszka, 2021; Stojczew, 2021; Vărzaru et al., 2021). In contrast, many companies have also accelerated the digitization of their services (Mouloudj et al., 2023), including hospitality services. Hence, "scientific research has intensified around strategizing the digitization of relations and the use of digital technologies" (Bovsh et al., 2023, p.134).

The term hospitality refers to "a segment of the service industry that includes hotels, restaurants, entertainment, sporting events, cruises, and other tourism-related services" (Terdpaopong, 2020, p.224). Globally, travel and tourism are the major job creation sectors, and global socio-economic and cultural development (McCabe & Qiao, 2020). However, the hospitality industry currently and due to the widespread COVID-19 is experiencing one of its most acute operational, commercial, and financial crises (Deri et al., 2023; Temelkov, 2022; Thams et al., 2020). In many destinations, cultural venues such as theatres and museums were closed, sporting events were postponed, the banning of access to beaches and national parks, and the closure of non-essential retail and hospitality outlets (Baum & Hai, 2020 ; Kilu et al., 2023). Therefore, companies were quick to find solutions in order to reduce losses. The COVID-19 crisis has compelled a new telecommuting reality incorporations and government organizations (Kilu et al., 2023). Several virtual markets have also emerged thanks to digital platforms, which led to the emergence of a wide gap between the digital and the real economy (Schilirò, 2021).

The characterization of the digital transformation phenomenon during the COVID-19 crisis should begin with an explanation of the meanings of the terms "transformation" and "digitization". Transformation is defined as a change or evolution and applies to elements both on the micro and macro scale (Mazurek, 2019). Digitization, on the other hand, is a concept defined as "a broadcasting digital form of various types of data or the conversion of ordinary, written and spoken language into digital" (Żabińska, 2016). Vial (2019, p.118) defines digital transformation as "a process that aims to improve an entity by triggering significant changes to its properties through combinations of information, computing, communication, and connectivity technologies". It refers to the use of new digital technologies in order to gain major business improvements (Liere-Netheler et al., 2018). The phenomenon of digital transformation is, therefore, a process of transformation and digitization of current solutions, markets and entities, and stuff. It is possible thanks to the development of information and communication technologies and the internet (Parviainen, 2017).

In terms of digitization, the greatest developed areas of the Polish economy include financial service, media, retail business, specialized and business service, along with the chemical and pharmaceutical industries (Ziółkowska, 2020). In 2020, 98.6% of Polish enterprises have access to the Internet and using a broadband Internet connection (fixed or mobile); 85.3% of enterprises using fixed broadband connection - DSL or other fixed (e.g. fiber optics technology (FTTP), cable technology); 71.3% enterprises having a website and more than 75% enterprises using a mobile broadband connection via portable devices (e.g. laptop, smartphone) (Statistics Poland, 2021a). According to Ziółkowska (2020), several Polish firms are until now in the phase of "automating single workstations" instead of creating "ecosystems devices" that work together without human being involvement (which characterizes Industry 4.0). In addition, notwithstanding Poland presently acquires the "fourth-largest digitally enabled talent pool in the EU, fewer people hold jobs that involve digital tasks than in Western Europe"(McKinsey, 2016).

The impacts of digital transformation on organizational and business performance (Chen et al., 2016; Mubarak et al., 2019; Popović-Pantić et al., 2019; Sousa-Zomer et al., 2020; Zhao et al., 2023), labor market (Dengler & Matthes, 2018), formal and informal organizational structures (Bonanomi et al., 2020), marketing activities (Ziółkowska, 2021), business process management (Pilipczuk, 2021), and project processes (Kozarkiewicz, 2020), brand promotion and positioning (Melović et al., 2020) have been widely discussed in the literature. However, little research focuses on digital transformation in the context of the global health crisis (e.g., Ahmad et al., 2021; Gabryelczyk, 2020; Kudyba, 2020; Soto-Acosta, 2020; Winarsih et al., 2021). In addition, previous research in Poland were mainly focused on assessing digital transformation in different sectors, such as energy sector (Chwiłkowska-Kubala et al., 2021; Chwiłkowska-Kubala et al., 2022; Siuta-Tokarska et al., 2022), banking sector (Czerwińska et al., 2021), and public sector (Krauze-Maślankowska, 2021). But the digital transformation in Poland' hospitality sector, is yet to be adequately investigated. Hence, the purpose of this chapter is to understand the factors that may drive or hinder digital transformation, especially in Poland, where few published papers on digital transformation in tourism and hotels have been done.

Based on the above, we try through our chapter to answer the following questions: Q1: What is the impact of the COVID-19 crisis on the hospitality industry in Poland? Q2: How the COVID-19 pandemic has impacted the hospitality industry's digital transformation? Q3: What are the challenges of adopting the hospitality industry's digital transformation during the COVID-19 pandemic in Poland? Therefore, the aim of our study is: (1) Identifying the compelling reasons, motivations, and incentives to adopt digital transformation in the hospitality industry; (2) Shedding light on the impact of the COVID-19 crisis on the hospitality industry in Poland; (3) Determining the impact of COVID-19 pandemic on hospitality industry's digital transformation; and (4) Identifying the main challenges and obstacles that could stymie hospitality industry's digital transformation during the COVD-19 pandemic in Poland.

To pursue the chapter purposes, a comprehensive review of literature pertaining to the topic has been conducted; the rest of the paper has been broken down into the following: the second part presents the background of the study by addressing digital transformation in the hospitality sector and accentuating the effect of digital transformation on hospitality business performance. The third part includes the chapter methodology, and the fourth part presents the results of the study. The fifth part presents the discussion and practical solution that may serve various stakeholders in Poland. Finally, the sixth part included future research directions and conclusion.

RESEARCH BACKGROUND

Digital Transformation in the Hospitality Industry

According to The results, in January 2020, the global human population was counting over 7.6 billion units. The Internet usage factor was around 57%, which means that over 4.3 billion people were connected to the Internet (Kemp, 2020). More than 5 billion people in the world use mobile devices and more than 3 billion people use social channels through them (approximately 93% of all users of social channels).

Digitization is not only associated with technology tools but also with the company's business philosophy (Ziółkowska, 2020). Although it is recognized that there is a digital gap between Poland and Western Europe, Poland has the potential to grow and expand in digitalization (McKinsey, 2016). According to the IAB Polska report, the number of Internet users continues to grow; in 2020, the number of people connected to the Internet was approximately 32 million people. 27.5 million people are connected to the Internet via mobile devices. Penetration of Internet access among households in Poland is 90.4% and is higher than in the previous year by 4 percentage points (Kolenda, 2020). Thanks to this data, we can see that potential target groups of digital service recipients have a great possibility of much more efficient and easier access to the customer.

Furthermore, digitalization affects the way and results of companies' business, as well as several dimensions such as circular supply chain management (Romagnoli et al., 2023), labor market (Sarabdeen & Alofaysan, 2023), and environmental performance (Shen et al., 2023). For instance, using a sample of 247 Chinese companies, the empirical evidence confirms that digital transformation positively affects companies' performance (Zhao et al., 2023). This is why late adoption of digitization tools can lead to the risk of losing position in fiercely competitive markets. Accordingly, digitization can affect the internal work environment of companies and bring new business opportunities. Consequently, the impact of digitization can be described from three different perspectives: (1) external opportunities, for example, the emergence of a new market such as online food delivery; (2) internal efficiency, for example, improving the quality of operations and reducing the work cost; and (3) disruptive change, for example, adopting cloud services, 6G technology, or augmented reality in tourism services (Parviainen et al., 2017, p.6).

On this basis, with the aim of protecting its competitive position and ensuring its survival in the markets, the hospitality industry increasingly employs technology in various operations. An ICT-based solutions approach can help tourism companies improve the quality of decisions and enrich the travel experience (Ammirato et al., 2015). Ongori and Migiro (2010) conclude that ICTs adoption and assimilation in SMEs is pivotal to boost their competitiveness. Nonetheless, reaping the benefits of digitization requires the company to share digital skills, recruit digital talent, provide an attractive work environment, and provide appropriate support (Schilirò, 2021). Besides, Chwiłkowska-Kubala et al. (2022) noted that enterprise size had no significant effect on the digitalization level. This suggests that small size of hospitality company does not prevent it from adopting digital solutions.

Digital Transformation and Hospitality Business Performance

The growth of digital technologies such as "social media platforms, the smartphone revolution, big data analytics, and the Internet-of-Things"has had a great influence on the growth of contemporary business models (Aloini et al., 2022; Ammirato et al., 2021) and hospitality industry (Terdpaopong, 2020). A recent study by Nikopoulou et al. (2023) demonstrated that "digital maturity of organizations, financial resource availability, and government regulations" are the significant antecedents of intention to adopt digital technologies in the hospitality industry. Ongori and Migiro (2010) argue that ICTs usage in SMEs will improve accessibility into the international markets. Previous studies dealt with the merits of adopting digital transformation, but the investigation of the effects of digital transformation on performance is still opaque and unexplored (Mitroulis & Kitsios, 2019). However, some literature has shown that digital transformation exerts positive effects on innovation, productivity, and financial performance (Popović-Pantić et al., 2020) and enhancing customer experiences (Jayawardena et al. et al., 2023; Terdpaopong, 2020). Bouarar et al. (2022, p.33) concluded that "successful digital transformation can speed up the pace of innovation, increase productivity, improve customer experiences and satisfaction, reduce costs, and improve business performance."

The digital transformation effects are associated with its advantages and the benefits it can provide to companies; as the adoption of digital technology leads to changing systems, processes, procedures, and work methods, which leads to improving the effectiveness and efficiency of providing products and services. In this way, the advantages of digital transformation include improving the customer experience and satisfaction, enhancing productivity, boost sales and profits, and increasing cost-saving (Gebayew et al., 2018). Popović-Pantić et al. (2020) found that product innovation mediates the effect of digital technology on financial performance. Therefore, digital transformation brings about several benefits not only to customers and the public hospitality companies as well, among these benefits is that it significantly helps to reduce costs and efforts, which results in improved operational efficiency, it improves quality facilitate procedures in providing services, and creates opportunities for a providing more creative and innovative services compared to conventional methods, digital transformation also helps hospitality companies to provide more services which are accessible to as much as possible of customers and public.

Nwankpa and Roumani (2016) found that digital transformation mediates the positive influence of IT capability on firm performance. Furthermore, the results show that digital transformation has a significant positive influence on both innovation and firm performance. Chen et al. (2016) found that the service-oriented portal function dimension "portal maintenance service, B2B function, and cloud computing" significantly influence SME's organizational performance. Popović-Pantić et al. (2019) found that digital transformation has a significant, positive, and strong impact on business performance (financial and non-financial). Mubarak et al. (2019) discovered that "big data, cyber-physical systems, and interoperability" have clear effects on business performance; however, the influence of the Internet of Things was found to be insignificant. Additionally, Sousa-Zomer et al. (2020) found that the building blocks of the digital transforming capability" i.e.; digital-savvy skills, digital intensity, and context for action and interaction," can drive the sustainability of corporate business performance in the digital economy context. Nonetheless, Aral and Weill (2007) found that firms' total IT investment is not related to performance. Also, Koski (2010) suggests that mobile connectivity does not significantly contribute to the firm' growth and profitability.

Based on the above discussion, we argue that effective digital transformation adoption has a significant effect on hospitality organizations' performance. However, the level of impact varies from one firm to another and from one context to another, undoubtedly a proper digital transformation is bound to reduce transaction costs and increase customer satisfaction, positively impacts firms' performance.

METHODOLOGY

The research methods include a critical review of existing literature. According to Torraco (2016), the literature review is different forms of scientific study that builds on existing research to generate new knowledge. Literature review as a research method is more relevant than ever (Snyder, 2019). The literature review is one of the scientific methodologies whereby researchers collect, analyze, and criticize high-quality manuscripts that have been written on a given topic, recent literature reviews have been widely employed in various disciplines.

In this study, the researchers employed theoretical analysis, and "the analysis scale was at the international level" (Bouarar et al., 2020; Mouloudj et al., 2020); with a particular focus on the Polish model. According to Mkwizu (2020), "literature review as a research methodology has been documented by previous and current scholars" (p.9).The literature reviewed in this chapter covers journals indexed in Scopus and Web of Science, reports, internationally recognized databases, and conference papers to explore the digital transformation in the context of the COVID-19 pandemic with a focus on the hospitality industry in Poland. By incorporating findings and perspectives from many empirical findings, literature reviews can discuss research questions in an unprecedented manner (Snyder, 2019).

FINDINGS AND ANALYSIS

The Impacts of the COVID-19 Crisis on Poland's Hospitality Industry

The COVID-19 pandemic crisis has been challenging for all companies in various sectors of activity, tourism, on the other hand, was one of the sectors of the economy most affected by the crisis (Abbas et al., 2021; Korinth & Ranasinghe, 2020; Manczak & Gruszka, 2021; Stojczew, 2021).Undoubtedly,epidemics and all threats associated with them lead to a decline in tourism demand in places where a certain infectious disease exists (Dębski et al., 2021). Tourist destinations have implemented emergency measures and restrictions that have influenced individual mobility worldwide (Vărzaru et al., 2021). While some industries experienced minor consequences, firms in the hospitality industry were closed for months (Baum & Hai, 2020). COVID-19 imposed lockdown and travel bans which abruptly halted the demand for hospitality service, forcing many hospitality businesses to temporarily or permanently close (Gursoy & Chi, 2020; Manczak & Gruszka, 2021).

Beaches and resorts were closed, cities were empty, people's movements were restricted, and travel among different places was sternly banned (Vărzaru et al., 2021). Many hotels, restaurants, and companies related to tourism activities have been forced to employ new internal working methods, and they were urged to provide products through digital channels. Besides, the low number of tourists caused a significant loss for hotel, restaurant, transport, entertainment, or souvenir sectors, and due to the multiplier, it indirectly impacted other cooperating sectors and other branches (Marjański & Sułkowski, 2021). COVID-19 has also affected employment (Kilu et al., 2023), as many companies, including hospitality companies, dismissal of part of their workers, and this had a negative impact on family income. Accordingly, hospitality companies were affected by two aspects: (1) reducing their workforce due to lack of demand; and (2) lower demand due to lower household incomes.

In Poland, upon the announcement of an epidemic emergency on14 March 2020, all activities associated with the operation of tourist accommodation establishments and short-stay accommodations, as well as health treatment, have been limited; and the imposed restrictions on the movement of people ensued a decrease in the number of tourists accommodated in tourist accommodation establishments (Statistics Poland, 2020). In 2020, both the occupancy and RevPAR in Polish top urban hotel markets plummeted to respectively 25% and 28% of the value of the corresponding day in 2019 and the lowest values of occupancy (18% of the value of the corresponding day in 2019) were registered in Lublin, Łódź, and Sopot when the first deaths announced by COVID-19 in Poland (Napierała et al., 2020). In March 2020, according to data presented by Statistics Poland (2020), the number of local tourists fell by 63% and of foreign ones by 69%. It is estimated that in comparison with March 2019, the number of tourists accommodated in hostels decreased the most (by over 70%), and the least in motels (by 55%) and health establishments (by 50%); and the number of tourists accommodated in establishments with 10 or more bed places was about 65% lower than in March 2019 (Statistics Poland, 2020). The small number of tourists visiting accommodation establishments was noticed in all Polish provinces, with the lowest decrease in "West Pomerania Province (by 34.4%) and Warmia-Masuria Province (by 36.6%), and the highest decrease in Mazovia Province (61.8%)" (Dębski et al., 2021).

Regarding the negative effects of COVID-19 on hotels' performance, restrictions in people's mobility and limited activity of the hotels caused a significant fall in the number of individuals visiting tourist accommodation establishments in Poland; and during the lockdown in April 2020, almost no tourist activity was registered, and the number of tourists in accommodation establishments was 28 times lower comparing to the previous year (Dębski et al., 2021).Moreover, Napierała et al (2020), who discussed the effect of the geographiclocation of the COVID-19 pandemic on the hotels' performances in Poland, noticed that hotel performance indicators witnessed a significant decrease in the period when the COVID-19 pandemic hit Europe mainly. They also found that the negative impact of Polish cases of COVID-19 is more significant in less visited destinations like "Lublin and Sopot".

According to Statistics Poland (2021b), in April 2021, 379.8 thousand tourists were hosted in tourist accommodation establishments that are 4 times more than a year ago. The number of overnight stays also increased to the same extent. An increase in the number of tourists and overnight stays was also noticed in May 2021. Compared to the same month of the previous year, both the number of tourists and the number of overnight stays increased three fold. In May 2021, the highest number of persons was accommodated in Mazowieckie Voivodship (143.0 thousand) and Małopolskie Voivodship" (133.3 thousand), and the lowest in Opolskie Voivodship (13.5 thousand). An increase in the number of tourists occurred in establishments located in all Voivodships, of which the largest increase occurred in Małopolskie Voivodship (over 4 times) (Statistics Poland, 2021b).

The COVID-19 Crisis' Impact on the Hospitality Industry's Digital Transformation

Companies witnessed a major change and, in a very short time, opted for solutions based on digital technologies. Meanwhile, it has become necessary to adjust management and collaboration models to guarantee that no one within organizations is less important and feels excluded from this digitization process. In states that a key element for the success of teamwork in COVID-19 time is the involvement of all team members in the company's major challenges. But this process has necessarily become more challenging, as the amount of information that naturally flowed in the same physical space is now becoming difficult to reach everyone in a rapid and effective way (Berger, 2020). The COVID-19 pandemic had a huge effect on the growth of online commerce, mainly due to trade closure and mobility restrictions. This action may have a wider extent. To suggest that the dematerialization of processes will substantially decrease some fixed costs for companies. Meanwhile, the younger generations will use their own behavior of consumption of products and services in the digital society.

The hospitality industry has been one of the sectors hit by the COVID-19 pandemic (Breier et al., 2020; Gursoy & Chi, 2020). The COVID-19 pandemic has, without doubt, brought about organizational changes, obliged the development of new business strategies, and, acted as a stimulator for digital transformation in many areas of the economy, healthcare, and education (Gabryelczyk, 2021). First theoretical and practical observations in the hospitality industry suggest that business model innovation might be an effective solution to successfully overcome the COVID-19 crisis (Breier et al., 2020; Kilu et al., 2023). Hence, COVID-19 has quickened the digital transition, and the adoption of digital technologies has become a must to assure the continuity of work, private, and social life (Schilirò, 2021).

During the COVID-19 crisis, the adoption of digital transformation in the tourism sector has increased. The hospitality industry has invested a lot of money to adopt new technologies for the improvement of tourism services. In this context and from the operational perspective, conventional business operations have known disruption caused by new tools, including increasing reliance on online distribution channels (Alrawadieh et al., 2021), such as online learning systems (Hauke et al., 2021; Mouloudj et al., 2021), online food delivery services (Bouarar et al., 2021), online travel agencies (Alrawadieh et al., 2021), and mobile health apps (Mouloudj et al., 2023). Accordingly, the transformation towards adopting digitization was a necessity (not an option) in order to adapt to a new situation.

From a strategic perspective, it is of vital importance to bring about a strategy related to digital transformation (Ziółkowska, 2020), where digital strategy prioritizes using technology to improve business performance. Heavin and Power (2018) argue that a strategic vision for digital transformation is beneficial, but the vision must be predicated upon a full understanding of customer needs and technical possibilities. In addition, the adoption of information and communication technologies and the digital transformation of businesses is very important in developing business strategies, boosting creativity and innovation, and improving competitiveness (Ongori & Migiro 2010). Diener and Špaček (2021) revealed that elements of strategy and management, technology and regulation, customers, and employees occupy a high level of attention within the digital transformation. According to Adamczewski (2018), digital transformation requires more than just the adoption of digital tools, but a comprehensive change of mindset throughout the organization. However, it is noteworthy that the precondition for benefiting from digital transformation is the company's ability to apply it innovatively (Popović-Pantić et al., 2020). Hence, digital innovation plays a significant role in the success of the digital transformation strategy. Abudaqa et al. (2022) also confirm that innovation is a pre-requisite for digital transformation. Digital innovation, according to Bouabdellah (2023, p.1), is "an ongoing strategy that has drastically transformed businesses, communications, economies, and other facets of daily life." In this context, Abudaqa et al. (2022) found that digital transformation strategies positively correlated with overall performance of small and medium firms; and they revealed that innovation can significantly moderates this relationship.

Challenges of Adopting the Hospitality Industry's Digital Transformation in Poland

Many researchers have tried to determine the key factors of successful digital transformation. Martin (2018) found that successful digital transformation depends on having the appropriate digital-knowledge leaders in place, developing capabilities for the workforce of the future, empowering people to work in new methods, providing a day-to-day tool a digital improvement, and frequently relating to traditional and digital methods. On the other hand, according to Gupta (2018), major hurdles to digital transformation regarding organizations are vague vision and objective of the digital transformation; poor management understanding, knowledge, and experience in digital transformation; little organizational agility; poor digital leadership skills "forward-looking, understanding of technology, open-mind, collaboration"; lack of organizational culture flexibility; the non-alignment of rewards and incentives along with digital transformation; opaque measurement and rewarding systems; lack of employee involvement and engagement; and employees' resistance to change. In addition, Bouarar et al. (2022, p.33) revealed that the important obstacles to adopting digital transformation are " lack of knowledge, lack of digital expertise, poor digital leadership, resistance to change, inflexible culture, unclear vision, and objective, lack of collaboration and alignment." Similarly, Jayawardena et al. (2023, p.1) argued that "the lack of technology acceptance strategy is one of the fundamental causes of failures observed in digital transformation".

De Bernardi et al. (2019) analyzed the ambit to which digitalization should be applied in museums' communication strategies, and to identify the logic preventing digital transformation in cultural heritage strategic management in Italy. They found that the level of digital readiness is still poor. In addition to the well-known systemic financial deficit of cultural institutions, other key factors hamper the integration of digitalization processes in cultural heritage management. The most known obstacles include institutional pressures and the lack of organizational and managerial harmony between different departments and functions of museums.

In Poland, the main obstacles that stymie a comprehensive digital transformation are the increased costs of its adoption and shortages of the necessary infrastructure (e.g., not enough broadband connections) (Ziółkowska, 2020). The Polish labor market contains employees from four generations, but not all of them acquire the necessary competencies of the knowledge-based economy and the 4.0 economy (Kryk, 2021). According to Müller and Hopf (2017, p. 1496), human knowledge faculties are the major factor for digital transformation. As claimed by Moroz (2018), that infrastructure is a primordial factor, but it is not enough. Along with the necessary digital infrastructure, a digital faculty of the workers is pivotal to a organizational changes to respond to this critical situation (Fonseca & Picoto, 2020). For local Polish companies, securing the necessary financial resources is the biggest challenge, alongside with lack of employee motivation from the managerial staff or a clear vision of how the changes should be applied (Ziółkowska, 2020). In the logistics service industry, Cichosz et al. (2020) found that the complexity of the logistics network and lack of resources are among the most important obstacles to adopting digital transformation. They also showed that the most important factor for the success of the digital transformation is "a leader having and executing a digital transformation vision, and creating a supportive organizational culture."

It has been noticed that senior leadership teams who lack digitalization experience are a huge hurdle to business transformation (Sawy et al., 2016). Therefore, managers with the know-how and experience of digital are prone to implement the constant renewal required in the digital era (Sousa-Zomer et al., 2020). According to Machado et al. (2020), the lack of knowledge among manufacturing companies is the major hurdle in the transition to Industry 4.0 in Sweden. Sari et al. (2020) think that the main obstacle to successfully implementing Industry 4.0 in the Turkish manufacturing industry is the absence of a collaborative strategy for digitalization. Cichosz et al. (2020) found that the most important factor for the success of the digital transformation is "a leader having and executing a digital transformation vision, and creating a supportive organizational culture." Accordingly, Polish hospitality companies have to overcome barriers to digitizing their business to achieve a successful digital transformation.

DISCUSSION

The COVID-19 has severely buffeted the tourism and leisure industry and has made it among the most damaged global industries (Abbas et al., 2021; Breier et al., 2020). The repercussions and pace of technological disruption in organizations are increasing and have been accelerated by COVID-19 (Kilu et al., 2023). The COVID-19 pandemic has quickened the digital transformation process, and because of the rapid digital transformation; economies around the world are going through severe changes in businesses, consumers, investments, trade, and government activities (Bouarar et al., 2022; Schilirò, 2021). In the era of economic changes, companies in Poland should be aware of how significantly their business is influenced by digital transformation. These changes also take place inside the organization and existing processes, such as customer service. Polish enterprises should take into account that customers in their strategy, the market has changed and requires constant attention. The existing processes are not enough to meet his needs, if only because the service process does not end there at the time of purchase.

In the future, these elements will be a key opportunity for Polish enterprises wishing to achieve market success. In such dynamic market conditions, the permanent nature of competitive advantages turns into a portfolio of temporary advantages. Business owners in Poland should constantly monitor customer service management processes both at home and at the competition in order to be able to react and maintain as soon as possible present position or strengthen it. We found that several factors stimulate organizations to urgently adopt digital transformation during the COVID-19 crisis, among these factors, for instance, management support, technology infrastructure, globalization, the increasing usage of smart devices, rivalry pressure, and customer awareness. However, digital transformation will be the only way to progress if it changes the traditional way of doing business and confers value by contributing to innovation (Popović-Pantić et al., 2020). In the critical period of transformation, clear leadership, and the commitment of all stakeholders, who will be involved in the whole process of transition is of vital importance (Heavin & Power, 2018).

The results also revealed that major barriers to digital transformation in the context of COVID-19 include lack of knowledge, lack of digital expertise, poor digital leadership, resistance to change, inflexible culture, unclear vision and objective, and lack of collaboration and alignment. Schilirò (2021) argues that the main success factors for a sound and inclusive digital transformation depend on the joint efforts of the state, business, and people. This means that flexible and joint strategic action can achieve better results. In addition, knowledge management and training programs can also contribute to overcoming organizational obstacles.

Governments are already developing strategies to boost their tourism economies in the aftermath of COVID-19 by help in sustain able tourism recovery, encouraging digital transformation and the transition to a greener tourism framework, and rethinking tourism for the future. In this context, an important catalyst for Polish enterprises to implement intelligent organizations lies in the dissemination of knowledge about intelligent organizations and the promotion of the good and successful experiences and practices of enterprises that already implemented the intelligent organization formula (Adamczewski, 2018). Besides, analyzing recovery measures utilized by European governments' shows that authorities are relatively aware of the dire need to boost tourists' sense of security, fulfill the increased need for information, and stimulate confidence. However, Napierała et al. (2020) argued that the hotel industry, "especially in the most internationalized, biggest Polish cities," is likely to recover only when problems resulting from the COVID-19 epidemic will be solved at the European level. This suggests that effective management of tourism crises during epidemics requires intensified international cooperation efforts.

Digitization of companies will increase the importance given to the digital channels of marketing and sales of companies. It will also encourage teleworking and consumption of technological products as more people will get involved in using hybrid communication mechanisms accessible from anywhere, and not exclusively in the physical environment of companies and their homes. The success of a company's digital transformation processes will depend to a great extent on the adoption of the community, namely its employees, suppliers, partners, and customers. Among the expected challenges is the adoption of telework and a distance working model that confer high interactivity and cooperation, in which the talent overcomes the geographical location of these people. The growth of e-commerce is another challenge as it is related to it a whole value chain that must be properly integrated to provide a differentiated shopping experience for the customer. The flexibility of digital products and services will also be a determining factor for their adoption. Hence, modern Polish enterprises willing to operate on the market effectively have to guarantee the flexibility of their organization and its ability to adopt innovative business models and reorganize logistics processes (Adamczewski, 2018).

Accordingly, digital transformation in the Polish hospitality industry is expected to witness an increase in costs in the short run. However, hospitality firms' performance and revenues are expected to improve in the medium and long run, especially with the end of the pandemic. Finally, we hope that the study findings may help managers and decision-makers to design effective programs, policies, and strategies that would illuminate and pave the way for Polish enterprises to adopt digital transformation.

FUTURE RESEARCH DIRECTIONS

This research primarily focused on exploring of digital transformation in the Polish hospitality industry in general. Hence, we believe that it would be beneficial for futures studies to address the following topics, (1) the challenges of implementation digital transformation strategy in Polish hotels; (2) the impact of digital transformation on Poland's hotels' performance; supply chain performance; or entrepreneurship performance, (3) the impact of adopting digital technology on the effectiveness of managing tourist complaints, and (4) determinants of adoption digital transformation in hospitality industry.

CONCLUSION

The hospitality industry in most countries of the world in general, and in Poland in particular, was greatly affected during the COVID-19 pandemic, as hotels, restaurants, airports, museums, and parks closed its doors to visitors. Accordingly, there was a decrease in the revenues of some hospitality companies, and other companies suffered financial losses, while some companies declared bankruptcy. This has led to layoffs of many employees in the tourism sector. Hence, many hospitality companies have had to adopt digital solutions and plan for digital transformation. The results revealed that successful adoption of digital transformation in the hospitality industry can reduce costs and improve profitability (in the medium and long term), and it also mitigates the negative effects of COVID-19 on hospitality companies. However, successful digital transformation remains a distant dream for some companies for several reasons, including lack of vision clarity, lack of infrastructure, lack of funding, lack of experience, lack of support, and employee resistance to change. Therefore, we believe that an accurate diagnosis of the nature and level of these barriers can provide insights into how to remove these barriers. For example, government support can overcome the financial difficulties that may face some hospitality companies that have been severely affected by the COVID-19 crisis.

ACKNOWLEDGMENT

This research received no specific grant from any funding agency in the public, commercial, or not-for profit sectors.

REFERENCES

Abbas, J., Mubeen, R., Iorember, P. T., Raza, S., & Mamirkulova, G. (2021). Exploring the impact of COVID-19 on tourism: Transformational potential and implications for a sustainable recovery of the travel and leisure industry. *Current Research in Behavioral Sciences*, *2*, 1–11. doi:10.1016/j.crbeha.2021.100033

Abudaqa, A., Alzahmi, R. A., Almujaini, H., & Ahmed, G. (2022). Does innovation moderate the relationship between digital facilitators, digital transformation strategies and overall performance of SMEs of UAE? *International Journal of Entrepreneurial Venturing*, *14*(3), 330–350. doi:10.1504/IJEV.2022.124964

Adamczewski, P. (2018). Intelligent organizations in digital age. *MEST Journal, 6*(2), 1-11. https://doi.org/. doi:1-11

Ahmad, A., Alshurideh, M. T., Al Kurdi, B. H., & Salloum, S. A. (2021). Factors impact organization digital transformation and organization decision making during COVID-19 pandemic. In: Alshurideh M., Hassanien A.E., Masa'deh R. (Eds), The Effect of Coronavirus Disease (COVID-19) on Business Intelligence. Springer, Cham. https://doi.org/ doi:10.1007/978-3-030-67151-8_6

Aloini, D., Latronico, L., & Pellegrini, L. (2022). The impact of digital technologies on business models. Insights from the space industry. *Measuring Business Excellence*, *26*(1), 64–80. doi:10.1108/MBE-12-2020-0161

Alrawadieh, Z., Alrawadieh, Z., & Cetin, G. (2021). Digital transformation and revenue management: Evidence from the hotel industry. *Tourism Economics*, *27*(2), 328–345. doi:10.1177/1354816620901928

Ammirato, S., Felicetti, A. M., Linzalone, R., & Carlucci, D. (2021). Digital business models in cultural tourism. *International Journal of Entrepreneurial Behaviour & Research*, *28*(8), 1940–1961. doi:10.1108/IJEBR-01-2021-0070

Aral, S., & Weill, P. D. (2007). IT assets, organizational capabilities, and firm performance: How resource allocations and organizational differences explain performance variation. *Organization Science*, *18*(5), 763–780.

Baum, T., & Hai, N. T. T. (2020). Hospitality, tourism, human rights and the impact of COVID-19. *International Journal of Contemporary Hospitality Management*, *32*(7), 2397–2407. doi:10.1108/IJCHM-03-2020-0242

Berger, R. (2020). *Digital workplace in the era of COVID-19*. Roland Berger. https://www.rolandberger.com/en/Point-of-View/Digitalworkplace-in-the-era-of-Covid-19.html (Accessed: May 5, 2020).

Bonanomi, M. M., Hall, D. M., Staub-French, S., Tucker, A., & Talamo, C. M. L. (2020). The impact of digital transformation on formal and informal organizational structures of large architecture and engineering firms. *Engineering, Construction, and Architectural Management, 27*(4), 872–892. doi:10.1108/ECAM-03-2019-0119

Bouabdellah, M. (2023). Digital Innovation in Healthcare. In A. Bouarar, K. Mouloudj, & D. Martínez Asanza (Eds.), *Integrating Digital Health Strategies for Effective Administration* (pp. 1–19). IGI Global., doi:10.4018/978-1-6684-8337-4.ch001

Bouarar, A. C., Mouloudj, K., & Mouloudj, S. (2020). The impact of coronavirus on tourism sector - an analytical study. *Journal of Economics and Management, 20*(1), 323–335.

Bouarar, A. C., Mouloudj, S., & Mouloudj, K. (2021). Extending the theory of planned behavior to explain intention to use online food delivery services in the context of COVID -19 pandemic. In C. Cobanoglu, & V. Della Corte (Eds.), Advances in global services and retail management (pp. 1–16). USF M3 Publishing.

Bouarar, A. C., Mouloudj, S., & Mouloudj, K. (2022). Digital transformation: Opportunities and challenges. In N. Mansour & S. Ben Salem (Eds.), *COVID-19's Impact on the Cryptocurrency Market and the Digital Economy* (pp. 33–52). IGI Global. doi:10.4018/978-1-7998-9117-8.ch003

Bovsh, L. A., Hopkalo, L. M., & Rasulova, A. M. (2023). Digital Relationship Marketing Strategies of Medical Tourism Entities. In A. Bouarar, K. Mouloudj, & D. Martínez Asanza (Eds.), *Integrating Digital Health Strategies for Effective Administration* (pp. 133–150). IGI Global. doi:10.4018/978-1-6684-8337-4.ch008

Breier, M., Kallmuenzer, A., Clauss, T., Gast, J., Kraus, S., & Tiberius, V. (2020). The role of business model innovation in the hospitality industry during the COVID-19 crisis. *International Journal of Hospitality Management, 92*, 1–10. doi:10.1016/j.ijhm.2020.102723 PMID:36919038

Chen, Y.-Y. K., Jaw, Y.-L., & Wu, B.-L. (2016). Effect of digital transformation on organizational performance of SMEs: Evidence from the Taiwanese textile industry's web portal. *Internet Research, 26*(1), 186–212. doi:10.1108/IntR-12-2013-0265

Chwiłkowska-Kubala, A., Cyfert, S., Malewska, K., Mierzejewska, K., & Szumowski, W. (2021). The relationships among social, environmental, economic CSR practices and digitalization in Polish energy companies. *Energies, 14*(22), 7666. doi:10.3390/en14227666

Chwiłkowska-Kubala, A., Malewska, K., & Mierzejewska, K. (2022). Digital transformation of energy sector enterprises in Poland. *Scientific Papers of Silesian University of Technology – Organization and Management Series, 162*, 101-120. http://dx.doi.org/ doi:10.29119/1641-3466.2022.162.5

Cichosz, M., Wallenburg, C. M., & Knemeyer, A. M. (2020). Digital transformation at logistics service providers: Barriers, success factors and leading practices. *International Journal of Logistics Management, 31*(2), 209–238. doi:10.1108/IJLM-08-2019-0229

Czerwińska, T., Głogowski, A., Gromek, T., & Pisany, P. (2021). Digital transformation in banks of different sizes: Evidence from the Polish banking sector. In I. Boitan & K. Marchewka-Bartkowiak (Eds.), *Fostering Innovation and Competitiveness With FinTech, RegTech, and SupTech* (pp. 161–185). IGI Global. doi:10.4018/978-1-7998-4390-0.ch009

De Bernardi, P., Bertello, A., & Shams, R. (2019). Logics hindering digital transformation in cultural heritage strategic management: An Exploratory Case Study. *Tourism Analysis*, *24*(3), 315–327. doi:10.3727/108354219X15511864843876

Dębski, M., Borkowska-Niszczota, M., & Andrzejczyk, R. (2021). Tourist Accommodation Establishments during the Pandemic – Consequences and Aid Report on a Survey among Polish Micro-enterprises Offering Accommodation Services. *Journal of Intercultural Management*, *13*(1), 1–25. doi:10.2478/joim-2021-0001

Dengler, K., & Matthes, B. (2018). The impacts of digital transformation on the labor market: Substitution potentials of occupations in Germany. *Technological Forecasting and Social Change*, *137*, 304–316. doi:10.1016/j.techfore.2018.09.024

Deri, M. N., Ari Ragavan, N., Niber, A., Zaazie, P., Akazire, D. A., Anaba, M., & Andaara, D. (2023). COVID-19 shock in the hospitality industry: Its effect on hotel operations within the Bono region of Ghana. *African Journal of Economic and Management Studies*. doi:10.1108/AJEMS-07-2022-0264

Diener, F., & Špaček, M. (2021). Digital Transformation in Banking: A Managerial Perspective on Barriers to Change. *Sustainability*, *13*(4), 1–26. doi:10.3390u13042032

Gabryelczyk, R. (2020). Has COVID-19 accelerated digital transformation? Initial lessons learned for public administrations. *Information Systems Management*, *37*(4), 303–309. doi:10.1080/10580530.2020.1820633

Gebayew, C., Hardini, I. R., Panjaitan, G. H. A., & Kurniawan, N. B., & Suhardi, (2018). A systematic literature review on digital transformation. *International Conference on Information Technology Systems and Innovation* (pp. 260-265). IEEE. https://doi.org/10.1109/ICITSI.2018.8695912

Gupta, S. (2018). *Organizational barriers to digital transformation. KTH Royal Institute of Technology.* School of Industrial Engineering and Management.

Gursoy, D., & Chi, C. G. (2020). Effects of COVID-19 pandemic on hospitality industry: Review of the current situations and a research agenda. *Journal of Hospitality Marketing & Management*, *29*(5), 527–529. doi:10.1080/19368623.2020.1788231

Hauke, J., Bogacka, E., Tobolska, A., & Weltrowska, J. (2021). Students of public and private universities in Wielkopolska region (Poland) facing the challenges of remote education during the COVID-19 pandemic. *Studies of the Industrial Geography Commission of the Polish Geographical Society*, *35*(4), 205–226. doi:10.24917/20801653.354.13

Heavin, C., & Power, D. J. (2018). Challenges for digital transformation: Towards a conceptual decision support guide for managers. *Journal of Decision Systems*, *27*(1), 38–45. doi:10.1080/12460125.2018.1468697

Jayawardena, C., Ahmad, A., Valeri, M., & Jaharadak, A. A. (2023). Technology acceptance antecedents in digital transformation in hospitality industry. *International Journal of Hospitality Management, 108*, 103350. doi:10.1016/j.ijhm.2022.103350

Kemp, S. (2020). Global internet use accelerates. Retrieved from https://wearesocial.com/digital-2020-global-internet-use-accelerates

Kilu, R. H., Sanda, M.-A., & Alacovska, A. (2023). Demystifying business models (shifts) among Ghanaian creative entrepreneurs in a COVID-19 era. *African Journal of Economic and Management Studies, 14*(2), 188–204. doi:10.1108/AJEMS-07-2022-0305

Kolenda, P. (2020). *Raport Strategiczny Internet 2019/2020*. ICAN Institute.

Korinth, B., & Ranasinghe, R. (2020). Covid-19 pandemic's impact on tourism in Poland in March 2020. *Geo Journal of Tourism and Geosites, 31*(3), 987–990. doi:10.30892/gtg.31308-531

Koski, H. (2010). Firm growth and profitability: The role of mobile IT and organizational practices. *Discussion Paper No. 1222*. The Research Institute of the Finnish Economy. https://www.etla.fi/wp-content/uploads/2012/09/dp1222.pdf

Kozarkiewicz, A. (2020). General and specific: The impact of digital transformation on project processes and management methods. *Foundations of Management, 12*(1), 237–248. doi:10.2478/fman-2020-0018

Krauze-Maślankowska, P. (2021). Open data and smart city initiatives for digital transformation in public sector in Poland. A survey. In: Wrycza, S., Maślankowski, J. (eds) *Digital Transformation*. Springer, Cham. https://doi.org/10.1007/978-3-030-85893-3_5

Kryk, B. (2021). Generations on the Polish labor market in the context of competencies needed in the economy based on knowledge and 4.0. [Economics and Law]. *Ekonomia I Prawo, 20*(1), 121–137. doi:10.12775/EiP.2021.008

Kudyba, S. (2020). COVID-19 and the acceleration of digital transformation and the future of work. *Information Systems Management, 37*(4), 284–287. doi:10.1080/10580530.2020.1818903

Liere-Netheler, K., Packmohr, S., & Vogelsang, K. (2018). Drivers of digital transformation in manufacturing. Proceedings of the 51st Hawaii International Conference on System Sciences, (pp. 3926-3935). Semantic Scholar. https://pdfs.semanticscholar.org/5783/7648a8ca127462f2ef35f2e4a6e3a4f7508e.pdf

Machado, C. G., Winroth, M., Carlsson, D., Almström, P., Centerholt, V., & Hallin, M. (2019). Industry 4.0 readiness in manufacturing companies: Challenges and enablers towards increased digitalization. *Procedia CIRP, 81*, 1113–1118. doi:10.1016/j.procir.2019.03.262

Madani, A., Boutebal, S. E., Benhamida, H., & Bryant, C. R. (2020). The impact of the COVID-19 outbreak on the tourism needs of the Algerian population. *Sustainability, 12*(21), 1–11. doi:10.3390u12218856

Manczak, I., & Gruszka, I. (2021). Averting the effects of the COVID-19 pandemic in tourism - a semantic field analysis. *Studies of the Industrial Geography Commission of the Polish Geographical Society, 35*(3), 164–176. doi:10.24917/20801653.353.10

Marjański, A., & Sułkowski, Ł. (2021). Consolidation strategies of small family firms in Poland during the Covid-19 crisis. *Entrepreneurial Business and Economics Review*, *9*(2), 167–182. doi:10.15678/EBER.2021.090211

Martin, J. F. (2018). *Unlocking success in digital transformations*. McKinsey & Company.

McCabe, S., & Qiao, G. (2020). A review of research into social tourism: Launching the annals of tourism research curated collection on social tourism. *Annals of Tourism Research*, *85*, 103103. doi:10.1016/j.annals.2020.103103

McKinsey & Company. (2016). *Digital Poland: Capturing the opportunity to join leading global economies*. McKinsey. https://www.mckinsey.com/~/media/McKinsey/Business%20Functions/McKinsey%20Digital/Our%20Insights/Digital%20Poland/Digital%20Poland.ashx

Melović, B., Jocović, M., Dabić, M., Vulić, T. B., & Dudic, B. (2020). The impact of digital transformation and digital marketing on the brand promotion, positioning, and electronic business in Montenegro. *Technology in Society*, *63*, 101425. doi:10.1016/j.techsoc.2020.101425

Mitroulis, D., & Kitsios, F. (2019). MCDA for assessing the impact of digital transformation on hotel performance in Thessaloniki. *Proceedings of the 8th International Symposium & 30th National Conference on Operational Research*, (pp. 53–57). HELORS. http://eeee2019.teiwest.gr/docs/HELORS_2019_proceedings.pdf#page=54 .

Mkwizu, K. H. (2020). Digital marketing and tourism: Opportunities for Africa. *International Hospitality Review*, *34*(1), 5–12. doi:10.1108/IHR-09-2019-0015

Moroz, M. (2018). Acceleration of digital transformation as a result of launching programs financed from public funds: Assessment of the implementation of the operational program digital Poland. *Foundations of Management*, *10*(1), 59–74. doi:10.2478/fman-2018-0006

Mouloudj, K., Bouarar, A. C., Asanza, D. M., Saadaoui, L., Mouloudj, S., Njoku, A. U., Evans, M. A., & Bouarar, A. (2023). Factors Influencing the Adoption of Digital Health Apps: An Extended Technology Acceptance Model (TAM). In A. Bouarar, K. Mouloudj, & D. Martínez Asanza (Eds.), *Integrating Digital Health Strategies for Effective Administration* (pp. 116–132). IGI Global. doi:10.4018/978-1-6684-8337-4.ch007

Mouloudj, K., Bouarar, A. C., & Fechit, H. (2020). The impact of COVID-19 pandemic on food security. Les cahiers du CREAD, 36(3), 159-184. https://doi.org/ doi:10.6084/m9.figshare.13991939.v1

Mouloudj, K., Bouarar, A. C., & Stojczew, K. (2021). Analyzing the students' intention to use online learning system in the context of COVID-19 pandemic: A theory of planned behavior approach. In W. B. James, C. Cobanoglu, & M. Cavusoglu (Eds.), Advances in global education and research (Vol. 4, pp. 1–17). USF M3 Publishing.

Mubarak, M. F., Shaikh, F. A., Mubarik, M., Samo, K. A., & Mastoi, S. (2019). The impact of digital transformation on business performance: A study of Pakistani SMEs. Engineering Technology & *Applied Scientific Research*, *9*(6), 5056–5061.

Müller, E., & Hopf, H. (2017). Competence center for the digital transformation in small and medium-sized enterprises. *Procedia Manufacturing, 11,* 1495–1500. doi:10.1016/j.promfg.2017.07.281

Napierała, T., Leśniewska-Napierała, K., & Burski, R. (2020). Impact of geographic distribution of COVID-19 cases on hotels' performances: Case of Polish cities. *Sustainability, 12*(11), 4697. doi:10.3390u12114697

Nikopoulou, M., Kourouthanassis, P., Chasapi, G., Pateli, A., & Mylonas, N. (2023). Determinants of digital transformation in the hospitality industry: Technological, organizational, and environmental drivers. *Sustainability, 15*(3), 2736. doi:10.3390u15032736

Nwankpa, J. K., & Roumani, Y. (2016). IT capability and digital transformation: A firm performance perspective. In *37th International Conference on Information Systems*, vol.5 (pp. 3839-3854). Dublin, Ireland.

Ongori, H., & Migiro, S. O. (2010). Information and communication technologies adoption in SMEs: Literature review. *Journal of Chinese Entrepreneurship, 2*(1), 93–104. doi:10.1108/17561391011019041

Parviainen, P., Tihinen, M., Kääriäinen, J., & Teppola, S. (2017). Tackling the digitalization challenge: How to benefit from digitalization in practice. *International Journal of Information Systems and Project Management, 5*(1), 63–77. doi:10.12821/ijispm050104

Pilipczuk, O. (2021). Transformation of the business process manager profession in Poland: The impact of digital technologies. *Sustainability, 13*(24), 13690. doi:10.3390u132413690

Poland, S. (2020). *Tourism in Poland in the face of COVID-19 pandemic.* Statistics Poland. https://stat.gov.pl/en/topics/culture-tourism-sport/tourism/tourism-in-poland-in-the-face-of-covid-19-pandemic,6,1.html

Poland Statistics. (2021a). *Enterprises having access to the Internet.* SWAID. http://swaid.stat.gov.pl/en/NaukaTechnika_dashboards/Raporty_predefiniowane/RAP_DBD_NTSI_10.aspx .

Poland Statistics. (2021b). *Occupancy of tourist accommodation establishments in Poland in April and May 2021.* Poland Statistics. https://stat.gov.pl/en/topics/culture-tourism-sport/tourism/occupancy-of-tourist-accommodation-establishments-in-poland-in-april-and-may-2021,5,31.html

Popović-Pantić, S., Semenčenko, D., & Vasilić, N. (2019). The influence of digital transformation on business performance: Evidence of the women-owned companies. *Ekonomika Preduzeća, 67*(7-8), 397–414. doi:10.5937/EKOPRE1908397P

Popović-Pantić, S., Semenčenko, D., & Vasilić, N. (2020). Digital technologies and the financial performance of female SMEs in Serbia: The mediating role of innovation. *Economic Annals, 65*(224), 53–82. doi:10.2298/EKA2024053P

Romagnoli, S., & Tarabu', C., MalekiVishkaei, B., & De Giovanni, P. (2023). The impact of digital technologies and sustainable practices on circular supply chain management. *Logistics, 7*(1), 1. doi:10.3390/logistics7010001

Sarabdeen, M., & Alofaysan, H. (2023). Investigating the impact of digital transformation on the labor market in the era of changing digital transformation dynamics in Saudi Arabia. *Economies, 11*(1), 12. doi:10.3390/economies11010012

Sari, T., Güleş, H. K., & Yiğitol, B. (2020). Awareness and readiness of Industry 4.0: The case of the Turkish manufacturing industry. *Advances in Production Engineering & Management, 15*(1), 57–68. doi:10.14743/apem2020.1.349

Sawy, O. A. E., Amsinck, H., Kræmmergaard, P., & Vinther, A. L. (2016). How LEGO built the foundations and enterprise capabilities for digital leadership. *MIS Quarterly Executive, 15*(2), 141–166.

Schilirò, D. (2021). Digital transformation, COVID-19, and the future of work. *International Journal of Business Management and Economic Research, 12*(3), 1945–1952.

Shen, Z., Chen, J., Bai, K., Li, Y., Cui, Y., & Song, M. (2023). The digital impact on environmental performance: Evidence from Chinese publishing. *Emerging Markets Finance & Trade.* doi:10.1080/1 540496X.2022.2164188

Siuta-Tokarska, B., Kruk, S., Krzemiński, P., Thier, A., & Żmija, K. (2022). Digitalisation of enterprises in the energy sector: Drivers—business models—prospective directions of changes. *Energies, 15*(23), 8962. doi:10.3390/en15238962

Snyder, H. (2019). Literature review as a research methodology: An overview and guidelines. *Journal of Business Research, 104*, 333–339. doi:10.1016/j.jbusres.2019.07.039

Soto-Acosta, P. (2020). COVID-19 pandemic: Shifting digital transformation to a high-speed gear. *Information Systems Management, 37*(4), 260–266. doi:10.1080/10580530.2020.1814461

Sousa-Zomer, T. T., Neely, A., & Martinez, V. (2020). Digital transforming capability and performance: A micro foundational perspective. *International Journal of Operations & Production Management, 40*(7/8), 1095–1128. doi:10.1108/IJOPM-06-2019-0444

Stojczew, K. (2021). Ocena wpływu pandemii koronawirusa na branżę turystyczną w Polsce. *Prace Naukowe Uniwersytetu Ekonomicznego we Wrocławiu, 65*(1), 157-172. https://doi.org/ doi:10.15611/pn.2021.1.09

Temelkov, Z. (2022). Financial performance of selected hotel groups and resorts during COVID-19 pandemic: 2019/2020 comparison. *Менаџмент у Хотелијерству и Туризму, 10*(1), 41–51.

Terdpaopong, K. (2020). Digital Transformation in the Hospitality Industry in an Emerging Country. In K. Sandhu (Ed.), *Leadership, Management, and Adoption Techniques for Digital Service Innovation* (pp. 223–243). IGI Global. doi:10.4018/978-1-7998-2799-3.ch012

Torraco, R. J. (2016). Writing integrative literature reviews: Using the past and present to explore the future. *Human Resource Development Review, 15*(4), 404–428.

Vărzaru, A. A., Bocean, C. G., & Cazacu, M. (2021). Rethinking tourism industry in pandemic COVID-19 period. *Sustainability, 13*(12), 6956. doi:10.3390u13126956

Winarsih, I. M., & Fuad, K. (2021). Impact of COVID-19 on digital transformation and sustainability in small and medium enterprises (SMEs): A conceptual framework. In: Barolli L., Poniszewska-Maranda A., Enokido T. (Eds), Complex, Intelligent and Software Intensive Systems. CISIS 2020. Advances in Intelligent Systems and Computing, (vol. 1194 pp. 471-476). Cham: Springer. https://doi.org/ doi:10.1007/978-3-030-50454-0_48

Żabińska, J. (2016). Cyfryzacja jako determinanta zmian w strukturze europejskiego sektora bankowego. *Zeszyty Naukowe Wydziału Zamiejscowego w Chorzowie Wyższej Szkoły Bankowej w Poznaniu.* Yadda. http://yadda.icm.edu.pl/yadda/element/bwmeta1.element.ekon-element-000171481860

Zhao, F., Meng, T., Wang, W., Alam, F., & Zhang, B. (2023). Digital transformation and firm performance: Benefit from letting users participate. [JGIM]. *Journal of Global Information Management, 31*(1), 1–23. doi:10.4018/JGIM.322104

Ziółkowska, M. (2020). Managers' decisions and strategic actions of enterprises in Poland in the face of digital transformation. [Economics and Law]. *Ekonomia I Prawo, 19*(4), 817–825. doi:10.12775/EiP.2020.053

Ziółkowska, M. J. (2021). Digital transformation and marketing activities in small and medium-sized enterprises. *Sustainability, 13*(5), 2512. doi:10.3390u13052512

KEY TERMS AND DEFINITIONS

Digital Hospitality: Is a multidisciplinary concept including smart hospitality and virtual hospitality that refers to the use of digital technologies to improve hospitality services performance.

Digital Transformation of Hospitality: Refers to planning to gradually adopt advanced digital technologies (e.g., 5G, virtual reality, blockchain, apps, artificial intelligence systems, and big data), in the work of hospitality companies, such as online reservations and digital payments, in order to improve the quality of the customers' experience.

Hospitality Industry: Is one of the most important industries in the service sector that includes a wide range of companies such as restaurants, hotels, resorts, museums, and parks, which provide various services such as food, beverage, lodging, and transportation services.

Chapter 3
Robotic Process Automation as an Emerging Technology in Tourism, Hotels, and Food Service

Sameera Khan

https://orcid.org/0000-0002-8724-6817

Vardhaman College of Engineering, Hyderabad, India

Dileep Kumar Singh

Narsee Monjee Institute of Management Studies, Hyderabad, India

ABSTRACT

A new wave of automation known as robotic process automation is revolutionizing company productivity and delivering excellent ROI. This book chapter examines the use of Robotic Process Automation (RPA) in the hospitality, travel, and food service industries. It focuses on how RPA handles routine activities, enhances operational effectiveness, and improves customer experiences. RPA streamlines operations and enables personalized client interactions in the travel and tourism sector. It automates front desk operations, inventory management, and back-office duties in hotels, optimizing resource allocation and enhancing visitor experiences. In the food service sector, RPA automates order processing, inventory management, and supply chain management, improving operational effectiveness and enabling personalized ordering experiences. The chapter covers popular RPA solutions, analyzes implementation difficulties, and emphasizes the future potential of RPA, including integration with cutting-edge technology. By automating mundane tasks and providing a basic framework for implementation, RPA offers significant advantages in industries like aviation, hotels, finance, tourism, and more. This chapter highlights the significance of RPA in enhancing productivity, reducing errors, and shaping the future of these industries.

DOI: 10.4018/978-1-6684-6985-9.ch003

1. INTRODUCTION

The globe has undergone a significant shift due to ongoing digital developments. Such transformations do not leave business operations unaffected. Robotic process automation is one such innovation. As per the definition given by IEEE (The Institute of Electrical and Electronics Engineers) Standards Association, Robotic Process Automation is "A preconfigured software instance that uses business rules and predefined activity choreography to complete the autonomous execution of a combination of processes, activities, transactions, and tasks in one or more unrelated software systems to deliver a result or service with human exception management." (Moffitt et al., 2018). According to a report published by Mckinsey (Lhuer, 2016), more than 81% of predictable physical work, 69% of data processing, and 64% of data-collection activities could feasibly be automated. With this huge capability for automation, a lot of time and human resources can be saved. Also, the error rates which include data entry errors, calculation errors, processing errors, compliance errors, communication errors, etc. can be significantly decreased. RPA finds its application in nearly all industries and service sectors. The extent of automation can vary from one domain to another.

Industries have been dramatically affected by the Industry 4.0 revolution(Ribeiro et al., 2021). An essential component of this shift has been digitalization. RPA has played a significant role in this area of technology. It could have a big impact on the profitability and level of competitiveness among competitors. With the introduction of RPA in industries and service sectors, the workforce now will just not be the manpower. It will also include a digital workforce created by RPA and usually called "bots"(Choi, R'Bigui, et al., 2021)

RPA can be used to automate a wide range of functions, including sending daily updates or details, front-office tasks, back-office tasks, end-to-end processes, etc. If such monotonous work is delegated to a digital workforce, human labor can be used for jobs that require greater intelligence and added value. The design and deployment of such bots can be done using a variety of RPA tools(Khan, 2020). One option is to list the steps that must be taken in order, and another is to employ recorders. Recorders are a lot simpler approach to creating a bot and do not require any programming expertise. These recorders employ artificial intelligence to pinpoint the task-related elements that a user wants to automate.

Some basic problems and challenges(Syed et al., 2020) Among the challenges an institution faces when implementing RPA are the instillation of employee insecurity, worry about losing sensitive data due to hacking or a system crash, frequent changes in the type of work a workforce undertakes, the need for proactive planning, the requirement that the entire process be managed under human supervision, etc. To name a few, some of the reasons why businesses use RPA to automate their business processes are to reduce human error, increase accuracy, cut down on the time required for humans to perform repetitive tasks, work around the clock, increase their digital workforce, and handle large data sets more effectively. A few of the benefits of using RPA for any company firm are given below (Siderska, 2020)

- Accuracy – RPA nearly eliminated typos and other human errors, boosting accuracy to an extremely high level. Additionally, it keeps the data collection method uniform.
- Free of biases: RPA's work cannot be hampered by biases or favoritism when collecting data or filtering data.
- Low Technical Barrier: Using RPA development tools, a person with little programming experience or technical expertise can easily configure a bot.

- Compliance: Bots completely follow instructions given to them. They also offer an audit trail history that may be used to examine a bot's operation.
- Consistency: Without growing weary or compromising performance, bots are capable of repeatedly performing the same activity with the same accuracy and efficiency for hours or even days at a time.
- Effective use of human resources: The workforce can be used more effectively. Employees who were performing repetitive jobs may be given some more beneficial tasks.
- Service quality – RPA may undoubtedly boost a company's service quality while also increasing efficiency and accuracy.
- Reliability: Bots may follow rules without error, carry out certain jobs flawlessly, and stick to their plan of action throughout various iterations.

A business must first identify tasks that are repetitive and need to be automated before implementing any RPA tools. The return on investment might be greatly hampered by improper process choices(Sobczak, 2022). A suitable instrument that satisfies the process need should be chosen after the process has been chosen. The top three RPA tools are BluePrism, UiPath, and Automation Anywhere. Each of them has a unique set of features that make them more suitable for particular industries. For mid-sized and big businesses in the banking, finance, IT, telecom, and healthcare industries. With a digital workforce that is centered on payment, purchase, and HR management, AA is a good fit for the banking, finance, and IT industries. BluePrism is intended for large-sized businesses. It offers robust support for back-office automation, making it more appropriate for businesses that produce media and those in the healthcare industry. Both mid-sized and large-sized businesses can use UiPath. Its extensive support for front- and back-office automation makes it valuable across a range of sectors, including infrastructure, insurance, and human resources.

2. RPA AS A SOLUTION TO THE MUNDANE TASK

Increasingly Robotic process automation (RPA) is a technology that uses software robots to automate repetitive, routine tasks usually performed by humans. The technology is gaining significant attention due to its potential to improve efficiency, reduce costs and minimize errors in business operations.

According to (Madakam et al., 2019), RPA technology and the use of robots in corporate processes are becoming standards in enterprises all over the world. The core business processes such as payroll, employee status changes, new hire recruitment and onboarding, accounts receivable and payable, invoice processing, inventory management, report creation, software installations, data migration, and vendor onboarding, to name a few, can all benefit immediately from robotic process automation. In addition, RPA has several uses in FMCG (Fast-moving consumer goods), real estate, telecom, energy and utilities, healthcare and pharmaceuticals, financial services, outsourcing, retail, and other industries. RPA's many allied technologies, including artificial intelligence, machine learning, deep learning, data analytics, HR (Human Resource) analytics, virtual reality, home automation, blockchain technologies, 4D printing, etc., operate in the background to help RPA be implemented properly in business operations. Additionally, it covers the material of many start-ups and established businesses. The implementation of RPA can result in significant time and cost savings, with an estimated average return on investment of 200% within the first year. The study also found that the use of RPA can lead to increased job satisfaction for employees, as they are freed from repetitive tasks and can focus on more value-adding activities.

RPA can improve overall organizational efficiency by automating manual and time-consuming tasks, reducing errors, and improving data accuracy. Due to the scalability and adaptability of RPA, it can be easily integrated into existing systems and processes. How businesses should manage RPA, though, is unknown. When a business builds up its RPA activities, the burden of maintenance in particular might grow significantly. Guidelines for establishing low-maintenance RPA implementations have been proposed by (Noppen et al., 2020) to prevent or reduce high-maintenance efforts.

Despite its numerous benefits, the implementation of RPA technology can also present certain challenges. According to a study by the Journal of Business and Economics Research, RPA implementation requires significant organizational changes, including the re-design of processes and the retraining of employees. The study also found that organizations need to carefully consider data privacy and security issues when implementing RPA technology.

Figure 1. Applications of RPA in various industries

Application of RPA in Various Industries

Financial Services & Banking	Healthcare	Telecommunications	Retail
Automate data validations, data migration between banking applications, customer account management, report creation, form filling, loan claims processing, updating loan data & backing up teller receipts	Patient data migration & processing, reporting for doctors, medical bill processing, insurance data automation & claim processing, claim status and eligibility automation, and patient record storage	Collecting and consolidating data from client phone systems, backing up information from client systems, uploading data, extracting data about competitor pricing, phone manufacturing information,	Extracting product data from manufacturers websites, automatically updating online inventory and product information, importing website and email sales.
Manufacturing	**Technology / Software**	**Hospitality**	**Consumer Goods**
ERP Automation, automation of logisticsdata, data monitoring, product pricing comparisons.	Hardware and software testing for functional, load and mobile performance.	Competitor pricing analysis, guest data processing, data verification, payment processing, user account creation.	Order processing, data entry, resolution consulting, claims processing, FTP automation, incentive claims processing.

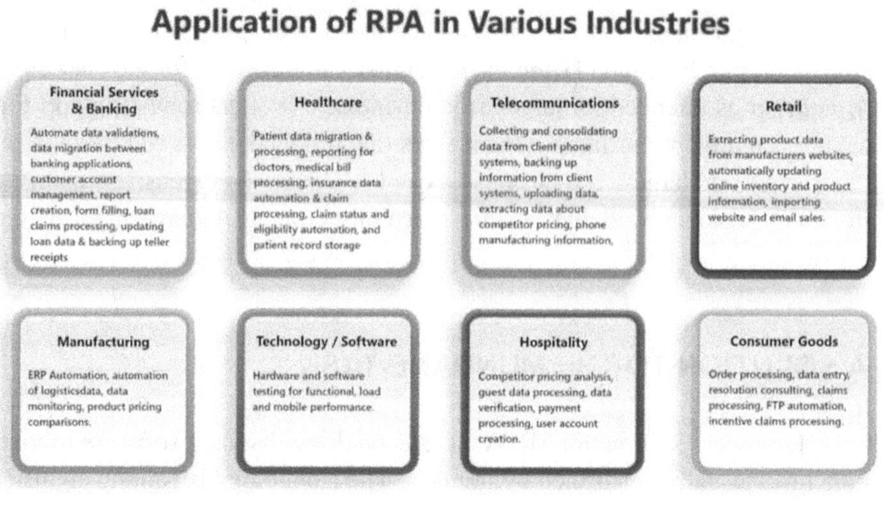

The typical requirements for processes that are suitable for RPA, according to (Fung, 2013), are:

- small cognitive requirements (it is difficult for multifaceted processes with countless complicated tasks to be handled by RPA).
- no necessity for access to numerous systems as RPA is functional on top of existing applications
- comparatively frequently performed processes and tasks are upright candidates for RPA implementation
- processes with a high probability of human error and liaising with external parties is also suitable.

In conclusion, RPA is a promising technology that can significantly improve business operations by automating mundane tasks and reducing errors. However, organizations must carefully consider the implementation of RPA and address any potential challenges to fully realize its benefits.

3. RPA IN TOURISM

RPA is an innovative technology that has the potential to revolutionize various industries, including the tourism industry. This industry is characterized by a high volume of repetitive and routine tasks, such as bookings, reservations, and customer service, which can be automated using RPA. This technology has the potential to significantly improve the efficiency and competitiveness of the hotel and tourism industry while providing numerous benefits to both customers and businesses(Sanjeev & Birdie, 2019).

One of the primary benefits of RPA in the tourism industry is increased efficiency and speed. Automating repetitive tasks using RPA can significantly reduce the time required to complete them, leading to faster processing times and reduced wait times for customers(Goyal & Singh, 2021). For example, RPA can be used to automate the process of booking flights, hotels, and other travel-related services. This can help to streamline the booking process and reduce the time customers spend waiting for their travel arrangements to be confirmed.

RPA also has the potential to improve customer satisfaction by providing a more seamless and personalized experience. For example, RPA can be used to automate the process of sending personalized travel itineraries to customers based on their travel preferences. This can help to improve the overall customer experience and increase customer loyalty. Additionally, RPA can be used to automate customer service functions, such as answering frequently asked questions, providing travel recommendations, and resolving customer complaints. This can help to improve the responsiveness and quality of customer service, leading to increased customer satisfaction.

Another benefit of RPA in the tourism industry is reduced costs. Automating routine tasks using RPA can help to reduce labor costs and increase cost efficiency. For example, by automating the process of booking flights and hotels, businesses can reduce the number of employees required to manage these tasks. This can help to reduce labor costs and increase cost efficiency, allowing businesses to pass these savings on to customers in the form of lower prices. Table 1 summarizes the comparison of the old process with the processes that employ RPA in the tourism industry.

However, implementing RPA in the tourism industry also presents certain challenges. One of the biggest challenges is ensuring data privacy and security. The tourism industry involves the collection and storage of sensitive personal and financial information, such as passport details and credit card information. Ensuring the security of this information is crucial for maintaining the trust of customers and avoiding data breaches. RPA vendors must implement robust security measures to ensure that sensitive information is protected at all times.

Another challenge of implementing RPA in the tourism industry is the potential for job loss. Automating routine tasks using RPA can result in the elimination of certain jobs, as they are no longer necessary. This can create a negative impact on employees and their families, leading to resistance to the adoption of RPA. To mitigate this impact, businesses must provide retraining and job placement support to affected employees to help them transition to new roles.

Table 1. Comparison of different manual and RPA-employed processes in the tourism industry

Name of Process	Old Process	New Process	Relevance	Limitations
Booking process	Manually search for different packages available. Wait for the dates on which the cost will be lower.	The bot will search different available packages and list them according to rates and dates available	Manually searching can be exhaustive and may not lead to an optimal result which is not the case with bots.	Initial deployment and the time taken by the system to stabilize will be time taking process.
PNR records and travel management	For travel agents, each update and cancellation has to be done manually. Also, there were chances of errors. Updating a lot of data was time taking process.	For travel agents, PNR information can be entered and updated using RPA bots, which can do so far more quickly and accurately. They can also effectively update ticket cancellations	To accurately handle all the PNR record management without any error, bots can be a good choice.	The process includes handling of personal information of clients. Lots of security has to be imposed while dealing with it.
Data Validation	Manually the documents were checked for verifying signatures. The chances of false acceptance are high.	Documents can be uploaded and a data capture mechanism powered by AI can be used to approve or reject the recommendation.	Frauds can be significantly reduced.	It can result in false rejection too.
Robot Chats	Usually, a person is employed in the helpdesk who will answer the queries.	A bot can be programmed to answer general queries. In case some specific queries are asked they can escalate it to the authorities	It can work 24 X 7 without making mistakes and getting tired. Also, manpower can be reduced.	Not all queries can be handled by bots.
Customer Insights	All the details have to be manually collected by the R&D team of the company	The digital workforce has access to customer insights and analytics from a variety of sources and may categorize customers based on data like demographics and average annual expenditure.	Varied information and analysis can be provided.	The reports generated can help make strategies but they cannot make strategies on their own.
Competitor Analysis	A person was specifically employed to keep an eye on schemes and prices of opponents which can cause a miss.	Bots can monitor rivals and their tactics, sending out notifications when their prices alter.	It will keep the competitive advantage up to date	Regular monitoring for adding emerging competitors is required.

In conclusion, RPA is a promising technology with the potential to significantly improve the efficiency, competitiveness, and customer satisfaction of the tourism industry. However, businesses must carefully consider the challenges of implementing RPA and take steps to mitigate any negative impacts, such as data privacy and security concerns and job loss. By taking a strategic approach to the implementation of RPA, businesses in the tourism industry can reap the benefits of this technology and remain competitive in a rapidly evolving market.

4. RPA IN HOTELS

The hotel industry has a vast application of RPA. It shares many of the applications with the tourism industry as well which has been discussed in the previous section like bookings and customer service. In addition to these, the use of RPA in the hotel industry is broad and covers a range of processes, from front-end activities like customer service and reservation management to back-end operations such as financial reporting and accounting. One of the key areas where RPA has made a significant impact is customer service. Automated chatbots have replaced traditional customer service representatives in many hotels. These chatbots are capable of answering common questions, providing recommendations, and even making reservations. This has not only improved the speed and efficiency of customer service but also reduced costs for hotels(Kali Durgampudi, 2022).

Another area where RPA has made a significant impact is in the management of reservations. The use of RPA has made it possible to automate the process of taking and managing reservations. This has not only reduced the workload of front-end staff but also improved the accuracy of reservations and reduced the risk of overbooking. Furthermore, RPA can be used to automate the process of generating reports, freeing up valuable time for hotel staff to focus on more critical tasks.

The use of RPA has also led to significant cost savings for hotels. By automating manual and repetitive tasks, hotels can reduce their labor costs while also improving efficiency. RPA also eliminates the need for manual data entry, reducing the risk of errors and increasing accuracy. Furthermore, RPA reduces the need for additional IT staff, as the technology is easy to use and does not require specialized skills.

RPA can play a significant role in inventory management for a hotel (Verma et al., 2020.). It takes a lot of time and breaks down due to several suppliers and travel services, which lowers or lowers the system's efficiency. Therefore, it is possible to use inventory process automation to automate a variety of business operations. The system for automating inventory procedures should be able to assign various tasks or business processes automatically using task search, filter/sort options, pagination, and open or finished tasks. In addition, the system for automating the inventory process must be able to automate the assignment of jobs to the back office hotel inventory team and the task of viewing ticket data, which automates the communication log. Once an inventory agent has performed a task, the inventory process automation system must be able to automatically allocate it to the quality check team.(Sharma & Guleria, 2021)

In addition to the above applications, RPA can also be employed to answer general inquiries on room availability, make and cancel reservations, facilitate check-ins and check-outs, marketing, competitive pricing, customer service, finances, report generation, manage customer and staff credentials, etc. (Bardia Eshghi, 2022).

In conclusion, RPA is transforming the hotel industry by reducing costs, improving efficiency, and enhancing customer service. The use of RPA has made it possible for hotels to automate manual and repetitive tasks, freeing up valuable time for front-end and back-end staff(Bardia Eshghi, 2022). The benefits of RPA in the hotel industry are undeniable, and this technology will continue to play an important role in the future of the industry. Table 2 summarizes the comparison of the old process with the processes that employ RPA in the hotel industry.

Table 2. Comparison of different manual and RPA-employed processes in the hotel industry

Name of Process	Old Process	New Process	Relevance	Limitations
System integration services	The availability of rooms was updated manually in the register, in ERP, on the website, and other platforms.	The hotel's website, third-party websites, or using RFP management platforms all can be updated simultaneously without any hassle. Also, real-time data is available for revenue, sales, CRM, and inventory	Saves lots of time and gives accurate information. Makes it easier to manage the system.	All the platforms must be linked otherwise inconsistent data will be reflected.
Competition reporting	A person was specifically employed to keep an eye on schemes and prices of opponents which can cause a miss.	Bots can monitor rivals and their tactics, sending out notifications when their prices alter.	It will keep the competitive advantage up to date	Regular monitoring for adding emerging competitors is required.
Optimized Pricing and rate utilization	Manually change the rates and update all other platforms	Bots can set your RMS to automatically modify which hotel rate levels are accessible when high demand levels are observed to encourage ideal pricing patterns and increase possible income, If not, provide discounts on days when demand is low.	An increase in revenue can be achieved	An attended bot is required means human intervention is required. The process is not 100% automated
Automated messaging and mailing	Guest list and the list of interested people who visited the hotel through any means are maintained manually and timely emails and messages were floated to them through a person.	A bot can timely update the guest list taking data from diverse sources and shooting emails and messages accordingly	Extra labor and time are saved	Sometimes human intervention is required
Report generation	Manually different formats of reports were generated daily, weekly, monthly, and annually. The probability of error was high and also may delay in delivery of reports	A bot can generate all the reports on a daily, weekly, monthly, and annual basis. Also, it is up-to-date, correct and timely	Accurate and timely data is generated	All the data must be in the same place to get the exact report

5. RPA IN FOODSERVICE

Robotic Process Automation (RPA) has been gaining significant traction in the food industry, as companies look to improve efficiency, reduce costs, and streamline their operations. In recent years, the food industry has been experiencing tremendous growth, and as a result, many companies are seeking innovative solutions to enhance their competitiveness and meet the changing demands of consumers. RPA is one of the most promising technologies that have the potential to revolutionize the way the food industry operates(Iqbal et al., 2017).

The food industry faces many challenges such as stringent regulations, changing consumer demands, and increasing competition. RPA can help companies overcome these challenges by automating routine and repetitive tasks, freeing up employees to focus on more critical and strategic activities. For example, in the food manufacturing process, tasks such as data entry, quality control, and inventory management can be automated, thus reducing the risk of human error and improving overall efficiency(Trienekens & Zuurbier, 2008).

In addition, RPA can also improve the speed and accuracy of operations. For instance, in the distribution process, RPA can automate the tasks of receiving and processing orders, which can lead to faster order fulfillment and improved customer satisfaction. Furthermore, RPA can also help companies manage their supply chain more efficiently by automating tasks such as tracking inventory levels, monitoring delivery schedules, and managing suppliers(Sharma & Guleria, 2021).

Moreover, RPA can also help companies improve their regulatory compliance. The food industry is heavily regulated, and companies are required to follow strict guidelines to ensure the safety and quality of their products. RPA can automate the tasks of tracking and monitoring compliance with regulations, thus reducing the risk of non-compliance and ensuring that companies are operating within the law. (Henson & Heasman, 1998)

In conclusion, the relevance of RPA in the food industry is clear. The technology has the potential to transform the way companies operate, from streamlining operations to improving efficiency and reducing costs. Furthermore, RPA can also help companies to comply with regulations, thereby improving their reputation and enhancing their competitiveness. In addition to various tasks which have been discussed earlier like booking, report generation, finance, marketing, and sales, some more tasks in the food industry can be automated(Bardia Eshghi, 2023). Table 3 summarizes the comparison of the old process with the processes that employ RPA in the food industry.

Table 3. Comparison of different manual and RPA-employed processes in the food industry

Name of Process	Old Process	New Process	Relevance	Limitations
Inventory Management	Manual and regular checks for inventory were required.	To set acceptable order levels when stock exceeds a minimum threshold, the bots can monitor client demand and prior orders.	Timely and relevant order placing. Reduces manual efforts	Human intervention is required
Supply chain Management	Manual tracking was required and was not accurate	Bots can be used in e-mail automation in addition to transport system integration to keep suppliers and customers up-to-date	Enhance accuracy and reliability	Human intervention is required
Automated alerts for cleaning and checking	No such mechanism exists	Alerts for managers and employees for scheduled work can help in enhancing the quality of services	Enhanced quality of services	NA
Customer Service	Waiters have to answer all the basic queries of the customer	Bots can make chats with the customer and answer all the basic queries	Relevant and timely information is provided to customers increasing the customer satisfaction	NA

6. RPA TOOLS

It's crucial to pick the appropriate jobs or processes to automate when implementing RPA technology to streamline business processes (Syed et al., 2020). If the incorrect procedures are not chosen for automation, RPA may result in slowed-down business operations rather than generating a return on investment. The processes that will be chosen for automation will include traits like being heavily rule-based, working with large amounts of data, having digitized input, being very manual and tedious, requiring little exception-handling, and being less complicated and well-documented. One needs access to any RPA technology to automate a business process. Three main parts make up RPA tools: a graphical modeling tool, an orchestrator, and a few supplemental tools. The modeling tool streamlines the design process and is simple enough for even inexperienced people to utilize. Bot administration and operation are handled by the orchestrator. Phases of development, testing, and production all make use of it. Vendor-specific additional tools are available. These may consist of an analytics tool, a scheduler, a collaborator, audit tools, or some AI elements. RPA software products must function in a variety of settings, including desktop, online, and Citrix environments. These tools must simplify the logical design and deployment of bots for certain purposes while requiring less programming language expertise. Data collection from a variety of sources must be allowed. RPA bots should typically be stored in cloud locations and used as needed.

Additionally, a management module is required to handle duties like versioning, keeping audit trails, scheduling, and collaboration, among other things.

Client-server-based architecture and web-based orchestrator are the two architecture types that RPA technologies have. Every node in a network that uses client-server architecture is either a client or a server. A web-based orchestrator, on the other hand, helps to connect automated actions into a workflow that is specified to achieve a specific goal. UiPath, Automation Anywhere, BluePrism, Power Automate and are some commonly used RPA software programs.

Automation Anywhere- Among RPA tools, Automation Anywhere is among the top service providers. The client-server foundation of AA's design. The Bot creator, Control room, and Bot runner are the three fundamental elements of AA. The creation of bots is made simple by the bot maker. Along with handling credentials, security concerns, client rights, and bot execution and scheduling, the control room also evaluates and manages security issues. The bot is run using a bot runner, and its analytics are recorded and reported back to the control room. Task bots, Meta bots, and IQ bots are the three categories of bot generation that AA enables. While meta bots are used to generate building blocks of bots that may be reused in other task bots, task bots are frequently used to automate rule-based and repetitive operations. IQ bots, on the other hand, have cognitive and intelligent capabilities that are utilized to process unstructured data. Three different types of recorders are offered by AA to capture user activity and turn it into a script for a bot. By imitating user activities, tools like web recorders, smart recorders, and screen recorders are utilized to automate the process. The analytics engine of AA BOT INSIGHTS, which enables the visualization of user data and business insight extraction from it, is one of the additional aspects of automation anywhere. Both BOT STORE, an online marketplace where a variety of plug-and-play bots are accessible, and BOT FARM, which enables businesses to purchase RPA tools on a usage basis unlike a capacity or licensing basis (*Automation Anywhere University*, n.d.).

UiPath- A platform for creating software bots to automate corporate activities is offered by UiPath, a multinational software provider. A web orchestrator-based architecture called UiPath was created using the.NET framework. UiPath Studio, UiPath Orchestrator, and UiPath Robots are its main components. The studio offers a variety of tasks and workspaces for creating and using user-defined bots. Since it employs a drag-and-drop methodology while working with activities, it is simpler to utilize. The orchestrator enables the user to upload, deploy, and manage resources for a bot in the cloud. Additionally, it controls provisioning, configuration, logging, and bot queues. Robots are employed to carry out human-like jobs. They come in both attended and unsupervised varieties. While unattended bots operate autonomously, attended bots require human participation to complete their job. UiPath offers five different types of recorders: basic recording, desktop recording, web recording, image recording, and Citrix recording. The basic recording is used for single activities; desktop recording is used for multiple actions that can be between different apps; web recording records web and browser activities; image recording and Citrix recording is used for virtual environments and is capable of image, text, and keyboard automation. One of UiPath's most vital parts is the orchestrator. It is utilized to control several bots in the setting. Assets have been included to enable communication amongst the bots. User credentials may also be stored in assets. In UiPath, queues are used to control workload. To monitor the activity of the bots, it also manages audit trails and records (*UiPath Academy*, n.d.).

BluePrism- Another well-known provider of RPA technologies and solutions is BluePrism. It provides a drag-and-drop approach for developing bots and is based on the Java &.NET foundation. Process diagram, process studio, object studio, and application modeler are the four key parts of BluePrism. Business workflows are represented by process diagrams, which are produced using fundamental programming ideas. These graphical process representations are employed in the development, evaluation, modification, and scalability of business capabilities. Process Studio offers a toolkit with a variety of drag-and-drop activities for creating process diagrams. Visual Basic objects are created using Object Studio and can communicate with other apps. Application Modeller is a feature of Object Studio that allows for the creation of application models. This makes a target application's UI Elements visible to the Blue Prism software. BluePrism introduces connected RPA, which collaborates with cognitive and artificial intelligence capacities. Additionally, BluePrism offers a control room for monitoring bot activity and audit trails. Additionally, it now includes cross-platform support for a variety of different cloud and AI technologies. Workload management involves using work queues to control the simultaneous execution of several bots. BluePrism also enables, to mention a few, robot screen capture, customized dashboards, multilingual interface support, and intelligent surface automation (*BluePrism University*, n.d.).

Power Automate- The intelligent cloud-based tool Microsoft Power Automate, formerly known as Microsoft Flow, gives you the ability to automate repetitive, manual, and time-consuming procedures with no coding and little work. It is a component of the Microsoft Power Platform and acts as a single solution that combines robotic process automation (RPA) and digital process automation (DPA), which are both further improved by artificial intelligence, to create a comprehensive solution (AI).

With the help of these tools, you can build automated workflows across apps and services in a connected ecosystem that includes hundreds of connectors and thousands of prebuilt templates. This makes it easier for your company processes to be more simplified, allowing you to increase productivity and save time. It also helps synchronize files, get notifications, and gather data.

Microsoft Power Automate can be used to perform a certain task, automate processes, send notifications on a real-time basis, collect data, synchronize files, etc. To do all this work, it facilitates the creation of five types of flows- cloud flow, instant cloud flow, automated cloud flow, scheduled cloud flow, and desktop flows. To automate daily office work, business process flow functionality is provided. Table 4 summarises the properties of tools used for RPA(Khan, 2020).

Table 4. Comparison of RPA tools

Comparison Criterion	Automation Anywhere	UiPath	BluePrism	Microsoft Power Automate
The architecture of the platform	Client-server architecture	Web-based orchestrator	Client-server architecture	Cloud-based
Tool popularity	more prevalent than other RPA programs but less well-known than UiPath and BP. The popularity of AA is rising daily.	RPA's most widely used tool Topping the charts for a long time.	popular but not as widely used as UiPath.	Lesser than others. It is gaining popularity with time
Product availability	The community edition is only usable with BotCreator rights, while the enterprise edition offers a one-month trial. Management and audit logs are unavailable. There are no features for the API. There is no repository access for the control room.	All users have access to the community edition, however, the generated bots cannot be shared. With 1 orchestrator, 10 licenses for UiPath Studio, StudioX, and Studio Pro, 10 Attended, 10 Unattended, 10 Tests, 2 AI Robots, 10 Action Centers, and 1 Insight, the Enterprise edition is available for a 60-day free trial.	offers a free trial of the product for a month. Only 15 processes and 1 digital worker are allowed. One digital worker and five processes are restricted under the learning edition's 180-day free license.	90 days free trial is provided. a single user can only have 6000 requests across all their Power Platform products
Usability	UI is intricate. More suited to developers and those with adequate coding knowledge.	Simple user interface. can also be utilized by beginners.	The UI is straightforward and makes it simple to create bots.	Easy to use. Non-technical users can also use it
Types of the process that can be automated	can be applied to front- and back-office automation.	can be applied to front- and back-office automation.	Can be applied to back-office automation.	Best for front office automation
Recorders	Smart, screen and online recorders are the three different varieties. These recorders are suitable for both desktop and web applications.	Basic, web, desktop, image, and Citrix recorders are among the five categories. UiPath makes it simpler to record human behaviors so that you may further emulate them thanks to its extensive selection of recorders.	There are no recorders available. A process must be created using drag-and-drop tools.	A desktop recorder is provided

continues on following page

Table 4. Continued

Comparison Criterion	Automation Anywhere	UiPath	BluePrism	Microsoft Power Automate
Cognitive capability	Medium cognitive capabilities	Medium cognitive capabilities	Low cognitive capabilities	Microsoft cognitive services are used to empower power automate. Good cognitive capability
The coding requirement	supports drag-and-drop methods as well as recordings. Thus, coding is not required.	supports drag-and-drop methods as well as recordings. Thus, coding is not required.	No recorders are used, but the built-in functionality and utilization of process diagrams make it simple to use. supports coding but does not require it	Supports mostly the drag-drop approach. So no coding requirement
Pricing	Cloud starter 9000$ (Customisable) Approx 20000$ annually	Customizable as per requirement Approx 18000$ annually	Around 15000$ annually	Provides different subscription plans by the user and by flows. Range from 15$ to 100$ per month. Also, have a license by run which costs from 0.6$ to 3 $ per run
Reliability & security	There is high security offered. AA offers a strongly protected credential vault for storing confidential user information.	User data that is private and sensitive is saved using a credential manager. The appropriate encryption has been included.	Sensitive data is stored securely using BluePrism Credential Manager. Users can select the key generation algorithm and storage location.	No specific credential vault is provided
Encryption algorithm	For encryption, RSA with a 2048-bit master key is employed. Data encryption employs the AES-256-bit key.	supports encryption techniques like Triple DES, RC2, DES, AES, and RC2.	Information about credentials is encrypted using ciphers. Obfuscating all source codes lowers the danger of assaults, reverse engineering, and patching.	RSA-CBC and MD5
Certifications	Available online	Available online	Available online	Available online

7. CHALLENGES IN RPA IMPLEMENTATION

RPA, which accelerates corporate growth by eliminating a lot of manual and repetitive work, has recently attracted a lot of interest from academia and a variety of industries. But there are still numerous obstacles in the way of RPA's implementation right now (Gao et al., 2019; Nawaz, 2019.). According to the global survey report (Protiviti, 2019), there are various challenges faced by industries on various levels like- organizational level, technological level, financial and regulatory levels. Fig-1 (extracted from (Protiviti, 2019)) explains the extent of these challenges on diverse levels. At an organizational level, it varies from 23-40%, at the technological level it ranges from 30-40%, whereas at the financial and regulatory level, it varies from 30-37%.

Figure 2. Obstacles in adopting RPA.

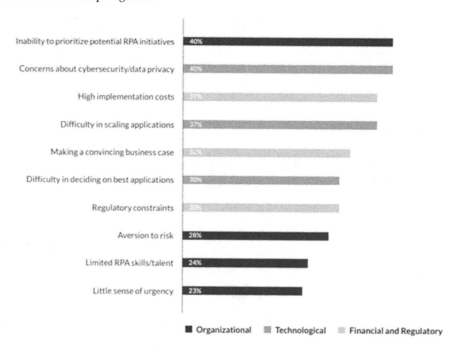

Thus, at the organizational level, the major challenge is to identify and prioritize RPA initiatives. In doing so, there is a huge risk involved. Various companies are not likely to take such risks. This is another challenge faced by RPA implementation. Another major criterion is the lack of skilled RPA developers. Since RPA is still an emerging technology, there is a lack of skilled RPA developers.

Concerning the technological barriers in the path of RPA implementation, security and data privacy plays the most significant role. Data privacy and security have always been seen as essential. User interface data, which might contain sensitive information, is mined as the foundation of RPA. Consequently, the creation of a secure RPA is required. Another technological aspect is the need for scalability. If the business scales up and RPA processes fail to manage the pace it can create a huge hindrance to the growth of the company. Deciding on which platform will be well suited for the particular business is an additional technological challenge. If the right application is not taken up it can result in inaccurate or expensive automation(Syed et al., 2020).

Implementation of RPA involves lots of financial constraints. Before automating a process, a study of anticipated Return on Investment (ROI) is required. Implementation of RPA and its stabilization can take a long time. In this course, both manual and automatic work should go on. Therefore, this process can be expensive. In a company, if there are not enough cases that can be automated then it would not be beneficial to invest in this technology. Lastly, RPA must meet all the regulatory constraints (Choi, R'bigui, et al., 2021). Apart from this, RPA can also result in the potential loss of jobs and unemployment.

Implementing RPA can present various challenges, such as identifying the right processes for automation, managing the change in workflows, and ensuring proper governance and security. Here are some ways to overcome these challenges:

- Develop a clear automation strategy: It is essential to have a clear understanding of the business processes that are suitable for automation and the expected outcomes. This will help in identifying the right processes and prioritizing them for automation.
- Conduct a feasibility study: Before implementing RPA, it is necessary to conduct a feasibility study to assess the potential impact of automation on the current workflows and processes. This will help in identifying any potential roadblocks and determining the feasibility of automation.
- Involve stakeholders: Involving stakeholders in the automation process, including employees, managers, and IT professionals, can help in identifying potential challenges and garnering support for the implementation.
- Ensure proper governance: Establishing governance and security policies is essential to ensure compliance with regulatory requirements and safeguard data privacy.
- Provide adequate training: Providing adequate training to employees on the new processes and technologies can help in reducing resistance to change and ensuring a smooth transition.
- Monitor and evaluate the results: It is important to monitor and evaluate the results of the RPA implementation regularly. This will help in identifying any issues and opportunities for improvement.

By following these steps, organizations can overcome the challenges of RPA implementation and successfully achieve the expected benefits.

8. SOME BUSINESS EXAMPLES OF RPA IMPLEMENTATION

RPA has been adopted by various businesses by now. Some of the examples of successful implementation of RPA in hotel, tourism and food industries are listed below-

- **Cargill-** Cargill, an American global food and agriculture company, successfully implemented RPA to automate its invoice processing and accounts payable (AP) operations. With the help of Automation Anywhere's RPA solution, Cargill was able to process invoices 5-10 times faster, while reducing processing errors by 95%. The RPA solution also helped Cargill to improve the accuracy of its financial data and reduce manual labour costs. As a result, Cargill was able to streamline its AP operations and improve its overall financial performance.
- **WAM-** The Italian multinational company WAM Group used RPA to streamline its accounts payable processes, which had been a manual and time-consuming task. The RPA solution automated invoice processing, resulting in a reduction of 60% in invoice processing time. Moreover, the solution reduced the risk of errors and improved the overall accuracy of financial data. As a result, the finance team was able to spend more time on higher-value tasks such as analysis and strategic planning. Overall, the RPA solution enabled WAM Group to improve its financial processes and better allocate its resources.
- **Foodstuffs-** Foodstuffs, a New Zealand-based grocery cooperative, used RPA to automate its manual and repetitive processes such as data entry, report generation, and invoice processing. With the RPA solution, Foodstuffs was able to process 500 invoices per day with 100% accuracy, resulting in a reduction of 65% in invoice processing time. The solution also improved the overall efficiency and accuracy of the finance team's work, allowing them to focus on more strategic

tasks. Overall, the RPA solution enabled Foodstuffs to improve its financial processes and achieve significant cost savings while delivering improved customer service.

- **Melia Hotels-** Melia Hotels International used RPA to automate its manual processes for updating reservation data across multiple systems, which was a time-consuming and error-prone task. The RPA solution resulted in a reduction of 80% in the time required for the task, allowing employees to focus on more strategic tasks. The solution also improved accuracy and reduced errors in the reservation data, resulting in a better experience for the customers. As a result, Melia Hotels International was able to improve its operational efficiency and better serve its customers.

9. FUTURE SCOPE

Robotic Process Automation (RPA) has the potential to greatly impact and improve various aspects of the hotel, tourism, and food industry. With the increase in technological advancements and growth in the culture of smart cities and smart villages, RPA can provide a new dimension to these industries. Implementation of RPA in the hotel, tourism, and food industry can significantly increase the ROI, customer satisfaction, customer procurement, customer retention, data accuracy, reliable ERP, effective supply chain management, and sales. On the other hand, it can reduce manual work, service failure and save time which can be utilized more fruitfully.

A rule-based automation system is better suited to the RPA technologies now available. Automation of very complex cognitive processes that frequently require exceptions and require logical thought is exceedingly challenging. The management of complicated and cognitive processes may one day be solved by intelligent or cognitive RPA. In the future, Artificial intelligence, machine learning, computer vision, or natural language processing can be used in association with RPA to provide an enhanced experience to users (Syed et al., 2020).

10. CONCLUSION

The tourism, hospitality, and food sectors are currently expanding at an unheard-of rate and significantly contribute to GDP. Additionally, it has worldwide employment implications. This paper has explored various areas in which these industries can be benefitted through RPA and reduce their operational complexity, and time taken for various processes and achieve enhanced accuracy. The application of RPA in the tourism, hospitality, and food sectors has been analyzed in this paper by identifying processes that can be automated, what was the old process to do the same, and how RPA can replace them. However, some challenges can arise during RPA implementation. Before automating any process or before employing RPA for a particular business, they must first determine which repetitive tasks may be automated. The return on investment might be greatly hampered by improper process choices. A suitable instrument that satisfies the process need should be chosen after the process has been selected. This paper also gives an insight into the most popular automation tools based on architecture, pricing, usability, popularity, reliability, security, the encryption algorithm used, availability, etc.

While RPA has the potential to bring many benefits to businesses, it can also have some negative impacts. One of the main concerns is job displacement, as the automation of repetitive tasks can lead to the loss of some jobs. This can create anxiety and resistance to change from employees, who may be concerned about job loss or the perception that the technology is unreliable. Additionally, overreliance on RPA can create vulnerabilities in the system, particularly if the technology fails or is hacked, leading to business disruptions and the potential loss of sensitive data. Implementing RPA can also be costly and complex, particularly for large organizations with multiple systems and processes. Ethical concerns can also arise, particularly in relation to privacy, data security, and bias. Organizations need to ensure that their use of RPA aligns with ethical and legal standards. Overall, while RPA can bring many benefits, organizations need to be aware of the potential negative impacts and take steps to mitigate them, including careful planning, effective communication with employees, and a commitment to ethical and responsible use of the technology.

11. REFERENCES

Automation Anywhere University. (n.d.). *About.* Automation Anywhere University. https://university.automationanywhere.com/

Eshghi, B. (2022). *Top 12 Use Cases of RPA in Hotels & Hospitality Industry in 2023.* AI Multiple. https://research.aimultiple.com/rpa-in-hospitality-industry/

Eshghi, B. (2023). *RPA in Food Industry: Top 11 Use Cases in 2023.* AI Multiple. https://research.aimultiple.com/rpa-in-food-industry/

BluePrism University. (n.d.). *Certification.* BluePrism University. https://university.blueprism.com/certification

Choi, D., R'bigui, H., & Cho, C. (2021). Robotic Process Automation Implementation Challenges. *Lecture Notes in Networks and Systems, 149,* 297–304. doi:10.1007/978-981-15-7990-5_29

Choi, D., R'bigui, H., & Cho, C. (2021). Candidate digital tasks selection methodology for automation with robotic process automation. *Sustainability (Basel), 13*(16), 8980. doi:10.3390u13168980

Fung, H. P. (2013). Criteria, Use Cases and Effects of Information Technology Process Automation (ITPA). *Advances in Robotics & Automation, 03*(03). doi:10.4172/2168-9695.1000124

Gao, J., van Zelst, S. J., Lu, X., & van der Aalst, W. M. P. (2019). Automated Robotic Process Automation: A Self-Learning Approach. *On the Move to Meaningful Internet Systems: OTM 2019 Conferences: Confederated International Conferences: CoopIS, ODBASE, C&TC 2019, Rhodes, Greece, October 21–25, 2019, Proceedings,* (pp. 95–112). Springer. 10.1007/978-3-030-33246-4_6

Goyal, N., & Singh, H. (2021). A Design of Customer Service Request Desk to Improve the Efficiency using Robotics Process Automation. *2021 6th International Conference on Signal Processing, Computing and Control (ISPCC),* (pp. 21–24). IEEE. 10.1109/ISPCC53510.2021.9609338

Henson, S., & Heasman, M. (1998). Food safety regulation and the firm: Understanding the compliance process1Financial support from the Ministry of Agriculture, Fisheries and Food (MAFF) is acknowledged. However, the views expressed in this paper reflect those of the author and not necessarily those of MAFF.1. *Food Policy, 23*(1), 9–23. https://doi.org/10.1016/S0306-9192(98)00015-3. doi:10.1016/S0306-9192(98)00015-3

Iqbal, J., Khan, Z. H., & Khalid, A. (2017). Prospects of robotics in food industry. Food Science and Technology (Brazil), 37(2), 159–165. https://doi.org/ doi:10.1590/1678-457X.14616

Durgampudi, K. (2022, November 21). How Robotic Process Automation Can Quietly Change The Way People Work. *Forbes*. https://www.forbes.com/sites/forbestechcouncil/2022/11/21/how-robotic-process-automation-can-quietly-change-the-way-people-work/?sh=3179e31a5fa1

Khan, S. (2020). Comparative Analysis Of Rpa Tools-Uipath, Automation Anywhere And Blueprism. *International Journal of Computer Science and Mobile Applications*, 8(11), 1–6. doi:10.47760/ijcsma.2020.v08i11.001

Lhuer, X. (2016). The next acronym you need to know about: RPA (robotic process automation). *Digital McKinsey, 17*, 1–5. https://www.mckinsey.com/business-functions/digital-mckinsey/our-insights/the-next-acronym-you-need-to-know-about-rpa%0Ahttps://usblearn.belpark.sun.ac.za/pluginfile.php/40119/mod_resource/content/0/2018MPhilFS_Robotic_process_automation_McKinsey_Company

Madakam, S., Holmukhe, R. M., & Kumar Jaiswal, D. (2019). The Future Digital Work Force: Robotic Process Automation (RPA). *Journal of Information Systems and Technology Management, 16*, 1–17. doi:10.4301/S1807-1775201916001

Moffitt, K. C., Rozario, A. M., & Vasarhelyi, M. A. (2018). Robotic Process Automation for Auditing. *Journal of Emerging Technologies in Accounting, 15*(1), 1–10. doi:10.2308/jeta-10589

Nawaz, N. (2019). Article ID: IJARET_10_02_057 Cite this Article Dr Nishad Nawaz, Robotic Process Automation for Recruitment Process. [IJARET]. *International Journal of Advanced Research in Engineering and Technology, 10*(2), 608–611. http://www.iaeme.com/IJARET/index.asp608http://www.iaeme.com/IJARET/issues.asp?JType=IJARET&VType=10&IType=02http://www.iaeme.com/IJARET/issues.asp?JType=IJARET&VType=10&IType=2. doi:10.34218/IJARET.10.2.2019.057

Noppen, P., Beerepoot, I., van de Weerd, I., Jonker, M., & Reijers, H. A. (2020). How to Keep RPA Maintainable? In D. Fahland, C. Ghidini, J. Becker, & M. Dumas (Eds.), *Business Process Management* (pp. 453–470). Springer International Publishing. doi:10.1007/978-3-030-58666-9_26

Protiviti. (2019). *Taking RPA to the Next Level - How companies are using robotic process automation to beat the competition*. Protiviti.

Ribeiro, J., Lima, R., Eckhardt, T., & Paiva, S. (2021). Robotic Process Automation and Artificial Intelligence in Industry 4.0 – A Literature review. *Procedia Computer Science, 181*, 51–58. doi:10.1016/j.procs.2021.01.104

Sanjeev, G. M., & Birdie, A. K. (2019). The tourism and hospitality industry in India: emerging issues for the next decade. In Worldwide Hospitality and Tourism Themes (Vol. 11, pp. 355–361). Emerald Group Publishing Ltd. doi:10.1108/WHATT-05-2019-0030

Sharma, A., & Guleria, K. (2021). A Framework for Hotel Inventory Control System for Online Travel Agency using Robotic Process Automation. *2021 International Conference on Advance Computing and Innovative Technologies in Engineering, ICACITE 2021*, (pp. 764–768). IEEE. 10.1109/ICACITE51222.2021.9404613

Siderska, J. (2020). Robotic Process Automation-a driver of digital transformation? *Engineering Management in Production and Services*, *12*(2), 21–31. doi:10.2478/emj-2020-0009

Sobczak, A. (2022). Robotic Process Automation as a Digital Transformation Tool for Increasing Organizational Resilience in Polish Enterprises. *Sustainability (Basel)*, *14*(3), 1333. doi:10.3390u14031333

Syed, R., Suriadi, S., Adams, M., Bandara, W., Leemans, S. J. J., Ouyang, C., ter Hofstede, A. H. M., van de Weerd, I., Wynn, M. T., & Reijers, H. A. (2020). Robotic Process Automation: Contemporary themes and challenges. *Computers in Industry*, *115*, 103162. doi:10.1016/j.compind.2019.103162

Trienekens, J., & Zuurbier, P. (2008). Quality and safety standards in the food industry, developments and challenges. *International Journal of Production Economics*, *113*(1), 107–122. https://doi.org/https://doi.org/10.1016/j.ijpe.2007.02.050. doi:10.1016/j.ijpe.2007.02.050

UiPath Academy. (n.d.). *About.* UiPath Academy. https://academy.uipath.com/

VermaK.ChahalA.KumarV.NayakP.SinghP. (n.d.). *3 RD INTERNATIONAL CONFERENCE ON INNOVATIVE COMPUTING AND COMMUNICATION (ICICC-2020) Automated Order Management using Robotic Process Automation.* SSRN. https://ssrn.com/abstract=3563094

Chapter 4
Determination and Evaluation of Parameters Affecting Tourism Revenue by Machine Learning Methods

Hande Mutlu Ozturk
(iD) https://orcid.org/0000-0002-4404-0106
Pamukkale University, Turkey

Ozgur Guler
Pamukkale University, Turkey

Olcay Polat
(iD) https://orcid.org/0000-0003-2642-0233
Pamukkale University, Turkey

ABSTRACT

The main focus of this chapter is to examine the tourism income of Türkiye as a case country, taking into account the structure of the tourism industry and relevant economic and social indicators. Statistical methods are used to investigate the factors that influence tourism income and to demonstrate the impact of these variables. The chapter aims to identify the key factors that should be considered when planning tourism-related activities and to explore the suitability of different models for future predictions. In addition, the chapter explores the use of machine learning models, such as artificial neural networks (ANN) and gradient boosted regression trees (GBRT), to compare their performance with the established multiple linear regression model. Furthermore, the chapter adds to the existing literature on tourism economics and forecasting methods by examining the performance of different models in predicting tourism income and highlighting the importance of factors such as the country's image, safety, and transportation opportunities in shaping tourism income in Türkiye.

DOI: 10.4018/978-1-6684-6985-9.ch004

INTRODUCTION

Tourism is one of the most important sources of income for some countries. In addition to its direct contribution to employment and foreign exchange, indirect contributions such as the personal expenditures of visitors and its impact on the auxiliary sectors that provide products and services to the tourism sector should also be taken into consideration. According to the economic impact report prepared by the World Travel and Tourism Council for 2019, the travel and tourism sector accounts for one in every four new jobs created in the last five years, making it the top employment generating sector for governments. Apart from this, the report also states that the travel and tourism sector constitutes 10.3% of the world's gross domestic product (GDP), 6.8% of total global exports and 28.3% of global service exports, amounting to $8.9 trillion (WTTC 2019).

Türkiye is among the most preferred countries in the tourism sector due to its cultural and historical heritage, natural beauty, sea-sand-sun tourism, and location. Considering the contribution of tourism income to both the balance of payments and employment in Türkiye, it is clear that the sector is one of the most important economic resources. Despite this, Türkiye cannot reach tourism income levels comparable to the countries with which it competes in terms of the number of tourists. The general report of the Turkish Ministry of Culture and Tourism (2019) states that although Türkiye ranks 6th in the world in the number of visitors for 2018, it has fallen to 15th in tourism income. In other words, while the average tourism income per capita in the world in 2019 was $1020, this figure was $580 for Türkiye, compared to $1050 for Germany and $1520 for Thailand, which comes after Türkiye in terms of the number of tourists. For the United States, the average tourism income per capita is $2700 (UNWTO 2020). Thus, it is clear that Türkiye, which has a satisfactory place in the world in the number of foreign visitors, cannot show the same success in tourism income.

The aim of this study is to determine the most important factors affecting tourism income with meaningful statistical methods and to show how these variables can describe tourism income. Thus, it seeks to identify which factors should be considered when planning activities related to tourism and which models can improve future forecasts. In this study, after conducting research on the variables affecting tourism income and determining the most important variables with classical statistical methods, the performance of the established model is compared with machine learning models. While determining the important explanatory variables, a multiple linear regression model supported by econometric models was used. The performance of the established multiple linear regression model was compared with Artificial Neural Network (ANN) and Gradient Boosted Regression Trees (GBRT) models. In the best model found, a five-year future prediction was made. In addition, the effect of the Covid-19 pandemic on foreign tourism income was analyzed. Thus, the pandemic process was evaluated and the performance of the model was tested.

This study differs from other studies in two main areas. In the literature, there are studies on the relationship between macroeconomic variables and tourism (Uysal and El Roubi 1999; Dritsakis 2004; Aydın et al. 2015; Pekmezci and Bozkurt 2016; Tengilimoğlu and Kuzucu 2019). In the determination of the parameters affecting income from tourism, which is the first stage of the study, 20 variables that can be divided into four groups as economy, tourism capacity, freedom, security and country development were studied and econometric methods were used while determining the important variables. A second point of differentiation is about measuring the performance of the relationship between the independent and dependent variables. In other studies, while machine learning methods such as ANN are applied in general for future predictions about tourism, only time series predictions are used without explanatory variables and seasonal decomposition (Çuhadar et al. 2009; Çuhadar 2013; Koutras et al. 2016; Keskin 2019; Höpken et al. 2020; Çuhadar 2020). Ma et al. (2016) identified the major characteristics of Chinese tourist arrivals in Australia between 1991 and 2015 and identified the patterns of Chinese tourist arrivals in Australia with the help of SARIMA time series models. Chu (2014) proposed an S-shape curve model that forecasts of tourist flows to Las Vegas with a logistic growth regression model that accounts for demand saturation patterns. Arbulu et al. (2021) developed an accurate methodology by using Monte Carlo simulation to analyse complex scenarios in situations of extreme uncertainty, such as the one presented by the unprecedented Covid-19.

In studies conducted after 2010, it has been observed that hybrid models, in which more than one model is combined, are more successful in forecasting than single models. These hybrid models are especially created by combining econometric models and artificial intelligence models (Song et al. 2019).

Cooper et al (1993) have defined tourism demand in various ways depending on the economic, psychological, geographical, and political perspectives. In the study, two types of demand curves are defined: the direct demand curve, which depends on the relationship between two variables such as price and 'quantity', and the inverse demand curve, which indicates that the amount of demand for tourism falls with the increase in tourism-related price and vice versa. Song and Turner (2006) Song and Turner (2006) stated that many factors come into play in tourism estimation and suggested that the effect and degree of all parameters are different. In the study, it was emphasized that the relations between the factors are variable, for example the role of tourism stakeholders, the visa policy of the government, the opening of a theme park can affect the number of tourists. Fleming and Toepper (1990), in their study, emphasized that one of the important issues in any tourism development effort is the total economic impact of the programs carried out by the state and local tourism offices and the private sector, and made a estimation about this. The study focuses on economic needs and discusses some basic approaches to measuring both positive and negative economic impacts. Bonn and Harrington (2008) examined the effect of Capacity Utilization Model (CUM), Regional Economic Models and Inc (REMI) models on the Impact Analysis for Planning (IMPLAN) model in their study. The results for total output, income and employment are compared.

Su et al (2021) emphasized the difficulties of making measurements related to tourism. The study evaluated current tourism flows and their effects on environmental, socio-cultural and economic macro scenarios and examined the main techniques used in the evaluation. Within the scope of the study, demand and supply assessments are discussed together with the latest techniques such as traditional time series and econometric models and artificial intelligence methods using the latest developments in computer science as well as the main national tourism statistical measurements. Li (2022) used the Sparse Principal Component Analysis Convolutional Neural Network Long-Short-Term Memory (SPCA-CNNLSTM) model to accurately predict the tourist traffic during the holidays for the city of Liuzhou, China.

In this study, GBRT models, which are not seen very often in the tourism forecasting literature, but which can give better results than classical regression trees, were modeled and compared with the multiple linear regression method and ANN methods with the help of explanatory variables. In addition, the data were divided into seasonal components and organized with the TRAMO/SEATS program, which is a stochastic seasonal decomposition method. Briefly, unlike the literature, this study is based on the creation of a regression model that can explain tourism income with 20 independent variables in various fields, providing secondary conditions such as stationarity, and comparing this model with machine learning methods and making future predictions with the best model. Also, ANN and GBRT methods were compared with the TRAMO/SEATS program to organize the data, complete the missing data, determine the seasonality factors and predict the future of some factors.

Tourism is one of the world's largest and fastest-growing industries (De Freitas, 2017), with millions of people traveling domestically and internationally every year. The sector has a significant impact on the global economy, generating billions of dollars in revenue and providing employment opportunities for millions of people. Tourism plays a crucial role in the economy of many countries, especially those that rely heavily on the industry. It brings in foreign currency, stimulates economic growth, and creates jobs in various sectors such as hospitality, transportation, and retail (Rasulovich, 2021). Moreover, tourism contributes to the preservation and promotion of cultural heritage and natural attractions. It provides opportunities for local communities to showcase their culture and traditions, which can foster cultural exchange and understanding among different nations (Ursache, 2015; Hosseini et al. 2021). Tourism also provides a platform for education and personal growth. Traveling to new places exposes individuals to different cultures, languages, and ways of life, broadening their horizons and enriching their understanding of the world. Tourism is a global phenomenon, with millions of people traveling to various destinations worldwide. According to UNWTO (2020), international tourist arrivals reached 1.5 billion in 2019, before the outbreak of the COVID-19 pandemic (Kumar and Nafi, 2020 and 2021). Some of the world's most popular tourist destinations include France, Spain, the United States, China, Italy and Türkiye, attracting millions of visitors every year. These destinations offer a variety of attractions, including historical sites, natural wonders, and cultural experiences (Isik et al. 2018).

The COVID-19 pandemic has had a devastating impact on the tourism industry, with international travel restrictions and declining consumer confidence leading to a significant drop in the number of tourists. According to UNWTO (2020), international tourist arrivals declined by 74% in 2020, resulting in a loss of $1.3 trillion in export revenues from tourism (Cretu et al. 2021). The pandemic has also led to job losses and business closures in the tourism sector, affecting millions of people and communities worldwide. Moreover, the pandemic has also had a negative impact on the preservation and promotion of cultural heritage and natural attractions, as many tourist destinations have been forced to close or limit access to protect public health. (Nhamo et al. 2020). The impact of the pandemic on the tourism sector is expected to be long-lasting. The World Tourism Organization predicts that it will take several years for the industry to recover to pre-pandemic levels. Moreover, the pandemic has accelerated the shift towards digitalization and online booking systems, which are likely to become a permanent feature of the industry. The pandemic has also changed consumer behavior, with many people opting for domestic travel over international travel, which could lead to a shift in the market towards more sustainable and local tourism (Atkinson, 2021).

Tourism is a vital industry with a significant impact on the global economy and society. Despite the challenges posed by the COVID-19 pandemic, the industry is expected to recover in the coming years as the world begins to emerge from the crisis. The importance of tourism cannot be overstated, and it remains an essential part of the global economy and a valuable source of income for millions of people and communities worldwide (Puah et al. 2018; Widiastuti et al. 2021). This study mainly focuses on the problem of examining tourism revenues by using the structure of the tourism industry and Türkiye's economic and social indicators. Within the scope of the study, the prediction models are specially emphasized, and machine learning models are discussed in great detail. However, the impact of Covid 19, which was effective for a very short time (2 years) in a long time period, but seriously affected the tourism sector, was discussed incompletely due to lack of data.

BACKGROUND

TRAMO/SEATS Method

TRAMO/SEATS, developed by Gomez and Maravall (1996), is a program that detects outliers, missing observations in the sample, and noisy values and errors, and establishes a regression model that can perform prediction and interpolation based on the ARIMA method. TRAMO/SEATS provides the error values of the series, seasonality factors, seasonally adjusted series, trend and cycle components of the series with signal extraction management in single time series.

In the Tramo stage of the model, a preliminary correction is made for the time series seasonal decomposition, and the aforementioned corrections are made. These corrections are realized by detecting and correcting outliers, missing observations and values that do not comply with seasonality. A regression model is constructed with the help of non-stationary ARIMA errors. Values such as economic crises and holidays in the study are determined automatically by the program and a dummy variable is assigned to the regression model for them. Outliers and forward-looking estimations are made using the least squares or maximum likelihood method. The program can automatically assign an ARIMA model without imposing restrictions on missing observations in the series. It detects various outliers and makes corrections for the linearization step. It calculates the optimal estimates for the series with mean squares of error. After this stage, the series is defined in the SEATS section and linearized to be estimated. In the Seats stage, the series is separated into the aforementioned seasonal parts by a signal extraction method. At this stage, which ARIMA model the data will work with is automatically determined by the program with TRAMO. Then, the effect of seasonality, change in level to the trend component, and outliers to the irregular component were studied. SEATS is used to filter the series and obtain residuals using the ARIMA estimation method (Gomez and Maravall 1996). Since the TRAMO/SEATS method does not follow a deterministic path like classical additive decomposition or multiplicative decomposition, it can produce performance results with a smaller error rate. Contrary to the aforementioned methods, while TRAMO/SEATS is creating seasonal components, it is not only able to obtain the same results for each period, but also seasonality factors that can change historically, plus it can make predictions. In addition, the ARIMA model provides an advantage over other models as it can determine whether the seasonality of the series should be additive or multiplicative, outliers internally, i.e. automatically.

Multiple Linear Regression

Regression is a classical statistical method that expresses the relationship between independent and dependent variables with a mathematical model. Regression models, which are used for different purposes, are generally used in the field of estimation. Regression can express the relationship between dependent and independent variables with linear or non-linear curves. If the model contains more than one independent variable, it is called multiple linear regression. In this context, a multiple linear regression equation is as follows (Eq. 1).

$$y = \beta_0 + \beta_1 x_1 + \beta_2 x_2 + \beta_3 x_3 + \ldots + \beta_n x_n + c \tag{1}$$

According to Eq. (1), the dependent variable y, which has n independent x variables, is explained by multiple linear regression. β_0 represents the constant coefficient, $\beta_{1+2+..+n}$ the coefficient of the independent variables and c the error amount. As can be understood from the equation, not every y observation value may be located on the regression line drawn. This equation aims to find a mathematical way to explain only x and y variables within the framework of a rule with optimum error. In this study, the least squares (LS) method was used for estimation in the regression model. In this method, it is desired that the sum of the squares of the error (c) be minimized.

In addition to the definition of regression, one of the concepts that should be emphasized is the long-circuit trend, which is used to explain the main trend of the time series in the long run. It should be noted that in order to obtain a significant relationship between more than one variable, the series used should not show a strong negative or positive trend. Otherwise, if there is a trend in the time series within the scope of the variables, an unrealistic regression occurs depending on the direction of the trend. Although there is a strong trend between two variables containing time series in econometric analyses, although the independent variable cannot explain the dependent variable in real life, the explanatory value of R^2 may be high. Granger and Newbold (1974) named this situation spurious regression in their study. Therefore, the stationarity of the time series is directly proportional to the power of a regression equation to reflect reality (Granger 2001). Time series that do not show a strong trend show a constant variance and mean feature, and whose covariance are expressed as stationary series.

Unit Root Analysis

Dickey and Fuller (1979) developed a test based on the distribution of the least squares estimator of the parameters. The regression equation for a start is given in Eq. (2).

$$y_t = p y_{t-1} + c_t \tag{2}$$

In Eq. (2), y_t is the series for which stationarity will be investigated, y_{t-1} is its one-period lag, and c_t stands for the regression error value. If the parameter p is equal to 1, y_t is a non-stationary series with a unit root (see in Eq. 3).

$$\Delta y_t = \delta y_{t-1} + c_t \tag{3}$$

where, the parameter δ will be equal to (p-1). During the unit root test, the equality of the δ parameter to 0 is investigated. Dickey and Fuller (1979) created the test based on three different models. In Eq. (4), a model without a constant and without a trend is expressed, in Eq. (5), only a fixed model is expressed, and in Eq. (6), both a fixed model and a trend are expressed. The unit root test is to test the lagged value of y_t with the t statistic. While doing this, critical values are used.

$$\Delta y_t = \delta y_{t-1} + c_t \tag{4}$$

$$\Delta y_t = \beta_1 + \delta y_{t-1} + c_t \tag{5}$$

$$\Delta y_t = \beta_1 + \beta_2 t + \delta y_{t-1} + c_t \tag{6}$$

If $H_0 = \delta \geq 0$, there is a unit root and the series is not stationary.

If $H_1 = \delta < 0$, there is no unit root and the series is stationary.

If the value of the calculated statistic as a result of the analysis is less than the determined critical value, the H_0 hypothesis will be rejected, and in this case, it can be said that the series is stationary. Later, Dickey and Fuller (1981) developed this test with a new study and included lagged values in the equation, and the Extended Dickey Fuller (ADF) test was created. The only difference between this test and the previous models is that it can include lagged values in the model over longer periods. Thus, autocorrelated error terms can be checked. In time series, cyclical changes are detected in certain periods, especially when long intervals are used. This situation is an indicator of both political and economic elements (terrorism, economic crisis, etc.) for countries in econometric data. In classical unit root analysis, this structural break is ignored. Perron (1989) suggested that a structural break that can be determined externally can be included in the model he developed. Zivot and Andrews (1992) suggested that exogenous determination of such a structural break may lead to the rejection of the unit root. In the proposed analysis, they wanted to estimate the reflection of the break in the name of the highest weight for the hypothesis with a stationary trend. The smallest t statistical value was selected. This test works just like the ADF test. A dummy variable was assigned for the internally determined structural break. Thus, the effect of structural break was eliminated and the standard unit root test was applied.

Neural Networks

Machine learning is a method that creates its own experience on the computer by making use of past data. It can make a model that agrees with this experience and can improve the model according to the experience it gains, namely the numbers and quality of data. Artificial Neural Networks (ANN), a sub-branch of artificial intelligence, is a machine learning method that models the learning system of the human brain by imitating it through a computer. The back propagation method, which was explained in detail by Rumelhart and McClelland in 1986, provides a great advantage over other methods and makes ANN applications very effective in solving problems. An Artificial Neural Network forms a structure by connecting artificial nerve cells (neurons). Artificial neurons provide the input data from the external environment, collect the data with an aggregation function, and process and transmit it with the activation function. At the end of the transfer, they output the information obtained to the external environment. ANNs, just like in real life, have connections between each other, and each of these connections has a weight. To summarize, artificial nerve cells have five stages: input, weight, summation function, activation function and output (Öztemel 2003).

An ANN is a structure that consists of layers and has neurons connected to each other in each layer. An ANN operation proceeds through three layers. In the input layer, external information is obtained and usually transmitted to the next layer without any processing. While the hidden layer may not be present in some structures, the number of neurons it contains may vary independent of other layers and is the place where information is processed. The output layer processes the information it receives from the hidden or input layer and exports the output it finds (Öztemel 2003). In the models, the number of neurons and the presence and number of hidden layers may vary according to the general conditions of the problem. In addition, the flow direction also changes according to the complexity of the problem and the type of data. In this study, an Multi-Layer Perceptron (MLP) with a hidden layer is used. It is known that more neurons or layers in the structure will complicate the computation and processing. Therefore, the complexity of the problem has to be considered when determining the number of neurons in the layers or the number of hidden layers. However, if the problem is not linear or very complex, a single hidden layer is considered sufficient (Öztemel 2003).

Backpropagation Algorithm

Resilient Backpropagation (RPROP) algorithm is a learning algorithm that is not fuzzy compared to other algorithms and provides direct adaptation based on gradient information. For faster learning with the algorithm an update value (Δij) is assigned for each weight update, which only determines the size of the weight update. This adaptive update value E develops a learning rule according to the local knowledge of the error function (Riedmiller and Braun 1993) (Eq. 7).

$$\Delta_{ij}^{(t)} = \begin{cases} n^+ * \Delta_{ij}^{(t-1)}, & if \ \dfrac{\delta E}{\delta w_{ij}}^{(t-1)} * \dfrac{\delta E}{\delta w_{ij}}^{(t)} > 0 \\[3mm] n^- * \Delta_{ij}^{(t-1)}, & if \ \dfrac{\delta E}{\delta w_{ij}}^{(t-1)} * \dfrac{\delta E}{\delta w_{ij}}^{(t)} < 0 \\[3mm] \Delta_{ij}^{(t-1)}, & other \end{cases} \tag{7}$$

$$0 < n^- < 1 < n^+$$

In their study, Riedmiller and Braun (1993) showed that when the partial derivative changes sign, the algorithm skips a local minimum value when the weight value (wij) takes a very large value and the update value should be reduced by n⁻. In cases where the sign of the derivative is preserved, the update value is increased to intensify the scan in that region. After this situation is repeated for each weight, if the derivative is positive, that is, if the error increases, the weight value is decreased by the update value. If the derivative is negative, the update value is added (Eq. 8).

$$\Delta_{ij}^{(t)} = \begin{cases} -\Delta_{ij}^{(t)}, & if \ \dfrac{\delta E}{\delta w_{ij}}^{(t)} > 0 \\[3mm] +\Delta_{ij}^{(t)}, & if \ \dfrac{\delta E}{\delta w_{ij}}^{(t)} < 0 \\[3mm] 0, & other \end{cases} \tag{8}$$

Here, the next weight value update is in Eq. (9).

$$w_{ij}^{(t+1)} = w_{ij}^{(t)} \Delta w_{ij}^{(t)} \tag{9}$$

However, in partial derivative change, if the previous step is too large and the minimum value is missed, the weight update is moved back one step to prevent a possible error. The sign of the derivative may change again in the next step due to the back-stepping process in Eq. (10). In order to prevent the update value from being subjected to the same operation twice, the update value is not adapted in the next step. An equation is used to ensure this. As a result, the sign of the partial derivative is used to provide both learning and convergence. Due to the exponential decrease in the sigmoid activation function, the weights further away from the output layer change more slowly, which slows the system down. In the RPROP algorithm, the whole network has an equal chance to learn, since the sign, not the magnitude, of the derivative is used. This results in both faster and stronger outputs than other algorithms (Riedmiller and Braun 1993).

$$\Delta w_{ij}^{(t)} = -\Delta w_{ij}^{(t-1)}, \quad if \quad \frac{\delta E^{(t-1)}}{\delta w_{ij}} * \frac{\delta E^{(t)}}{\delta w_{ij}} < 0 \tag{10}$$

Gradient Boosted Regression Trees

Gradient incremental trees, which is a machine learning algorithm introduced by Friedman (1999), is based on the principle of improving the model by using the data obtained from the error functions of a weakly produced model at the beginning. To put it another way, it uses the predictions produced by weak learners to create strong learners. This model can also be evaluated on regression. Here, it is desired to create an up-to-date tutorial with the highest correlation with the negative gradient found in the loss function. The Mean Squares Error (MSE) method is used to determine the error rate (Zhou et al. 2017). In the Gradient Boosted Regression Trees (GBRT) method, the predictions are made sequentially, not independently, and are included in the ensemble algorithms class. In the first iteration, a function is created for the prediction. The function that calculates the MSE from the deviations between the estimation and the desired value is combined with the estimation function, and the deviations are calculated again and an attempt is made to reduce the error rate to zero. In the loss function (Eq. 11), y_i is the desired value, and $\hat{y}_i$ is the predictive value. A new formula is created by including the learning rate (α) in this equation.

$$MSE = \Sigma(y_i - \hat{y}_i)^2 \tag{11}$$

Thus, in Eq. (12), the learning rate is updated throughout the iterations so that the value of the loss function approaches the minimum (Keleş et al. 2020).

$$\hat{y}_i = \hat{y}_i - \alpha * 2 * \Sigma(y_i - \hat{y}_i) \tag{12}$$

The main purpose of the GBRT algorithm is to add a regression tree to the difference between the desired value and the predicted values calculated by the weak learners at each step. In other words, a fixed-size decision trees algorithm is chosen as weak learners here. The GBRT algorithm tries to find the function with the sum of the weighted values obtained from the weak learners. In Eq. (13), N is the number of weak learners, β is the weight value for each learner, h(x; a_n) is a small regression tree defined by the parameter set a_n. Similar to other augmentation algorithms, GBRT creates an additional model as in Equation (14).

$$F\left(x\right) = \sum_{n=1}^{N} \beta_n h\left(x; a_n\right) \tag{13}$$

$$F_n\left(x\right) = F_{n-1}\left(x\right) + \beta_n h\left(x; a_n\right) \tag{14}$$

Thus, in order to create a better model at each stage, the $h(x; a_n)$ estimator is added on top of the previous function. The decision tree and its weight value β_n are trained to minimize the value of the loss function. L(y, F(x)) is the missing function here, and since Eq. (15) is a difficult optimization problem, it should be made easy with the gradient descent algorithm. To implement the gradient descent algorithm, the $h(x; a_n)$ training set is trained using $\{x_i, g_n(x_i)\}$. Here $g_n(x_i)$ is the negative gradient of the loss function calculated in Eq. (16) by replacing the right side of the equation with MSE (Dabiri and Abbas 2018).

$$\left[h\left(x; a_n\right), \beta_n\right] = \operatorname{argmin} \sum_{i=1}^{M} L(y_i, F_{n-1}\left(x_i\right) + \beta_n h\left(x_i; a_n\right) \tag{15}$$

$$g_n\left(x_i\right) = \left[\frac{\partial L\left(y_i, F\left(x_i\right)\right)}{\partial F\left(x_i\right)}\right]_{F(x)=F_{n-1}(x)} = \left[y_i - F_{n-1}\left(x_i\right)\right] \tag{16}$$

COMPILATION

The data set was created from quarterly data from 2003-2019. As the dependent variable, foreign (excluding citizens living abroad) tourism income data were obtained from TURKSTAT (2020). For the independent, explanatory variables, economy (GDP, exports, imports, dollar, euro, gold, foreign investment, inflation, tourism expenditure, unemployment), tourism capacity (international flight traffic, bed capacity, number of agencies, incentive amounts), freedom and 20 other variables have been determined for four main headings: for national security, (1) terrorism and press freedom indices and (2) the number of detainees, and for the development level of the country, (3) the urban population ratio and (4) human development and human capital indices. These variables are used by the Organization for Economic

Cooperation and Development (OECD), the Central Bank of the Republic of Türkiye Electronic Data Distribution System, TURKSTAT, the T.C. Ministry of Culture and Tourism, T.C. the Ministry of Industry and Technology, the State Airports Authority, the University of Groningen, the University of California-Davis, the United Nations Development Program, Reporters Without Borders, the Institute for Economics and Peace and the World Data Bank. In the analysis of the Covid-19 epidemic period, since the 2020 data of the global terrorism index has not been shared yet, the number of terror-related deaths to be used for the 2020 estimation of this index is from the data of the National Terrorism and Terrorism Response Studies Consortium (START) for the years 2015-2020. The data between the two countries were arranged through the International Crisis Group.

Some data were obtained annually. The reason for this is that the global terrorism index, human development index, press freedom index and other data are published only once a year. Such data are used as the quarters of a year. The data were seasonally adjusted with the TRAMO/SEATS program, and missing data were extrapolated. When using econometric methods, the logarithm of the data was taken in order not to be affected by unit differences and to obtain more flexible results. During the comparison of machine learning models, the normalization process, which is a requirement of Artificial Neural Networks, was applied to all models with the min-max normalization method. For prediction, decimal normalization was used as the normalization method.

REGRESSION MODEL

First of all, normality analysis and Pearson correlation analysis were performed to determine the variables that did not have a linear relationship, using the SPSS 25.0 program. The fact that the skewness and kurtosis values are between +1.5 and -1.5 is considered an indication that the data is normally distributed (Tabachnick and Fidell 2013).

The EURO, USD and INCENTIVE values are outside the critical values. Therefore, they were excluded from the system because they did not satisfy the normality assumption, which is a condition of the regression model. As a result of the Pearson correlation analysis, the only value with a t statistical probability value above 0.05 with the dependent variable of tourism income was UNEMPLOYMENT (0.227). This shows that there is no significant relationship between foreign tourism income and unemployment data. In the model established with the Eviews 10.0 program, first the stationarities of the variables were tested with the Extended Dickey Fuller (ADF) and Zivot-Andrews unit root tests that allow structural breaks. The null hypothesis is based on the existence of a unit root, that is, the series is not stationary. Table 1 shows the statistical values of the data as a result of the ADF unit root test, which first ignores the structural breaks. If any of the values is less than the determined critical values, it does not contain a unit root, that is, it is stationary (Dickey and Fuller 1981).

Since the LOGIHRACAT, LOGITHALAT, LOGTUTUKLU and LOGHDI variables are less than the critical values at the 0.05 significance level, the data are stationary. When the first-order differences of all variables are taken into account, stationarity is observed. The Zivot-Andrews unit root test, which allows a single structural break in order not to be affected by structural breaks, is given in Table 2. In this

Table 1. ADF unit root test results

Models	Level			First Order Difference		
Variables	Constant	Constant and Trend	No Constant and No Trend	Constant	Constant and Trend	No Constant and No Trend
LOGTGLR	-2.28	-2.55	1.89	-6.68	-6.67	-6.45
LOGACENTE	2.95	-1.79	13.60	-8.00	-8.49	-1.38
LOGALTIN	0.48	-2.12	4.37	-7.59	-7.52	-5.95
LOGBASIN	-1.54	-2.39	0.29	-8.02	-7.96	-8.06
LOGDIŞHAT	-1.99	-1.98	5.88	-6.66	-6.80	-4.92
LOGGSYİH	-1.16	-2.02	5.35	-7.17	-7.21	-5.37
LOGHCI	3.70	-1.60	3.22	-2.31	-5.74	0.19
LOGKENTN	-2.88	2.10	-0.53	-0.06	-7.77	-4.68
LOGTGDR	-2.76	-2.85	0.53	-10.52	-10.69	-10.51
LOGTERORİT	-1.49	-2.00	-0.29	-8.02	-7.96	-8.06
LOGTÜFE	1.76	0.02	13.11	-8.02	-8.45	-0.73
LOGTYATAK	-2.09	-0.90	14.5	-8.00	-8.49	-1.38
LOGYATIRIM	-1.76	-0.73	4.64	-8.18	-8.49	-6.28
LOGİHRACAT*	-0.47	-4.07	3.76	-9.28	-9.20	-9.34
LOGİTHALAT*	-2.24	-4.40	-1.33	-5.13	-5.10	-4.91
LOGTUTUKLU*	-0.65	-4.27	2.29	-2.75	-2.60	-1.47
LOGHDI*	-0.40	-1.63	-7.13	-8.15	-8.10	-1.39
Critical values (0.05 p)	-2,90	-3,48	-1,94	-2,90	-3,48	-1,94

Table 2. Zivot-Andrews unit root test results

Models	Fixed			Fixed and Trending		
Variables	t value	pp	Breakdown Period	t value	pp	Breakdown Period
LOGACENTE	-2.76	0.03	2014Q1	-3,53	0,53	2009Q4
LOGALTIN	-4.29	0.00	2013Q1	-3,9	0,00	2013Q1
LOGBASIN*	-4.65	0.00	2009Q1	-5,16	0,00	2009Q1
LOGDIŞHAT*	-6.88	0.00	2016Q2	-5,64	0,00	2016Q2
LOGGSYİH	-3.46	0.00	2008Q2	-3,48	0,02	2008Q2
LOGHCI	-2.59	0.27	2006Q3	-11,16	0,00	2010Q1
LOGKENTN	0.016	0.00	2005Q1	-1,78	0,82	2012Q2
LOGTGDR	-4.41	0.00	2007Q1	-5,23	0,10	2009Q3
LOGTERORİT*	-5.01	0.00	2015Q1	-5,42	0,00	2015Q1
LOGTUFE	-2.76	0.057	2012Q3	-3,1	0,22	2017Q2
LOGTYATAK	-2.53	0.00	2017Q1	-4,1	0,00	2014Q1
LOGYATIRIM	-2.92	0.00	2008Q4	-4,9	0,00	2008Q4
Critical values (0.05 p)	-4.93			-5.08		

test, in addition to the ADF test, structural breaks are included in the model using internally determined dummy variables. Thus, the effect of structural breaks on the existence of a unit root is eliminated (Zivot and Andrews 1992).

As seen in Table 2 above, the variables LOGBASIN, LOGDIŞHAT and LOGTERORIT do not contain a unit root at the 0.05 significance level, and the stationarity of these variables is accepted. The break in the terrorism index occurred in the first quarter of 2015, the break in the press freedom index in the first quarter of 2009, and the break in international flight traffic in the second quarter of 2016. The determined stationary data are inserted into a regression model and screened by backward sorting. During the process, the values with the highest t statistic p probability value are eliminated one at a time, and the process continues until there is no meaningless variable. Meanwhile, multiple connections in the equation are determined by the Variance Inflation Factor (VIF) and excluded from the system. If the

Table 3. Zivot-Andrews unit root test results

Independent variables	coefficients	Std. Errors	Statistical Value (t)	Probability Value	VIF VALUE
DIŞHATLAR	17.27206	0.779749	22.15078	0.000	1.691
TERORİT	935446.3	90220.45	10.36845	0.000	1.691
C	-3689561	548164.8	-6.730752	0.000	

VIF value is above 10, the variables are considered to be overly related (Büyükuysal and Öz 2016). As a result of the backward sorting process and VIF analyses, the regression values with significant F and t statistics and non-multilinearity are as shown in Table 3.

As seen in Table 3, the constant coefficient C on the regression model affects the model negatively, while the Inverted Global Terrorism Index (TERORITE) and International Flight Traffic (DIŞHATLAR) have positive coefficients in the model. The accuracy of the established model was confirmed by performing the normal distribution test for the residuals of the established regression model, the Breusch-Pagan-Godfrey test of variance, and the Wallis autocorrelation test, which is suitable for quarterly data. In the regression model that was modeled and provided the optimum conditions, TERORITE and INTERNATIONAL were chosen as the explanatory variables. The regression equation (Eq. 17) established in this direction is as follows. When the Eq. (17) and seasonally adjusted real tourism revenues and forecast values are compared, the graph on Figure 1 is obtained.

$$tglr = 17,27206 dishat + 935446,3 terorit - 3689561 \tag{17}$$

MACHINE LEARNING MODELS

The Knime 4.2.1 program was used for machine learning models. Training data and test data should be optimally partitioned so that the data does not suffer from over-learning problems when working on datasets and can be tested on a satisfactory dataset. In order to be able to compare other machine learning methods in partitioning, linear sample selection was used instead of random selection. As a partitioning method, 10-fold cross-validation is usually preferred to just partition a large dataset into 70% for training and 30% for testing. With the variables previously determined in the multiple linear regression model, the parameter values of the Multilayer Artificial Neural Networks (ANN) model, namely the Multi-Layer Perceptron (MLP) method and the GBRT algorithm, each of which is a machine learning method, have been determined and compared. The data were normalized between 0-1 values in order to meet the normalization condition, which is a requirement of ANN. The normalization process was carried out by the min-max method. The Mean Absolute Percent Error (MAPE) criterion was used to measure the performance of the models. In the MAPE calculation, a proportional inference is made by dividing each error value by the expected value. This ratio for each value is called the Absolute Percent Error (APE). The MAPE value was found by dividing the sum of these values by the total number of values, that is, by the average. MAPE prevents the problems that could otherwise be caused by unit differences and is not affected by the extreme changes created by structural breaks. In short, it makes the errors of a very

Figure 1. Estimation graph of tourism income with multiple linear regression.

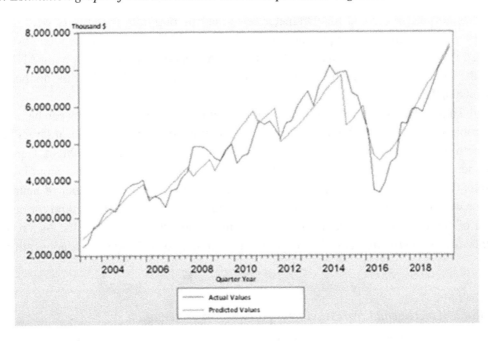

large value and a very small value comparable. For this reason, MAPE was chosen as the most suitable criterion for the comparison of different values such as index, flight capacity and currency. In addition, the R^2 value will be considered the second performance criterion.

Table 4. Artificial neural networks parameter representation.

Parameter	Parameter Information
Number of Neurons in Input Layer	2
Number of Hidden Layers	1, 2
Number of Neurons in the Hidden Layer	3, 5, 7, 10
Number of Neurons in Output Layer	1
Iteration	2500
Activation Function	Logistics
Weight Update	Feed Forward, Back Propagation
Algorithm Used	RPROP

Establishment of Artificial Neural Networks Model

Since there is no precise method for parameter selection, the optimum parameters were determined and the intervals in which the other parameters has no effect on performance were found as a result of repeated testing of the parameters on the analyses. Preferred parameters while applying ANN vary according to the variety, size and complexity of the data. The parameters selected for the application were used as shown in Table 4.

Although single hidden layers are sufficient in time series, the number of layers can be increased in complex problems. However, it should not be forgotten that increasing the number of layers has disadvantages, both in terms of computational performance and the over-learning problem. Generally, the number of neurons in the hidden layer is higher than the number of neurons in the input layer, giving better results (Kirk 2017). Therefore, the number of hidden layers and the number of neurons in the hidden layer were determined as in Table 4 above. Since there is no specific method for determining the number of neurons, this number was tested according to its effect of the results. As a result of the experiments, the optimum number of iterations was determined to be 2500 for this application.

Table 5. Gradient incremental regression trees parameter representation.

Parameter	Parameter Information
Maximum Tree Depth	2, 3, 4, 5
Learning Rate	0.1
Maximum Number of Decision Trees	100
Outlier Identification Rate (1- Alpha)	0.05

Establishing the Gradient Increasing Regression Trees Model

As a result of the analyses made with different values while determining the parameters at this stage, it was seen that the other parameters, with the exception of tree depth, were singular and the optimum. One of the biggest disadvantages of machine learning methods is that they do not have an optimal parameter range, and the trial-and-error method must be used in order for these values to reach optimum results. Table 5 shows the parameters determined by this method.

COMPARISON OF MODELS

In the multiple linear regression model R^2 and MAPE are found as 0,86 and 6,1%, respectively. It is seen that the R^2 value (0.86) in this scenario reveals the estimation has only a 2% difference from that (0.88)

Table 6. GBRT test results

Tree Depth	R^2	MAPE	Tree Depth	R^2	MAPE
2	0.892	5.7%	4	**0.899**	**5.3%**
3	0.891	5.5%	5	0.896	5.5%

Table 7. ANN test results

MODEL	R^2	MAPE	MODEL	R^2	MAPE
YSA(2,3,1)	0.919	4,9%	YSA(2,3,3,1)	0.923	5.1%
YSA(2,5,1)	**0.933**	**4,4%**	YSA(2,5,5,1)	0.913	5.2%
YSA(2,7,1)	0.922	4,7%	YSA(2,7,7,1)	0.915	5.6%
YSA(2,10,1)	0.911	5,3%	YSA(2,10,10,1)	0.936	4.4%

of the mathematically calculated equation. In this direction, it has been pointed out that the machine learning methods established in the application also contain explanatory values close to optimum. Table 6 shows the results obtained for the GBRT model. Since there are only two variables in the GBRT model (70% LN) and the results give progressively lower results after four tree depths, a maximum of five was used. The GBRT model gives better results than the multiple linear regression model. Under the same parameters, the best model is the model established with four tree depths, a MAPE value of 5.3% and the R^2 value being 0.899. In Table 7, the Multilayer Perceptron (MLP) method (70% LN), which is one of the ANN models, is used. Although one hidden layer is sufficient, the test parameters are also tested for two hidden layers in order not to miss the extreme results that two hidden layers may cause.

The best parameters are from the scenario with one hidden layer and five neurons in the hidden layer. According to these results, the model that gave the best results in the comparative model was the

Figure 2. Comparison of the predicted values of the models

Figure 3. Comparison graph of absolute percentage errors of the models

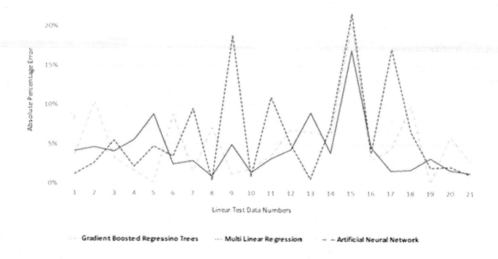

MLP method with a MAPE of 4.4% and an R^2 value of 0.933 [2:5:1] using the RPROP algorithm. As mentioned before, using a single hidden layer in general, except for very complex, non-linear models, provides higher performance results. In Figure 2, the models established as a result of the training data were estimated in the test data and compared with the desired, that is, real data. In the comparison made in Figure 2, it can be seen visually that the machine learning models follow the desired values curve

Figure 4. Estimation graph of tourism income with the best model

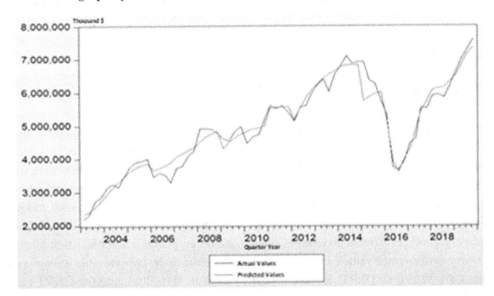

Table 8. ANN and GBRT models test results

MODEL	R²	MAPE	MODEL	R²	MAPE
YSA(2,3,1)	0.89	5.7%	GBRT(2 Depth)	0.872	7.1%
YSA(2,5,1)	0.944	4.5%	GBRT(3 Depth)	0.863	7.1%
YSA(2,7,1)	0.932	5.8%	GBRT(4 Depth)	0.884	7.0%
YSA(2,10,1)	0.918	4.8%	GBRT(5 Depth)	0.931	5.8%
YSA(2,3,3,1)	0.947	5.0%	GBRT(6 Depth)	0.944	5.0%
YSA(2,5,5,1)	0.849	8.1%	GBRT(7 Depth)	**0.949**	**4.3%**
YSA(2,7,7,1)	0.856	7.1%	GBRT(8 Depth)	0.935	4.8%
YSA(2,10,10,1)	0.86	6.9%	GBRT(9 Depth)	0.934	4.8%

better than the regression model. However, since the APE values of the models established for each test data provide a better visual comparison, those values were compared in Figure 3. The established ANN model made less erroneous predictions than other models. The estimation of all series for the ANN [2:5:1] model with the RPROP algorithm, which was determined to be the best model (Figure 4).

As is known, machine learning algorithms with linearity, normality, homogeneity of variance, etc. are methods that can work without requiring conditions. Accordingly, in Table 8, the MLP and GBRT methods with RPROP algorithms were compared using all the variables that could not be used due to the requirements of multiple linear regression. Since 20 variables were used during the comparison, parameters such as tree depths and neuron numbers were retested.

As can be seen in Table 8 above, the machine learning method that gave the best results was the GBRT model with a tree depth of seven. The MAPE value in the model was 4.3% and the R^2 value was 0.949. Compared to the other models, including the multiple linear regression model, MLP with the RPROP algorithm gave better results in the modeling using two parameters, while the GBRT method gave better results in the scenario where all parameters were included. In addition, when we look at our performance measurement values, MAPE and R^2, it is seen that increasing the number of variables does not significantly affect the accuracy of the result. The conclusions that can be drawn from here are that the Least Squares (LS) method, which has certain requirements and works with a classical mathematical equation, can give more meaningful and optimum results in parameter selections to be used in machine learning methods in order to use resources more effectively and to make specific determinations. In short, the use of statistical methods based on solid foundations and conditions, together with machine learning methods that give high performance estimation results, will yield more specific and more efficient results. Apart from this, it is thought that GBRT models can produce better results than ANN with a large number of parameters. For this reason, it can perform well as an alternative to the ANN model, which is frequently used in the current literature, especially in cases that progress as horizontal sections. It is known in the literature that if the MAPE value is greater than 50%, such models may be set up incorrectly, models with a MAPE value of 20% are acceptable, and models with a MAPE value of 10% give very good performance values (Çuhadar 2020). In general, it is clear that all results have very satisfactory performance values. In addition, the best model in the two-variable scenario is the MLP method with 4.4% MAPE and 0.933 R^2 values [2:5:1] RPROP algorithm, and the GBRT method with 4.3% MAPE and 0.949 R^2 values, which also uses 20 variables, are satisfactory enough.

FUTURE FORECASTING

Since the largest data to be estimated are not precise when predicting the future, the decimal normalization method was used instead of the min-max normalization method for ANN. A seasonally adjusted tourism income estimation was made using the parameters of the best model. With the best model found, the [2:5:1] MLP method with the RPROP algorithm, a five-year estimation of 20 data sets was created. In order to create these estimates, the independent variables first had to be estimated. Seasonally adjusted international flight traffic and terrorism index data for the years 2020-2024 were estimated with the TRAMO/SEATS method, which we used as a supporting program for our chosen models and built in time series forecasts. Then, with the ANN model established with these values, the seasonally adjusted future values of tourism income were estimated. Finally, the seasonality factor, which was estimated by the TRAMO/SEATS method for tourism income, was processed with the ANN estimation and the final estimation results were obtained. In Table 9, the estimation values described above are collected in a table.

The TRAMO/SEATS method was used for the seasonally adjusted estimations of the independent variables. The best model found estimated the seasonally adjusted foreign tourism income with seasonally adjusted variables. Afterwards, seasonal components of foreign tourism income were found, as TRAMO/SEATS can predict variable seasonal factors. Here, the program determined that the seasonal component of foreign tourism income is multiplicative. Therefore, five-year estimates of foreign tourism income were obtained by multiplying the forecast values in the seasonality factor with the seasonally adjusted foreign tourism income as a percentage. In Figure 5, a graph is given for these estimates to compare with the actual values.

As seen in Figure 5, growth was expected in Türkiye's foreign tourism revenues after 2020. Although there are data that could be used in 2020, one of the reasons why the sample size was chosen up to the end of 2019 is undoubtedly the Covid-19 epidemic, which affects all sectors and parameters. Crises and pandemic can cause sharp deviations in the curves showing the variables. Therefore, this can also affect

Table 9. Future forecast values

PERIODS	SEASONAL ADJUSTED FORECASTS			SEASONALITY FACTOR FORECAST	T. INCOME FORECAST
	TERORİT-TRAMO/SEATS	INTERNATIONAL-TRAMO/SEATS	TGLR-[2:5:1] MLP	TGLR_SF- TRAMO/SEATS	TGLR_SF x TGLR(MLP) /100
2020-Q1	4.2322498	446374.8642	7550584.213	55.91923398	4222228.853
2020-Q2	4.3718676	456504.6118	7715165.724	98.90723729	7630857.27
2020-Q3	4.5126193	466864.241	7868740.625	157.1550748	12366125.21
2020-Q4	4.6543568	477458.9655	8011151.28	85.96427553	6886728.159
2021-Q1	4.7969515	488294.1193	8142494.294	55.99647145	4559509.493
2021-Q2	4.9402914	499375.1573	8263095.516	98.99910646	8180390.727
2021-Q3	5.0842791	510707.6646	8373464.684	157.3144527	13172670.14
2021-Q4	5.2288301	522297.3371	8474242.819	85.96427544	7284821.438
2022-Q1	5.3738709	534150.0214	8566151.648	55.99647087	4796742.612
2022-Q2	5.5193374	546271.6843	8649949.421	98.99910614	8563372.608
2022-Q3	5.665174	558668.4337	8726395.883	157.3144488	13727881.59
2022-Q4	5.8113325	571346.4977	8796225.947	85.96427449	7561611.818
2023-Q1	5.9577708	584312.2754	8860131.866	55.99647126	4961361.194
2023-Q2	6.1044523	597572.2962	8918751.918	98.99910629	8829484.691
2023-Q3	6.2513453	611133.2405	8972664.737	157.3144527	14115298.42
2023-Q4	6.3984222	625001.8651	9022387.323	85.96427514	7756029.862
2024-Q1	6.545659	639185.3004	9068377.298	55.99647117	5077971.279
2024-Q2	6.6930347	653690.5639	9111034.231	98.99910646	9019842.478
2024-Q3	6.8405313	668524.9989	9150705.499	157.3144544	14395382.43
2024-Q4	6.9881329	683696.0804	9187690.53	85.96427433	7898131.492

the rate at which dependent variables can be explained by arguments and the error rate. In addition to the independent variables, the dependent variable is also affected by the structural breaks in that time period. In line with all this, the forecast values seen in Figure 6 above actually show us a scenario of what kind of tourism income we would have if the Covid-19 epidemic had not occurred. On the other hand, we also have data for 2020, i.e., during the epidemic period. In the continuation of this study, in order to examine the loss of tourism income during the epidemic period and to evaluate the performance of the independent variables we selected and the model due to the effect of Covid-19, a forecast of four numbers was made for the year 2020 in which the independent variables were selected from the realized values.

In this estimation process, 2020 data for international flight traffic were obtained from the same source, while 2020 data for the global terror index could not be found. For this reason, the best model we found, the [2:5:1] MLP method with the RPROP algorithm, estimated the global terror index for the year 2020, in which terror-related deaths were assigned as an independent variable for the years 2003-2020. The reason for estimating the independent variable is to obtain a value as close as possible to the epidemic process. In this modeling, the data were separated linearly as 70% training and 30% testing, and the decimal normalization method was used. Since the training data used were changed, the performance

Figure 5. Foreign tourism income future forecast

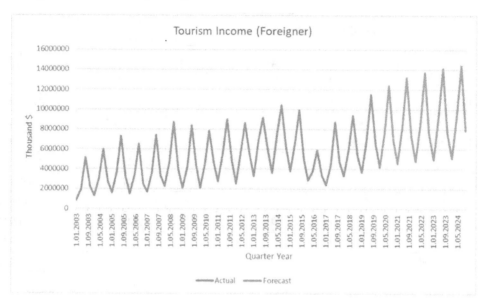

Table 10. Future forecast values due to the effect of Covid-19

PERIODS	SEASONAL ADJUSTED VALUES			SEASONALITY FACTOR FORECAST	T. INCOME FORECAST
	TERORITE (MLP FORECAST)	INTERNATIONAL - SA (ACTAL VALUE)	TGLR-[2:5:1] MLP	TGLR_SF-TRAMO/ SEATS	TGLR_SF x TGLR(MLP) /100
2020-Q1	4.201271	359522.10	6702890.502	55.91923398	3748205.024
2020-Q2	4.324619	166189.59	3278993.512	98.90723729	3243161.893
2020-Q3	4.448995	134501.14	2730531.212	157.1550748	4291168.367
2020-Q4	4.573994	170743.82	3698577.909	85.96427553	3179455.704

values of the model were checked again. Accordingly, the global terror index for 2020 was estimated as 5,613, with an R^2 value of 0.978 and an error rate of 2.4%. Then the terrorism index and international flight traffic values were estimated using the MLP method with the RPROP algorithm, which was seasonally adjusted by going through the same process as the previous forecast. As can be seen in Table 10, the seasonally adjusted foreign tourism income found as a result of the estimation was processed with the seasonal components previously found with the TRAMO/SEATS method, and an estimate for 2020 was made taking into account the effect of the pandemic period. In order to better observe the predicted values of the Covid-19 effect in Table 11, it has been visualized with the graph in Figure 6.

The margin of error between the real epidemic period tourism income values and the 2020 forecast values using independent variables during the pandemic period, that is, the MAPE values, was 24%. Even with these results, it is clear that the model is acceptable. However, as can be clearly seen in Figure 6, the largest error in the comparison of the three values occurred in the 3rd quarter estimation, when

Figure 6. Future forecast of foreign tourism income under the influence of Covid-19

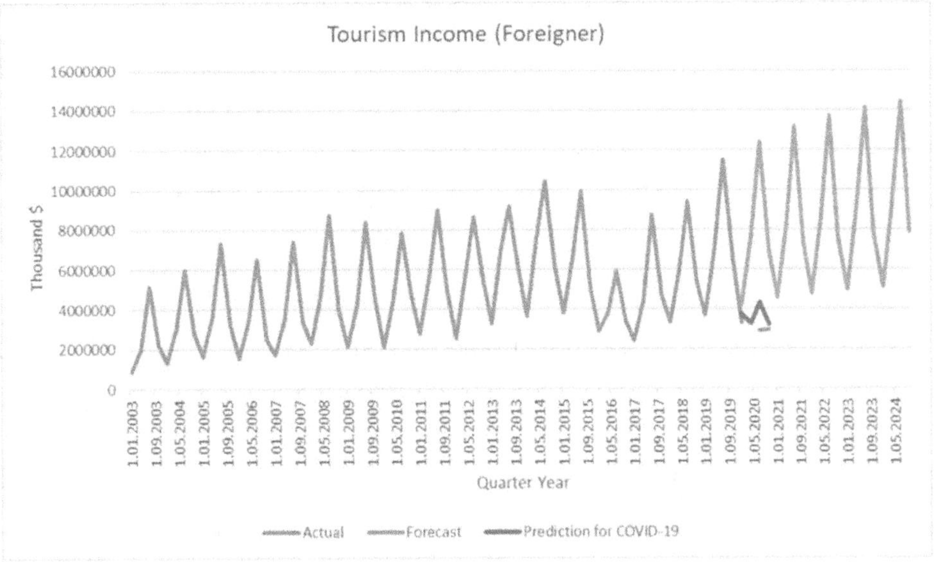

Table 11. Absolute error percentage values of future prediction under the effect of Covid-19

Periods	Actual Tourism Income	Epidemic Period Forecast	APE Value
2020-Q1	3292351	3748205.024	13.8%
2020-Q2	-	3243161.893	-
2020-Q3	2875002	4291168.367	49.25%
2020-Q4	2929765	3179455.704	8.5%

Table 12. Absolute error percentage values of the future forecast with the effect of Covid-19 and with reduced seasonality

Periods	Actual Tourism Income	Epidemic Period Forecast	APE Value
2020-Q1	3292351	3748205.024	13.8%
2020-Q2	-	3243161.893	-
2020-Q3	2875002	2730531.212	5%
2020-Q4	2929765	3179455.704	8.5%

the seasonality factor was the highest. For this reason, the APE ratios of the values were also examined in Table 11.

When we looked at the APE values, it was confirmed that the 3rd quarter was the period that is responsible for the main difference. The 1st and 4th quarters, where the seasonal factors were less effective,

Figure 7. Future projection of foreign tourism income under the effect of Covid-19 and seasonality reduced

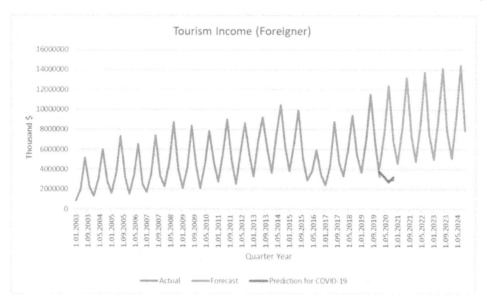

were estimated with a good error rate. It can be deduced from this that the only value with a very high seasonality factor, the 3rd quarter value, was not included in the process with the seasonality factor, and its seasonal effect was ignored. In this case, the value was changed using only ANN prediction. The APE value is given in Table 12.

When we removed the seasonality factor and therefore the effect of the seasonality factor of the 3rd quarter forecast, which has the highest APE value, the MAPE value dropped to 9%, making the model a very high performance model. As can be seen visually in Figure 7, removing the seasonality factor from the 2020 3rd quarter data brought our model closer to the real values.

The reason why only the 3rd quarter value was adjusted for seasonality is that the only seasonality factor that increases the seasonally adjusted forecast values produced as a result of ANN forecasts is in the 3rd quarter. In the epidemic process, it was seen that the closures that took place all over the world removed the positive seasonal factor perception in the tourism sector.

CONCLUSION AND EVALUATION

In the first stage of the study, an equation was formulated using the classical regression model, which has too many requirements. In this equation, econometric methods such as unit root analysis were also used. As a result of these analyses, two out of the 20 selected variables were found to affect tourism revenues significantly and at high rates. These variables are international flight traffic and the global terrorism index, with an explanatory rate of approximately 0.88 R^2.

The multiple regression equation in the second stage of the application was compared with machine learning methods. First of all, the same data were divided into training and test series on the Knime 4.2.1 program to see whether the bivariate linear regression equation with high R^2 value could preserve its accuracy in practice. Using the test data, the R^2 value was approximately 0.86 and the MAPE value was approximately 6%. This shows that with the established multiple linear regression model, a very successful result was obtained in terms of performance criteria which was very close to the statistical results found in the previous stage. Then, this model, whose independent variables were determined to be the terror index and international flight traffic, was compared with machine learning models in terms of performance. The model that best explains tourism income with two variables was ANN with a Multi-layered Flexible Backpropagation (RPROP) algorithm, where 70% of the data was reserved for education, with a single hidden layer and five hidden layer neurons. The MAPE value obtained with this model was 4.4% and the R^2 value was 0.933. Then, the machine learning models, which do not have the requirements of the regression models, were compared by including all 20 variables, and it was seen that the best model was the GBRT model with a tree depth of seven. The best result obtained with this model had a 0.949 R^2 value and a 4.3% MAPE error value, not far from the numbers obtained from the ANN model established with two variables.

It has been observed that good solutions can emerge when econometric methods, regression and machine learning methods are combined. Making detailed use of classical statistical methods such as regression and econometric models while determining the variables to be used in machine learning models would be very useful, both in terms of more efficient resource use and in reflecting the solid foundations brought by classical statistical models to machine learning methods. In addition, the comparison between machine learning models has shown that Gradient Boosted Regression Trees (GBRT) models can give better results in datasets with large cross-sections, and Artificial Neural Network (ANN) models in small datasets. Therefore, it is thought that it would be beneficial to use GBRT models in multi-country comparisons based on panel data in future studies in terms of tourism revenues, and to use ANN models in time series models with few or no explanatory variables.

To make a general evaluation in terms of the factors affecting tourism income, it has been realized that creating an image of a country with safer and easier transportation opportunities on a global scale can affect tourism income on a large scale. Contrary to widely accepted thought, it has been shown that economic factors do not have a significant effect on tourism revenue. The short-term economic effects, which are thought to affect tourism, are believed to be caused by the high number of middle-income tourists coming to Türkiye. In this regard, it is thought that research on how the changes in tourist profiles can affect tourism income and which studies can be effective in ensuring high-income tourist flow could be beneficial. When we look at the estimations we made with the model we obtained, in the scenario that is assumed to be free of the epidemic for 2020, the annual foreign tourism income was calculated to be $31.1 billion. This value shows that if there were no epidemic, there would have been an increase of approximately 8% in foreign tourism income compared to the previous year. On the other hand, the same value shows that Türkiye's foreign tourism income decreased by 71% due to the epidemic. To put it another way, foreign tourism income lost due to the epidemic in 2020 corresponds to approximately 3% of Türkiye's GDP that year. According to the estimation results, although the epidemic eliminated the increase in the seasonality factor in tourism income, the acceptable error rate in our estimation is a testament to the success of the independent variables found and the model established with them.

Due to the increase in the seasonally adjusted value of the seasonality factor because of Covid-19, this high seasonality factor in the 2020 3rd quarter forecast, the performance decreased during the pandemic period forecast. The reason for this is that the seasonality component of the 3rd quarter, which coincides with the summer season, may affect the forecasts in years without crises and epidemics. It would be a mistake to think that the tourism revenues in 2020, when the whole world was closed, would be positively affected by the seasonality factor. It was thought that the seasonality value of this period should be removed and the model re-evaluated. In the results obtained, removing only the 3rd quarter seasonality factor from the process reduced MAPE, our performance measure, from 24% to 9%. This confirms the situation we just brought up. In addition, the results of the comparison of these values with the actual values showed how strong our model and our independent variables, international flight traffic and the global terror index, are in explaining foreign tourism revenues. On the other hand, although there is no real data for the 2nd quarter of 2020, which coincides with the full closure period around the world, it is thought that the 2nd quarter forecast value, where the seasonality factor is the least effective, has a low error rate and this will push us to talk about the forecast. If tourism had been fully realized in April, May and June during the epidemic period of 2020, the estimated total foreign tourism revenue would have been $3.24 billion. This value represents a 53% decrease compared to the same period of 2019. It is seen in the forecast scenarios that, whether the Covid-19 effect is included or not in the model, the decrease in tourism revenues has deeply affected the Turkish economy.

REFERENCES

Arbulú, I., Razumova, M., Rey-Maquieira, J., & Sastre, F. (2021). Measuring risks and vulnerability of tourism to the COVID-19 crisis in the context of extreme uncertainty: The case of the Balearic Islands". *Tourism Management Perspectives*, *39*, 100857. doi:10.1016/j.tmp.2021.100857 PMID:34580625

Atkinson, J. (2021). The times they are a-changin': But how fundamentally and how rapidly? Academic library services post-pandemic. In *Libraries, digital information, and COVID* (pp. 303–315). Chandos Publishing. doi:10.1016/B978-0-323-88493-8.00019-7

Aydin, A., Darici, B., & Taşçi, H. M. (2015). Uluslararası turizm talebini etkileyen ekonomik faktörler: Türkiye üzerine bir uygulama. *Erciyes Üniversitesi İktisadi ve İdari Bilimler Fakültesi Dergisi*, *0*(45), 143–177. doi:10.18070/euiibfd.85938

Bonn, M. A., & Harrington, J. (2008). A comparison of three economic impact models for applied hospitality and tourism research. *Tourism Economics*, *14*(4), 769–789. doi:10.5367/000000008786440148

Büyükuysal, M. Ç. ve Öz, İ. İ. (2016). Çoklu doğrusal bağıntı varlığında en küçük karelere alternatif yaklaşım: Ridge regresyon. *Düzce Üniversitesi Sağlık Bilimleri Enstitüsü Dergisi*, *6*(2), 110–114.

Chu, F. L. (2014). Using a logistic growth regression model to forecast the demand for tourism in Las Vegas. *Tourism Management Perspectives*, *12*, 62–67. doi:10.1016/j.tmp.2014.08.003

Cooper, C. (1993). The tourist destination-Introduction. In C. Cooper, J. Fletcher, D. Gilbert, & S. Wanhill (Eds.), *Tourism:principles and practice* (pp. 77–79). Pitman Publishing.

Cretu, R. C., Stefan, P., & Alecu, I. I. (2021). Has tourism gone on holiday? Analysis of the effects of the COVID-19 pandemic on tourism and post-pandemic tourism behavior. *Scientific Papers. Series Management, Economic, Engineering in Agriculture and Rural Development*, *21*, 191–197.

Çuhadar, M. (2013). Türkiye'ye yönelik dış turizm talebinin MLP, RBF ve TDNN yapay sinir ağı mimarileri ile modellenmesi ve tahmini: Karşılaştırmalı bir analiz. *Journal of Yasar University*, *8*(31), 5274–5295.

Çuhadar, M. (2020). Türkiye'nin dış aktif turizm gelirlerinin alternatif yaklaşımlarla modellenmesi ve tahmini. *Ankara Hacı Bayram Veli Üniversitesi Turizm Fakültesi Dergisi, 23*(1), 115–141. doi:10.34189/tfd.23.01.006

Çuhadar, M., & Güngör, İ. ve Göksu, A. (2009). Turizm talebinin yapay sinir ağları ile tahmini ve zaman serisi yöntemleri ile karşılaştırmalı analizi: Antalya iline yönelik bir uygulama. *Süleyman Demirel Üniversitesi İktisadi ve İdari Bilimler Fakültesi Dergisi, 14*(1), 99–114.

Dabiri, S., & Abbas, M. (2018). Evaluation of the gradient boosting of regression trees method on estimating car-following behavior. *Transportation Research Record: Journal of the Transportation Research Board, 2672*(45), 136–146. doi:10.1177/0361198118772689

De Freitas, C. R. (2017). Tourism climatology past and present: A review of the role of the ISB Commission on Climate, Tourism and Recreation. *International Journal of Biometeorology, 61*(S1), 107–114. doi:10.100700484-017-1389-y PMID:28647761

Dickey, D. A., & Fuller, W. A. (1979). Distribution of the estimators for autoregressive time series with a unit root. *Journal of the American Statistical Association, 74*(366), 427–431. doi:10.2307/2286348

Dickey, D. A., & Fuller, W. A. (1981). Likelihood ratio statistics for autoregressive time series with a unit root. *Econometrica, 49*(4), 1057–1072. doi:10.2307/1912517

Dritsakis, N. (2004). Cointegration analysis of German and British tourism demand for Greece. *Tourism Management, 25*(1), 111–119. doi:10.1016/S0261-5177(03)00061-X

Fleming, W. R., & Toepper, L. (1990). Economic impact studies: Relating the positive and negative impacts to tourism development. *Journal of Travel Research, 29*(1), 35–42. doi:10.1177/004728759002900108

Friedman, J. H. (1999). Greedy function approximation: A stochastic boosting machin. Technical Report. Department of Statistics Stanford University.

Gomez V. & Maravall A. (1996). *Programs TRAMO and SEATS: instructions for the user*. Banco de España. Servicio de Estudios.

Granger, C. W. J. (2001). Spurious regressions in econometrics. In B. H. Baltagı (ed.), A Companion to Theoretical Econometrics. Oxford: Blackwell. doi:10.1017/CCOL052179207X.006

Granger, C. W. J., & Newbold, P. (1974). Spurious regressions in econometrics. *Journal of Econometrics, 2*(2), 111–120. doi:10.1016/0304-4076(74)90034-7

Höpken, W., Eberle, T., Fuchs, M., & Lexhagen, M. (2021). Improving tourist arrival prediction: A big data and artificial neural network approach. *Journal of Travel Research, 60*(5), 1–20. doi:10.1177/0047287520921244

Hosseini, K., Stefaniec, A., & Hosseini, S. P. (2021). World Heritage Sites in developing countries: Assessing impacts and handling complexities toward sustainable tourism. *Journal of Destination Marketing & Management, 20*, 100616. doi:10.1016/j.jdmm.2021.100616

Isik, C., Dogru, T., & Turk, E. S. (2018). A nexus of linear and non-linear relationships between tourism demand, renewable energy consumption, and economic growth: Theory and evidence. *International Journal of Tourism Research, 20*(1), 38–49. doi:10.1002/jtr.2151

Keleş, M. B., & Keleş, A. ve Keleş A. (2020). Makine öğrenmesi yöntemleri ile uçuş fiyatlarının tahmini. *Eurosia Journal of Mathematics, 7*(11), 72–78.

Keskin, H. İ. (2019). Using the seemingly unrelated regression model in the estimation of tourism demand of Türkiye. *Journal of Tourism Theory and Research, 5*(2), 182–190. doi:10.24288/jttr.526021

Kirk, M. (2017). *Thoughtful Machine Learning with Python: A Test-Driven Approach*. O'reilly.

Koutras, A., & Panagopoulos, A. ve Nikas, I. A. (2016). Forecasting tourism demand using linear and nonlinear prediction models. *Academica Turistica-Tourism and Innovation Journal, 9*(1), 85–98.

Kumar, S., & Nafi, S. M. (2020). *Impact of COVID-19 pandemic on tourism: Recovery proposal for future tourism. GeoJournal of Tourism and Geosites, 33*.

Kumar, S., & Nafi, S. M. (2020). *Impact of COVID-19 pandemic on tourism: Perceptions from Bangladesh*. SSRN 3632798.

Li, W. (2022). Prediction of Tourism Demand in Liuzhou Region Based on Machine Learning. *Mobile Information Systems, 2022*, 1–9. doi:10.1155/2022/9362562

Ma, E., Liu, Y., Li, J., & Chen, S. (2016). Anticipating Chinese tourists arrivals in Australia: A time series analysis. *Tourism Management Perspectives, 17*, 50–58. doi:10.1016/j.tmp.2015.12.004

Nhamo, G., Dube, K., Chikodzi, D., Nhamo, G., Dube, K., & Chikodzi, D. (2020). Impacts and implications of COVID-19 on the global hotel industry and Airbnb. *Counting the Cost of COVID-19 on the global tourism industry,* 183-204. Research Gate.

Öztemel, E. (2003). *Yapay Sinir Ağları*. Papatya Yayıncılık.

Pekmezci, A., & Bozkurt, K. (2016). Döviz kuru ve ekonomik büyüme: Türk turizm sektörü için bir analiz. *SDÜ Sosyal Bilimler Enstitüsü Dergisi, 2*(24), 98–110.

Perron, P. (1989). The great crash, the oil price shock and the unit root hypothesis. *Econometrica, 57*(6), 1361–1401. doi:10.2307/1913712

Puah, C. H., Jong, M. C., Ayob, N., & Ismail, S. (2018). The impact of tourism on the local economy in Malaysia. *International Journal of Business and Management, 13*(12), 151–157. doi:10.5539/ijbm.v13n12p151

Rasulovich, K. A. (2021). The role of agro-tourism in the development of socio-economic infrastructure in rural areas. *Наука и образование сегодня, 3*(62), 13-14.

Riedmiller, M., & Braun, H. (1993). A direct adaptive method for faster backpropagation learning: The RPROP algorithm. *International Conference On Neural Networks*. IEEE. 10.1109/ICNN.1993.298623

Rumelhart, D. E., & McClelland, J. L. (1986). *Parallel Distributed Processing. Explorations in the Microstructure of Cognition* (Vol. 1). MIT Press. Cambridge. doi:10.7551/mitpress/5236.001.0001

Song, H., Qiu, R. T. R., & Park, J. (2019). A review of research on tourism demand forecasting: Launching the Annals of Tourism Research Curated Collection on tourism demand forecasting. *Annals of Tourism Research*, *75*, 338–362. doi:10.1016/j.annals.2018.12.001

Song, H., & Turner, L. (2006). Tourism demand forecasting. *International handbook on the economics of tourism*, 89-114.

Su, Y., Cherian, J., Sial, M. S., Badulescu, A., Thu, P. A., Badulescu, D., & Samad, S. (2021). Does tourism affect economic growth of China? A panel granger causality approach. *Sustainability (Basel)*, *13*(3), 1349. doi:10.3390u13031349

Tabachnick, B. G., & Fidell, L. S. (2013). *Using Multivariate Statistics*. Pearson.

Tengilimoğlu, E. ve Kuzucu, S. C. (2019). Döviz kuru oynaklıkları turist başına düşen ortalama harcamayı etkiler mi? 2003-2018 yılları arasında bir nedensellik analizi. *International Human and Civilization Congress, From Past to Future, Ines*, (pp. 148-156). IEEE.

TURKSTAT. (2020). *İstatiksel Tablolardan ve MEDAS Veritabanından Derlenmiş bilgiler*. TURKSTAT. https://www.tuik.gov.tr/

UNWTO (UN World Tourism Organization). (2020). *A compilation of data on inbound tourism by country*. UNWTO. https://www.unwto.org/country-profile-inbound-tourism

Ursache, M. (2015). Tourism–significant driver shaping a destinations heritage. *Procedia: Social and Behavioral Sciences*, *188*, 130–137. doi:10.1016/j.sbspro.2015.03.348

Uysal, M., & El Roubi, M. S. (1999). Artificial neural networks versus multiple regression in tourism demand analysis. *Journal of Travel Research*, *38*(2), 111–118. doi:10.1177/004728759903800203

Widiastuti, I. A. M. S., Astawa, I. N. D., Mantra, I. B. N., & Susanti, P. H. (2021). The Roles of English in the Development of Tourism and Economy in Indonesia. *SOSHUM: Jurnal Sosial Dan Humaniora*, *11*(3), 305–313. doi:10.31940oshum.v11i3.305-313

WTO. (2007). *A practical guide to tourism destination management*. World Tourism Organization.

WTTC (World Travel and Tourism Council). (2019). *Economic Impact Reports*. WTTC. https://wttc.org/Research/Economic-Impact

Zhou, C., Yu, H., Ding, Y., Guo, F., & Gong, X.-J. (2017). Multi-scale encoding of amino acid sequences for predicting protein interactions using gradient boosting decision tree. *PLoS One*, *12*(8), 1–18. doi:10.1371/journal.pone.0181426 PMID:28792503

Zivot, E., & Andrews, D. W. K. (1992). Further evidence of the great crash, the oil price shock and the unit root hypothesis. *Journal of Business & Economic Statistics*, *10*, 251–270.

Chapter 5
Perspective of the Application of New Technologies in the Business Activities in Rural Tourism:
Case of the Republic of Serbia

Predrag Miroslav Vuković

Institute of Economic Sciences, Belgrade, Serbia

Marija Mosurović Ružičić

Institute of Economic Sciences, Belgrade, Serbia

ABSTRACT

A characteristic of rural tourism is the physical distance between supply, which is located in rural areas, and demand, which is located in cities. The role of tourist intermediaries is to connect supply and demand. Tourism distribution channels are seen as a "bridge" between supply and demand and the role of intermediaries is precisely to find ways to facilitate the exchange process. The lack of ICT application can lead to a decrease in demand, a decrease in economic activities, and thus can negatively affect the development of certain areas. The authors will specify the necessary steps which will enable the improvement of business activities and the promotion of rural tourist destinations, both at the micro level, and at the macro level. The expectation is that this approach will increase the income and stop the negative trends burdening the life in rural areas (depopulation, population migration to urban centres, the decline in macroeconomic indicators, etc.) that are present not only in Serbia but also in other countries.

DOI: 10.4018/978-1-6684-6985-9.ch005

INTRODUCTION

Tourism is currently one of the most promising industries worldwide, and there is an urgent need to understand better the innovation in this sector (Carvalho & Costa, 2011; Vučetić, 2017). The authors Carvalho & Costa (2011) refer to the United Nation's confirmation of the importance of the service sector for the economies of developing countries: by improving the service sector, national competitiveness is promoted in a broad sense; the development of a knowledge-based economy underlines the importance of the growing share of services in a large number of activities; and the usage of ICT can facilitate service trade as well.

Rural tourism, as one specific form of tourism, has the potential to accelerate economic recovery. It has various positive effects because it connects economic and non-economic activities and thus enables additional employment and investment in rural areas. In this way, it manages to mitigate negative demographic, economic, sociological and other trends in rural areas. Due to rural tourism development in some countries, those rural households shifted from performing their primary agricultural production to being exclusively engaged in rural tourism. Nowadays, farmers in many EU countries use rural tourism as an additional source of income in their farms. This is the case in Germany, Austria, France, Italy, Slovenia, and so on. Following these positive examples, rural tourism has become a trend in many countries. Today in Europe, a large number of member state governments invest in rural tourism development and expect the effects of the investment. Such strategic commitment is motivated by the slogan "tourism is as efficient as it is invested in". Having in mind that the Republic of Serbia is in the process of joining the EU, the proposed topic is relevant, current, and timely.

Furthermore, by using SWOT analysis the authors (Krasavac-Chroneos et al., 2018) displayed Serbia's perspectives for rural tourism development, underlining the potential of synthesis of agriculture and tourism. Nowadays the application of the information and communication technology (ICT) is present in all branches of the economy, including the sector of (Vidas-Bubanja & Bubanja, 2017). It is predicted that ICT will mark the development of business and economies worldwide in the first half of the 21st century (Pedersen & Wilkinson, 2018). There is no agreement on what the most important success factors in IT applications are. The research conducted at the Technical University of Eindhoven (Nederland) aimed to establish the critical success factors in the application of ICT. It was concluded that technically the emphasis is on the information system, while from the organizational aspect these are: business processes, organizational culture, and structure (Stankić, 2014). Moreover, there is a link between long-term growth and ICT competencies (Antlová, K., Popelínský, L., & Tandler, 2011), especially when it comes to small and medium-sized enterprises and the development of entrepreneurship (Antlova, 2014; Irefin, I. A. et al., 2012). Many authors (Geoffrey, 2007; Goffi & Cucculelli, 2019; Niavis & Tsiotas, 2019; Salinas Fernández et al., 2020) explore the multidimensional nature of tourism and underline the influence of a large number of different factors on destination competitiveness. They highlight the economic, social, cultural, political and technological dimensions of competitiveness.

Increasing number of tourist destinations on the market is present worldwide. Each of the destinations strives to secure its own competitive position. On daily basis, a large amount of information is eagerly presented by tourist organisation to meet ever-increasing demands of tourists.

Through marketing, efforts are made to stimulate tourist demand through promotion, that is, to introduce as many tourists as possible to a brand-new destination's offer. Those destinations that succeed in animating as wide a circle of tourists as possible are assumed to be able to make greater profits and be more competitive in the market. For this reason, the interest of tourist organisations to improve and develop ICT is growing.

Family firms are a dominant way to do business in rural tourism (Kallmuenzer & Peters, 2018). The authors (Paunović et al. (2022) indicated that the appearance of business process innovations in Serbian family firms positively influenced customer satisfaction and customer retention rate.

The desk-research method was utilised in the paper and the primary sources of the data were the papers and publications from scientific journals and other professional literature.

The aim of this chapter is to emphasise the importance of ICT for rural tourist area development. This research will underline: the importance of the application and utilization of ICT for the development of Serbia's rural tourism within the frame of concrete, measurable results; the importance of the education of the local rural population for ICT application; steps to enable the improvement of business activities in rural tourism, and the ways to improve the promotion of rural tourism destinations in Serbia both at the micro and macro level.

The lack of ICT application can lead to a decrease in demand, a decrease in economic activities, and thus can negatively affect the development of certain areas, i.e. it can lead to a decrease in investment (Nguyen et al., 2015). Rural areas are nowadays burdened with numerous problems (depopulation, population migration to urban centres, reduction of macroeconomic indicators, etc.). To stop these negative trends, the development of rural tourism is encouraged and the reason for this is that tourism, with its synergistic character, connects a large number of economic and non-economic activities. This approach has proven to be successful in a numerous of countries (UN FAO, 2021). This is why the EU promoted a multifunctional concept of agricultural development in its Common Agricultural Policy, which implies that farms, in addition to primary agricultural production, should be developed into rural tourism.

POSSIBILITIES FOR THE APPLICATION OF ICT IN RURAL TOURISM BUSINESS

ICT implementation is essential since it has a significant supporting business function linked with innovation activities that affect business process innovation. The changes that occurred in the business environment also affected the development of methodological approach for measuring innovation. One of the reasons for publishing the latest, fourth edition of the "Oslo Manual", the guidelines for innovation measurement was, among other things, the ever-increasing implementation of information and communication technologies (ICT) in all sectors to facilitate the measurement of the digital transformation process. This enables better understanding of innovation processes and their impact (OECD/Eurostat, 2018). In the beginning, the concept of innovation was mainly associated with technological innovations and capital-intensive industries. The field of innovation and innovation activities should be observed within the broader innovation concept where, in addition to technological changes, non-technological changes occupy a significant place and affect business development (Mosurović & Kutlača, 2011). Technology and ICT innovations, besides changes in the economy on micro and macro levels, cause changes in human behaviour, sociology, psychology, etc.(Gössling, 2021). Lately, the development of ICT has had a significant impact on the tourism sector, and the innovations in the service sector have become increasingly important (Decelle, 2004). Besides that, ICT implementation in rural tourism implies the development of the whole tourism sector (Vučetić, 2017).

Nowadays, information represents the key to a successful business. Timely access to information is one of the conditions for the survival and preservation of the competitiveness of economic entities (Chae et al., 2018). The need to exchange information between supply and demand affects the strengthening and rise of the IT sector. This process is present with varying intensity in all economy sectors. The tourism industry is certainly one of those in which the development of IT is a condition for survival. A characteristic of rural tourism is the physical distance between supply, which is located in rural areas, and demand, which is located in urban city centres. The role of tourist intermediaries is to connect supply and demand. Tourism distribution channels are seen as a "bridge" between supply and demand (Pearce & Schott, 2005), and the role of intermediaries is precisely to find ways to facilitate the exchange process (Clerides et al., 2006).Tourism industry is often perceived as a less innovative sector compared to production and capital-intensive sectors (Alsos et al., 2014). Bearing in mind the specificity of the products and services in the tourism sector, legal entities are forced to continuously improve their prod-ucts, processes and procedures to gain sustainable competitive advantage (Alsos et al., 2014; Carvalho & Costa, 2011; Tüzünkan, 2017). Sustainable development is main progressing force for developing countries and incorporates environmental, social, and economic performance (Ružičić et al., 2021). The large and intense changes in the tourism sector also resulted in redefining the strategies of the legal enti-ties to improve their competitiveness (Carvalho & Costa, 2011). Moreover, the process of globalization resulted in the growth of tourist demand. With the development of ICT, there was a change in the profile of tourists since the information became more accessible to them. The concept of sustainability should be an integral part of the strategy, because the legal entities operating in the tourism sector are supposed to achieve a synergy of economic, social, and ecological dimensions (Tüzünkan, 2017).

Tourism is becoming an important element of various regional policies which aim to mitigate the effects of uneven socio-economic development. In particular, this has become an obligatory element of rural development policies, along with creative thinking, an innovative approach to the introduction and implementation of new technologies, as well as the application of the proven concepts in an innovative way (Roblek et al., 2021). The literature underlines that product differentiation through innovation further encourages the residents of rural areas to engage in tourism through entrepreneurship (Tang et al., 2022).

The implementation of ICT in some countries has shown that ICT can both play a key role in business activities (for example, in small and medium-sized enterprises and entrepreneurship), and simultane-ously influence the development of certain regions (rural or urban), or encourage the development of the local economy (Premkumar & Roberts, 1999). Furthermore, the implementation of ICT can also be particularly beneficial for the tourism sector (Akca et al., 2007).

The implementation of digital innovation, i.e. ICT, is especially present today in the so-called mass tourism. The application of Computer Reservation System (CRS) and Global Distribution System (GDS) has entirely changed the role recently played by intermediaries in the tourism business - travel agencies/tour operators (UNWTO, 2007). On the other hand, the application of ICT in rural tourism is lagging behind (Buhalis & Law, 2008). This resulted in rural tourism falling behind other types of tourism. This is present to a greater or lesser extent in all countries where rural tourism is developing and can be perceived as a general problem in rural tourism.

The typical feature of rural tourism is the physical distance between supply, located in rural areas, and demand, located in urban city centres. The role of tourist intermediaries is to connect supply and demand. Tourism distribution channels are seen as a "bridge" between supply and demand (Pearce & Schott, 2005), and it is the role of intermediaries to find the ways to facilitate the exchange process (Clerides et al., 2006). The lack of ICT implementation can lead to a decrease in the demand for the holidays in rural destinations, a decrease in economic activities; and this can, consequently, negatively affect rural development, that is, it can lead to a decrease in future investments in rural areas (Premkumar & Roberts, 1999). To make the offer of rural tourism available to the urban tourist demand, one of the important conditions is the education of the local rural population to use and implement ICT. This requires special training programs. Actually, the progress in ICT development has enabled a wide range of tools and services that can improve business activities in rural tourism (Akca et al., 2007).

The intensity of rural tourism development varies from country to country. The level of rural tourism development depends on: the natural geographical characteristics of the rural area, the presence of natural and social attractiveness in these areas, the achieved level of economic development of that area, and the local community's awareness of the importance that rural tourism can have on the development of the area.

It is expected that ICT will be used extensively in business-related activities such as management, organization, product packaging, marketing, distribution, and business networking (Benckendorff et al., 2019).

The process of digitalization and computerization of business nowadays is present in tourism, such as airline companies, hotel chains, travel corporations, etc. The use of advanced technologies includes new products, microprocessors, different types of media, and communication technologies. It is estimated that about 1/7 of the world's population is directly or indirectly involved in the tourism business. Nowadays, it is practically impossible to imagine doing business in tourism (tourist activities) without the application of modern ICT (Awad Alhaddad, 2015).

Table 1. Distribution of digital travel sales worldwide from 2014 to 2019, shares in % by region

Region	2014	2015	2016	2017	2018	2019
North America	34.0	35.8	33.8	31.8	30.2	28.9
Western Europe	29.8	26.2	24.8	23.5	22.3	21.2
Asia, Pacific	24.7	28.4	31.5	34.3	36.3	38.5
Latin America	5.3	4.6	4.9	5.1	5.3	5.5
Middle East and Africa	4.3	3.8	3.9	4.1	4.4	4.7
Central and Eastern Europe	1.2	1.2	1.2	1.2	1.2	1.2

Source: Statista, https://www.statista.com/statistics/499760/forecast-ofonline
-travel-sales-share-by-region/ (retrieved: 14th November, 2022).

Note: Due to the presence of the corona virus (COVID - 19) in the period 2020-2023, there was a decrease in the volume of the business of tourism in all regions of the world. For this reason, the data for these years are not relevant.

Based on the data in Table 1 which provides the insight into the regional distribution of the volume of digital travel, it can be concluded that the highest turnover was achieved in the North American region in 2015 (35.8%) out of the total travels sold online worldwide. However, a slight decline can be seen in the following years. Therefore, in 2019 it amounted to 28.9%, whereas Asia Pacific region recorded rise. The smallest and almost unchanged participation is recorded by the region of Central and Eastern Europe at 1.2%, which was constant during all the period.

Travel agencies use computerized reservation systems (CRS) to obtain information and make booking. Not only do hotel use the technologies to integrate their front office, back office, and food and beverage departments, but they use it for almost every aspect of their operations, from schedule generation to flight planning and analysis (Benckendorff et al., 2019).

Authors Rowley & Shirley (2019) state that the application of latest IT has made the usual distance that exists between tourists and destinations irrelevant. Such a business principle is also applied by small family farms that offer their tourist services to the market, and in this way they become as competitive as organized entrepreneurial firms. Starting from this point of view, it is acceptable to believe that if modern IT is not applied in business in rural areas, it could lead to negative consequences such as: decrease in tourist demand, decrease in economic activity, lag in economic development, and decrease in investments.

Akca et al. (2007) gives the following opportunities to rural residents:

- to promote rural tourist products,
- to use benefits of e-trade,
- to make transfer knowledge from urban to rural areas and vice-versa,
- to apply in official state procedures - for example, the applications for funds from various state funds, such as the funds intended to stimulate the development of rural tourism, etc.,
- to use geographical information system (GIS) for the management of natural resources, etc.

What presents an obstacle for local shareholders to apply the latest ICT to encourage practical business and continuous progress is the low population density of rural areas (Salemink et al., 2017). AlBar & Hoque (2017) argued that effective communication is a presumption for adopting new innovative knowledge and technologies. ICT has multiple functions and it can help overcome physical barriers, restructure the economy and develop rural areas in the long run.

UNWTO (2007) presented an e-business system which can be used in tourist destination management. E-business enables destination organizations to develop and implement a wide range of relationships with consumers/tourists, product suppliers, and market intermediaries. Accessing the system through one or more channels (such as: PC, handheld device, interactive TV, kiosks, in-car systems, etc.), can help users to select from a wide variety of services (or applications) that will draw, in turn, expand on the appropriate databases.

ICT IN RURAL TOURISM DEVELOPMENT IN SERBIA

Tourism as an economy sector has always been insufficiently developed "even though more than 85% of Serbia's territory is rural, especially if we are talking about the rural tourism product" (Gajić et al., 2023).

In Serbia, the intensive development of rural tourism began after the Second World War, as in most other countries. The following are the initial factors that contributed to the development:

1) The accelerated process of industrialization that led to an increase in the number of residents in urban city centres, which led to the strengthening of the urban population's demand for vacations in rural areas;
2) The development of road traffic and the increasing availability of cars as a means of transport, which have become available to a large number of people. In this way, various rural tourist destinations have become more accessible to citizens;
3) Higher standard of living of the population;
4) Strengthening of global trends towards the return of natural and anthropogenic values possessed by rural areas (rural tourist destinations).

In this way, as in most other countries, we can talk about the phased development of rural tourism. The first one who expressed his views on this matter was Zdorov (2009). Accepting his views and observing the character and dynamics of the development of rural tourism in Serbia, it can be concluded that rural tourism in Serbia has gone through two phases - the phase of independent establishment and the phase of dedicated development. Moreover, it had almost identical characteristics, but it is possible to determine the time frames in the development of both phases based on this study (Vukovicć et al., 2016; Vuković & Subić, 2016).

Namely, the first phase of dates from the seventies until 2006 followed by rapid development which lasted until 2008. Such a precise timing of the second phase of dedicated development begins is due to the fact that at that time the Ministry of Agriculture, Forestry and Water Management allocated about 92 million dinars or approximately one million euros in order to stimulate the development of rural tourism in all regions of Serbia (Ministry of Agriculture and Water Management, 2009). After this decision, the number of villages and municipalities engaged in rural tourism increased significantly. Nowadays, almost all regions in Serbia have recorded certain level of the development of rural tourism (Vuković et al., 2016).

Since then, rural tourism has started to develop evenly in all regions. The pioneer villages experiencing this development were Sirogojno, Seča Reka, and Devići. The time frame for the beginning of this phase in development stems from the fact that at that time the biggest tourist agencies in Serbia ("Putnik" and "Yugoturs") included rural tourism and these villages in their tourist arrangements (Vuković, 2017).

Todorović & Bjelac (2009) provide the following data "According to the 1992 Serbian Tourist Association, in the Knić municipality there were about 35,000 foreign tourists from 21 countries. The largest number of tourists was recorded from Great Britain, Germany, Russia, and Italy".

Strengths in the development of rural tourism characteristic of the first phase of development relate to: preservation and abundance of natural resources, rich cultural and historical heritage, abundance and diversity of rural settlements, wealth of local traditions, traditional hospitality, and diversification of tourist products. Weaknesses characteristic of the first phase of development refer to: inadequate rural infrastructure, "archaic" tourism products, underdeveloped information system, unsatisfactory level of quality of various services, lack of educational programs for farmers, lack of experience, lack of motivation, and lack of awareness in rural areas about economic and other benefits of rural tourism development (Milojević, 2004).

The basic weaknesses observed in the second phase of development relate to the underdevelopment of accommodation capacities and the unsatisfactory level of the quality of existing accommodation capacities, insufficient utilization of existing capacities, incomplete offer of basic catering services, economy of small volume and low prices, under-development of additional services, low investment capacity of households, and slow trend of development.

The expansion of rural tourism in Serbia has been recorded since 2009 when the number of households and villages engaged in rural tourism increased. The region that experienced the greatest expansion of development is Central Serbia.

The main advantages of Serbia, when it comes to rural tourism, are 'preserved nature, mild climate, clean air, rich flora and fauna, unpolluted rivers and lakes, a tradition in the production' (Krasavac-Chroneos et al., 2018). As one of the results resulting from the process of EU accession, Serbia accepted the Concept of Common Agricultural Policy, which promotes multifunctional agriculture and rural development model based *on rural development, sustainable food, environmental protection, organic farming, rural tourism*, etc(Vuković & Subic, 2016). The authors (Ristić et al., 2020) showed in their research that the multifunctionality of agriculture would be an important factor in further sustainable rural development in the Republic of Serbia. However, this concept is still underdeveloped and should be supported at all decision-making levels.

The problems with which the further development of rural tourism in Serbia is burdened are related to the insufficient education of interested farmers regarding treating tourists; insufficient number of tourist attractions in villages and weak connection with the municipal, and insufficient and inadequate social and road infrastructure.

Bearing in mind that rural tourism began to develop rapidly in all parts of the year, in 2011 the document "Master plan for sustainable development of rural tourism in Serbia" was published, in which the data on the dimensions of rural tourism development were presented.

For now, this is the only valid document in which the data on the dimensions of rural tourism in the Republic of Serbia are presented. Bearing in mind that in the period up to 2019, when the "Covid-19" virus pandemic began, there were no major changes, we will present the data from this document.

The document specifies that 106 local tourist organizations operate. Rural tourism encompassed 2.7 million overnight stays, which is the sum of individual overnight stays in rural tourism (145,354). This data comes from the municipalities and LTOs. As pointed out in this document, "no central institution is in charge of collecting the data, except for the Council of each municipality or the local tourist organizations'. The number of tourists overnight stays usable for rural tourism (2,556,128). Rural tourism provides more than 32,000 beds (registered and unregistered), where more than 10,000 beds are in the countryside. The total number of beds is estimated to bring more than 5 billion RSD annually in income and 5 bn. RSD in direct income to the tourism sector. The income of 10 bn. RSD does not include visitors who stay for a night or stay with their friends or family (although they also spend money and use other services during their stay) and it does not include the indirect contribution to the local economy in the sense of income and employment. The income of 10 bn. RSD is 16% of direct GDP from travel and tourism, as calculated by the *World Council for Travel and Tourism in Serbia* for the year 2010, which is 64.2 bn. RSD.

Based on this issue, we can conclude that rural tourism today has an upward trend in development.

What appears to be a problem in tourism business activities is the lack of ICT implementation. This is characteristic of the organization and management of rural tourism at all levels, that is, from local, regional to national. The use of the Internet in tourism has not reached the level that could be considered satisfactory. One of the following tasks is the development of the network economy, which would be based on the adopted Information Society Development Strategy.

(Vidas-Bubanja & Bubanja, 2017) argued that tourism is information intensive and creates necessary predictions for electronic commerce which can play a significant role in tourism service realisation. According to these authors basic components of tourist products are information and confidence. 'These two elements are the base of every e-commerce transaction, making the transition of tourism as a traditional service activity to e-tourism phase a logical forward development step. That is, the implementation of ICT in tourism industry can further improve the quality of tourism information and raise confidence of potential consumers, as basic competitive factors of tourism product. The online travel industry is primarily made up of travel e-commerce sites and review sites. Travel e-commerce sites specialize in the selling of travel products such as flights, hotels, and rental cars. These can be either purchased directly through a travel company's website, such as the Lufthansa website, or through an online travel agency (OTA), such as Expedia'.

Today, the term emerging distribution systems is widely used. For example, the World Tourism Organization (UNWTO) under this term implies all the systems that use Internet technology, and the goal is to improve business and make a wide range of information about different tourist destinations and products available. This term also includes various "tools" offered by modern ICT (UNWTO, 2007, p. 73).

In Serbia, there is no single database of rural tourism accommodation offers. The reason lies in the fact that there are numerous limitations in the field in collecting information about accommodation facilities.

The biggest database of accommodation facilities in rural tourism in the Republic of Serbia is in the National Association "The Rural Tourism of Serbia". This association was co-founded by nine associations in 2002 and today has more than 500 members. In 2004, the association set up an official Internet presentation (www.selo.rs), which was conceived as a unique database on the offer of rural tourism. The association is an active member and representative of Serbia in the European Federation of Rural Tourism (EUROGITES). Over the years, the Internet site has recorded continuous rise of a number of visitors.

This association is in charge of the activities and projects supported by the European Union, the Ministry of Trade, Tourism and Telecommunications of the Republic of Serbia, the Ministry of Agriculture, Forestry and Water Management of the Republic of Serbia, the World Tourism Organization (UN WTO), the United Nations Development Program (UNDP) for developing rural tourism, etc.

Up to now, there has been no central server at the global level for the needs of rural tourism, with the databases from all countries in which the data collected from all national servers monitoring tourism supply or rural tourism would be combined. Moreover, for the time being, there is no such database that could be expected to exist in the near future. The idea is that a suitable company that is a leader in global distribution systems should be connected to such a database to provide services for the promotion and sale of rural tourism offers/products.

In the Republic of Serbia, one of the leading systems for reserving accommodation facilities is Booking.com. The idea is that the offer of rural tourism that is presented on the website of the National Association "The Rural Tourism of Serbia" can be linked to one of these websites or simply enable the reservation and sale of tourist arrangements directly through the Internet presentation of this association. In this way, tourist offers can be brought closer to the tourist demand, i.e. sales would be enabled. Furthermore, the idea is that the offer of the "EUROGITES" association can also be connected to the leading national reservation systems and thus promote the development of rural tourism. Certainly, for this type of business, the education of local stakeholders in rural areas is necessary through appropriate training, which would promote the development of rural tourism.

It is expected that ICT implementation will contribute to rural tourism development, the increased income of the local population, and the increased competitiveness of rural households/farms as small legal entities on the tourist market, which is precisely one of the characteristics of increased competition. This approach is in line with the principles of multifunctional agriculture insisted by the EU in its well-known Common Agricultural Policy (CAP), which is presented in a large number of different documents, the UN FAO (UN FAO, 2021), etc.

EDUCATION OF THE LOCAL RURAL POPULATION (RESIDENTS) FOR THE APPLICATION OF NEW ICT

Based on the abovementioned, it is obvious that modern society requires an appropriate level of digital literacy and skills in order for a person to keep up with new trends and be able to seize all the opportunities that technology provides. The author (Bradić-Martinović, 2022) highlights that the process of improving the digital skills of citizens has been taking place in Serbia - "In 2007 there were more than 70% of citizens who didn't have any information skills, whereas in 2019 almost 60% displayed advanced knowledge in this area.

It can be stated that almost all rural areas in the world have a problem with the application of latest technologies and that they hardly gain the benefits from the application of latest technologies. In this sense, various programs of education of the local population are implemented in order to eliminate these problems. Unfortunately, the rate of literacy and the rate of broadband penetration in rural areas is still at a low level (Halili & Sulaiman, 2018).

Hosseini et al. (2009) list several challenges that rural communities are facing when using ICT:

- *Organisational* - lack of interest and expertise in using ICT, concerns about the risks of using ICT, low quality of the services provided by the service centre, and the lack of interest by the private sector to participate in the development of ICT for rural areas.
- *Technical* - the use of low-bandwidth, hardware shortages, lack of appropriate infrastructure, lack of software and telecommunications systems, and weak expertise in rural areas in using ICT.
- *Financial* - involves high costs for the purchase of hardware and software, high costs for internet access, the cost of maintaining the system, the cost for upgrading systems, and the lack of investors to invest in rural areas from both the public and private sectors.
- *Social* - includes the aspect of rural 'technophobes' not wanting to use ICT since rural communities do not fully understand the benefits and advantages of using ICT.

All these factors slow down the process of regulation due to the lack of support from the parties involved, insufficient equipment, and old or inappropriate strategic places to develop ICT activities in rural areas. Furthermore, rural communities also have the lowest level of knowledge of using ICT, lack of expertise in handling ICT equipment, and great doubts about the advantages of ICT. All these factors clearly show the challenges and problems faced by rural communities and influence the acceptance of using ICT.

Ghavifekr & Ibrahim (2015) argued that the development of ICT has changed the teaching and learning process from traditional methods to a technology-based approach. Since the advent of internet technology, there has been a change in the pattern of life of communities and society around the world in education, economics, politics, and social life in general.

- There are two systems of education regarding tourism: a formal and an informal one. *Formal education.* In Serbia there is a system for educating mangers for the needs of tourism taught at faculties, high schools, secondary schools, and private schools and universities. Special attention is paid to the wide range of education (lectures, consultations, exercises, professional practice, professional school trips, company visits, visits of specialists, and so on). The model of "complete separation" has mostly been used in Serbia. Market orientation in doing business, that and the process of privatization, which represents long-term orientation of our economy, require inclusion and usage of other models in the educational system, where the emphasis needs to be on practice and training of the so-called "informal educational system".
- *Informal education system.* The education of the local population for the providing of the services of the appropriate quality to the tourists has not been performed enough in Serbia. This role should be given and performed by the local tourist organization, regional chambers of commerce, units of the local government, non-governmental organizations, etc. Most often, the trainings are intended for different groups of local stakeholders.

The document 'Report on digital inclusion in the Republic of Serbia for the period from 2014 to 2018' contains the following data (Ožegović, 2019):

- According to the research of the Republic Institute of Statistics "Usage of ICT in the Republic of Serbia in 2018", 72.1% of households in Serbia own a computer, which represents an increase of 4% compared to 2017, and 6.3% compared to 2016. Nevertheless, this is still a weaker result compared to the European Union average, which was 84% in 2017;
- The upward trend of computer users is evident. It increased by 3.3% compared to 2017, by 4.5% compared to 2016, and by 5.9% compared to 2015;
- The biggest gap regarding the usage of computers in households is visible in the structure of households according to monthly income: computers are mostly owned by households with a monthly income exceeding 600 euros (87.9%), while the share of households with an income of up to 300 euros is only 54.8%;
- The stratification according to the criteria of education, employment and age is also noticeable. Among computer users, 59.7% have secondary education, 17.7% of users have less than secondary education, and 22.6% have higher education. Computers are used far more often by persons with a university degree and higher education than persons with less than secondary education (93% versus 41.1%);
- In 2018, 86.8% of employees used a computer compared to 74.6% of the unemployed and 44% of others, including pensioners. Over 95% of young people aged 16-24 use a computer at least once every three months in contrast to about 83% of people aged 25-54). It is noticeable that men aged 55-74 use a computer more than women of the same age (45.6% compared to 36.6%).

In order to be able to increase the efficiency of the application of modern ICT, it is necessary to increase the level of information literacy of local stakeholders in rural areas in the Republic of Serbia. Therefore, it is necessary to constantly conduct education, training processes, and appropriate events. All stakeholders in the field of business are interested in the organisation of training activities in the application of modern ICT, from the local administration, the economy, the tourist organization of Serbia as a representative of the national tourism organization, local chambers of commerce, farmers, but also actual and potential tourists. This problem is not only characteristic of the Republic of Serbia, but it is characteristic of almost all countries where rural tourism is in a developing process.

CONCLUSION

The paper has showed that there is a close connection between ICT and the development of rural tourist areas with huge potential for tourism and economy progress. The accelerated development of rural tourism has been noted since the seventies of the twentieth century, first in the countries of Western Europe and then in North America. The positive results achieved by these countries, solving a large number of problems burdening rural areas (unemployment, depopulation, migration of the local population to urban city centres, macroeconomic indicators, etc.) influenced other countries to follow the positive example, striving to use their natural and social potentials for the development of rural tourism.

The analysis conducted in this paper indicates the potential of rural tourism as an important development factor, especially for developing countries such as Serbia. The features of rural areas in Serbia represent a significant predisposition for the development of tourism. Nevertheless, the main obstacle is insufficiently developed infrastructure, so these areas are not visible enough to potential tourists. In this paper, special attention is paid to insufficient application of ICT as a significant infrastructural deficiency.

In Serbia, rural tourism has been developing with varying intensity, variety and characteristics since the seventies of the twentieth century. The developing process depends on:

- geographical characteristic of a local area (region);
- presence of natural and social/anthropogenic resources (potential and actual tourist attractions);
- achieved level of development of local economy;
- awareness of local population about the role that rural tourism could have on the development of rural community;

The issue of competitiveness of rural destinations is a current issue nowadays. In order to achieve the competitiveness of tourist destinations, it is necessary to use appropriate management techniques. To achieve the adequate level of developing rural tourism, it is necessary to use different marketing and management activities. In this way tourist destinations meet tourists' demands and encourage people to stay in the rural area. The management of the tourist destination must constantly work on improving the quality and quantity of the so-called tourist offer. Different accommodation and non-pension facilities and different activities which could be practiced in rural areas could raise competitiveness and create an image of the destination. In order to ensure competitiveness, it is necessary to follow the current trends in the economy and the market, as well as theoretical studies, in order to be able to react in a timely manner, that is, to respond to current challenges with appropriate solutions.

A characteristic of rural tourist destinations is the physical, social and cultural distance from consumer needs urban cities centres. The problem of exchange is difficult and can be further complicated if there is a lack of adequate "business skills" in order to bring rural supply closer to tourist demand. Tourism distribution channels are seen as a "bridge" between supply and demand, and the role of intermediaries is precisely to find the ways to facilitate the exchange process. In order to improve competitiveness in business in rural tourism, it is necessary to apply modern ICT.

Information and communication technology enables residents of rural areas to:

- advertise rural tourist products;
- use e-commerce;
- transfer the knowledge from urban to rural areas and vice versa;
- apply the official state procedure;
- apply geographic information system in tourism management with natural and human resources, etc.

The official evidence about the number and type of accommodation facilities used in rural tourism in Serbia currently does not exist. The official evidence does not exist with the competent ministry (the Ministry of Trade, Tourism and Telecommunications), the Tourist Organization of Serbia (as National Tourism Organization), nor with the Chamber of Commerce of Serbia. The existing lists of accommodation units are incomplete and not updated regularly. Each of this institution possesses its own database of rural accommodation capacities. For this reason, management and monitoring of the development is confronting obstacles. However, there are associations that allow their members to present their rural tourism offer. National Association "The Rural Tourism of Serbia" represents one of the most famous, but the membership is not the condition which is regulated by law or by any other administrative regulation.

It is expected that in the future the request for the creation of a single and official database on all accommodation capacities of rural tourism offer will be approved. This way it would make it easier to:

- monitor the development of rural tourism;
- apply advanced marketing and management techniques in planning, organization, coordination, and development;
- apply modern ICT, which would bring the rural tourist offer closer to the urban city population.

In modern business conditions, the use of the Internet and other ICTs is replacing the role of tourist intermediaries. Such trends are particularly present in the so-called "mass tourism". The basic idea presented in the paper is that a suitable company that is a leader in global distribution systems (GDS) should be connected to such a database and provide the services for the promotion and sale of rural tourism products of different rural tourist destinations. If one of these databases would be connected to one of the GDS, the supply of rural tourism products would be made available not only to domestic but also foreign markets. The results would be:

- the higher number of tourist visits,
- the increase of tourist balance of trade,
- the higher number of overnight stays in tourist accommodation, etc.

When we talk about Serbia and the application of modern ICT in rural tourism business, all stakeholders at all levels of management need to go through formal and informal education systems (different programs of education and training.). These educational programs depending on the type and activity are engaged in in rural tourism. The competent ministries (Ministry of Tourism, Ministry of Agriculture, Forestry and Water Management), Tourist Organization of Serbia as a national organization which promotes rural tourism; local administrations; farmers as providers of tourist services should be involved in the organization of such activities.

Globally, as well as in the Republic of Serbia, there is no single database of rural tourist destinations and their tourist products. The associations that monitor the development of rural tourism at the regional or continental level (such as EUROGITES) usually refer potential tourists to the websites of national tourist organizations that monitor the development of rural tourism. The suggestion of this paper is to form a unique database and connect it to the global distribution systems (GDS).. This will lead to the global development of rural tourism and its popularization.

ACKNOWLEDGMENT

Paper is a part of research financed by the Ministry of Science, Technological development and Innovation of the Republic of Serbia, and agreed in decision no. 451-03-47/2023-01/200009 from 03.02.2023.

REFERENCES:

Akca, H., Sayili, M., & Esengun, K. (2007). Challenge of rural people to reduce digital divide in the globalized world: Theory and practice. *Government Information Quarterly*, *24*(2), 404–413. doi:10.1016/j.giq.2006.04.012

AlBar, A. M., & Hoque, M. R. (2017). Factors affecting the adoption of information and communication technology in small and medium enterprises: a perspective from rural Saudi Arabia online: 21 Oct 2017. *Information Technology for Development*, *25*(4), 715–738. doi:10.1080/02681102.2017.1390437

Alsos, G. A., Eide, D., & Madsen, E. L. (2014). Introduction: innovation in tourism industries. In Handbook of Research on Innovation in Tourism Industries (pp. 1–24). Edward Elgar Publishing. doi:10.4337/9781782548416.00006

Antlova, K. (2014). Main Factors for ICT Adoption in the Czech SMEs. Springer. doi:10.1007/978-3-642-38244-4_7

Antlová, K., Popelínský, L., & Tandler, J. (2011). Long Term Growth of SME from the View of ICT Competencies and Web Presentations. *E+M. Ekonomie a Management*, *14*, 125–139.

Awad Alhaddad, A. (2015). The Effect of Advertising Awareness on Brand Equity in Social Media. *International Journal of E-Education, e-Business, e-. Management Learning*, *5*(2), 73–84. doi:10.17706/ijeeee.2015.5.2.73-84

Benckendorff, P. J., Xiang, Z., & Sheldon, P. J. (2019). *Tourism information technology*. Cabi international Australia.

Bradić-Martinović, A. (2022). *Digitalne veštine građana Srbije*. Research Gate.

Buhalis, D., & Law, R. (2008). Progress in information technology and tourism management: 20 years on and 10 years after the Internet—The state of eTourism research. *Tourism Management, 29*(4), 609–623. doi:10.1016/j.tourman.2008.01.005

Carvalho, L., & Costa, T. (2011). Tourism innovation–a literature review complemented by case study research 23-33. *Tourism & Management Studies*, 23–33.

Chae, H.-C., Koh, C. E., & Park, K. O. (2018). Information technology capability and firm performance: Role of industry. *Information & Management, 55*(5), 525–546. doi:10.1016/j.im.2017.10.001

Clerides, S., Nearchou, P., & Pashardes, P. (2006). Intermediaries as Quality Assessors: Tour Operators in the Travel Industry. SSRN *Electronic Journal*. doi:10.2139/ssrn.505282

Decelle, X. (2004). A conceptual and dynamic approach to innovation in tourism. *Maître de Conférences*. https://www.oecd.org/cfe/tourism/34267921.pdf

Gajić, T., Đoković, F., Blešić, I., Petrović, M. D., Radovanović, M. M., Vukolić, D., Mandarić, M., Dašić, G., Syromiatnikova, J. A., & Mićović, A. (2023). Pandemic Boosts Prospects for Recovery of Rural Tourism in Serbia. *Land (Basel), 12*(3), 624. doi:10.3390/land12030624

Geoffrey, C. (2007). *Modelling destination competitivness – A szrvet abd abakzsis of the impact of competitivness atributes*. CRC for Sustainable Tourism Pty Ltd.

Ghavifekr, S., & Ibrahim, M. S. (2015). Effectiveness of ICT integration inMalaysian schools: A quantitative analysis. *International ResearchJournal for Quality in Education, 2*(8), 1–12. /ijres.net/index.php/ijres/article/view/79/43

Goffi, G., & Cucculelli, M. (2019). Explaining tourism competitiveness in small and medium destinations: The Italian case. *Current Issues in Tourism, 22*(17), 2109–2139. doi:10.1080/13683500.2017.1421620

Gössling, S. (2021). Tourism, technology and ICT: A critical review of affordances and concessions. *Journal of Sustainable Tourism, 29*(5), 733–750. doi:10.1080/09669582.2021.1873353

Halili, S. H., & Sulaiman, H. (2018). Factors influencing the rural students' acceptance of using ICT for educational purposes. *Kasetsart Journal of Social Sciences*. doi:10.1016/j.kjss.2017.12.022

Hosseini, S. J. F., Niknami, M., & Chizari, M. (2009). To determine thechallenges in the application of ICTs by the agricultural extensionservice in Iran. Journal of Agricultural Extension and Rural Development,1(1), 292e299. *Journal of Agricultural Extension and Rural Development, 1*(1), 292–299.

Irefin, I. A., Abdul-Azeez, I. A., & Tijani, A. A. (2012). An Investigative Study of the factors Affecting the Adoption of Information and Communication Technology in Small and Medium Enterprises In Nigeria. *Australian Journal of Business and Management Research, 02*(02), 01–09. doi:10.52283/NSWRCA.AJBMR.20120202A01

Kallmuenzer, A., & Peters, M. (2018). Entrepreneurial behaviour, firm size and financial performance: The case of rural tourism family firms. *Tourism Recreation Research, 43*(1), 2–14. doi:10.1080/02508281.2017.1357782

Krasavac-Chroneos, B., Radosavljević, K., & Bradić-Martinović, A. (2018). SWOT analysis of the rural tourism as a channel of marketing for agricultural products in Serbia. *Ekonomika Poljoprivrede*, *65*(4), 1573–1584. doi:10.5937/ekoPolj1804573K

Milojević, L. (2004). The Social and Cultural Aspects of Rural Tourism. In Rural Tourism in Europe: Experiences, Development and Perspectives (pp. 115–121). UN WTO.

Ministry of Agriculture and Water management. (2009). *Analysis of the budgetary support to the development of rural tourism in Serbia and diversification of economic activities in the countryside*. MAWM. http://www.minpolj.gov.rs/?script=lat

Mosurović, M., & Kutlača, D. (2011). Organizational design as a driver for firm innovativeness in Serbia. *Innovation (Abingdon)*, *24*(4), 427–447. doi:10.1080/13511610.2011.633432

Nguyen, T. H., Newby, M., & Macaulay, M. J. (2015). Information Technology Adoption in Small Business: Confirmation of a Proposed Framework. *Journal of Small Business Management*, *53*(1), 207–227. doi:10.1111/jsbm.12058

Niavis, S., & Tsiotas, D. (2019). Assessing the tourism performance of the Mediterranean coastal destinations: A combined efficiency and effectiveness approach. Journal of Destination Marketing & Management, 14, 100379. doi:10.1016/j.jdmm.2019.100379

Ožegović, J. (2019). *Report on digital inclusion in the Republic of Serbia for the period from 2014 to 2018*. Government of the Republic of Serbia.

Paunović, M., Lazarević-Moravčević, M., & Mosurović Ružičić, M. (2022). Business Process Innovation of Serbian Entrepreneurial Firms. *Economic Analysis*. doi:10.28934/ea.22.55.2.pp66-78

Pearce, D. G., & Schott, C. (2005). Tourism Distribution Channels: The Visitors' Perspective. *Journal of Travel Research*, *44*(1), 50–63. doi:10.1177/0047287505276591

Pedersen, J. S., & Wilkinson, A. (2018). The digital society and provision of welfare services. *The International Journal of Sociology and Social Policy*, *38*(3/4), 194–209. doi:10.1108/IJSSP-05-2017-0062

Premkumar, G., & Roberts, M. (1999). Adoption of new information technologies in rural small businesses. *Omega*, *27*(4), 467–484. doi:10.1016/S0305-0483(98)00071-1

Ristić, L., Despotović, D., & Dimitrijević, M. (2020). Multifunctionality of Agriculture as a Significant Factor for Sustainable Rural Development of the Republic of Serbia. *Economic Themes*, *58*(1), 17–32. doi:10.2478/ethemes-2020-0002

Roblek, V., Petrović, N. N., Gagnidze, I., & Khokhobaia, M. (2021). Role of a Digital Transformation in Development of a Rural Tourism Destinations. *VI International Scientific Conference Challenges of Globalization in Economics and Business,* (pp. 297–305). Research Gate.

Rowley, T. D., & Shirley, L. P. (2019). Removing rural development barriers through telecommunications: Illusion or reality. In *In Economic Adaptation* (pp. 247–264). Routledge. doi:10.4324/9780429041082-14

Ružičić, M. M., Miletić, M., & Dobrota, M. (2021). Does a national innovation system encourage sustainability? Lessons from the construction industry in Serbia. *Sustainability (Basel)*, *13*(7), 3591. doi:10.3390u13073591

Salemink, K., Strijker, D., & Bosworth, G. (2017). Rural development in the digital age: A systematic literature review on unequal ICT availability, adoption, and use in rural areas. *Journal of Rural Studies, 54*, 360–371. doi:10.1016/j.jrurstud.2015.09.001

Salinas Fernández, J. A., Serdeira Azevedo, P., Martín Martín, J. M., & Rodríguez Martín, J. A. (2020). Determinants of tourism destination competitiveness in the countries most visited by international tourists: Proposal of a synthetic index. *Tourism Management Perspectives, 33*, 100582. doi:10.1016/j.tmp.2019.100582

Stankić, R. (2014). Key Success Factors for Implementation of Business Information System University in East Sarajevo, Faculty of Economic. *Bčko*, (November), 18–25.

Tang, G. N., Ren, F., & Zhou, J. (2022). Does the digital economy promote "innovation and entrepreneurship" in rural tourism in China? *Frontiers in Psychology, 13*, 979027. doi:10.3389/fpsyg.2022.979027 PMID:36312131

Todorivić, M., & Bjelac, Ž. (2009). Rural tourism in Serbia as a Concept of Development in Undeveloped Regions, journal. *Acta Geographica Slovenica, 49*(2), 453–473. doi:10.3986/AGS49208

Tüzünkan, D. (2017). The Relationship between Innovation and Tourism: The Case of Smart Tourism. *International Journal of Applied Engineering Research: IJAER, 12*(23), 13861–13867. http://www.ripublication.com

UNFAO. (2021). *Cultivating Our Futures - Issues Paper: The Multifunctional Character of Agriculture and Land, webl.* UNFAO. https://www.fao.org/3/x2777e/x2777e00.htm

UNWTO. (2007). *A Practical Guide to Tourism Destination Management.* UNWTO.

Vidas-Bubanja, M., & Bubanja, I. (2017). The importance of ICT for the competitiveness of tourism companies. *Tourism in Function of Development of the Republic of Serbia*, (pp. 470–489). TISC.

Vučetić, Š. (2017). The importance of using ICT in the rural tourism of the Zadar county. *DIEM, 3*(1), 176–187. hrcak.srce.hr/187378

Vuković, P. (2017). Character and dynamics of development rural tourism in the Republic of Serbia, *Ekonomika, 63*. http://www.ekonomika.org.rs/sr/PDF/ekonomika/2017/Ekonomika-2017-4.pdf

Vuković, P., Simonović, Z., & Kljajić, N. (2016). Complementarity of Multifunctional Agriculture and Rural Development with Rural Tourism and Possibilities for their Implementation in the Republic of Serbia, *International Journal of Scientific and Technology Research*, 195-212.

Vuković, P., & Subić, J. (2016). Sustainable Tourism Development of Rural Areas in Serbia as a Precondition to Competitiveness. In *Global Perspectives on Trade Integration and Economies in Transition* (pp. 342–361). IGI Global. doi:10.4018/978-1-5225-0451-1.ch017

Zdorov, A. B. (2009). Comprehensive development of tourism in the countryside. *Studies on Russian Economic Development, 20*(4), 453–455. doi:10.1134/S107570070904011X

Section 2

Innovation, Experiences, and Service Quality in the Hotel and Food Service Industry

Chapter 6
Technology Paradox in the Tourism Industry:
Technostress Perspective

Nilgün Demirel III
Tourism Faculty, Igdir University, Turkey

Onur Çelen
Harmancık Vocational School, Bursa Uludag University, Turkey

ABSTRACT

The aim of this chapter is to reveal technology paradox, and technostress concepts' effects in tourism industry. In the research the impact of the technology on the tourism industry employees was determined through qualitative research steps. Descriptive results were obtained using phenomenological research design. 20 participants were interviewed face-to-face and videoconferenced, and the audio and video files obtained were deciphered. Three main themes and seven sub-themes were identified, and the data were classified by content analysis. Analyses performed with percentage values, total and frequency graphs, and descriptive results. It is emphasized that the negative effects of technology on employees are greater. It is a fact that technology creates stress. No matter how competent one is in the use of technology, the unhindered development of technology has a negative impact on employees in the tourism industry, as it is thought to lead to a decrease in the labor force and unemployment.

INTRODUCTION

Our daily activities, experiences, and occupations are all surrounded by modern technology. Computer technology is extensively integrated into job processes, particularly in the office, and it has broken down the barriers between work and life. The demand for computer-based services is increasing quickly, placing a greater burden on employees to keep up with the rapidly changing technology (Shu, Tu, & Wang, 2011).

DOI: 10.4018/978-1-6684-6985-9.ch006

The human resource components of hotel innovation have received a lot of attention in studies about technological development. A new operating philosophy for managing hotels known as smart hotel design has resulted from the use of big data and developing artificial intelligence (AI) technologies. AI, cloud computing, and the internet of things support smart hotels' ability to compete and appeal to travelers by offering a sophisticated digital environment. Hotels can choose from a variety of options thanks to smart technology, which helps the establishment run more smoothly while also wowing visitors (Wu & Cheng, 2018). But, after COVID-19 pandemic process, digital wellbeing concept has appeared, and its framework is extending from articulations that assert that it could be a state that's accomplished through the mindful utilize of innovation to those that state that it is security of the wellbeing of individuals who work with innovation (Debasa, 2022). The research findings show that technological stress has a negative impact on both employee performance and wellbeing (including engagement and general wellbeing). The relationship between technological stress and performance is mediated by employee wellbeing. Employee performance and wellbeing are moderated and offset by organizational learning. Particularly, organizational learning dramatically worsens the connection between technological stress and engagement while significantly reducing the connection between technological stress and performance. A higher guest experience is promised through increased efficiency and convenience. A smart hotel attempts to satisfy visitors' needs by offering high-tech amenities including scene control, rapid feedback, efficient check-in processing, and guest identification as a result, a rising number of hotels are modernizing. Technology installed in hotel rooms can make stays more enjoyable for visitors. Smart hotels can better accommodate especially young visitors who grew up with the rapid development of technology with a comfortable and entertaining atmosphere (Choi, Mehraliyev & Kim, 2020).

In the digital world it is accepted that more technology means more satisfaction. On the other hand employees' commitment to their organizations and productivity would both be impacted by technostress (Tarafdar, Pullins, & Ragu- Nathan, 2015). But what is overlooked is how tourism employees feel when providing services? In fact, quality service is related with internal customer satisfaction. How reasonable is it to ignore employees when providing and implementing technological advances? According to literature technostress has impact on employees wellbeing. However, in tourism industry maintain the quality of businesses thanks to labor force. So, how is digitalization and technological development planned in torism industry? Should labor force ignore? How can digitalization be managed in torism industry? What do employees think about technological development in their departments? The research questions is designed to find an answer technology paradox.

There are some limitations in the research. The most important limitation is assumed to be the social world of the research and the fact that the population of the research includes only employees. It is thought that a qualitative study on tourists may provide different and striking results. In addition, the data collection process was carried out in a short period of time and some of the interviews were conducted via videoconferencing. It is recommended to prefer face-to-face interviews especially in qualitative research.

Technology, Technology Paradox, and Tourism Industry

The post-COVID-19 transformation necessitated technology and steps to improve the fragile structure of tourism started to change thanks to technology. This process brought tourism and leisure services and global travel to a halt and led to a complete decline in tourism revenues. The tourism industry, which is vulnerable to political, environmental and socio-economic risks, is trying to compensate for the negative processes through tourism flows. All sectors of the tourism industry have taken a series of measures to recover, whether economic, environmental, political or strategic, and have seen changes in consumption patterns. One of these is the introduction of online experiences such as virtual entertainment, reversal of the food and beverage service cycle, travel in virtual environments, or the introduction of robots in the service delivery process to minimize human interaction. Technology is at the heart of solutions for tourism's post-COVID-19 recovery and reopening of the economy. There are applications that are integrated into tourism, such as robotized artificial intelligence with contactless service delivery, big data for fast and real-time decision-making, social distancing and crowd control technologies, humanoid robots that measure body temperature. The concept of e-tourism is at the forefront of the applications developing only in tourism. It is the best example of the use of technology to revitalize tourism. In addition, applications such as artificial intelligence and robotic developments, smart destinations and smart tourism applications have been developed and accelerated during the fight against COVID-19. This is where the paradox of technology in tourism begins. On the one hand, there is the improvement of the economy and the tourism industry, on the other hand, the purposes of using technology and questioning the psychological structure of human beings while using it should be examined (Sigala, 2020).

The term -job resources- refers to the social, physical, and organizational components of a job that are useful in achieving work objectives, lowering workload demands, and/or promoting individual learning, growth, and development (Bakker & Demerouti, 2007). Digitalization is the key factor for development of this resources and work objectives. During the COVID-19 pandemic, the switch to digital platforms has caused "technostress" for a number of employees (Tuan, 2022a). This concept is newly discussed in tourism industry and it is related to digital well being. In tourism industry users who use information systems in an organizational setting may experience technostress (Tarafdar, Pullins, & Ragu- Nathan, 2015).

If someone talks about the good aspects of the use of technology in tourism; it has highly changed the travel and tourism sector, and has a significant impact on the competitiveness of travel agencies and destinations (Buhalis, 2020). So, tourism industry has a chance thanks to technology to raise the standards of the services it provides. It is an important competitive differentiator in the hotel industry is the way management is evolving to focus more on the needs of guests and provide superior service. The goal is to make sure customers are satisfied enough to choose to return to the same business (or organization), as is the case with all other service industries. With the help of technology, it is possible to track the interactions between a hotel and a customer, to keep track of a specific guest's preferences, and to cater to those preferences when a customer makes a subsequent reservation. So, small practices of service can be crucial in the service sector (Baines, 1998). Particularly in light of recent developments in the information and communication technology and the use of artificial intelligence techniques in a range of industries, including the hotel business, smart technology is becoming more and more significant in the tourism industry (Yasin, Abdelmaboud, Saad, & Qoura, 2021).

Robotics and artificial intelligence (AI) technological advancements are predicted to have a significant impact on many facets of life. All strategic and operational management functions are dominated by information technologies (ITs). ITs present potential and difficulties for the tourism industry because information is the industry's vitality. There is a risk that different innovation process will enter the market and threaten the position of the current process unless the current tourism industry increases its competitiveness by utilizing the developing ITs and innovative management methods. Only creative and inventive suppliers will be able to stay in business in the new era (Buhalis, 1998). The value of creativity in the growth of a healthy person must be recognized by society. Many of the ideas related to the growth of fully developed citizens are closely related to creativity. The great advancements being made through the use of technology in areas like information access, openness and exposure to new ideas, and interaction with significant others must not be minimized by a deficiency in the development of the human-computer interface. Only when a system is in place to support creativity's development can it be encouraged. It is essential to design all programs with the elements in mind that have been proven to foster creativity. For instance, human direction should be used for computer interaction. There should be enough time to consider a variety of options. To encourage the creation of novel associations, solutions should have some restrictions (Edwards, 2001). In the tourism industry, practices and procedures are being transformed into smart technologies and environments that have a significant impact on service process innovation, strategic management and marketing processes, as well as on competitive processes. In the future, smart processes will be used more intensively for competitive advantage and more emphasis will be placed on business dynamics, agility, artificial intelligence and neuromarketing, body language perception and human computer interaction (Buhalis, 2020).

As to negative and paradoxical aspect of technology; technology-related disturbances occurring in consumers' feelings and thoughts are evaluated within the framework of the technology paradox. It is important to consider the attitudes and reactions of consumers towards technology. Because these developments, which are imposed on consumers as innovations, may cause them to make mistakes in their purchasing processes and to start recovery (compensation) processes and not to prefer the business again due to these disruptions. For example, self-service cash registers are easy to control and can create chaos at the same time. It can be inefficient as well as efficient. Especially in shopping processes, technology can speed up the process, but the fact that technology does not accept human assistance when errors occur leads to the complete cancellation of the process. For consumers who do not master technological processes or can not use any technological tools, it can make them feel a sense of incompetence. It can be said that young and technologically inclined individuals can accept innovation more easily in this case, but individuals who are not interested in technology or older individuals cannot purchase products because they cannot follow innovations (Bulmer, Elms & Moore, 2018). On the other hand, consumers who benefit from a good or service exhibit contradictory behaviors when evaluating the benefits they derive from their experiences. While sometimes the experiences obtained from a product are positive, sometimes they may be negative. Similarly, although consumers perceive the benefits of technology, technological developments may harm human life and people may be disturbed by excessive technology (Johnson, Bardhi, & Dunn, 2008).

Digital Wellbeing Versus Technostress

Information may now be efficiently accessed and transmitted across spatial and temporal boundaries thanks to technology breakthroughs (Rennecker & Godwin, 2005). Growing public worries about technostress and the negative consequences of excessive use of information and communication technologies (ITC) on people's mental and physical health as well as other elements of life, such as the environment and society, gave rise to the notion of digital wellbeing. A situation of personal wellbeing that is experienced via the healthy and responsible use of digital technology is what is commonly referred to as "digital wellbeing" (Marsden, 2020). Digital wellbeing is the subjective wellbeing of people in a world where digital media are pervasive. A broad framework is created to incorporate empirical studies into a body of knowledge about how using digital media affects wellbeing. People's digital behaviours are shaped in the socio-technical environment. At the same time, in this relationship network, which emerges as benefit or wellbeing, the individual's behaviour and interaction with technology involves a cause and effect relationship that is linked to the welfare outcome (Büchi, 2021).

The word "digital wellbeing" as a whole has significant value, both for the people who should be able to experience it and for the organizations involved in its provision. Hence, one of the definitions of digital wellbeing that is frequently quoted states that it is a framework that takes care of one's own health, safety, relationships, and work-life balance in digital contexts (Beethem, 2016). The conditions that lead to digital wellbeing are adopting safe and responsible digital use. Digital wellbeing can be achieved if digital workload, distraction and digital stress are well managed. Correct use of digital media is associated with political and societal actions. The use of digital data is mandatory for digital wellbeing. If all these are provided, it is unlikely that digital wellbeing will be disrupted (Beethem, 2016). Technology plays a significant role in consumers' lives, and digital wellbeing intersects with aspects of general wellbeing such as physical, emotional, intellectual, spiritual, social, occupational, environmental, and financial wellbeing. The continuum of digital wellbeing could guide a new research agenda on the opportunities and difficulties in digital wellbeing in the travel and tourism industry. Certain nations have recently implemented a number of digital wellbeing programs, and unofficial groups formed to discuss these topics have grown in popularity and public awareness (Stankov & Gretzel, 2021).

Digital wellbeing in tourism has become an element that facilitates business processes and increases profitability for businesses, consumers, government agencies and technology providers facing technology-driven challenges. In the tourism industry, business processes for digital prosperity and the scope of new roles and responsibilities with technology are not defined. Undefined business processes and uncertainties create stress on employees (Stankov & Gretzel, 2021). Although it is known that technology affects employment, labor force and employee psychology, this occurs in two ways. By removing or relocating employees from the tasks they previously performed or by increasing the number of labor force to increase productivity, it reveals the positive or negative impact of the labor force (Yasin, Abdelmaboud, Saad & Qoura, 2021). In other words, it can be said that technology applied in line with digital wellbeing and its healthy, correct and responsible use increases personal wellbeing (Marsden, 2020).

Digital wellbeing refers to how the usage of digital media is related to wellbeing (Büchi, 2021). The pervasive use of computers in our lives is poised to fundamentally enhance and hamper the growth of creativity in society and the individual, so society will definitely suffer (Edwards, 2001). The effects of technological improvements on the workplace have frequently been contradictory, increasing disruptions and unpredictability while promoting accessibility and efficiency (Hoeven, Van Zoonen, & Fonner, 2016). Users of new technologies and experiences meet both benefits and drawbacks, leading to contention and contrasting points of view. People are more likely to adopt innovations when they are feeling positively than when they are feeling negatively. Technostress, or the stress a person feels as a result of not being able to handle the demands of employing information technology, has apparently been experienced by employees as a result of digitalization during the COVID-19 (Ayyagari, Grover & Purvis, 2011).

According to the Transactional Theory of Stress (TTS), the effects of stressors are determined by how people cope and evaluate their situation. The use of information and communication technologies brings about processes such as emotion-focused coping, challenging technology or not accepting technology (Tarafdar, Pullins & Ragu- Nathan, 2015; Zhao, Xia & Huang Wayne, 2020). *"The concept of technostress summarizes this situation and is expressed as a cognitive response that includes attitudes and feelings of sadness and demoralization to technology exposure".* Also, technology is barrier to productivity (Tarafdar, Pullins & Ragu- Nathan, 2015).

Technostress is relevant to people's intention to take digital-free tourism. In other words, technostress is a type of stress that users experience while using technologies (Liu & Hu, 2021). Also it refers to the *"negative impact on thoughts, attitudes, behaviors, or body physiology that is caused either directly or indirectly by technology"* (Shu, Tu, & Wang, 2011: 923). High levels of stress reactions induced by the current health crisis lead to negative active responses to online technology use (Panisoara, Lazar, Panisoara, Chirca, & Ursu, 2020). Three key ideas make up technostress: technological stressors, psychological effects, and behavioral responses. Overexposure to technology can lead to a condition known as techno-exhaustion, which is a psychological response to technological pressures (Ayyagari, Grover, & Purvis, 2011). Besides, technostress can act as a difficult stressor that encourages those who are motivated by promotions to advance (Tuan, 2022b).

This research examines the technology paradox from a technostress perspective in the tourism industry under three headings and their subheadings. The first is technological competence. Technological competence refers to employees' level of knowledge about technology and their use of technology. Competence; the of feeling competent in the use of technological tools and equipment and software programs. Incompetence; the situation of feeling incompetence in the use of software programs of technological tools and equipment. The second is technology and tourism industry. In the relationship between technology and tourism, it refers to the stress caused by technology in the emergence of disruptions in the use of technology and the positive aspects of technology use. In this theme, which is divided into three sub-headings, positive and negative situations are used together. Failures about technological equipment; in the service delivery process, the problems related to technological equipment or software in tourism businesses. Technology creates stress; the stress caused by technological developments and the use of technology in tourism businesses on employees. Technology is the best way; the positive impact of technology on service processes and advocating that technology should be a must in tourism enterprises. The third; the impact of technology on wellbeing. This dimension reveals whether technology affects people positively or negatively. Positive impact; the positive effects of technology on human and digital wellbeing. Negative impact; the negative effects of technology on human and digital wellbeing.

Research Methodology

The method of the research was designed within the framework of qualitative research methods. Qualitative research includes research methods adopted by many scientists in the field of social sciences. The design of this research method, which is adopted for the purpose of the study, aims to reveal all aspects of the subject under consideration in depth (Bayyurt & Seggie, 2017). Qualitative research has a large number of approach (pattern) options according to its purpose. A pattern is a logical string that connects the initial research questions of a study with the scope of the research and the results associated with the questions (Yin, 2009). In this research, the phenomenological research pattern from the five qualitative research approach proposals of will be carry out. The phenomenological qualitative research pattern describes the lived experiences of people about a phenomenon or concept by examining them (Creswell, 2007). In qualitative research, the concept of "social world" is used instead of the "universe/population" used in quantitative research. Social worlds refer to individuals who have experienced similar experiences and the groups formed by these individuals. The social world of the research consists of different sectors of tourism and employees working in Turkey and selected from the department that is intertwined with technology. In all types of qualitative research, sometimes all of the data are collected through interviews. An interview is defined as "*a process in which a researcher and participant participate in a conversation focused on questions related to the topic in a study*" or "conversation for a purpose" (Merriam, 2009). In this context, in order to obtain information about "how tourism employees interpret their use of technology and how they feel when they are exposed to technology". This research was supported by a semi-structured interview as a data collection technique. In qualitative research, open-ended questions provide the researcher with the opportunity to flexibly address the phenomenon who wants to study. In order to obtain qualitative data the following open-ended questions were determined according to literature review with the researches of Hoeven, Van Zoonen and Fonner, 2006; Stankov & Gretzel, 2020; Debasa, 2022; Tuan, 2022a; Tuan, 2022b:

- What is your competence about technological equipment at work?
- How do you feel when you are exposed to technological equipment?
- Can you provide information about your social media usage and inclusion in whatsapp groups?
- How do you feel when you receive news and information about new technological applications for the industry in whatsapp groups in your destination?
- What are the problems related to the technical equipment in your department?
- What do you think about robotization and becoming a technology business?
- What are the negative effects of the development of technology on your business?

During the interviews, which will be conducted in the form of video conferences and face-to-face interviews, a voice recording will be taken and a consent form will be signed, provided that the participants have permission. A total of 20 participants will be interviewed and the data will be deciphered and read in detail. As a result of the readings, the main theme and sub-themes will be created in the light of both literature and data. Main themes sub-themes and qualitative data will be defined in the MAXQDA software program and analyzes will be performed. In content analysis, researchers examine written documents based on social communication and recorded verbal communication materials. Content analysis will be applied in accordance with the research procedure. Content analysis is a technique used to make valid inferences through texts (Krippendorff, 2004).

Results From Qualitative Research on Tourism Industry Employees

Before the content analysis is carried out, the main themes' supporting themes are outlined from data. For qualitative research, delineating the sub-themes is crucial since it improves validity and reliability. The explanations, which serve as the coders' road map, are established by the coders' collective agreement by reading the data one at a time. They are made to have the same meaning when the data is read. Given that perception and understanding are not always compatible, it is conceivable for meanings to be similar.

After being outline the themes from data, firstly demographic characteristics of participants were summarized. Then, frequencies of sub-themes and sum of main themes were resulted. The percentages of sub-themes were showed with graphics and participants comments were listed.

Among the participants of the research, 11 participants are female and 9 participants are male. The average age of the participants was 32.50 years old. The participants in the research are mostly from the F&B department. Since the interview process was difficult and the participants did not accept to be interviewed, equality could not be achieved on the basis of departments. For this reason, interviews were carried out mostly in hotel businesses. 7 participants from Istanbul, 4 participants from Ankara, 4 participants from Antalya, 3 participants from Bursa, and 1 participant each from Muğla and Kocaeli contributed to obtaining the data of the research.

Table 1. Explanation of themes

Technological Competence	
Competence	It is expressed as the state of feeling competent in the use of technological tools and equipment and software programs.
Incompetence	It is expressed as the situation of feeling incompetence in the use of software programs of technological tools and equipment.
Technology and Tourism Industry	
Failures About Technological Equipment	In the service delivery process, it refers to the problems related to technological equipment or software in tourism businesses.
Technology creates stress	It refers to the stress caused by technological developments and the use of technology in tourism businesses on employees.
Technology is the best way	It refers to the positive impact of technology on service processes and advocating that t echnology should be a must in tourism enterprises.
The Impact of Technology on Wellbeing	
Positive Impact	It refers to the positive effects of technology on human and digital wellbeing.
Negative Impact	It refers to the negative effects of technology on human and digital wellbeing.

Note: Constituted by authors.

As a result of the content analysis conducted through main and sub-themes in line with the data obtained, it was determined that the most frequently mentioned main theme was "Technology and Tourism Industry" (85). It was determined that the sub-theme "Technology creates stress=51" under this main theme was the most mentioned sub-theme with a rate of 60.00%. According to this result, it is possible to say that technology creates stress on employees in the tourism industry in Turkey. On the other hand, there are also those who say that technology is the best way (27). It is stated that technology has good aspects, and makes important contributions to the tourism industry. At the same time, it was revealed that there are not many problems related to technological equipment, and when they do occur, they disrupt the service process. The second most mentioned main theme was "The Impact of Technology on Wellbeing= 61" and the most mentioned sub-theme within this main theme was "Negative Impact= 45". In this case, it is possible to say that technology has more negative effects on wellbeing. The main theme with the least number of comments is "Technological Competence=20". Participants generally stated that they were at a good level in terms of their ability to learn, apply and use technological equipment at a good level.

In the research, visualizations of the percentages of qualitative data were made with MAXQDA software. Although the colors do not have any importance on the visuals, there are summaries of the frequencies and percentages specified in Table-3. In addition, the participants' comments on each sub-theme are quoted verbatim without any changes.

Table 2. Demographics about participants

Participant	Department	Professional Experience	Sector	Age	Gender	City
P1	Front Office	13 Years	Hotel	34 Years	Female	Istanbul
P2	F&B	5 Years	Hotel	27 Years	Female	Ankara
P3	Cabin Crew	10 Years	Airline	34 Years	Female	Istanbul
P4	Tour Guide	15 Years	Tour Guiding	39 Years	Female	Antalya
P5	Front Office	10 Years	Hotel	29 Years	Male	Ankara
P6	F&B	7 Years	Hotel	31 Years	Male	Istanbul
P7	Guest Relations	20 Years	Hotel	38 Years	Female	Ankara
P8	Travel Agency	11 Years	Travel Agency	34 Years	Female	Istanbul
P9	Accounting	4 Years	Hotel	24 Years	Male	Muğla
P10	Front Office	2 Years	Hotel	22 Years	Female	Kocaeli
P11	General Manager	4 Years	Hotel	26 Years	Female	Ankara
P12	Cabin Crew	8 Years	Airline	31 Years	Female	Istanbul
P13	F&B	24 Years	Hotel	45 Years	Male	Antalya
P14	F&B	25 Years	Hotel	45 Years	Male	Bursa
P15	Front Office	7 Years	Hotel	27 Years	Female	Istanbul
P16	F&B	17 Years	Hotel	39 Years	Male	Antalya
P17	Front Office	30 Years	Hotel	50 Years	Male	Istanbul
P18	F&B	2 Years	Hotel	21 Years	Male	Antalya
P19	F&B	2 Years	Hotel	22 Years	Female	Bursa
P20	Accounting	14 Years	Travel Agency	37 Years	Male	Bursa

Note: Constituted by authors.

Table 3. Sum of main themes and frequencies- percentages of sub- themes

Main Themes	Sub-Themes	20	100%
Technological Competence	Competence	18	90%
	Incompetence	2	10%
		85	**100%**
Technology and Tourism Industry	Failures about technological equipment	7	8,24%
	Technology creates stress	51	60,00%
	Technology is the best way	27	31,76%
		61	**100%**
The Impact of Technology on Wellbeing	Positive impact	16	26,23%
	Negative impact	45	73,77%
Sum		**Σ 166**	

Note: Constituted by authors via MAXQDA.

Figure 1. Percentages of technological competence sub-themes
Note: *Constituted by authors via MAXQDA.*

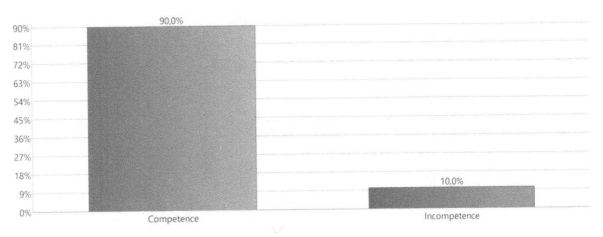

Competence: Employees consider themselves competent in technology. If they do not consider themselves competent, they act with the obligation to learn technological tools and equipment and make efforts to learn. The desire to follow innovation and constantly update oneself due to commitment to the job is among the motivations of the employees. Desire to succeed and being committed to business processes pave the way for employees to acquire all skills. If necessary, they are willing to volunteer to receive training for the processes of keeping up with innovation and the use of technology.

P20 *"We have to do it, we need to know the central reservation system well."*

P7 *"Our job is to collect the demands of people with the phone and computer and to realize their wishes, I have to use all kinds of equipment for this."*

P19 *"Developments in working life have led to a need for more competent employees. Therefore, the development of competent individuals in working life should be open to innovations by giving importance to formal and informal education, turning their work into a lifestyle and constantly improving themselves. By doing this, you can achieve success. I think I have all of these traits."*

Incompetence: The sub-theme "incompetence", which is the sub-theme with the least number of comments by the participants, is a reflection of the predisposition to technological developments. It is an indication that good training is provided in the tourism industry in Turkey and orientation process is applied in the businesses. Therefore, the participants are able to keep up with technological developments and do not consider themselves incompetence.

P9 *"At first I had a hard time using the computer, but now I'm competent."*

P16 *"I can use simple tools. Frankly, I am afraid of devices that are too complicated for me"*

Figure 2. Percentages of technology and tourism industry sub-themes
Note: Constituted by authors via MAXQDA.

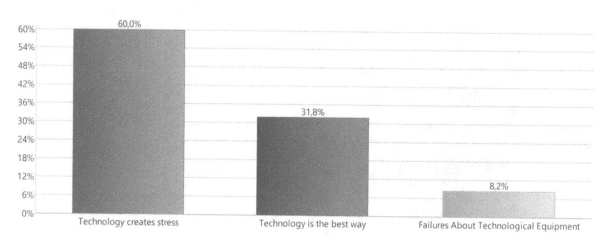

Technology creates stress: The tourism industry, which is integrated with technology, has brought many innovations. Employees are exposed to many messages, information and technological equipment with these innovations. This requires them to constantly update themselves, be alert and have problem solving skills. It is their most important task to continue the service delivery process at the point where technology fails. This is an important responsibility in ensuring service quality. They are also exposed to a lot of information or meetings outside of working hours. This situation causes the work process and family relationship to be intertwined. As a result of the interviews conducted with the participants, being exposed to messages, e-mails or meetings time off creates stress on the participants and they have difficulty focusing on work the next day. It was revealed that especially older workers think that they will lose their jobs with the development of technology. They even stated that in the future, with the further development of technology, the labor force will be completely eliminated. When they put themselves in the shoes of customers, they stated that they would not like to see robots in front of them.

P13 *"As people get older, sometimes they get more and more nervous about it. Previously, we used to write customers' orders on bills. I was stressed when I switched to electronic orders and cash registers. At first, it's about making a deficit and making mistakes."*

P12 *"I don't think positively at all, I want to see people in front of me.. I can't go to a robot and order and tell my problem..."*

P8 *"Even though you are a corporate company, a lot of information comes in, sometimes even an incoming e-mail, even when you say what is this, delete it, but the administrator sends it to the whatsapp group, asking if anyone knows about it, so I get stressed. Adapting is sometimes difficult..."*

Technology is the best way: It has been revealed that the employees who think about customer orientation and service quality are the ones who argue that technology is the right way. They even stated that those who are interested in technology are affected by technology and that technological developments excite them. According to them, technology is useful and actually the only way for the tourism industry. There are comments that those who figüre out technology quickly can stay in the industry for a long time. Those who think that technology is the only way, state that people make the service delivery process difficult and that mistakes are caused by employees or managers. Employee dissatisfaction leads to customer dissatisfaction. Therefore, the more technology is used, the more quality services will be provided. In addition, service quality will be standardized.

P9 *"I think about technology in general in order to provide the best service to our holiday guests. But because we generally support quality service and smiling faces, it is important for us to relax people and keep them away from troubles. Therefore, I think positively about technological innovations that are beneficial. The less people the more peace- supporting person I am."*

P6 *"Technological developments excite me and I try to learn as much as I can by being a part of it. I am aware that learning these developments early and improving myself will put me in front of many people."*

P10 *"I'm very excited because you learn that you can do something that you have a hard time with very easily... this is a good thing..."*

    Failures About Technological Equipment: Although there are not many failures related to technological equipment, there are situations that will make service delivery processes difficult. This leads to financial losses of businesses and customers are aggrieved. The role of human beings in the tourism industry is important when it is considered that service recovery processes will be carried out through labor force in order to lose and gain the aggrieved customer.

P15 *"There is usually a problem like the system is locked and customers react because they don't like to wait, which negatively affects our motivation."*

Figure 3. Percentages of the impact of technology on wellbeing sub-themes
Note: Constituted by authors via MAXQDA.

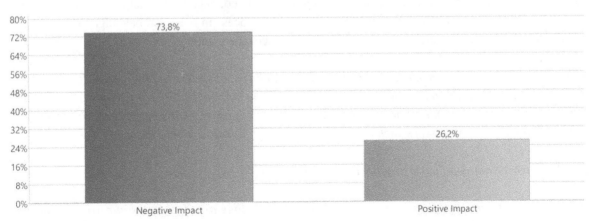

P16 *"Everyone avoids taking responsibility even when miscalculation is a problem."*

P11 *"Unfortunately, we are having problems with our computers, mainboard default and printers."*

Negative Impact: The tourism industry in Turkey is feeling the negative effects of technology. This is particularly due to concerns about the loss of labor force. Turkey is an industry that generates income from tourism and has a high economic cycle due to the high number of people working in tourism. For this reason, employees' concerns about not being able to earn income from tourism are among the negative effects of technology. The negative effects of technology are especially noticeable in service offerings that require one-to-one expression and involve emotional processes. Failure to listen to the guide during the tour process, disconnection in the reservation process, which leads to an increase in e-complaints. This failures, which results in financial losses of businesses, reveals the necessity of continuous monitoring of technology. The updates and monitoring technology can be made by humans and the maintenance of technological tools, are proof that humans will never be able to break away from the tourism industry.

P13 *"With the increasing age, people sometimes become more regressed in this regard. Previously, we used to write customers' orders on bills. I was stressed when I switched to electronic orders and cash registers. At first it's also about making a deficit and making mistakes in the safe."*

P4 *"While you are telling something, the people on the tour do not listen to you, they take photos or videos with their phones and share them on social media, they do not pay attention to you, they say that we did not understand something because they did not fully listen...So you are not important for the tour..."*

P15 *"Online reservation and feedback systems affect us negatively, even if you provide full and complete service, the consumer immediately evaluates you negatively, which causes us to be held accountable for the behavior we did not do to the management."*

Positive Impact: Although the negative effects of technology are mostly mentioned in the research, the positive effects of technology are not ignored. The aspect of technology that supports the tourism industry and facilitates business processes has not gone unnoticed. Technology helps to minimize errors, especially in the front office department. It speeds up the process in food and beverage preparation. It is thought to have more positive effects as long as it reduces labor force.

P19 *"With the developing technology, the use of the common robot system has increased. It necessitates a new adaptation in the tourism sector. These robots provide help and support to people in businesses. In this way, tourism has gained a new dimension. In addition, it has a positive effect on tourism in terms of facilitating the work on issues such as food and beverage businesses in tourism."*

P6 *"I feel more confident, more practical and comfortable."*

P2 *As long as it does not affect us, that is, our work, and as long as it facilitates our work, it does not have any negative effects...*

128

CONCLUSION AND DISCUSSION

Recently tourism industry has undergone tremendous change as a result of the application of information technology. The technology has gained advantageous to this industry both tactical and strategic tasks. So, competitiveness and inventive practices and ideas are getting important for changing competitive conditions (Khatri, 2019). Digital technology and devices have improved our lives in terms of convenience and efficiency, but they are also linked to social, psychological, and physical issues (Jiang & Balaji, 2021). This view constituted the main purpose of the research and the research was designed accordingly. As a result of the interviews conducted to prove the idea, it was proved that technology creates stress in the tourism industry in Turkey. It was supported through qualitative data that the technology paradox has a negative aspect for the employees. Although technological developments have a positive impact on the tourism sector, considering the labor-intensive nature of tourism, the data revealed that technological developments create an intense source of stress for employees.

Although the aspect of tourism that is intertwined with technology puts employees under stress, applying the requirements of digital transformation and catching up with the age provides a competitive advantage. The use of technology is inevitable in the stages of increasing the profitability of businesses and implementing sustainability. Any innovation made to meet demand is the key to the success of the business. Therefore, it is necessary to implement the right innovation management processes. Considering the mental and psychological processes of the employees in the process of adopting innovations means that business processes are not disrupted. For this purpose, the suggestions put forward for implementation are as follows;

- Adaptation of employees to technological developments should be ensured,
- Technology-related trainings should be provided during recruitment processes,
- In out-of-hours communication processes or meetings, the rights of employees should be respected and compensation should be adopted by emphasizing that time is precious,
- Instead of dismissing employees due to technological developments, they should be assigned elsewhere,
- Considering that technological developments in the tourism industry will replace human beings, it should be ensured that new employment branches are created,

Technological advances should not mean the revolution of human beings. For this reason, every proposal to be implemented will be a good example for the creation of a healthy society. Due to the fact that part of the research was conducted through a remote data collection process, the sample did not include participants equally representing sectors and departments. This may affect the data obtained from the research. The most important recommendation for future studies is that equal level departments or sectors that can represent the social world should be preferred. For example, the differences between hotel businesses exposed to technological processes working in the front office department can be measured.

REFERENCES

Ayyagari, R., Grover, V., & Purvis, R. (2011). Technostress: Technological antecedents and implications. *Management Information Systems Quarterly*, *35*(4), 831–858. doi:10.2307/41409963

Baines, A. (1998). Technology and tourism. *Work Study*, *47*(5), 160–163. doi:10.1108/00438029810370492

Bakker, A. B., & Demerouti, E. (2007). The job demands-resources model: State of the art. *Journal of Managerial Psychology*, *22*(3), 309–328. doi:10.1108/02683940710733115

Bayyurt, Y., & Seggie, F. N. (2017). *Nitel araştırma yöntem, teknik, analiz ve yaklaşımları*. Anı Yayıncılık.

Beethem, H. (2016, February 13). *From digital capability to digital wellbeing: Thriving in the network.* Open Research Online. http://oro.open.ac.uk/72433/1/A%20From%20digital%20capability%20to%20 digital%20wellbeing%20thriving%20in%20the%20network%20Helen%20Bee.pdf

Büchi, M. (2021). Digital wellbeing theory and research. *New Media & Society*, 1–18.

Buhalis, D. (1998). Strategic use of information technologies in the tourism industry. *Tourism Management*, *19*(5), 409–421. doi:10.1016/S0261-5177(98)00038-7

Buhalis, D. (2020). Technology in tourism-from information communication technologies to eTourism and smart tourism towards ambient intelligence tourism: A perspective article. *Tourism Review*, *75*(1), 267–272. doi:10.1108/TR-06-2019-0258

Bulmer, S., Elms, J., & Moore, S. (2018). Exploring the adoption of self-service checkouts and the associated social obligations of shopping practices. *Journal of Retailing and Consumer Services*, *42*, 107–116. doi:10.1016/j.jretconser.2018.01.016

Choi, Y., Mehraliyev, F., & Kim, S. S. (2020). Role of virtual avatars in digitalized hotel service. *International Journal of Contemporary Hospitality Management*, *32*(3), 977–997. doi:10.1108/IJCHM-03-2019-0265

Creswell, J. W. (2007). *Qualitative inquiry research design choosing among five approaches*. Sage Publications.

Debasa, F. (2022). Digital wellbeing tourism in the fourth industrial revolution. *Journal of Tourism Sustainability and Well-being*, *10*(3), 227–237.

Edwards, S. M. (2001). The technology paradox: Efficiency versus creativity. *Creativity Research Journal*, *13*(2), 221–228. doi:10.1207/S15326934CRJ1302_9

Hoeven, C. L., Van Zoonen, W., & Fonner, K. L. (2016). The practical paradox of technology: The influence of communication technology use on employee burnout and engagement. *Communication Monographs*, *83*(2), 239–263. doi:10.1080/03637751.2015.1133920 PMID:27226694

Jiang, Y., & Balaji, M. S. (2021). Getting unwired: What drives travellers to take a digital detox holiday? *Tourism Recreation Research*, *47*(5-6), 453–469. doi:10.1080/02508281.2021.1889801

Johnson, D. S., Bardhi, F., & Dunn, D. T. (2008). Understanding how technology paradoxes affect customer satisfaction with self-service technology: The role of performance ambiguity and trust in technology. *Psychology and Marketing*, *25*(5), 416–443. doi:10.1002/mar.20218

Khatri, I. (2019). Information technology in tourism & hospitality. *Journal of Tourism & Hospitality Education Industry: A Review of Ten Years'. Publications, 9*, 74–87.

Krippendorff, K. (2004). *Content analysis an introduction to its methodology*. Sage Publications.

Liu, Y., & Hu, H. (2021). Digital-free tourism intention: A technostress perspective. *Current Issues in Tourism, 24*(23), 3271–3274. doi:10.1080/13683500.2021.1883560

Marsden, P. (2020, July 17). *What is digital wellbeing? A list of definitions*. Digital Wellbeing: https://digitalwellbeing.org/what-is-digital-wellbeing-a-list-of-definitions/

Merriam, S. B. (2009). *Qualitative research a guide to design and implementation*. Jossey-Bass- John Wiley & Sons.

Panisoara, I. O., Lazar, I., Panisoara, G., Chirca, R., & Ursu, S. A. (2020). Motivation and continuance intention towards online instruction among teachers during the Covid-19 pandemic: The mediating effect of burnout and technostress. *International Journal of Environmental Research and Public Health, 17*(21), 1–28. doi:10.3390/ijerph17218002 PMID:33143180

Rennecker, J., & Godwin, L. (2005). Delays and interruptions: A self-perpetuating paradox of communication. *Information and Organization, 15*(3), 247–266. doi:10.1016/j.infoandorg.2005.02.004

Shu, Q., Tu, Q., & Wang, K. (2011). The impact of computer self-efficacy and technology dependence on computer-related technostress: A social cognitive theory perspective. *International Journal of Human-Computer Interaction, 27*(10), 923–939. doi:10.1080/10447318.2011.555313

Sigala, M. (2020). Tourism and COVID-19: Impacts and implications for advancing and resetting industry and research. *Journal of Business Research, 117*, 312–321. doi:10.1016/j.jbusres.2020.06.015 PMID:32546875

Stankov, U., & Gretzel, U. (2021). Digital wellbeing in the tourism domain: Mapping new roles and responsibilities. *Information Technology & Tourism, 23*(5), 5–17. doi:10.100740558-021-00197-3

Tarafdar, M., Pullins, E. B., & Ragu-Nathan, T. S. (2015). Technostress: Negative effect on performance and possible mitigations. *Information Systems Journal, 25*(2), 102–132. doi:10.1111/isj.12042

Tuan, L. T. (2022a). Employee mindfulness and proactive coping for technostress in the COVID-19 outbreak: The roles of regulatory foci, technostress, and job insecurity. *Computers in Human Behavior, 129*, 1–9. doi:10.1016/j.chb.2021.107148 PMID:34975214

Tuan, L. T. (2022b). How and when does hospitality employees' core beliefs challenge foster their proactive coping for technostress?: Examining the roles of promotion focus, job insecurity, and technostress. *Journal of Hospitality and Tourism Management, 22*, 86–99. doi:10.1016/j.jhtm.2022.05.017

Wu, H.-C., & Cheng, C.-C. (2018). Relationships between technology attachment, experiential relationship quality, experiential risk and experiential sharing intentions in a smart hotel. *Journal of Hospitality and Tourism Management, 37*, 42–58. doi:10.1016/j.jhtm.2018.09.003

Yasin, E. S., Abdelmaboud, A. E., Saad, H. E., & Qoura, O. E. (2021). Side and the light side of technostress related to hotel innovations: Transforming the hospitality industry or threat to human touch. *International Journal of Tourism. Archaeology and Hospitality, 1*(1), 44–59.

Yin, R. K. (2009). *Case study research: Design and methods*. Sage Publications.

Zhao, X., Xia, Q., & Huang Wayne, W. (2020). Impact of technostress on productivity from the theoretical perspective of appraisal and coping processes. *Information & Management, 57*(8), 1–11. doi:10.1016/j.im.2020.103265

Chapter 7
Measuring the Service Quality of Artificial Intelligence in the Tourism and Hospitality Industry

Jeganathan Gomathi Sankar

BSSS Institute of Advanced Studies, India

Arokiaraj David

St. Francis Institute of Management and Research-PGDM, India

ABSTRACT

The tourism industry is rapidly adopting artificial intelligence (AI) to enhance customer experiences and improve service delivery. However, the adoption of AI has raised concerns about concierge chatbots, digital assistance, proactiveness, anthropomorphism, and security, and its impact on overall customer satisfaction. Therefore, the aim of this research is to measure the service quality of AI in the tourism industry, with primary data collected in Pondicherry. This study is a quantitative research study that utilized a survey method to collect primary data. A total of 350 respondents were targeted, with 307 valid responses obtained. The data were analyzed using confirmatory factor analysis and structural equation modeling. The study highlights that AI technology has a significant positive impact on the service quality of the tourism industry. This study contributes to the literature by providing empirical evidence to service quality in the context of AI technology and the importance of AI technology to enhance service quality and customer satisfaction.

INTRODUCTION

Artificial intelligence (AI) technology has advanced significantly in recent years, and more and more applications are now leveraging this technology (Huang & Rust, 2021). The utilisation of technology like machine learning, huge data, interpreting, and processing natural language is referred to as artificial

DOI: 10.4018/978-1-6684-6985-9.ch007

intelligence (AI) (Poole & Mackworth, 2010). These software agents are created to carry out operations that ordinarily need human intellect, such as problem-solving, in a methodical manner in order to obtain the optimal result given the facts at hand (Russell & Norvig, 2021). There is a commonly held notion that AI will have large and far-reaching impacts on the services industry, acting as a primary driver of its expansion, according to several studies (Huang & Rust, 2021; Noor et al., 2022). The potential advantages of AI for organizations and suppliers of services include increased revenue through better support for business and marketing decisions and decreased operational costs through automation (Davenport et al., 2020; Neuhofer et al., 2021; Prentice et al., 2020). Makadia (2020) points out that market value growth projections for AI agents additionally highlight their potential. As a result, it is widely believed that AI will be essential to the expansion and advancement of the services sector in the years to come. One of the most significant aspects of AI is its capacity to offer customers customer service that is comparable to that of a human. Comparatively to other types of technology-based self-service (Wirtz et al., 2018). Additionally, there is evidence to support the idea that AI is superior to people and non-AI self-service technologies in several service-related tasks. Because AI is not limited by human flaws like unintentional biases and relative inefficiency, it can perform some parts of service more efficiently than human service workers (Wirtz et al., 2018). Also, compared to non-AI self-service systems, which are typically pedantic in following specified interaction rules, AI can adapt and provide greater possibility for tailored social engagement with personalisation to consumers in service encounters (Wirtz et al., 2018).

According to industry analysts, AI is becoming more significant in the services sector (Mustak, et al., 2021). The application of AI in marketing research has attracted more interest in recent years (Feng et al., 2021; Mustak et al., 2021). Most recent research in the field of services marketing has concentrated on customer acceptance of AI and its continuous use (Kong et al., 2022; Gursoy et al., 2019; Wirtz et al., 2018; Xu et al., 2020). Recently, conceptual studies investigating the potential effects of AI on services have been published (Mustak et al., 2021). According to Wirtz et al., (2018), research into the use of service robots at the micro, meso, and macro levels is advised. In addition, Huang and Rust (2020) did research on the effects of various forms of Artificial intelligence on societal norms, services, and consumer behaviour. Moreover, Dwivedi et al., (2019) put forth frameworks that take public policy issues into account and offer instructions for reducing security and social desirability worries connected to AI applications. There is currently no research on how the usage of AI might affect consumer assessments of service quality, which is a vital topic for study, even if these studies lay the groundwork for additional research in the quickly developing field of AI.

THEORETICAL BACKGROUND

Service Robots

A service robot is defined as "a robot that performs helpful activities for humans or equipment" by the International Federation of Robotics (Tung & Law, 2017). The two main kinds of service robots are personal and professional service robots. Robots used for personal or non-commercial chores are referred to as personal service robots, whilst those used for professional reasons are referred to as professional robots. In fields including medicine, agriculture, and logistics, professional robots have already demonstrated a significant impact (Cheng Hong et al., 2019). For example, medical robots are worth more than $6 billion in 2018 (Tian et al., 2019). The market for professional service robots is anticipated to expand as they demonstrate their financial viability (Wirtz et al., 2018). In example, it is anticipated that sales of professional service robots would rise by 20% to 25% yearly and total $19 billion by 2020 (Tung & Law, 2017; Thummula et al., 2019). The hotel business has included professional service robots into numerous aspects of its operations due to the significant rise of robotic technology. One of the most well-known instances is the Henn Na Hotel in Japan, which was the first hotel to employ robots and has plans to hire eight more as a method to save labour expenses (Baird et al., 2018; David et al., 2019[a]). Two service robots were employed by the Singaporean hotel Jen Orchard Gateway to provide for guests' requirements, such as providing amenities (Kim et al., 2017; Ganeshkumar & David, 2022; Ravi et al., 2018.). Another hotel adopting service robots is Yotel, which uses them for monotonous jobs like keeping baggage and providing services (Bhimasta & Kuo, 2019). At hotels, robots have been deployed in a variety of capacities to assist staff members and provide services to guests (Ivanov et al., 2017). In light of Hilton's investment in Connie, an AI-powered robot concierge, the hotel sector has been expressing strong interest in robot concierge services where direct engagement is required due to their physical presence, such as their movability and embodiment.

System-based, autonomous, and adaptable interfaces known as service robots may engage with customers, communicate with them, and provide services on their behalf. They can be real or virtual, humanoid (i.e., anthropomorphic), or not (Wirtz et al. 2018). Chatbots are interactive agents that communicate with people in a narrowly defined area or on a specific issue using natural language phrases, according to Huang & Rust (2021). They are typically used on the Internet to seek out information, site navigation, and other services. Lester et al. (2004) defined chatbots as systems that make use of natural language to engage users in task- and information-focused text-based discussions for a variety of applications. Different chatbots may have varying degrees of intelligence. Therefore, Huang and Rust (2018) distinguish four types of intelligences: mechanical, analytical, intuitive, and empathic, depending on the nature of the service. The simplest kind of intelligence is mechanical intelligence, which is the capacity to carry out routine tasks effortlessly without the need for special instruction (Huang & Rust 2018). Because it is rule-based in this instance, the chatbot is unable to comprehend its surroundings. To digest information, solve problems, and learn from it is what is meant by analytical intelligence (Sternberg 2005; Huang & Rust 2018). Mass-personalization based on big data is made possible by machine learning and data analytics approaches, which enable technology to learn from data and identify insights without being programmed. Mechanical and analytical intelligences, which mimic intelligence but lack intuition, are still regarded as "weak AI" (Huang & Rust 2018; Ganeshkumar et al., 2020). At the ultimate degree of complexity, there are two types of intelligence: empathic intelligence and intuitive intelligence. Empathetic intelligence refers to the capacity of the AI to recognise and comprehend other people's emotions, to impact them, and to respond accordingly (Huang & Rust 2018). Though they are the most developed generation of AI, these two intelligences are still a long way from becoming practical. According to Wirtz et al., (2018), the way service agents differ from humans is through their particular (limited) strengths, perceptions, and flaws. To give varied, individualised services, human personnel must have a thorough awareness of their clients and the service processes, and learning is necessary for this. Also, setting up staff connections to Customer Relationship Management (CRM) systems takes time and effort (David et al., 2018; Sudhakar et al., 2017). Service workers need extensive training at considerable expenses but may later provide a competitive edge. Better hiring, selection, training, motivation, and organisational practises can differentiate services. Chatbots, on the other hand, swiftly and thoroughly gather knowledge using AI and CRM systems to identify the best answers. Moreover, chatbots offer homogenous services in a very dependable way since they lack human error and weariness, act uniformly, do not exhibit heterogeneity, and are free from human mistake (Huang & Rust 2018; Nihmathullah et al., 2022). They also don't actually experience or communicate true emotions. Indeed, a mechanical and/or analytical degree of design is used in the majority of chatbots. They have the benefits of being exceedingly consistent, constantly accessible, and cost- and time-effective, but they may not please customers and have a poor potential for competitive advantage. In fact, there's a chance that their pre-programmed scripts won't handle user requests correctly, which might irritate and annoy the user. To satisfy client expectations, it is crucial to take into account a chatbot's efficiency and competencies. The best uses of service robots are seen to be in the areas of information providing, processing, reservations, and payments (Ivanov & Webster, 2019).

Chatbot

A chatbot is a computer software created to mimic human conversations utilising voice techniques or text written in natural language (Pillai & Sivathanu, 2020). AI-based chatbots are described by Ramach-andran et al., (2018) as chat robots that converse with people via an underlying computer programme and AI technology. According to Adam et al. (2021), AI-based chatbots are created for turn-by-turn user dialogues based on textual input. Chatbots combine intelligent backend systems with an interface that can be accessible by a variety of gadgets, including Google, Siri, Amazon Alexa, laptops, and smart-phones (Guzman & Pathania, 2016; Madaan et al., 2021). With the aid of clever backend systems that simplify communication, chatbots can engage people via text or speech (Sheehan, 2018). The use of chatbots and digital intelligent assistants is growing, providing conversational system capabilities for use in marketing, sales, and customer service (Cocca et al., 2018; Choudhary et al., 2021). Chatbots can give dialogues that are more appealing to consumers because of machine learning and clever software algorithms. The most recent AI-based chatbots are far more intelligent, potent, and competent than the older iterations, which enhances human-technology connection (Rajan & Saffiotti, 2017; Jeganathan & David, 2022). Chatbots have found a number of uses in the hotel and travel sectors, including travel booking, customer booking, customer service, and giving advice to customers on travel-related con-cerns. Tourism businesses can now provide 24/7 customer assistance, boost income prospects, improve engagement, automatically gather leads, save overhead expenses, gain an edge over their competitors, and save time thanks to chatbots (Sheehan, 2018). Making use of chatbots for client assistance are travel agencies like Makemytrip, Expedia, Kayak, Skyscanner, and Cheapflights (Pillai & Sivathanu, 2020). And around 44% of consumers said they prefer utilising AI-based chatbots over human customer service representatives, according to a survey by Aspect research software (Sweezey, 2020). The acceptance of chatbots by consumers is crucial to effectively promote tourist businesses because technology influ-ences all generations (Sweezey, 2020). Yet, there are currently few studies in the tourist industry that examine how novel technology like artificial intelligence and robots are altering the way that tourism is conducted (Tussyadiah, 2020).

Recent studies in the field of tourism have concentrated on the use of robotics, automation, and artificial intelligence (AI), with the major focus being on current applications and potential future consequences (Ivanov and Webster, 2019; Murphy et al., 2019). The adoption and acceptance of these technologies for technology-mediated interactions, however, as well as the impact of perceived trust (Chien & Hassenzahl, 2020; Arokiaraj et al., 2020[a]; Tussyadiah, 2020), application of the uncanny valley theory (Yu, 2020), and human replacement (Yu, 2020; Ivanov et al., 2019). To properly comprehend the possible effects of robotics, automation, and AI technologies on the tourism business, more research is thus necessary. Academic research has not focused much on the adoption of AI-based chatbots in the tourism and hospitality sectors (Brandtzaeg & Følstad, 2017; McLean et al., 2020). There is a dearth of study analysing the elements that lead to the adoption of this technology because it is still in its infancy (Io and Lee, 2019; Sheehan, 2018; Lakshman & David, 2023). Consequently, the goal of this study is to carefully examine the variables that influence the acceptance of chatbots used for tourism. Prior research has emphasised the significance of elements including perceived value, perceived usability, trust, and pleasure in influencing the adoption of technology (Ivanov et al., 2017; Io and Lee, 2019; Ganeshkumar et al., 2022). These considerations are pertinent in the area of chatbots used for tourism planning since consumers are more likely to adopt technology, they find beneficial, simple to use, and reliable. Moreover, user experience and personalization have been underlined in research as important factors in chatbot acceptance (Sheehan, 2018; Candello et al., 2017). If a chatbot offers a tailored experience that caters to their unique wants and tastes, users are more inclined to use it. A good user experience can also boost user happiness and technology confidence, which can encourage technology adoption.

Chatbots in the Tourism and Hospitality Industry

The usage of chatbots in the travel and tourism sector can have a number of benefits, including better customer service and greater accessibility to a business's offerings. By including chatbots into these channels, which are becoming more and more popular due to apps like Facebook Messenger and WhatsApp, customers will find these channels to be easier to use during the presale and after-sale processes (Singh, 2018). Chatbots can carry out a variety of functions, including creating reminders, booking tickets, sharing traffic or weather information, and organising appointments (Albayrak et al., 2021; David, 2020). Chatbots have a lot of promise, and they can help a number of tourism-related industries, including lodging, dining, vehicle rentals, travel agencies, and visitor information centres. In order to promote efficiency and boost customer happiness, chatbots can assist with tasks like booking bookings, responding to frequently asked queries, and giving recommendations. However, there are possible drawbacks to using chatbots, such as the possibility of technical failures, the inability of complex questions to be understood, and the absence of the personalised touch that certain clients may desire (Albayrak et al., 2021; Kalburgi et al., 2023 Ganeshkumar et al., 2023$_a$). However, the advantages of adopting chatbots in the tourist sector exceed any potential drawbacks, and businesses can investigate numerous chatbot use cases to enhance their services and acquire a competitive edge. Lasek and Jessa's (2013) research supports the economic benefit of hotel chatbots by showing that growing the proportion of online bookings has a favourable effect on sales growth. Other businesses that have made use of Facebook's technology to build their own chatbots include Expedia and Marriott Hotels. These chatbots provide fundamental services like reserving rooms and additional extras like spa services and dinner (Ukpabi et al., 2018; Feleen & David, 2021). Chatbots can improve the experience for visitors before their arrival and collect useful information that may be utilised to offer tailored services (Bhargava, 2017).

The potential advantages of chatbots have also been observed in the restaurant sector. In 2016, Taco-Bell introduced its own TacoBot, which streamlines food ordering, makes suggestions, and responds with humour. With their own proprietary chatbots, other eateries including Burger King, Pizza Hut, and Dominos have followed suit (Ukpabi et al., 2018; David et al., 2022$_b$). Soon, customers will be able to request deliveries using social media platforms like Facebook and WhatsApp, and chatbots will eventually be able to process payments, as shown by MasterCard's Masterpass app. Chatbot implementation can lower costs for both users and companies. Consumers no longer need to contact, which lowers their communication costs, and businesses no longer require hiring customer care agents or contracting out answering services to call centres (Ukpabi et al., 2018). Businesses are able to provide individualised services while also obtaining important data to enhance their offers by integrating chatbots into the client journey. Gamanyuk (2017) contends that chatbots have advantages beyond only speeding up ordering and delivery. These benefits include enabling users to complete chores without downloading mobile apps. Customers can utilise chatbots, for instance, to look up and explore restaurant reviews, menus, prices, and available tables. Users may simply reserve, modify, cancel, or reschedule tables as needed to manage their restaurant bookings while they are on the road. Customers can search for and select restaurants using chatbots based on a variety of factors, including party size, date, time, desired cuisine, pricing, or distance (Jeganathan et al., 2022; Gamanyuk, 2017).

Service Quality

Service quality is viewed by customers as a long-term overall assessment of service performance and is quantified as an attitude, according to Parasuraman et al. (1994a) and Cronin Jr. and Taylor (1994). Many studies have been conducted on the factors that affect service quality, and they have been confirmed in a variety of service contexts. Despite the fact that some academics see service quality dimensions as antecedents, most think of them as parts of the multidimensional service quality concept (Luo et al., 2019; Arokiaraj et al., 2020$_b$). Although the majority of the literature on the development of service quality scales suggests that the service quality construct is reflective, there is ongoing discussion among scholars as to whether it actually consists of pivotal higher orders (Martnez & Martnez, 2010; Parasuraman et al., 2005). Yet, several academics advise caution when thinking about the fundamental specification (Hair et al., 2018).

A substantial body of knowledge has been produced through research on service quality, including well-developed structures and models that have been improved, expanded, and verified in a variety of service situations (Seth et al., 2005). Most scholars agree that service excellence is an overarching assessment and attitude that goes beyond judgements unique to particular transactions (Zaibaf et al., 2013; Parasuraman et al., 1994$_b$). One viewpoint on service quality is provided by the disconfirmation hypothesis, which compares consumers' expectations with actual performance of services (Torres et al., 2014). This comparison includes both expressive (psychological) and instrumental (functional) performance outcomes to assess service quality (Grönroos, 1984). Creating a tolerance range between intended and minimal expectations, within which service delivery is deemed adequate, is another method of gauging expectations (Parasuraman et al., 1993). The measurement and management of service quality have benefited greatly from the inclusion of these ideas and metrics in service quality research.

The subjectivity of service assessments underlined the need to enhance our knowledge of service quality for face-to-face service environments, according to Zeithaml, Parasuraman, and Berry's 1985 study. Because of improvements in service innovation utilising technology, the majority of research on the quality of human services was undertaken in the 1990s, whereas research on the quality of services enabled by technology was conducted in the 2000s (Huang & Rust, 2018; David et al., 2022$_a$). The service management literature was blended with concepts from the information systems literature to construct scales that manage self-service technologies and applications that run on distributed infrastructures like the internet (Ding et al., 2011; Ganeshkumar et al., 2023$_b$). Studies that assess consumer perceptions of technology-based service environments and human service environments (Manzoor et al., 2019; Rosenbaum & Russell-Bennett, 2020) continue to use the seminal SERVQUAL scale developed by Parasuraman, Zeithaml, and Berry in 1988. Service quality research is still relevant in the literature (Xiao & Kumar, 2019). AISA-related SERVQUAL has also been modified in recent empirical investigations (Meyer-Waarden et al., 2020).

Based on the qualities of the service agent, the technique of service delivery, and the general service environment, customers' assessments of service quality may vary (Rust & Oliver 1993). Given this, SERVQUAL was not well adapted for the online service environment. As a consequence, additional qualities (such "system availability") that were not included in SERVQUAL but were important for website-based services were introduced (Rita, Oliveira, & Farisa, 2019).

A thorough analysis of the literature reveals that AI may be used in a variety of technological and social service contexts. AISA's ability to provide individualised help with a human touch may also be assessed using service quality measures that include human service agents like Mittal and Lassar (1996). Moreover, AISA and contact centres both need dynamic replies to a range of voice service demands (Burgers et al. 2000).

The performance of AI service agents cannot be evaluated with the same service quality metrics that are often used for human-delivered services (AISA). Recent research on the customer service quality of chatbots and robots in cafés has revealed that even SERVQUAL, a commonly used metric for customer service, is unable to fully capture the customer service performance of AISA (Meyer-Waarden et al., 2020). This is because AISA is distinctive, which has altered how services are provided and the atmosphere for all services, resulting in diverse customer assessments of service quality (Rust & Oliver, 1993). AISA employ several techniques, such as voice recognition, natural language processing, and machine learning, to achieve intelligence (Jordan & Mitchell, 2015). AISA has made significant strides in the performance of well-defined, automated operations (Davenport et al., 2020), and has the potential to carry out tasks that are more intuitive and empathetic in the future (Huang & Rust, 2018; Sudhakar et al., 2017; David et al., 2019[b]). Anthropomorphism is a crucial component of AISA that may be observed in chatbots, virtual assistants, and humanoid robots (Goudey & Bonnin, 2016; Moussawi, 2016). Consumers' feelings of social presence can be influenced by their interactions with anthropomorphic AISA, which can boost their trust and enjoyment of service interactions (Noor et al., 2020). It is clear that AI service agents (AISA) may deliver client experiences that fall halfway between those obtained from human-based services and technology-based service systems, claim Bock, Wolter, and Ferrell (2020) and Lu et al. (2020). To what degree these service quality characteristics are significant for consumers when assessing AISA is unclear, as is which service quality aspects, typically used to evaluate human- or technology-based service experiences, are crucial for customers. In addition, new service quality characteristics that are unique to the AISA service environment and not present in conventional service settings may exist because of the distinctive characteristics of AISA, such as intelligence and anthropomorphism (Bock et al., 2020).

AI Service Quality

Integrating AI technologies into the service industry, this essay argues that AI services must be viewed as a component of the entire level of service quality since doing so will have financial consequences for the company. The degree of service being provided to meet and surpass customers' expectations is frequently determined by how customers see a company's service offerings. (Parasuraman et al., 1995). One of the key indicators of customer loyalty and satisfaction, two ideas that are widely acknowledged to be highly correlated, is service quality (Shi et al., 2014). Studies have not yet examined how a service affects an organization's performance as shown by customer-related results or the inclusion of a service in studies that evaluate service quality (satisfaction and loyalty behaviours).

Despite the fact that service companies (like hotels) provide AI services to boost operational performance and customer satisfaction, a survey of the pertinent literature reveals that relatively little academic study has tried to understand how consumers/customers react to such services. Users' responses are probably affected by their views on and experiences with technology, given that technology is essential to the quality of AI services. Technology-based services may impact client happiness and loyalty, according to a prior research. Self-Service Technology (SST) caused both good and bad things to happen for different customers, according to Mueter et al., (2000) investigation using the critical incident approach. SST performed better than service workers, was easy to use, and was time and money efficient. The research also highlighted customers' dissatisfaction with subpar designs and technological malfunctions. Another technology-based approach is customer relationship management (CRM), which is used to retain customer and service encounter data for the best economic outcomes (such as purchasing and loyalty behaviour) (Kumar et al., 2010). Applications and CRM software have gotten a lot of funding. Nonetheless, research (like that of Mithas et al., 2005) shows that such an investment is financially positive as CRM systems are favourably related with customer happiness, which encourages customer loyalty. Innovation services were taken into account when McKecnie, (2011) evaluated the service quality of the banking business.

RESEARCH METHODOLOGY

The purpose of this research is to investigate the use and perception of AI-powered services in the tourism and hospitality industry for this data were collected from Pondicherry, India. This research collected data from customers who have used such services from an online portal to book hotels, restaurants, and other tourist services. The participants of this study will be customers who have used AI-powered tourism and hospitality services from the online portal. The inclusion criteria for participants are as follows: (a) they must be over the age of 18; (b) they must have used the online portal to book tourism and hospitality services such as hotel or restaurant reservations; (c) they must be able to understand and complete the questionnaire in English. 350 respondents were approached to collect data for this research. Among the 350 gathered responses, 307 were found valid. The data collected from the questionnaire have been analyzed using descriptive statistics and inferential statistics. Descriptive statistics was used to summarize the characteristics of the participants and their responses to the survey questions. Inferential statistics, such as confirmatory factor analysis and structural equation modelling, have been used to examine the relationships between the variables in the study. This study will go with the ethical standards of research, which include getting participants' informed permission, protecting their privacy, and preventing individuals from suffering harm as a result of the study. AI services differ depending on the research circumstances. As this study is focused on hotels, the items that were used to assess AI service quality were adopted from He et al. (2017), Prentice et al. (2020) and Noor et al. (2021). This metric reflects the services respondents received from online travel and hotel service providers that used AI. The clients that participated in this study used the following AI services, which were also included in the study: concierge chatbots, digital assistance, proactiveness, anthropomorphism, and security. In the framework of this study, a total of 15 items were created, including 6 items pertaining to customer satisfaction and word-of-mouth marketing. Participants were asked to judge how happy they were with various services offered by the AI technologies (1 = strongly disagree to 5 = strongly agree).

AI service quality was measured by adopting Prentice et al.,'s (2020) two factors namely concierge chatbots and digital assistance. Concierge chatbots are designed to provide a personalized experience for customers. These chatbots act as personal assistants to customers by understanding their preferences, providing recommendations, and answering queries. A concierge chatbot should have the ability to understand natural language and respond in a way that is human-like, empathetic, and helpful. Additionally, the chatbot should be able to handle complex queries, redirect the customer to a live agent if necessary, and provide seamless handoffs between the chatbot and human agents. Digital assistance refers to the ability of an AI system to provide support to customers through various digital channels such as chat, email, social media, and phone. An AI system should be able to respond to queries quickly and accurately, providing customers with the information they need. Additionally, it should be able to learn from previous interactions and use this knowledge to improve future interactions. A digital assistance system should also have the ability to integrate with other systems and tools, providing a seamless experience for customers.

Noor et al., (2021) identified two factors, proactiveness and anthropomorphism, and a construct from He et al. (2017), security, to measure AI service quality were also been employed. Proactiveness refers to the ability of an AI system to anticipate and address customer needs before they even express them. A proactive system should be able to use customer data and previous interactions to suggest solutions, provide relevant information, and make personalized recommendations. The system should also be able to anticipate potential issues and offer proactive solutions to prevent them from occurring. This can lead to increased customer satisfaction and loyalty. Anthropomorphism refers to the degree to which an AI system resembles human characteristics, behaviors, and emotions. A system with high anthropomorphism can create a more human-like interaction with customers, which can lead to greater engagement and satisfaction. However, the system should be careful not to create false expectations or overpromise on its abilities, as this can lead to disappointment and frustration. Security refers to the measures taken by an AI system to protect sensitive customer data, ensure privacy, and prevent unauthorized access. A system that prioritizes security can create a sense of trust and confidence in customers, leading to greater loyalty and satisfaction. The system should also be transparent about its security measures and communicate them clearly to customers. Using three measures adapted from Bogicevic et al., (2017), customer satisfaction with AI and staff service quality were assessed. In addition to measuring customer satisfaction, the researchers also aimed to measure word-of-mouth communication. They did this by adopting variables from a study conducted by Ismagilova et al., (2020). Word-of-mouth communication is when customers talk about a product or service to others, which can have a significant impact on a company's reputation and success.

Postulated Model and Hypotheses

The relationship between service quality and customer satisfaction is widely studied in the field of marketing and has been shown to be positive and significant. Service quality is defined as the customer's overall evaluation of the superiority or excellence of the service provided, while customer satisfaction refers to the customer's subjective assessment of their experience with the service. Several studies have found that higher levels of service quality lead to higher levels of customer satisfaction (Koc, 2020; Nilashi et al., 2021). Thus, the following hypothesis was developed which as shown in the above figure 01.

Figure 1. Postulated model

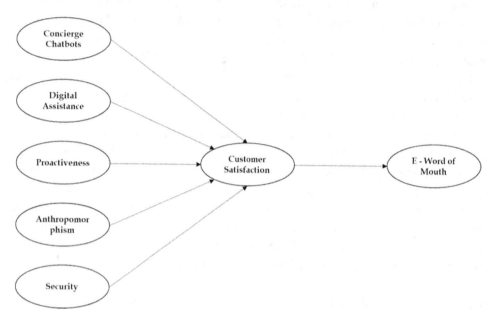

- H1: AI service quality positively influences customer satisfaction.

Tourism businesses rely heavily on customer satisfaction to build brand loyalty and drive repeat business. Studies have shown that satisfied tourists are more likely to engage in positive e-wom communication, which can help to increase brand awareness and drive new business to tourism destinations (Hung & Petrick, 2011; Kankhuni, & Ngwira, 2022). Thus, the following hypothesis was developed

- H2: There is a significant relationship between customer satisfaction and electronic word of mouth communication.

The relationship between service quality and electronic word-of-mouth (e-wom) communication in tourism is an important area of research. Studies have shown that service quality has a significant impact on eWOM communication in the tourism industry (Lai et al., 2021). When tourists perceive high levels of service quality, they are more likely to engage in positive e-wom communication and share their positive experiences with others (Kankhuni, & Ngwira, 2022). Conversely, when tourists perceive low levels of service quality, they are more likely to engage in negative e-wom communication and share their negative experiences with others (Mukhopadhyay et al., 2022)

- H3: AI service quality has a significant impact on electronic word of mouth communication

ANALYSIS AND RESULTS

Measurement Model

The measurement characteristics for the latent constructs, reflecting in nature and having many indicators, were determined for the final model. Internal consistency and high levels of reliability are verified by composite reliability (CR) scores (Nunnally, 1978). The discriminant validity is established by comparing the inter-correlations of the factors with the AVE off-diagonal values (Fornell & Larcker, 1981).

Figure 2. Measurement model

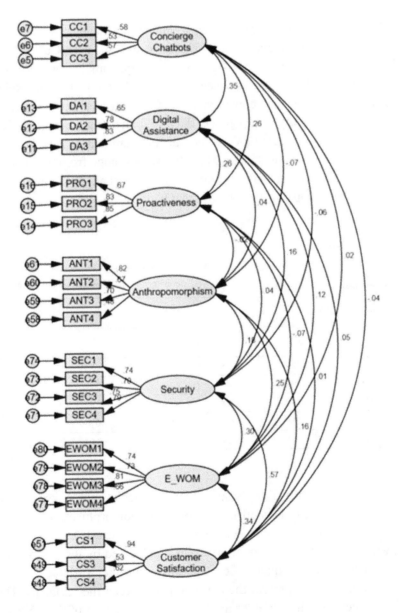

Table 1. Psychometric properties

Factor	Composite Reliability	Average Variance Extracted
Concierge Chatbots	0.778	0.513
Digital Assistance	0.800	0.574
Proactiveness	0.830	0.621
Anthropomorphism	0.767	0.560
Security	0.834	0.557
Customer Satisfaction	0.751	0.517
E-Word of Mouth	0.824	0.540

The table 1 provides information on the psychometric properties of seven constructs, namely Concierge Chatbots, Digital Assistance, Proactiveness, Anthropomorphism, Security, Customer Satisfaction, and E-Word of Mouth, that have been portrayed in figure 02. The psychometric properties evaluated are the Composite Reliability (CR) and Average Variance Extracted (AVE). Composite Reliability measures the internal consistency of the items in a construct, and its values range from 0 to 1. In this table 1, all constructs have Composite Reliability values greater than 0.7, indicating good internal consistency (Hair et al., 2017). Average Variance Extracted (AVE) measures the amount of variance that is captured by the construct relative to the measurement error. The AVE values range from 0 to 1, with values greater than 0.5 indicating adequate convergent validity consistency (Hair et al., 2017). All constructs have AVE values greater than 0.5, which confirms their convergent validity.

Table 2. Discriminant validity

	Customer Satisfaction	Concierge Chatbots	e-wom	Security	Anthropomorphism	Proactiveness	Digital Assistance
Customer Satisfaction	0.719						
Concierge Chatbots	-0.042	0.660					
e-wom	0.338	0.024	0.735				
Security	0.567	-0.064	0.304	0.746			
Anthropomorphism	0.155	-0.072	0.254	0.181	0.678		
Proactiveness	0.014	0.266	-0.067	0.037	-0.019	0.788	
Digital Assistance	0.045	0.354	0.121	0.161	0.036	0.262	0.757

Table 2 shows the discriminant validity results for the variables: customer satisfaction, concierge chatbots, e-WOM (electronic word-of-mouth), security, anthropomorphism, proactiveness, and digital assistance. The values on the diagonal represent the square root of the average variance extracted (AVE) for each variable, which measures the amount of variance explained by the variable itself (Hair et al., 2017). The values should be higher than the correlations with other variables to show discriminant validity. The off-diagonal values represent the correlations between the variables. The table 2 shows that all the correlations between the variables are below the square root of the AVE for each variable, indicating good discriminant validity. The CMIN/DF ($\chi2$/df) value is 2.088, which is below the recommended value of 3 for an acceptable model fit based on absolute fit indices. The fit indices suggest that the measurement model has an acceptable fit. However, it is important to note that the $\chi2$ test is sensitive to sample size and may be significant even when the model fit is adequate. Therefore, other fit indices such as the RMSEA, CFI, and AGFI should also be considered to assess the model fit (Kline, 2015). The Root Mean Square Error of Approximation (RMSEA) is 0.060, which is below the recommended value of 0.09 for an acceptable model fit based on absolute fit indices. The Comparative Fit Index (CFI) value is 0.904, which is above the recommended value of 0.9 for an acceptable model fit based on incremental fit indices. The Adjusted Goodness-of-Fit Index (AGFI) value is 0.850, which is above the recommended value of 0.8 for an acceptable model fit based on parsimony fit indices. The obtained values for the various fit indices are compared to the recommended values from Hair et al. (2017) to assess the adequacy of the model fit.

The relationship between AI Service Quality and Customer Satisfaction has been tested by structural model depicts in figure 03. The path coefficient value is 0.580, indicating a positive and significant relationship between AI Service Quality and Customer Satisfaction. The p-value of 0.001 is less than the conventional significance level of 0.05, indicating that the relationship is statistically significant. Therefore, we can conclude that higher AI Service Quality leads to higher levels of Customer Satisfaction. The obtained values for CMIN/DF, RMSEA, CFI, and AGFI are 2.468, 0.069, 0.981, and 0.885, respectively. According to Hair et al. (2017), a CMIN/DF value less than 3, RMSEA less than 0.09, CFI greater than 0.9, and AGFI greater than 0.8 indicate an acceptable model fit. The obtained values for all fit measures meet or exceed the recommended values.

Figure 3. Structural model

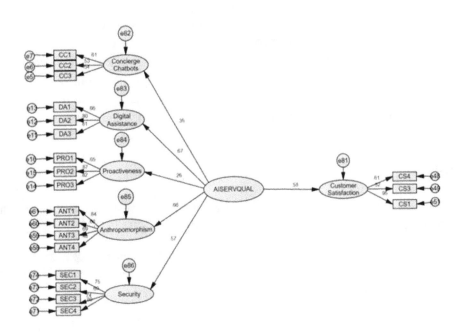

DISCUSSION

Artificial Intelligence (AI) is rapidly transforming the tourism and hospitality industry. It has become an essential tool for hotels to enhance customer experience, improve efficiency, and increase revenue. However, little research has been conducted on how AI service quality affects customer satisfaction and electronic word-of-mouth in the hospitality industry. This study has identified five dimensions of AI service quality in the hospitality industry: concierge chatbot, digital assistance, proactiveness, anthropomorphism, and security. The study found that these dimensions significantly explain overall service quality, and digital assistance had the most significant effect on customer satisfaction and e-wom. For those who work in the hotel industry, the study's conclusions have significant practical ramifications. While users may find AI-powered technologies convenient, they may not signify a competitive advantage that would encourage positive customer feedback. Customer satisfaction and overall service quality rating still heavily rely on employee services. Customers tend to love dealing with robots, though, and when personnel services are excluded, AI services play a significant role in determining the overall quality of a business. To improve AI service quality in the tourism and hospitality industry, hotels must ensure that their AI-powered tools are secure, proactive, and offer personalized services. Hotels must also ensure that their AI-powered tools are not perceived as impersonal or dehumanizing, as this can negatively affect customer satisfaction and e-wom.

This study presents a number of limitations that must be acknowledged. First off, the study was limited to hotels in Pondicherry, that might limit the applicability of our conclusions. As a result, care must be exercised while interpreting the findings of this research. Second, the AI service quality scale used in this study was based on practitioner perspectives and non-academic sources. As a result, we recommend that this scale be cross validated for future applications. We are aware of this drawback and emphasise how crucial it is to confirm this measurement in other settings. Future study should overcome these constraints in order to properly understand the connection between AI services, organisational performance, and customer-related outcomes. Future research should, for instance, examine how consumers' views and actions towards data privacy relate to their trust in AI services as well as other AI services that may influence customer reaction. More understanding of the advantages and disadvantages of AI may be gained by analysing these problems. In conclusion, we recognize the limitations of our study and advise caution when interpreting our findings. We hope that our research will stimulate further exploration into the implications of AI services for organizational performance and customer-related outcomes.

REFERENCES

Adam, M., Wessel, M., & Benlian, A. (2021). AI-based chatbots in customer service and their effects on user compliance. *Electronic Markets*, *31*(2), 427–445. doi:10.100712525-020-00414-7

Albayrak, T., Caber, M., & Sigala, M. (2021). A quality measurement proposal for corporate social network sites: The case of hotel Facebook page. *Current Issues in Tourism*, *24*(20), 2955–2970. doi:10.1080/13683500.2020.1854199

Arokiaraj, D., Ganeshkumar, C., & Paul, P. V. (2020a). Innovative management system for environmental sustainability practices among Indian auto-component manufacturers. *International Journal of Business Innovation and Research*, *23*(2), 168–182. doi:10.1504/IJBIR.2020.110095

Arokiaraj, D., Ramyar, R. A., Ganeshkumar, C., & Gomathi Sankar, J. (2020b). An empirical analysis of consumer behaviour towards organic food products purchase in India. *Calitatea Qual Access Success, 21*.

Baird, K., Su, S., & Munir, R. (2018). The relationship between the enabling use of controls, employee empowerment, and performance. *Personnel Review*, *47*(1), 257–274. doi:10.1108/PR-12-2016-0324

Bhargava, V. (2017). Are social chatbots the Future of hassle-free travel? *The chatbot magazine*.

Bhimasta, R. A., & Kuo, P. Y. (2019, September). What causes the adoption failure of service robots? A Case of Henn-na Hotel in Japan. In Adjunct proceedings of the 2019 ACM international joint conference on pervasive and ubiquitous computing and proceedings of the 2019 ACM international symposium on wearable computers (pp. 1107-1112). ACM.

Bock, D. E., Wolter, J. S., & Ferrell, O. C. (2020). Artificial intelligence: Disrupting what we know about services. *Journal of Services Marketing*, *34*(3), 317–334. doi:10.1108/JSM-01-2019-0047

Bogicevic, V., Bujisic, M., Bilgihan, A., Yang, W., & Cobanoglu, C. (2017). The impact of traveler-focused airport technology on traveler satisfaction. *Technological Forecasting and Social Change*, *123*, 351–361. doi:10.1016/j.techfore.2017.03.038

Brandtzaeg, P. B., & Følstad, A. (2017). Why people use chatbots. In *Internet Science: 4th International Conference, INSCI 2017, Thessaloniki, Greece, November 22-24, 2017* [Springer International Publishing.]. *Proceedings*, *4*, 377–392.

Burgers, A., de Ruyter, K., Keen, C., & Streukens, S. (2000). Customer expectation dimensions of voice-to-voice service encounters: A scale-development study. *International Journal of Service Industry Management*, *11*(2), 142–161. doi:10.1108/09564230010323642

Candello, H., Pinhanez, C., & Figueiredo, F. (2017, May). Typefaces and the perception of humanness in natural language chatbots. In *Proceedings of the 2017 chi conference on human factors in computing systems* (pp. 3476-3487). IEEE. 10.1145/3025453.3025919

Cheng, H., Jia, R., Li, D., & Li, H. (2019). The rise of robots in China. *The Journal of Economic Perspectives*, *33*(2), 71–88. doi:10.1257/jep.33.2.71

Chien, W. C., & Hassenzahl, M. (2020). Technology-mediated relationship maintenance in romantic long-distance relationships: An autoethnographical research through design. *Human-Computer Interaction*, *35*(3), 240–287. doi:10.1080/07370024.2017.1401927

Choudhary, N., David, A., & Feleen, F. (2021). Employee Engagement and Commitment in Service Sector. *Wesleyan Journal of Research, 13*(4.7), p107-112.

Cocca, P., Marciano, F., Rossi, D., & Alberti, M. (2018). Business software offer for industry 4.0: The SAP case. *IFAC-PapersOnLine*, *51*(11), 1200–1205. doi:10.1016/j.ifacol.2018.08.427

Cronin, J. J. Jr, & Taylor, S. A. (1994). SERVPERF versus SERVQUAL: Reconciling performance-based and perceptions-minus-expectations measurement of service quality. *Journal of Marketing*, *58*(1), 125–131. doi:10.1177/002224299405800110

Davenport, T., Guha, A., Grewal, D., & Bressgott, T. (2020). How artificial intelligence will change the future of marketing. *Journal of the Academy of Marketing Science*, *48*(1), 24–42. doi:10.100711747-019-00696-0

David, A. (2020). Consumer purchasing process of organic food product: An empirical analysis. [QAS]. *Journal of Management System-Quality Access to Success*, *21*(177), 128–132.

David, A., Ganesh Kumar, C., & Jeganathan, G. S. (2022a). Impact of Food Safety and Standards Regulation on Food Business Operators. In *Au Virtual International Conference* (pp. 355-363). SSRN.

David, A., Kumar, C. G., & Paul, P. V. (2022b). Blockchain technology in the food supply chain: Empirical analysis. [IJISSCM]. *International Journal of Information Systems and Supply Chain Management*, *15*(3), 1–12. doi:10.4018/IJISSCM.290014

David, A., Nagarjuna, K., Mohammed, M., & Sundar, J. (2019a). Determinant Factors of Environmental Responsibility for the Passenger Car Users. *International Journal of Innovative Technology and Exploring Engineering*, 2278-3075.

David, A., Ravi, S., & Reena, R. A. (2018). The Eco-Driving Behaviour: A Strategic Way to Control Tailpipe Emission. *International Journal of Engineering & Technology*, *7*(3.3), 21-25.

David, A., Thangavel, Y. D., & Sankriti, R. (2019b). Recover, recycle and reuse: An efficient way to reduce the waste. *Int. J. Mech. Prod. Eng. Res. Dev*, *9*, 31–42.

Ding, D. X., Hu, P. J. H., & Sheng, O. R. L. (2011). e-SELFQUAL: A scale for measuring online self-service quality. *Journal of Business Research*, *64*(5), 508–515. doi:10.1016/j.jbusres.2010.04.007

Dwivedi, Y. K., & Wang, Y. (2022). Guest editorial: Artificial intelligence for B2B marketing: Challenges and opportunities. *Industrial Marketing Management*, *105*, 109–113. doi:10.1016/j.indmarman.2022.06.001

Feleen, F., & David, A. (2021). A Comparative Study of Work From Home vs Work From Office: Preference of Women Employees in IT Industry. *Design Engineering (London)*, *7*(1), 5763–5775.

Feng, C. M., Park, A., Pitt, L., Kietzmann, J., & Northey, G. (2021). Artificial intelligence in marketing: A bibliographic perspective. *Australasian Marketing Journal*, *29*(3), 252–263. doi:10.1016/j.ausmj.2020.07.006

Fornell, C., & Larcker, D. F. (1981). *Structural equation models with unobservable variables and measurement error: Algebra and statistics*. Sage.

Gamanyuk, A. (2017). Restaurant table reservation chatbot for Facebook bootmaker.

Ganeshkumar, C., & David, A. (2022, August). Digital Information Management in Agriculture—Empirical Analysis. In *Proceedings of the Third International Conference on Information Management and Machine Intelligence: ICIMMI 2021* (pp. 243-249). Springer Nature Singapore.

Ganeshkumar, C., David, A., & Jebasingh, D. R. (2022). Digital transformation: artificial intelligence based product benefits and problems of Agritech industry. In *Agri-Food 4.0*. Emerald Publishing Limited. doi:10.1108/S1877-636120220000027010

Ganeshkumar, C., David, A., Sankar, J. G., & Saginala, M. (2023a). Application of Drone Technology in Agriculture: A Predictive Forecasting of Pest and Disease Incidence. In *Applying Drone Technologies and Robotics for Agricultural Sustainability* (pp. 50–81). IGI Global.

Ganeshkumar, C., Prabhu, M., Reddy, P. S., & David, A. (2020). Value chain analysis of Indian edible mushrooms. *International Journal of Technology*, *11*(3), 599–607. doi:10.14716/ijtech.v11i3.3979

Ganeshkumar, C., Sankar, J. G., & David, A. (2023b). Adoption of Big Data Analytics: Determinants and Performances Among Food Industries. [IJBIR]. *International Journal of Business Intelligence Research*, *14*(1), 1–17. doi:10.4018/IJBIR.317419

Goudey, A., & Bonnin, G. (2016). Must smart objects look human? Study of the impact of anthropomorphism on the acceptance of companion robots. [English Edition]. *Recherche et Applications en Marketing*, *31*(2), 2–20. doi:10.1177/2051570716643961

Grönroos, C. (1984). A service quality model and its marketing implications. *European Journal of Marketing*, *18*(4), 36–44. doi:10.1108/EUM0000000004784

Gursoy, D., Chi, O. H., Lu, L., & Nunkoo, R. (2019). Consumers acceptance of artificially intelligent (AI) device use in service delivery. *International Journal of Information Management*, *49*, 157–169. doi:10.1016/j.ijinfomgt.2019.03.008

Guzmán, I., & Pathania, A. (2016). *Chatbots in customer service*. Accenture. http://bit. ly/Accenture-Chatbots-Customer-Service

Hair, J. F., Harrison, D., & Risher, J. J. (2018). Marketing research in the 21st century: Opportunities and challenges. *Brazilian Journal of Marketing-BJMkt. Revista Brasileira de Marketing–ReMark*, (Special Issue), 17.

Hair Jr, J. F., Sarstedt, M., Ringle, C. M., & Gudergan, S. P. (2017). *Advanced issues in partial least squares structural equation modeling*. Sage publications.

He, D., Ai, B., Guan, K., García-Loygorri, J. M., Tian, L., Zhong, Z., & Hrovat, A. (2017). Influence of typical railway objects in a mmWave propagation channel. *IEEE Transactions on Vehicular Technology*, *67*(4), 2880–2892. doi:10.1109/TVT.2017.2782268

Huang, M. H., & Rust, R. T. (2018). Artificial intelligence in service. *Journal of Service Research*, *21*(2), 155–172. doi:10.1177/1094670517752459

Huang, M. H., & Rust, R. T. (2021). Engaged to a robot? The role of AI in service. *Journal of Service Research*, *24*(1), 30–41. doi:10.1177/1094670520902266

Hung, K., & Petrick, J. F. (2011). Why do you cruise? Exploring the motivations for taking cruise holidays, and the construction of a cruising motivation scale. *Tourism Management*, *32*(2), 386–393. doi:10.1016/j.tourman.2010.03.008

Io, H. N., & Lee, C. B. (2019). Understanding the Adoption of Chatbot: A Case Study of Siri. In *Advances in Information and Communication Networks: Proceedings of the 2018 Future of Information and Communication Conference (FICC)*, Vol. 1 (pp. 632-643). Springer International Publishing. 10.1007/978-3-030-03402-3_44

Ismagilova, E., Slade, E. L., Rana, N. P., & Dwivedi, Y. K. (2020). The effect of electronic word of mouth communications on intention to buy: A meta-analysis. *Information Systems Frontiers, 22*(5), 1203–1226. doi:10.100710796-019-09924-y

Ivanov, D., Dolgui, A., & Sokolov, B. (2019). The impact of digital technology and Industry 4.0 on the ripple effect and supply chain risk analytics. *International Journal of Production Research, 57*(3), 829–846. doi:10.1080/00207543.2018.1488086

Ivanov, S., & Webster, C. (2019b). What should robots do? A comparative analysis of industry professionals, educators, and tourists. In *Information and Communication Technologies in Tourism 2019: Proceedings of the International Conference in Nicosia,* (pp. 249-262). Springer International Publishing.

Ivanov, S. H., Webster, C., & Berezina, K. (2017a). Adoption of robots and service automation by tourism and hospitality companies. *Revista Turismo & Desenvolvimento (Aveiro), 27*(28), 1501–1517.

Jeganathan, G. S., & David, A. (2022). Determination of Hospitality Services Quality and Customer Satisfaction-A Holserv Approach. In *Au Virtual International Conference* (pp. 325-334).

Jeganathan, G. S., David, A., & Ganesh Kumar, C. (2022). Adaptation of Blockchain Technology In HRM. *Korea Review of International Studies,* 10-22.

Jordan, M. I., & Mitchell, T. M. (2015). Machine learning: Trends, perspectives, and prospects. *Science, 349*(6245), 255–260. doi:10.1126cience.aaa8415 PMID:26185243

Kalburgi, N. K., David, A., & Muralidhar, L. B. (2023). Understanding the Perceptions of Students towards YouTube as a Learning Tool-An Empirical Approach. *Central European Management Journal,* 2336-2693.

Kankhuni, Z., & Ngwira, C. (2022). Overland tourists' natural soundscape perceptions: Influences on experience, satisfaction, and electronic word-of-mouth. *Tourism Recreation Research, 47*(5-6), 591–607. doi:10.1080/02508281.2021.1878653

Kim, D. H., Park, G. M., Yoo, Y. H., Ryu, S. J., Jeong, I. B., & Kim, J. H. (2017). Realization of task intelligence for service robots in an unstructured environment. *Annual Reviews in Control, 44,* 9–18. doi:10.1016/j.arcontrol.2017.09.013

Kline, R. B. (2015). *Principles and practice of structural equation modeling.* Guilford publications.

Koc, E. (2020). Do women make better in tourism and hospitality? A conceptual review from a customer satisfaction and service quality perspective. *Journal of Quality Assurance in Hospitality & Tourism, 21*(4), 402–429. doi:10.1080/1528008X.2019.1672234

Kong, X., Wu, Y., Wang, H., & Xia, F. (2022). Edge Computing for Internet of Everything: A Survey. *IEEE Internet of Things Journal, 9*(23), 23472–23485. doi:10.1109/JIOT.2022.3200431

Kumar, B. K., Vijayalakshmi, G., Krishnamoorthy, A., & Basha, S. S. (2010). A single server feedback retrial queue with collisions. *Computers & Operations Research, 37*(7), 1247–1255. doi:10.1016/j.cor.2009.04.019

Lai, I. K. W., Liu, Y., & Lu, D. (2021). The effects of tourists' destination culinary experience on electronic word-of-mouth generation intention: The experience economy theory. *Asia Pacific Journal of Tourism Research*, *26*(3), 231–244. doi:10.1080/10941665.2020.1851273

Lakshman, K., & David, A. (2023). *Senior Citizens' Perceptions on E-banking Services*. Exceller Books.

Lasek, M., & Jessa, S. (2013). Chatbots for customer service on hotels' websites. *Information Systems Management*, 2.

Lester, R. K., & Piore, M. J. (2004). *Innovation—The missing dimension*. Harvard University Press. doi:10.4159/9780674040106

Lu, V. N., Wirtz, J., Kunz, W. H., Paluch, S., Gruber, T., Martins, A., & Patterson, P. G. (2020). Service robots, customers and service employees: What can we learn from the academic literature and where are the gaps? *Journal of Service Theory and Practice*, *30*(3), 361–391. doi:10.1108/JSTP-04-2019-0088

Luo, J., Wong, I. A., King, B., Liu, M. T., & Huang, G. (2019). Co-creation and co-destruction of service quality through customer-to-customer interactions: Why prior experience matters. *International Journal of Contemporary Hospitality Management*, *31*(3), 1309–1329. doi:10.1108/IJCHM-12-2017-0792

Madaan, G., Swapna, H. R., Kumar, A., Singh, A., & David, A. (2021). Enactment of sustainable technovations on healthcare sectors. *Asia Pacific Journal of Health Management*, *16*(3), 184–192. doi:10.24083/apjhm.v16i3.989

Makadia, J., Pashchapur, R., & Dhulasawant, T., & PY, D. R. (2020, July). Autonomous Flight Vehicle Incorporating Artificial Intelligence. In *2020 International Conference on Computational Performance Evaluation (ComPE)* (pp. 419-426). IEEE. 10.1109/ComPE49325.2020.9200061

Manzoor, F., Wei, L., Hussain, A., Asif, M., & Shah, S. I. A. (2019). Patient satisfaction with health care services; an application of physician's behavior as a moderator. *International Journal of Environmental Research and Public Health*, *16*(18), 3318. doi:10.3390/ijerph16183318 PMID:31505840

Martínez, J. A., & Martínez, L. (2010). Some insights on conceptualizing and measuring service quality. *Journal of Retailing and Consumer Services*, *17*(1), 29–42. doi:10.1016/j.jretconser.2009.09.002

McKechnie, S. (2011). Consumer confidence in financial services after the crunch: New theories and insights. *International Journal of Bank Marketing*, *29*(2). Advance online publication. doi:10.1108/ijbm.2011.03229baa.001

McLean, G., Osei-Frimpong, K., Wilson, A., & Pitardi, V. (2020). How live chat assistants drive travel consumers' attitudes, trust and purchase intentions: The role of human touch. *International Journal of Contemporary Hospitality Management*, *32*(5), 1795–1812. doi:10.1108/IJCHM-07-2019-0605

Meuter, M. L., Ostrom, A. L., Roundtree, R. I., & Bitner, M. J. (2000). Self-service technologies: Understanding customer satisfaction with technology-based service encounters. *Journal of Marketing*, *64*(3), 50–64. doi:10.1509/jmkg.64.3.50.18024

Meyer-Waarden, L., Pavone, G., Poocharoentou, T., Prayatsup, P., Ratinaud, M., Tison, A., & Torné, S. (2020). How service quality influences customer acceptance and usage of chatbots? *SMR-Journal of Service Management Research*, *4*(1), 35–51. doi:10.15358/2511-8676-2020-1-35

Mithas, S., Krishnan, M. S., & Fornell, C. (2005). Why do customer relationship management applications affect customer satisfaction? *Journal of Marketing*, *69*(4), 201–209. doi:10.1509/jmkg.2005.69.4.201

Mittal, B., & Lassar, W. M. (1996). The role of personalization in service encounters. *Journal of Retailing*, *72*(1), 95–109. doi:10.1016/S0022-4359(96)90007-X

Moussawi, S. (2016). *Investigating personal intelligent agents in everyday life through a behavioral lens*. City University of New York.

Mukhopadhyay, S., Pandey, R., & Rishi, B. (2022). Electronic word of mouth (eWOM) research–a comparative bibliometric analysis and future research insight. *Journal of Hospitality and Tourism Insights*.

Murphy, A., & Liszewski, B. (2019). Artificial intelligence and the medical radiation profession: How our advocacy must inform future practice. *Journal of Medical Imaging and Radiation Sciences*, *50*(4), S15–S19. doi:10.1016/j.jmir.2019.09.001 PMID:31611013

Mustak, M., Salminen, J., Plé, L., & Wirtz, J. (2021). Artificial intelligence in marketing: Topic modeling, scientometric analysis, and research agenda. *Journal of Business Research*, *124*, 389–404. doi:10.1016/j.jbusres.2020.10.044

Neuhofer, B., Magnus, B., & Celuch, K. (2021). The impact of artificial intelligence on event experiences: A scenario technique approach. *Electronic Markets*, *31*(3), 601–617. doi:10.100712525-020-00433-4

Nihmathullah, Z., Ramasamy, R., & Raj David, A. (2022). *Event Impact Assessment: A Case of Puducherry*. Book Rivers.

Nilashi, M., Abumalloh, R. A., Alghamdi, A., Minaei-Bidgoli, B., Alsulami, A. A., Thanoon, M., Asadi, S., & Samad, S. (2021). What is the impact of service quality on customers' satisfaction during COVID-19 outbreak? New findings from online reviews analysis. *Telematics and Informatics*, *64*, 101693. doi:10.1016/j.tele.2021.101693 PMID:34887617

Noor, N., Hill, S. R., & Troshani, I. (2022). Developing a service quality scale for artificial intelligence service agents. *European Journal of Marketing*, *56*(5), 1301–1336. doi:10.1108/EJM-09-2020-0672

Nunnally, J. C. (1978). An overview of psychological measurement. *Clinical diagnosis of mental disorders: A handbook*, 97-146.

Parasuraman, A., Berry, L. L., & Zeithaml, V. A. (1993). More on improving service quality measurement. *Journal of Retailing*, *69*(1), 140–147. doi:10.1016/S0022-4359(05)80007-7

Parasuraman, A., Zeithaml, V. A., & Berry, L. L. (1985). A conceptual model of service quality and its implications for future research. *Journal of Marketing*, *49*(4), 41–50. doi:10.1177/002224298504900403

Parasuraman, A., Zeithaml, V. A., & Berry, L. L. (1994). Alternative scales for measuring service quality: A comparative assessment based on psychometric and diagnostic criteria. *Journal of Retailing*, *70*(3), 201–230. doi:10.1016/0022-4359(94)90033-7

Parasuraman, A., Zeithaml, V. A., & Malhotra, A. (2005). ES-QUAL: A multiple-item scale for assessing electronic service quality. *Journal of Service Research*, *7*(3), 213–233. doi:10.1177/1094670504271156

Parasuraman, R., Greenwood, P. M., & Alexander, G. E. (1995). Selective impairment of spatial attention during visual search in Alzheimer's disease. *Neuroreport: An International Journal for the Rapid Communication of Research in Neuroscience.*

Pillai, R., & Sivathanu, B. (2020). Adoption of AI-based chatbots for hospitality and tourism. *International Journal of Contemporary Hospitality Management, 32*(10), 3199–3226. doi:10.1108/IJCHM-04-2020-0259

Poole, D. L., & Mackworth, A. K. (2010). *Artificial Intelligence: foundations of computational agents.* Cambridge University Press. doi:10.1017/CBO9780511794797

Prentice, C., Dominique Lopes, S., & Wang, X. (2020). The impact of artificial intelligence and employee service quality on customer satisfaction and loyalty. *Journal of Hospitality Marketing & Management, 29*(7), 739–756. doi:10.1080/19368623.2020.1722304

Rajan, K., & Saffiotti, A. (2017). Towards a science of integrated AI and Robotics. *Artificial Intelligence, 247*, 1–9. doi:10.1016/j.artint.2017.03.003

Ramachandran, K. K., Mary, A. A. S., Hawladar, S., Asokk, D., Bhaskar, B., & Pitroda, J. R. (2022). Machine learning and role of artificial intelligence in optimizing work performance and employee behavior. *Materials Today: Proceedings, 51*, 2327–2331. doi:10.1016/j.matpr.2021.11.544

Ravi, S., David, A., & Imaduddin, M. (2018). Controlling & calibrating vehicle-related issues using RFID technology. *International Journal of Mechanical and Production Engineering Research and Development, 8*(2), 1125–1132. doi:10.24247/ijmperdapr2018130

Rita, P., Oliveira, T., & Farisa, A. (2019). The impact of e-service quality and customer satisfaction on customer behavior in online shopping. *Heliyon, 5*(10), e02690. doi:10.1016/j.heliyon.2019.e02690 PMID:31720459

Rosenbaum, M. S., & Russell-Bennett, R. (2020). Service research in the new (post-COVID) marketplace. *Journal of Services Marketing, 34*(5), I–V. doi:10.1108/JSM-06-2020-0220

Russell, S., & Norvig, P. (2021). Artificial intelligence: a modern approach. Global Foundations, 19, 23.

Rust, R. T., & Oliver, R. L. (Eds.). (1993). *Service quality: New directions in theory and practice.* Sage Publications.

Seth, N., Deshmukh, S. G., & Vrat, P. (2005). Service quality models: A review. *International Journal of Quality & Reliability Management, 22*(9), 913–949. doi:10.1108/02656710510625211

Sheehan, K. B. (2018). Crowdsourcing research: Data collection with Amazon's Mechanical Turk. *Communication Monographs, 85*(1), 140–156. doi:10.1080/03637751.2017.1342043

Shi, Y., Prentice, C., & He, W. (2014). Linking service quality, customer satisfaction and loyalty in casinos, does membership matter? *International Journal of Hospitality Management, 40*, 81–91. doi:10.1016/j.ijhm.2014.03.013

Singh, A. (2018). Facebook, WhatsApp, and Twitter: Journey towards Education. *SOSHUM: Jurnal Sosial Dan Humaniora, 8*(2), 139–149. doi:10.31940oshum.v8i2.987

Sternberg, R. J. (2005). Creativity or creativities? *International Journal of Human-Computer Studies*, *63*(4-5), 370–382. doi:10.1016/j.ijhcs.2005.04.003

Sudhakar, B. D., Kattepogu, N., & David, A. (2017). Marketing assistance and digital branding-an insight for technology up-gradation for MSME's. *International Journal of Management Studies & Research*, *5*(1), 2455–1562.

Sweezey, M. (2020). *The Context Marketing Revolution: How to Motivate Buyers in the Age of Infinite Media*. Harvard Business Press.

Thummula, E., Yadav, R. K., & David, A. (2019). A cost-effective technique to avoid communication and computation overhead in vehicle insurance database for online record monitoring. [IJMPERD]. *International Journal of Mechanical and Production Engineering Research and Development*, *9*(2), 711–722.

Tian, W., Liu, Y. J., Liu, B., He, D., Wu, J. Y., Han, X. G., Zhao, J., & Fan, M. (2019). Guideline for thoracolumbar pedicle screw placement assisted by orthopaedic surgical robot. *Orthopaedic Surgery*, *11*(2), 153–159. doi:10.1111/os.12453 PMID:31025807

Torres, E. N., Adler, H., & Behnke, C. (2014). Stars, diamonds, and other shiny things: The use of expert and consumer feedback in the hotel industry. *Journal of Hospitality and Tourism Management*, *21*, 34–43. doi:10.1016/j.jhtm.2014.04.001

Tung, V. W. S., & Law, R. (2017). The potential for tourism and hospitality experience research in human-robot interactions. *International Journal of Contemporary Hospitality Management*, *29*(10), 2498–2513. doi:10.1108/IJCHM-09-2016-0520

Tussyadiah, I. (2020). A review of research into automation in tourism: Launching the Annals of Tourism Research Curated Collection on Artificial Intelligence and Robotics in Tourism. *Annals of Tourism Research*, *81*, 102883. doi:10.1016/j.annals.2020.102883

Ukpabi, D., Karjaluoto, H., Olaleye, S. A., & Mogaji, E. (2018). Dual perspectives on the role of artificially intelligent robotic virtual agents in the tourism, travel and hospitality industries. In *EuroMed Academy of Business Conference Book of Proceedings*. EuroMed Press.

Wirtz, J., Patterson, P. G., Kunz, W. H., Gruber, T., Lu, V. N., Paluch, S., & Martins, A. (2018). Brave new world: Service robots in the frontline. *Journal of Service Management*, *29*(5), 907–931. doi:10.1108/JOSM-04-2018-0119

Wuenderlich, N. V., & Paluch, S. (2017). *A nice and friendly chat with a bot: User perceptions of AI-based service agents*. Semantic Scholar.

Xiao, L., & Kumar, V. (2021). Robotics for customer service: A useful complement or an ultimate substitute? *Journal of Service Research*, *24*(1), 9–29. doi:10.1177/1094670519878881

Xu, Y., Shieh, C. H., van Esch, P., & Ling, I. L. (2020). AI customer service: Task complexity, problem-solving ability, and usage intention. *Australasian Marketing Journal*, *28*(4), 189–199. doi:10.1016/j.ausmj.2020.03.005

Yu, C. E. (2020). Humanlike robots as employees in the hotel industry: Thematic content analysis of online reviews. *Journal of Hospitality Marketing & Management*, 29(1), 22–38. doi:10.1080/1936862 3.2019.1592733

Zaibaf, M., Taherikia, F., & Fakharian, M. (2013). Effect of perceived service quality on customer satisfaction in hospitality industry: Gronroos' service quality model development. *Journal of Hospitality Marketing & Management*, 22(5), 490–504. doi:10.1080/19368623.2012.670893

Chapter 8
Territorial Identities and Gastronomy Tourism in the South Danube Region:
Case Study of Fish Soup Brewing Tradition

Mihály László Vörös
 https://orcid.org/0000-0003-3471-5998
Edutus University, Hungary & HELIA Research Group, Hungary

Aleš Gačnik
Faculty of Tourism Studies, University of Primorska, Slovenia

ABSTRACT

The research study comprised by the book chapter investigates territorial identities and diversified features of the South-Danube region and presents a gastronomy tourism case study of an old culinary tradition of brewing and eating fish soup. This is not only a dish consumed with great frequency in the diet of local and regional residents but creates food offering and cooking demonstration of a summer gastronomy tourism festival organized every year in the town of Baja. In addition, this fish-dish is one of the most popular meals offered in almost every local and neighborhood restaurant's menu for tourist guests. The study covers short analysis on the healthiness of this dietary custom and highlights that this culinary feast and gastronomy tourism attraction also became a brand and cultural heritage which can contribute to enhance the image of the place and to promote sustainable development of gastronomy tourism.

INTRODUCTION

The research study comprised by the book chapter investigates territorial identities and diversified features of the South-Danube region and presents a gastronomy tourism case study of an old culinary tradition of brewing and eating fish soup. This is not only a dish consumed with great frequency in the diet of local and regional residents but creates food offering and cooking demonstration of a summer gastronomy

DOI: 10.4018/978-1-6684-6985-9.ch008

festival organized every year in the town of Baja. In addition, this fish-dish is one of the most popular meal offered in almost every local and neighborhood restaurant's menu for tourist guests. The research highlights that this culinary feast and gastronomy tourism attraction also became a brand which can contribute to enhance the image of the place and to drive sustainable development of gastronomy tourism. Based on reviewing international literatures the study also covers short analysis on the beneficial health effects of the nutrition components and healthiness of this dietary custom and culinary feast which belongs well-being of people living in the region and of tourists as well.

The chapter is divided on four main parts. After a short introduction the first part presents the material and methodology of the research. The second part contains a short theoretical introduction to the concepts of territoriality and territorial identity then demonstrates the diversified features and characteristics of South Danube region in connection with the case study theme and location. The third, the main part, presents the case study on the healthy pepper fish-dish brewing tradition and festival. Finally, the main attributes new findings and novelty of the research are summarized in a fourth conclusion section.

Material and Methodology

This research study aims exploring food and gastronomy traditions as relevant cultural heritages and tourism attractions to reveal the diversified relations and synergy between territory, territorial and local food identities and sustainable gastronomy tourism.

The methodology of the research includes qualitative methods of independent observation with participation (ethnographic/ethnological method) jointly with case study research method utilizing own ethnographic experiences and observations achieved in field research as well as the technique of creative thinking supported by economic and regional science methodologies.

The case study created by the research demonstrates the history, brewing traditions and consumption culture of a typical pepper fish-dish, called fisherman's soup, fish soup, or fish paprikás originating from and existing in the South – Danube region.

In the exploration of the relevant territorial features which have strong links with the regional or local food and gastronomy heritage and culture the research followed the guidelines of the "active territoriality" concept (Pollice, 2003). The selection of the region and topic of the case study research and applying ethnographic methods (Banini[1], 2017).was based on a broad and thorough analysis of the literature

This research doesn't contain comparison between selected gastronomy tourism examples but the authors, partners from Hungary and Slovenia, generated new knowledge and experiences by implementing joint field research on fish-dish gastronomy traditions and published joint research findings on comparing agricultural products, achieved different domestic or EU quality brands, utilized as components of gastronomic offers and heritage (Gačnik &Vörös 2018).

TERRITORIAL IDENTITIES AND CHARACTERISTICS OF SOUTH-DANUBE REGION

The defining element of the region's geography is the River Danube, the second longest river in Europe, after the Volga, rising in the Black Forest mountains of western Germany and flows to its mouth on the Black Sea in Romania. On its 2,872-kilometer way it runs singularly in a horizontal manner, in the direction of West to East. Along its course it passes through Germany, Austria, Slovakia, Hungary,

Croatia, Serbia, Bulgaria, Romania and at its lower end it creates the border of Moldova and Ukraine. The Hungarian part of South-Danube region includes Tolna, Baranya and Bács-Kiskun counties which are among Hungary's nineteen counties established after World War II in 1950 under the control of the administrative territorial reform forced by USSR in East European countries. Town Baja, the location of the gastronomy tourism case study belongs to Bács-Kiskun county. Going to the south crossing the Hungarian border there are two sub-regions: Osijek-Baranja County in Croatia located in the east from Danube and the West Bačka District which is one of seven administrative districts of the autonomous province of Vojvodina in Serbia with its administrative seat of the municipality Sombor. Going along the Danube from Baja and Mohács we found further populated settlements (Bezdan, Apatin) in Serbia.

The Concept of Territorial Identities

The essence of territory is partially physical substrate (soil, arable land, water, forestry, biodiversity, renewable energy, not renewable resources, landscape, buildings, infrastructures, etc.), on the other hand it has also different historical, cultural, traditional as well as economic and social aspects.

"The territory can be regarded as that portion of geographical space which reflects a given community and represents the community's individual and collective actions. The territory's specificity – regarded as the difference from the geographic surroundings – comes from the process of interaction between this community and the environment" (Pollice, 2003 p. 108).

"Territoriality can be defined as a responsible, active and participatory way of inhabiting places, so that territorial identity can be considered an 'active territoriality' aware of the material and relational value of the territory, directed to the well-being of both local communities and ecosystems" (Banini, 2017 p. 18). The same research study defines the concept of territorial identity: "…as a process of social construction, open and dynamic, through which the communities settled in a given territory, choose the distinctive features of the territory they inhabit or where they act in, shaping shared values, solutions, actions, and future trends" (Banini, 2017 p. 18).

The thoughts of Jacinthe Bessière (1998) illustrate and explain below very well the relationship between territories and local historical cultural identities highlighting also how local cuisine becomes a culinary heritage:

"Local economic development also raises the problem of defining territories. Indeed, as actors define their heritage and combine efforts in relation to the resources available, a territorial construction often develops on the basis of terroirs. The term terroir refers to a specific area with an outspoken cultural and historical identity. It includes the accumulation and transmission of local know-how. This is how we come to speak of local cultural produce and local cuisine (....).Turning to local development as a territorial construction process endowed with both local co-operation and a collective legacy, culinary heritage may be used as a means to boost development" (Bessière, 1998 p. 31).

These findings also provide a fairly instructive and multi-faced interpretation of brewing traditions as specific resources and its utilization for building the image of a place and local development (Bessière, 1998; Bannini & Pollice 2015). Beyond that it brings us close to understanding the term gastronomy tourism which finds its origin in the traveller's or tourists desire to discover, among other attractions, different tastes, flavours of gastronomy products created by local ingredients and recipes and tasting locally sourced typical quality or *origin food* products, as special values of a territory belonging to certain rural region:

"origin food (…) serves to define any foodstuff that people perceive to have some added value(s) because of its place of origin. The main actors of food originality are as follows:

Typicity: place-specific peculiarities of the production process and the final product;

Territoriality: degree of physical connection with the place of origin;

Traditionality: rootedness of an origin food history in its place of origin, including eating and gastronomic culture;

Communality: shared experience and practices, reflected in the presence of multiple producers (farmers, processors) and their collaboration." (van der Meulen, 2007 p.5).

We can conclude that the identification of a country or a rural territory or region in a county is associated with the name of a food or a special dish, as a result of local traditional gastronomy. Local food production is supported by the fact that regional and local characteristics such as certain raw materials and dishes, tastes, as well as food traditions and food culture might always be closely tied to particular regions creating and maintaining the identity of these regions. Local food chains can be considered a relevant pillar of territorial management and promotion (Vörös & Gemma 2011).

The short study of Duhart, F. (2020) defines and categorizes the territorial food identities, recommended to the attention of gastronomy actors, might also be interesting to consider: "A territorial food identity characterised by a set of cultural markers that the majority of the inhabitants of a territory consider as indigenous to their culture and the foreigners identify as typical of this part of the world or frankly exotic. These markers are various: products, recipes, culinary techniques, consumption patterns, table manners, food preferences or representations." (Duhart, 2020 p. 15). In the context of gastronomy tourism, we can divide the territorial food identity structure into the following categories:

(1) *District identity*, characterised by an absolute singularity marker e.g. a freshly baked bread or pastry with its unique taste which has been regularly produced and sold by a local baker-master in a certain district of a city or a town for customers buying these products every day because of its local special quality marker;

(2) *Local identity*, characterised by local unifying markers e.g. a traditional pepper fish-dish closely related to the dietary tradition and culture of a small town's inhabitants who have been preserved and honor this gastronomy tradition and insist on them for long time (see Case Study Section in this book chapter).

(3) *Regional identities*, characterised by intermediary unifying elements e.g. 'Kalocsai' Red Ground Pepper a regional seasoning as processed food product in Hungary which was added to the collection and database entitled "Traditions-Tastes-Regions" (TTR)[2] and the product achieved also a TTR trademark (HMOARD 2000; Török, 2019)2. Raw materials for processing this product are grown in areas of Danube and Tisza connections and in the near Danube territories of Sárköz;

(4) *National* (macro-regional or supra-national) *identities*, characterised by extremely federative elements e.g. The Collection of Hungarikum in Hungary (Hungarikum)[3] which has separate categories of intellectual achievements and material assets of the agriculture - including the field of forestry, fishing, hunting and animal health - with a special attention to agricultural products (food, winery, animal and plant species) as well as of tourism and hospitality heritages (Duhart, 2020; Vörös & Gemma 2015; Gačnik &Vörös 2018)

At the most local levels of identity, strong originality markers play a crucial role. At superior levels, federative elements are much more important. It can be emphasized that this territorial food identity structure allows to act more effectively when it comes to support local development by gastronomic initiatives or to protect and enhance a food heritage. This can be enforced at all identity levels.

Du Rand and Heath (2006) determined the components of *food tourism* as follows:

"The roots of food tourism lie in agriculture, culture and tourism (…) All three components offer opportunities and activities to market and position food tourism as an attraction and experience in a destination. Agriculture provides the product, namely, food; culture provides the history and authenticity; and tourism provides the infrastructure and services and combines the three components into the food tourism experience. These three components form the basis for the positioning of food tourism as one of the components in the tourism paradigm." (Du Rand &Heath, 2006 p. 209)

Du Rand and Heath (2006) also observed that food tourism globally remains a form of *niche or alternative tourism* generally linking to cultural or heritage tourism and although it forms an important component of tourism, is still very much a less promoted attraction in many countries.

Based on the above research findings and other studies (Mackenzie, 2019) we can continue to conclude that gastronomy tourism represents a strategic cross-sectoral intersection among agriculture, culture and tourism infrastructure. Agriculture provides special, typical, quality or origin food and terroir-origin wine, many of them are special culinary ingredients which are utilized in local gastronomy to provide authentic hospitality service for tourists. Culture comprises history, food production and processing traditions which guarantee and add history and authenticity to the gastronomy tourism services. Tourism resources ensure the proper infrastructure, supplementary services and management to offer opportunities and activities to be marketed as well as combine and aggregate the three components and attractions into a special, outstanding gastronomy tourism experience. Sustainable gastronomy tourism does not mean conservation the place and time, but rather that inhabitants suppliers and tourists as well can coexist with the heritage of gastronomy. Sustainable gastronomy tourism represents the culture of holistic coexistence with the heritage of gastronomy in the contemporary world (Gačnik, 2012).

The UNWTO jointly with its affiliate member, the Basque Culinary Centre (BCC) created guidelines for the development of gastronomy tourism: "Gastronomy tourism forms an integral part of local life and has been forged by the history, culture, economy and society of a territory (…) Gastronomy tourism is therefore based on a concept of knowing and learning, eating, tasting and enjoying the gastronomic culture that is identified with a territory. It is not possible to talk of gastronomy tourism without also talking about the culinary identity of the terroir as a distinguishing feature." (UNWTO – BCC, 2019 pp. 8-9)

Relaying on analogy the territory can be considered an "open book" which has been written by an "invisible ink" not readable in usual way (Ciani &Vörös 2020). A reasonable solution must be found to make this "book" readable for everyone based on a sophisticated, innovative and smart approach and vision. The territorial, regional and tourism research as well as territorial or regional planning and programming activities must be strongly promoted and supported by an appropriate and widespread use of advanced and innovative information and communication technologies (ICT). By structuring and content formulating the description of territory with its diversified values, heritages we can achieve bringing it "to come to life", namely showing and representing (attracting, speaking, touching, or even singing, smelling, tasting flavors etc.) in order to inspire and stimulate feelings, catalyzing creativity and arouse inventiveness.

Demonstration of the Regional Identities and Features

The waters of the Danube river reflect 3,000 years, in which the histories of the peoples at the river, of their societies and culture are testified by sources. Concerning prehistory and early history of Danube historians are mainly based on archaeological sources but for the Roman period and especially for the modern and contemporary age a chronologically growing abundance of written documents and evidence are available (Weithmann, 2011 pp. 9-10). At the beginning of the modern age the Danube was the main waterway of the freight transport network in the Carpathian Basin and gave support booming agricultural sector along with industrial and military facilities (Horvath, 2011 p. 86). Evidence of built-in heritage, monuments and historical buildings, as relevant tourism attractions can be found along the river (e.g. Vienna, Bratislava, Budapest and Belgrade). The natural potential determines the similarities between the Upper and Lower part of Danube but there is a considerable difference between the level of their economic and regional development.

An important feature of the Lower part of Danube and South-Danube region is the colourful ethnic composition. During the 17th–18th centuries huge areas in the Danube valley and in Pannonian plain had been devastated and depopulated because of the prolonged war between the Habsburg monarchy and the Ottoman Empire. After the end of Hungary's Ottoman Turkish occupation, a number of German settlers from different parts of Germany emigrated to Hungary in scope of an organized and supported resettlement program.The majority of the first settlers were Swabians boarded boats in Ulm, Swabia, and travelled to their new destinations down the Danube River. These boats called "Ulmer Schachtel" in German. Despite differing origins, the new immigrants were all referred as Swabians by their neighbours (Croats, Serbs, Hungarians, and Romanians).

The Danube Swabians (in German language called "Donau Schwaben") mostly lived in isolated and small rural communities and had an occupation in agriculture, fishing or shipping as well as in water milling or grain grinding in Danube river. The main territory of German settlers before World War I and II called the historical Bács-Bodrog County. Recently the north part of it belongs to Hungary where Baranya, Tolna and Bács-Kiskun counties are located. The south part belongs either to Croatia with its Osijek-Baranja County or to Serbia with the West Bačka District. The history of German/Swabian settlers came to a very sad fate during and after World War II. Great majority of this very creative and active ethnic group was either killed or deported to labour camp. And then after the conclusion of peace treaties they were expelled from the country (Kentish, 2019). Only small minority of them were lucky to escape but despite this the ethnic minority is still present even in the "shrunk" Hungarian counties. Therefore, in order to make our case study, based on partly ethnographic sources (Bredetzky, 1807; Erdei, 1971; Solymos, 1997; Szilágyi, 1997; Pusztai, 2007; Sümegi, 2019; Fejes, 2019), more understandable and more established it is important to note, that their occupation, lifestyle, food consumption patterns and table manners linked to the Danube river during the history have played a relevant role in the development of local food and gastronomic heritages (e.g. pepper sausages and fish- dishes) in all over South-Danube region.

The development potentials and competitiveness of tourism along the Middle and Lower Danube are very different and highly depending on several internal and external factors. Limited or lacking funds for investments have worsened mainly the conditions of the Eastern part of the region. An essential problem among the limitations of regional and inter-regional tourism development can be traced back to the area distribution and unfavorable political conditions aroused by the situation after World War II resulting in isolation and lack of regional cooperation between South-Danube countries concerned. This problem has been wisely realized by the EU politicians and decision makers who initiated to elaborate and implement the EU Strategy for the Danube Region (EUSDR) across national borders throughout the European territory since June 2011 (Datourway, 2011; EC-EUSDR, 2019). Concerning our case study research this strategy comprises relevant objectives to provide opportunity to promote cross-border cooperation between sub-regions, states, counties and communities. Thereby to support the integration of South-Eastern countries into the EU.

Tolna county is located in central Hungary along the west bank of the Danube. In the east, the Danube forms its common border with Bács-Kiskun county. Concerning its ethnic composition, the majority of the inhabitants are Hungarians (90.6%) and the main minorities are the Germans (4.7%) and Roma (4.1%). The most significant tourist attractions and destinations are around its county seat town of Szekszárd, with its historical wine region, and along the Danube. An important ecotourism attraction is the Germenc-Gyulaj wildlife reserve within the Danube-Dráva National Park and the fishing sites in the oxbows of the Danube. The natural assets in the region are partly form a basis for developing eco-tourism and partly exploited by hunting tourism. A unique monument, the Basilica of the Holy Blood can be found in municipality Báta located in southernmost part of the county on the west bank of the Danube. The neo-Romanesque style church with a single tower design is the only Holy Blood Church in today's Hungary. The Benedictine Abbey of Báta was built by King Saint László on a prominent point in the Trans-Danube hills overlooking the Danube basin, to a terrace already established by the Celts, which could only be approached from a single point. A relevant fact rarely mentioned in tourist guides is that the most southern part of the Abbey of Báta had been located in Apatin. The church had been one of the most important pilgrimage sites and monasteries of the medieval Kingdom of Hungary (Sümegi, 2019). Other important cultural, wine and gastronomy tourism attractions can be found in Tolna and Szekszárd.

Bács-Kiskun county is the largest county of Hungary located along the east bank of the Danube. Its ethnic composition is almost homogeneous, Hungarians create majority (93.9%), the ratio of minorities are much smaller: besides some Germans (2.0%) and Roma (2.3%) there are only few Croats and Serbs mainly living around Hajós (a "cellar village" with a number of vineyards and wineries owned by Swabian settlers). The county's tourism potentials are favourable owing to the product groups highlighted in the national strategy: medical, wellness and active tourism, cultural and heritage tourism, rural tourism, as well as gastronomy and wine tourism. Guest traffic is realised mainly in Bugac area located in the centre of the county near to Kecskemét and proximity of the Danube. The "Family-friendly Fish Soup Festival of Baja" a summer gastronomy event held in every second weekend of July has been becoming the most well-known festival not only among domestic tourists but more and more for international guests as well (see in Case Study section). Kecskemét and Baja both have sufficient receiving capacity in case of festival programmes that require high hotel capacity. Other popular target areas are nature conservation areas, thermal baths and sightseeing tours in Kecskemét, Kalocsa and Baja.

The availability of quality food and wine products produced and sold locally or utilized as culinary ingredients in restaurants and catering services have very important role to improve authenticity and sustainability of gastronomy tourism and enhance gourmet experience everywhere and in this region as well. Based on successful operation of TTR program (see in concepts section and endnote 2) in Hungary (Vörös&Gemma 2015;Török, 2019) there are a number of wine, spirits and food products produced or grown in the areas of these counties achieved certain EU quality protection brand[4] after they passed domestic certification.

The Town Baja is built on the banks of Danube river on the south-western corner of the county Bács-Kiskun, located on the border of the Great Plain and the Trans-Danube territories, owning a mixture of their geographical characteristics. Its regional historical and cultural proximity with Tolna county also has importance in the contexts of our case study. The tributary Sugovica (Kamarás-Danube) and the artificial Ferenc Canal have always been determined by the presence of natural water. Opposite of the central main square of the town Petőfi Island is located which separated by the Ferenc Canal from Nagypandúr Island. One of the most important feature of Baja's cultural and historical heritage is the colourful ethnic composition. According to local ethnographic sources (Solymos, 1997, Pusztai, 2007) during Turkish invasion period (16th–17th centuries) Serbs were settling in the place, then organised population movements brought new inhabitants: Catholic Bunjevci (in Hungarian: bunyevác) and German, Swabian settled down here. On 24th December 1696 the town status of Baja had been promulgated as a result of the request of the Serbs which created majority of the population that time. Commemorating this day, the coat of arms of the town depicts Adam and Eve. The multi-ethnic composition of the town changed in the course of the 19th century: Hungarians made up 75% of the population at the end of the century and Serb population fell to less than 1%. According to historical sources referred (Solymos, 1997) the explanation of the disappearance of Serbs from Baja might be a reason of the decline of commercial activities. During its history Baja was the centre of the northern part of the historical Bács-Bodrog county. After the World War I and Paris Peace Treaty this northern part of the historical county remained part of Hungary and the town of Baja became the seat of the so called 'broken' Bács-Bodrog county. Then the period after the World War II brought further changes in the position of the town: Baja and the remains of Bács-Bodrog county was then merged into the new county of Bács-Kiskun and Baja became the second largest town of the county after the county seat, Kecskemét. The border of Serbia in the south is very near and it is only 75 km distance for traveling to Apatin.

In scope of the EU pilot project entitled "Béda-Karapancsa" (Datourway 2011) thorough and detailed analyses has been performed to explore the ecotourism development potentials focusing on Béda-Karapancsa national protection area in Baranya county as well as on North-Eastern Croatia and North-Western Serbia of South-Danube region. With its more than 10.000 hectares territory Béda-Karapancsa is located at the Southern part of the Danube section of Duna-Drava National Park in the proximity of the town of Mohács. It consists mainly river floodplain forests, water bodies and marshes. The more than one 120-year old oak woods and the wetland habitats are particularly valuable parts of the area providing home for a number of strictly protected birds, resting places for migrating birds and in the water bodies there are many protected fish species and orchid species. This Danube territory is valuable scenery and has unique potential for ecotourism purposes jointly with gastronomy tourism which has close relationship with the local ecosystem and we are convinced that there must be considerable joint interest between the target groups of eco-tourists and gastronomy tourists. Concerning ecotourism potentials of cross-border sub-regions in Croatia and Serbia some important arguments and conclusions can be summarized as follows (Datourway, 2011)

In North-Eastern Croatia, the study area around Kopački Rit nature park covered one town area and four municipalities in Osijek-Baranja county. Three different types of environment can be consisted here: The town of Osijek with its close surroundings as populated urban area and the remaining area covered by the four municipalities of Draž, Kneževi Vinogradi, Bilje and Erdut, which is partly agricultural and rural and partly preserved its natural condition. Natural areas are mainly located near two big rivers Danube and Drava, with the largest area covered by Kopački rit nature park. Drava river creates border river between Croatia and Hungary, dividing two historical provinces Baranya county in the north and Slavonia in the south.

Concerning North-Western part of Serbia, the Gornje Podunavlje Special Nature Reserve is located in the West Bačka District in province of Vojvodina, along the left bank of the Danube. It comprises numerous meanders, dead river branches, canals, ponds, swamps and marshes formed as a result of continuous river dynamics, which is preserved in large part of reserve. It is bordered with two other protected areas: with Kopacki Rit in Croatia and Danube-Drava National park in Hungary with which it forms the almost 70,000 ha large Central Danube Floodplain Area. The natural reserve is very near to the state borders and easily accessible for tourist visitors. In close surroundings of the reserve, main urban centres Sombor and Apatin are in Serbia, Osijek in Croatia and Baja in Hungary.

The Transboundary UNESCO Biosphere Reserve according to "Mura-Drava-Danube" concept (Trisic et al., 2022) defines about 300,000 hectares of core and buffer zones and around 700,000 hectares of transition zones (Figure 1).

This large, primarily marshy area is rich in natural and cultural heritage. This area is called "the Amazon of Europe" and represents the world's first biosphere reserve among five countries. This biosphere reserve is important for the preservation of the natural and cultural values of the entire region. Especially rare plant and animal representatives live in this unique geographical unity. Some species inhabit only this area. Considering historical-cultural and environmental background of neighbouring cross-border regions of Hungary, Croatia and Serbia in South-Danube area it would be possible to integrate the tourism resources based on mutual exploitation of advantages, mobilizing resources and creating a recognisable tourism brand, a river destination in the Béda-Karapancsa region as well (Trisic et al., 2022).

CASE STUDY: FISH SOUP BREWING TRADITION AS A GASTRONOMY TOURISM ATTRACTION AND LOCAL HERITAGE

By exploring older literary sources, it was revealed that the chapter written on Tolna in Bredetzky's topography (Bredetzky, 1807) by Károly Unger mentioned fisherman's soup as a special fish-dish at first in the ethnographic literature. In the year of 1800 he had attended on a sturgeon (in Hungarian: viza) fishing evening in Tolna (the settlement Tolna not the Tolna county) and on this accession he tasted the carp fish soup made by a German (or Swebian, author's note) fisherman on the banks of the Danube. He summarized his gastronomic experiences as follows: "The first course was made of carp, which was cooked in pepper soup, the people living here call this "Halászly" (in Hungarian) and Turkish pepper (paprika) is used to season the fish parts. The fish slices tasted excellent to me, but my taste was not receptive enough to eat such a hot pepper soup." Being a German from the Highlands, Károly Unger had known Hungarian well, also the name "paprika", but writing for Viennese he uses the terms "Türkischer Pfeffer" and "Pfeffersuppe" in the texts written in German (Fejes, 2019).

Figure 1. Mura-Drava-Danubes UNESCO Biosphere Reserve, study areas.
(Trisic, I. – Privitera, D. – Stetic, S. – Petrovic. M.D. – Radovanovic, M.M. – Maksin, M. - Simicevic, D. – Jovanovic, S. S. & Lukic, D. 2022)

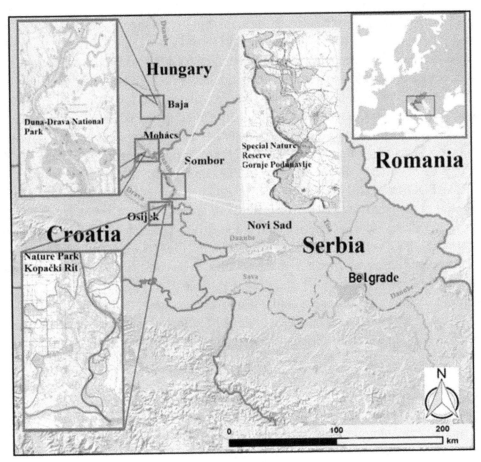

Knowledge passed down from father or grandfather to son can be extremely important in generating cuisine or gastronomy expertise. The author of this study having a Swabian fisherman's family origin, is aware of all skills and tricks to create an authentic fisherman's soup from killing, cleaning, slicing and salting the carp to preparing and mixing culinary ingredients then putting a fire under the caldron and brewing carefully in traditional way. His fisherman grandfather (János Witzmann 1897 – 1975) was born in Apatin and started fishing in Danube river as a little child. Than he moved with his family to Somogy then Fejér county and had been employed as a lake fish breeding master long time by the company located in Rétimajor until his retirement (the successor firm recently is Aranyponty, in English Gold Carp, Stock Company operating in the same place). His name and professional activity was well known and highly regarded not only in the Hungarian lake fishing community but Dr. Elek Voynarovich (Nagy, 2010), the world famous Hungarian fish biologist considered him as his main professional support in carp fry research (Purgel & Szabó 2019 pp. 36-37).

The Traditional Fisherman's Soup Cuisine and Dietary Tradition

The traditional fisherman's soup cuisine has been originating from South-Danube region and widespread along the river from the town Apatin (recently in Serbia) in the south, only 75 km distance from Baja, to Paks/Tass in the north in Hungary. The Hungarian, German, Swabian, Croatian, Bunyevci, Sokac, Serbian, Jewish and Roma residents of the multi-ethnic, multi-religious region with a rich gastronomy tradition all cook same type of soup, regardless of nationality or religion. It is very popular dish cooked at least once a week alike on both sides of the river in villages and towns in family households and restaurants as well (Vörös & Gačnik, 2018; Fejes 2020; Szilágyi 1997). All of the ingredients required to the fish soup are available locally and most of them has EU quality brands designation: the five basic culinary components are: water; fish ("Akasztói" Sikh Carp /in Hungarian "Sziki ponty"/ achieved PDO in 2020;salt; red onion ("Makói" Onion with PDO brand, 2009); red ground pepper („Kalocsai" Ground Paprika, PDO brand, 2011). Without high quality red ground paprika there is no real fish soup. The raw material of red pepper is a staple agricultural product of South-Danube region. The main traditional fish soup cooking areas and paprika cultivation are closely related. But it is mandatory to serve this gastronomy specialty with the so called "matchstick noodle" (in Hungarian "gyufatészta") which is prepared and cooked separately. Therefore, the noodle can be considered as the sixth components of the authentic South-Danube fish soup.

Concerning the cuisine expertise one should take into consideration all small steps and their sequences during the process of brewing operation which is similar to the way as invented by the Swabian fisherman and water-millers lived on Danube band from Apatin to Baja. However, it is worth to pay attention to small details and various hidden tricks.

"One can therefore argue that gastronomy is a complex, interdisciplinary activity….an inter-related branch of art and science that has a direct relation with chemistry, literature, biology, geology, history, music, philosophy, psychology, sociology, medicine, nutrition, and agriculture. The term 'culinary' often used in the context of gastronomy that describes a country's or region's dishes, foods, and food preparation techniques, which give rise to the country's or region's distinctive cuisine" *(Kivela & Crotts, 2006 pp.354-355).*

There is a legend, goes back in time to Danube Swabians, about how this custom of eating fish soup with noodles came about (Erdei 1971). It said that it is an innovation of Swabian water mill masters and owners in Apatin having close working relationship and cooperation with fishermen. They were obliged to give their miller boys a bowl of hot meal every day. Yes, but the stingy employers gave only a small bowl of fish soup, which was not enough for the lads. That's why they came up with the dough, which included flour from their own mill. This family history as well as ethnography background strengthen author's opinion that the denomination of "fishermen's soup" is more authentic than simple call it fish soup. Fishermen in Apatin (or in Baja) had prepared and cooked, using authentic culinary ingredients grown and produced in the most natural way, in open air in the natural environment on Danube banks close to the ecosystem blending into the landscape. This ethnographic origin legend might be easily a believable truth.

After World War II by the territorial division of historical Bács-Bodrog county the Daube part of region with Apatin was separated from Hungary. Therefore, in pepper fish soup cuisine 'Bajai' Fish Soup has become the main brand cooked in typical pear-shaped caldrons, with wood burning underneath, which are identified also 'Bajai' by the locals. The smoky aroma is part of the taste experience. 'Bajai' simultaneously means place, tradition, practice, cooking and serving methods. It differs from all other fish soups cooked in other regions of Hungary (e.g. at river Tisza or lake Balaton). The basic difference is that Bajai' fish soup made authentic fresh in which fish is not pasteurized and doesn't contain pre-made parts. The freshly kneaded "matchstick noodle" cooked separately in salted water and only soup served firstly with the noodle, because the soup is the main dish, and then not filleted fish slices as the second course.

Among Hungarians the town Baja considered as "the Capital of Fish Soup" not only because of its popular summer festival but the outstanding level of fish consumption: 30-35 kg/capita that is five to six times higher compared to the national average. The international tourism website of Taste Atlas under the title "The 3 Most Popular Fish-dishes in Hungary (What to eat in Hungary)" (Taste Atlas, 2023a) besides Fish Paprikás promotes also this fish-dish: "Every region has their own version, but the two most popular are Szeged halászlé, made with four types of fish, and Baja halászlé, made mainly with carp and served with homemade pasta called gyufatészta." (Taste Atlas 2023a). This brief presentation still acceptable but the website has a great mistake: it doesn't recommend any restaurants from Baja?

The obsessed followers of fish soup cult say so that "carp spoils in the third water," so no one can brew and eat fish soup without wine (Erdei, 1971;Vörös&Gačnik 2018;Szilágyi, 2019). Therefore, the wine is organic part of a fish soup feast. The rule of paring the soup and the wine appropriately: not white wine but red wine is the best choice because it is a pepper dish. Excellent quality red wines produced in the sub-region which are available and match well to the fish soup: Cabernet from Hajós – Baja wine region (PDO brand achieved in 2006) or Kadarka, Cabernet Franc, Merlot from Szekszárd wine region including Báta (with PDO brand, 2006).

Healthiness of fish soup feast based on literature survey

During this research it was raised that adequate answers should be given to the following questions based on trusted sources: "Can we justify by trusted scientific sources the healthiness of eating fisherman's soup and drinking red wine with it?" "Does this traditional dietary belong to the healthy category and how can we estimate at what extent?" We summarize the major findings of investigating and analysing selected international literature.

Concerning the beneficial health effects of red ground pepper, the most important basic seasoning component of fish soup the result is obvious: it has a number of natural compounds which is not only a useful addition to the diet but it can contribute to the treatment of certain diseases. Among its chemical composition the following main substances can be distinguished („htgetrid.com/en": https://htgetrid.com/en). (1) Vitamins (more than 10 items, including B,C and E). (2) Micro and macro elements. (3) Acids (including Omega-3 fatty acid and PUFAs Poly Unsaturated Fatty Acids). (4) Proteins (including fiber, essential oils and capsaicin which gives a specific sharpness). The author can proof that eating ground chilli red pepper proved to be effective against colds and to stimulate blood circulation.

Concerning the healthiness of the fish Leech (2019)[5] published an article entitled „11 Evidence-Based Health Benefits of Eating Fish" in a website which contains useful arguments of beneficial health effect of fish-dishes underlined by trusted sources. The research findings of Hei (2020) highlighted that fish contain many biomolecules and elements that are required in basic nutrition of growth, development and enhance performance ability physically and mentally and prevent and reduce the risks of many diseases. Fish provides all the essential amino acids, essential Omega-3 fatty acids, vitamins and minerals. Fish protein is highly digestible and of high biological value enhancing physical and mental performance ability which contributes human well-being. Olgunoglu (2017) made similar research and reflected that fish oil is considered to have the highest amounts of Omega-3 PUFAs (Poly Unsaturated Fatty Acids), which play important roles in the human body, such as in the synthesis of specific active compounds in, among others, the prevention of coronary heart disease. The research confirmed that in content of the fatty acids there might be significant variations and differences between fish species.

According to observational studies mental decline is a frequent old age disease. Morris et al., (2006) confirmed that people who eat more fish have slower rates of mental decline. Referring findings of large observational studies Virtanen et al (2007) established that, unsurprisingly, it is justified that people who eat fish regularly have a lower risk of heart attacks, strokes, and death from heart disease. In one study in more than 40,000 men attended in the United States showed that those who regularly ate one or more servings of fish per week had a 15% lower risk of heart disease. Studies of Raji et al., (2014) also revealed that people who eat fish every week have more grey matter - our brain's major functional tissue - in the parts of the brain that regulate emotion and memory.

Gourmands and chefs are convinced, researchers, based on sound scientific investigations, assumed that attending at an authentic fish soup feast including moderate red wine consumption belongs to a healthy dietary custom. Concerning beneficial health effects of moderate red wine consumption several research studies justified that chemical compounds which have antioxidant properties are good for health because they protect cells from oxidative stress. On the other hand, the existence of antioxidant phenolic compounds in red wine are justified by several laboratory survey. Mikolajková et al., (2021) established that "The more phenolic compounds, and thus resveratrol, a wine contains, the higher its antioxidant capacity. Its content may vary in the wine depending on the soil, grape variety, technique used, ripening and many other biotic and abiotic factors.". This article refers to the result of a laboratory survey accomplished by Švajdová, Z. in Czech Republic in 2011 (Tomas Bata University in Zlin) who analysed the level of resveratrol content in wine samples from 24 countries and confirmed that wine samples from Portugal and Hungary proved to be the leading varieties.

Certain variety of wines might be more beneficial than others. A famous Hungarian grape breeder and wine researcher (Csizmazia, 2001) justified high resveratrol content of an inter-species variety of red grape of Nero invented by him which became an authorised variety in Hungary in 1993 and then in EU. The resveratrol content of Nero shown an outstanding result: three-four times higher than other Hungarian grape varieties. Yoo et al., (2010) and several other researchers surveyed the health-promoting components of red wine. The findings of Lippi et al., (2010) strongly supported the cardiovascular benefits of red wine on cardiovascular health and that it would make a promising therapeutic supplement. The research also reveals that while there is a widespread agreement on the powerful antioxidant effects of polyphenols, the biological role of resveratrol in wine is still controversial in different reason? All these issues should be definitely cleared based on future research.

Finishing literature review on wine effects we refer to a research based on a new approach on analysing wine index of salubrity and health (WISH) Rossi & Fusco (2019) established that:

„Moderate ethanol consumption reduces stress and increases feelings of happiness and well-being and may reduce the risk of coronary heart disease. The debate as to whether the positive effect of red wine on longevity and health is only due to its content in ethanol or whether it has additional benefits linked with its no-alcoholic content remains open. We can affirm that moderate wine drinkers should not be discouraged and physicians can propagate the statement that moderate wine consumption may have beneficial effects" (Rossi,A. & Fusco, F. 2019 p.32)

Fish soup as a brand, gastronomy tourism attraction and cultural heritage in Baja

The small town, Baja with recently nearly forty thousand inhabitants represents an attractive destination for increasing domestic and foreign tourist and it is also known as the "Capital Town of Fish Soup". This regional and local gastronomy speciality is celebrated every year with a special family and local community event organized as the "Family-friendly Fish Soup Festival of Baja". This soup is prepared for family and group of friends in big caldrons in the centre of the city and nearby. "The importance of local residents in the developing process of gastronomy tourism is becoming one of the key elements. They are sources and validators of local history and cultural heritage, both tangible and intangible. Gastronomy can be an important inducer of new knowledge and relationships, boosting the economy while using sustainable principles as a guide for local development (Emmendoerfer et al., 2023 p. 59).

At the start of festival period entertaining the families has been the main role but during the time it became more and more a gastronomy tourist attraction. During festival period the downtown is filled with caldrons, almost the whole town is involved in cooking. The photo in *Figure 2* shows the row of caldrons set up in the main square in Baja under cooking operation at one of the fish soup festivals. The former residents are usually returning back to the "mother-land" for this weekend. The number of total guest participants attending in this widely known summer festival event all together can reach 20-30 thousand people.

Figure 2. Fish soup brewing festival in the main square in Baja

In 2015 'Bajai' fish soup achieved a relevant domestic quality label it was included in the Hungarikum collection (Hungarikum)[6] . Then a civil organization, Baja Fish Soup Champions Association (BFSCA) applied for adding fish soup brewing tradition to the UNESCO National Register of the Intangible Heritage (Fejes, 2020) and in 2021 the application entitled "Living Tradition of Baja Fish Soup Brewing"

was accepted. A fan group in FACEBOOK entitled "I like 'Bajai' Fish Soup" established to support this initiative. It has now 4,500 members visiting and posting photos and comments daily justifies well their keen interest and that "fish soup cult" in Baja is a living reality.

"What is fish-soup like as a brand? First of all, it is far more than a simple fish-dish. As a real brand it represents a concrete complexity of values and feelings. A good brand guarantees prominent quality without proof. Quality we became used to. Moreover, it represents values we can identify with and rely on." (Pusztai, 2007)

Therefore, the 'Bajai' fish soup became part of the cultural heritage of Baja and the region closely related to the traditions of people living next to the river Danube.

Sources of research also reflected that town Apatin, the native place of the grandfather of the author, which had been completely transformed in terms of ethnic composition, as Swabian-Hungarians almost eliminated from residents, became popular place for tourists owing to its proximity to the Danube and the forests. From forty-fifty years it is gathering place for recreational fishermen on event called "Apatinske Ribarske Večeri (Apatin Fishermen Nights) taken place yearly on the first day of July, and lasts five or seven days. Among the tourism attraction the so called Fiš Paprikaš should have an important role supported also by the Tate Atlas website: "Where to Eat the Best Fiš Paprikaš in the World?" (Taste Atlas, 2023b) Armed with fishermen's soup cuisine expertise we can establish that it is essentially the same traditional authentic version of fish-dish what had invented by Swabian fishermen and mill masters of Apatin. Concerning cross-border cooperation examples are rather few but it is worth to mention the twinning between Baja and Bezdán where a fish soup cooking event called "Vojvodina Gold Caldron" held every year with the participation of a local government delegation and best chefs of BFSCA from Baja.

CONCLUSION

The research study focused on exploring diversified relationship of territorial identities and gastronomy tourism attractions in South-Danube region and created a case study of an old local fish-dish brewing tradition transformed to a gastronomy tourism attraction. It highlights that this culinary feast and gastronomy tourism attraction also became a brand and cultural heritage contributes to enhance the image of the place and to promote sustainable rural development in South-Danube region.

Exploration of South-Danube region is based on adapting the "active territoriality" concept jointly with case study research and qualitative ethnographic/ethnological methodology utilizing own ethnographic experiences and advanced regional science approach which indicates novelty of this research study. Its goal is to reveal the main territorial features of surrounding of the case study location to highlight connection of these identities and features with local food and gastronomy resources.

Reviewing relevant literature sources the research justifies that consumption of fishermen's soup as a special culinary product jointly with moderate good quality red wine consumption belongs to healthy dietary customs and must have relevant role in well-being and quality of life of residents and tourists as well. The authenticity of the fish-dish basically depends on cuisine expertise, experience and high quality freshly prepared culinary ingredients grown and produced in the most natural way mainly locally. The atmosphere of cooking also contributes to the authentic fish soup feast: cooking and serving in open air in caldron with wood burning underneath and enjoying natural environment around with blending into the landscape.

Concerning economic sustainability and competitiveness of the hotels, guesthouses, apartments, restaurant and cafe shop businesses of the town of Baja, it plays a very important role how to attracts tourists, guests, visitors. Through designing and implementing a sophisticated marketing promotion program which operates before-after and under the summer festival guest have to convince to return back and visit continuously the town throughout the year. Therefore, it is not enough to focus only the fish soup festival event but it should also to be able to prevail comprehensively at town and regional level.

During the process of case study research we found and visited the website: "The 3 Most Popular Fish-dishes in Hungary (What to eat in Hungary)" (Taste Atlas, 2023a) and it was rather surprising, unfortunate and quite thought-provoking to reveal that this doesn't recommend to visitors any fish soup restaurant from Baja, known as the Capital Town of Fish Soup in Hungary? Glaring contradiction that its food description contains the name of the town to distinguish 'Bajai' and 'Szegedi' fish soup, nevertheless Taste Atlas offers only to visit a fishing inn from Szeged? One can come to an even more regrettable conclusion if click on another Taste Atlas website entitled "Where to Eat the Best Fiš Paprikaš in the World?" (Taste Atlas, 2023b) searched by international tourism visitors very often and widely from different part of the world recommends 15 restaurants from Croatia and Serbia located very near to the Hungarian border and to the town Baja. Based on own experiences one should observe that this is essentially the same fish-dish served also with noodle but called in different name *Fiš Paprikaš* (in English: Pepper Fish Stew; in Hungarian *Halpaprikás)* in Serbia and Croatia. Considering these finding, in order to evaluate the effectiveness of recent level of gastronomy tourism marketing communication and promotion strategy of the town Baja, the author, also a local patriot, should make the conclusion that it is still a lot to be done in the future by both the local government and local tourism entrepreneurs jointly with civil organizations as well.

Major findings of "Béda-Karapancsa" pilot project studies and literature (Datourway, 2011; Trisic et al., 2022) jointly with our case study research considerably support the conclusion that communities and decision-makers in cross-border regions have to focus their actions to accelerate cooperation towards a more specific, concrete and comprehensive direction, in order to integrate the tourism resources of neighbouring countries. It would be possible to exploit advantages and to create a recognisable tourism brand by mobilizing resources in order to transform South Danube region in an authentic and sustainable perspective. Our study highlights an opportunity to develop a new tourism product: Danube territory provides a valuable scenery a unique potential for ecotourism purposes jointly with gastronomy tourism. There must be considerable joint interest between the target groups of eco-tourists and gastro-tourists. An interesting research tasks might be in the future to compare the event "Apatinske Ribarske Večeri" (Apatin Fishermen Nights) in Apatin with the "Family-friendly Fish Soup Festival of Baja" held yearly in July for initiating cross-border cooperation in gastronomy tourism development.

REFERENCES

Bannini, T. (2017). Proposing a theoretical framework for local territorial identities: concepts, questions and pitfalls. *Territorial Identities and development, 2.* http://doi.org/ doi:10.23740/TID2201722

Bannini, T., & Pollice, F. (2015). Territorial identity as a strategic resource for the development of rural areas. *Semestrale di Studi e Ricerche di Geografia Roma – 27*. https://www.academia.edu/en/16373583/

Bessière, J. (1998). *Local Development and Heritage: Traditional Food and Cuisine as Tourist Attractions in Rural Areas*. Blackwell Publishers.

Bredetzky, S. (1807). *Digitale - Sammlungen*. BSBD (Bayerische Staatsbibliothek Digital). https://reader.digitale-sammlungen.de/de/fs1/object/display/bsb10009138_00204.html

Ciani, A., & Vörös, M. L. (2020). Rural Tourism and Agrotourism as Drivers of the Sustainable Rural Development – a Proposal for a Cross-Border Cooperation Strategy, In Burkiewicz, L. & Knap- Stefaniuk, A. (eds.) Management Tourism Culture. Studies and Reflections on Tourism Management. Ignatianum University Press Kraków.

Csizmazia, D. J. (2001). To protect our health. *Manuscript.*, *28*(October), 1–4.

Datourway. (2011). *The development of sustainable tourism in the Béda-Karapancsa area - Tourism development survey, strategy and action plan - Short version*. Béda-Karapancsa Pilot Project (Hungary-Croatia-Serbia). https://www.eubusiness.com/topics/food/door

Du Rand, G. E., & Heath, E. (2006). Towards a Framework for Food Tourism as an Element of Destination Marketing. May 2006. *Current Issues in Tourism*, *9*(3), 206–234. doi:10.2167/cit/226.0

Duhart, F. (2020). Territorial Food Identities Tips for Gastronomy Actors. WGI Global Report 2020. WGI.

EC - EUSDR. (2019). European Commission. European Union Strategy for the Danube Region (EUSDR). Danub Economics. https://www.danubecommission.org/dc/en/danube; https://www.interreg-danube.eu/about-dtp/eu-strategy-for-the-danube-region

Emmendoerfer, M., Chagas de Almeida, T., Richards, G., & Marques, L. (2023). Co-creation of local gastronomy for regional development in a slow city. *Tourism & Management Studies*, *19*(2), 51–60. doi:10.18089/tms.2023.190204

Erdei, F. (1971). Ethnographic cuisine (in Hungarian). Osiris, Budapest.

Fejes, A. (2019). *Newer data on the domestic spread of Paprika and the usage of Paprika pálinka* (Manuscript in Hungarian).

Fejes, A. (2020). Living tradition of making fish soup in Baja. Nomination data sheet for admission to the national register of intangible cultural heritage in Hungary. (Manuscript in Hungarian) p.10.

Gačnik, A. (2012). Gastronomy heritage as a source of development for gastronomy tourism and as a means of increasing Slovenian's tourism visibility. *Academica Turistica: Tourism & Innovation Journal*, 39-60.

Gačnik, A., & Vörös, M. (2018). Protected Food & Wine Products as a Driving Force for Creativity and Innovation of Gastronomy Tourism Development: Case of Slovenia and Hungary. *Agriculture, 15*(1-2), 19-34. https://www.agricultura-online.com/portal/index.php/issues/issue-21

Hei, A. (2020). Fish as a Functional Food in Human Health. *Diseases and Well – Being, 107th Indian Science Congress*. Banglore.

HMOARD. (2000). *Hungarian Ministry of Agriculture and Rural Development. Collection of Hungary's Traditional and Local Agricultural Products. Traditions-Tastes-Regions (in Hungarian HÍR). 2000.* CD-ROM.

Horváth, N. (2011). From Development to Sustainability? The EU Strategy for the Danube Region, In Tarrósy, I.; Milford, S. (eds.) *EU Strategy for the Danube Region Perspectives for the future.* EU.

Hungarikum. (n.d.). *Collection of Hungarikums.* Hungariankum. https://hungarikum.hu/

Kentish, P. (2019). The Danube Swabians: A story of cultural loss and revival. *Emerging Europe.* https://emerging-europe.com/after-hours/the-danube-swabians/a-story-of-cultural-loss-and-revival

Kivela, J., & Crott, J. C. (2006). Tourism and Gastronomy: Gastronomy's influence on how tourism experience influence destination. *Journal of Hospitality & Tourism Research (Washington, D.C.), 30*(3), 354–377. doi:10.1177/1096348006286797

Leech, J. (2019). 11 Evidence-Based Health Benefits of Eating Fish. *Healthline.* https://www.healthline.com/nutrition/11-evident-based-health-benefits-of-fish

Lippi, G., Massimo, F., & Guidi, G. C. (2010). Red wine and cardiovascular health: The "French Paradox" revisited. *International Journal of Wine Research, 2,* 1–7. doi:10.2147/IJWR.S8159

Mackenzie, R. (2019). Tranforming the terrior into a tourism destination. In S. K. Dixit (Ed.), *The Routledge Handbook of Gastronomic Tourism. Routledge* (pp. 70–78). doi:10.4324/9781315147628-10

Mikolajková, M., Ladicka, N., Janusova, M., Ondrova, K.,Mikulaskova, H.K. & Dordevic, D. (2021). *Resveratrol content in wine – resveratrol biochemical properties.* MASO INTERNATIONAL. doi:10.2478/mjfst-2022-0005

Morris, M.C.,Evans, A.D., Tangney, C.C., Bienias, J.L & Wilson, R.S. (2005). Fish consumption and cognitive decline with age in a large community study. *Arch Neurol, 62.* doi:10.1001/archneur.62.12.noc

Nagy, S. A. (2010). Elek Woynarovich 95 years (in Hungarian), In: Pisces Hunagarici 4.

Olgunoglu, I. A. (2017). Review on Omega-3 (n-3) Fatty Acids in Fish and Seafood. *Journal of Biology, Agriculture and Healthcare. 7*(12), 37-45.

Pollice, F. (2003). *The role of territorial identity in rural development processes.* Research Gate. https://www.researchgate.net/publication/242122046.

Purgel, I., & Szabó, I. (2019). Hey fishermen, anglers, and their activities in Fejér County (in Hungarian). Ministry of Agriculture.

Pusztai, B. (2007). Identity, Canonisation and Branding at the Baja Festival, In: *Tourism, Festivals and Local Identity: Fish Soup Cooking in Baja,* 83-101. Research Gate. https://www.researchgate.net/publication/329686693

Raji, C.A., Erickson, K.I., Lopez, O.L.,Kuller, L.H., Gach. M. H. Thompson, P.M., Riverol,, M. &Becker, J.T. (2014). Regular fish consumption and age-related brain gray matter loss. *Am J Prev Med 47*(4).444-51. . doi:10.1016/j.amepre.2014.05.037

Rossi, A., & Fusco, F. (2019). Wine Index of Salubrity and Health (WISH). An evidence-based instrument to evaluate the impact of good wine on well-being. *International Journal of Wine Research.*, *2019*(11), 23–37. doi:10.2147/IJWR.S177394

Solymos, E. (1997). Nationalities in Baja in the 18th-19th century. In J. Bárth (Ed.), *Mirror-images on the Sugovica* (pp. 101–103). (in Hungarian)

Sümegi, J. (2019). The foundation and early history of the Benedictine abbey in Báta (in Hungarian: A bátai bencés apátság alapítása és korai története). [Szekszárd, Hungary.]. *Year Book of the Wosinsky Mor County Museum.*, *XLI*, 2019.

Szilágyi, M. (1997). Fish foods as man foods. In *Variations of the food culture in the 18th-20th centuries (in Hungarian: A halételek, mint férfiételek, In: A táplálkozáskultúra változatai a 18-20. században)* (pp. 117–132). Kalocsa.

Taste Atlas. (2023a). The 3 Most Popular Fish-dishes in Hungary (What to eat in Hungary). *Taste Atlas.* https://www.tasteatlas.com/most-popular-fish-dishes

Taste Atlas. (2023b). Where to Eat the Best Fiš Paprikaš in the World? *Taste Atlas.* https://www.taste-atlas.com/fish-paprikas/wheretoeat

Török, Á. (2019). The recognitions and the embeddedness of the TTR trademark among the Hungarian consumers. Nutrition Marketing, 6, 81-97.

Trisic, I., Privitera, D., Stetic, S., Petrovic, M. D., Radovanovic, M. M., Maksin, M., Simicevic, D., Jovanovic, S. S., & Lukic, D. (2022). Sustainable Tourism to the Part of Transboundary UNESCO Biosphere Reserve "Mura-Drava-Danube". A Case of Serbia, Croatia and Hungary. *Sustainability (Basel)*, *2022*(14), 6006. doi:10.3390u14106006

UNWTO - BCC. (2019). *Guidelines for the Development of Gastronomy Tourism.* International Tourism Organization UNWTO and Basque Culinary Center., doi:10.18111/9789284420957

Van der Meulen, H. S. (2007, March). A normative deðnition method for origin food products. *Anthropology of Food*, *S2*(S2). doi:10.4000/aof.406

Virtanen, J. K.,Mozaffarian, D..Chiuve, S. E. & Rimm, E.B. (2007). Fish consumption and risk of major chronic disease in men. *Am J Clin Nutr.* *88*(6), 1618–1625. . doi:10.3945/ajcn.2007.25816

Vörös, M., & Gačnik, A. (2018). Gastronomy tourism enterprises in the digital economy: Case study on fish soup as a tourism attraction – Comparing fish-dish cuisine of 'Gold Carp' Fishing Inn in Rétimajor and brewing tradition in South-Danube region. Edutus College, Budapest.

Vörös, M., & Gemma, M. (2011). Current Status, Future Trends & Real-life Cases from Japan. In M. Bourlakis, V. Zeimpekis, I. Vlachos, (eds.) Intelligent Agrifood Chains and Networks. Wiley-Blackwell Publishing.

Voros, M., & Gemma, M. (2015). Promotion of Local Agricultural and Food Products by Using Geographical Indications and Traditional Specialties Guaranteed Schemes in Hungary, In: *The Proceedings for the International Farm Management.* International Farm Management Association (IFMA), Laval University. http://ifmaonline.org/proceedings/20th-vol1/

Weithmann, M. (2011). The Danube: A European River as Street, Bridge and Frontier. An Introductory Essay. In: Tarrósy, I. – Milford, S. (eds.) EU Strategy for the Danube Region Perspectives for the future. Pécs, Hungary.

Yoo, Y. J., Anthony, J., Saliba, A. J., & Prenzler, P. D. (2010). *Should Red Wine Be Considered a Functional Food? Comprehensive Reviews in FoodScience and FoodSafety* (Vol. 9). Institute of Food Technologists. doi:10.1111/j.1541-4337.2010.00125

ENDNOTES

[1] "…local territorial identity cannot be studied „from above and outside" on the basis of tangible and intangible aspects (landscape and urban forms, cultural and linguistic features, and so on), but through a complex and articulated field research, mainly with *ethnographic methods*, aimed at exploring representations and self-representations, local actors and relationship networks, strategies and practices, territorial ties and attitudes of local collectivities." (Banini, 2017 p. 19).

[2] Traditions-Tastes-Regions TTR programme launched in 1998, based on best practices of EU's Euroterroirs program, for promoting and economically stimulating the traditional and local food products of Hungary. It is a collection and computerized database of traditional and regional food products, landraces and indigenous animal species containing the origin, history and description of regional food markers by products.

[3] In 2012 the Hungarian Parliament adopted the Act XXX on Hungarian national values and Hungarikums with the aim of establishing appropriate legal framework for the identification, collection and documentation of national heritage values for their protection and making them available for the widest possible audience.

[4] The EU quality protection schemes for agricultural and food products adopted in European Union in 1992 and entered into force in Hungary from 2004, the year of EU accession: Protected Designation of Origin (PDO); Protected Geographic Indication (PGI); Traditional Specialities Guaranteed (TSG) (see in DOOR Database)

[5] Joe Leech is an Australian dietitian with a master's degree in nutrition and dietetics.

[6] Baja Fish Soup in Hungarikum Collection: https://hungarikum.hu/en/content/baja-fish-soup

Chapter 9
Mediterranean Diet and Tourism Innovation, Experiences, and Sustainability:
The HoST Lab Case Study

Alexandra Rodrigues Gonçalves
https://orcid.org/0000-0003-3796-1801
Universidade do Algarve, Portugal

Célia M. Q. Ramos
https://orcid.org/0000-0002-3413-4897
Universidade do Algarve, Portugal

Carina Viegas
https://orcid.org/0000-0002-8134-6817
Universidade do Algarve, Portugal

ABSTRACT

HoST Lab is an integrative approach that aims to innovate based on Mediterranean diet (MD) creating new products, services, and experiences, involving producers and distributors. The lab research also aims to know the emotions and sensations associated with eno-gastronomic experiences of the Mediterranean diet and their welcome among visitors-tourists, using a triangulation of traditional methods (interviews, focus groups, workshops), but also developing a digital solution for sharing results (webpage, digital survey, sentiment analysis). A set of reference indicators and a nutrition economic label will be developed and used to evaluate and monitor research results, both in a laboratory and in a real environment. The HoST Lab pretends to be a sharing and learning research space between the academy, the tourism, and the hospitality sector, in which proposals are tested, results transferred to the community, and well-being promoted among the local population and visitors, aspiring for a growing sustainable destination.

DOI: 10.4018/978-1-6684-6985-9.ch009

1. INTRODUCTION

The Algarve is located in the southernmost part of Portugal's mainland. The city of Faro, home to the University of Algarve public institution (location of HoSTLab - Hospitality, Sustainability and Tourism Experiences Innovation Laboratory) and the international airport serving the region, serves as the administrative hub for Algarve. Two zones—one to the west (Barlavento) and another to the east—are separated by the area, which is located in Faro District (Sotavento). The foundation of the Algarve's economy is tourism along with other significant activities such as fisheries industries (Ortega et al., 2013).

In 2017, the Regional Commission for the Mediterranean Diet of the Algarve highlighted in the *Plan of Activities to Safeguard the Mediterranean Diet* in the Algarve (2018-2021) the enhancement of local products, as a strategic differentiator of the Portuguese regions and other countries practising Mediterranean Diet (Freitas et al., 2022). The concept of gastronomy tourism has evolved to encompass the ethical and sustainable values of the territory, the landscape, the sea, local history and cultural heritage (Kiráľová & Malec, 2021).

The topic of "innovation" linked to the Mediterranean Diet is emerging in the research area. This study aims to contribute to filling the scarcity of scientific literature and to contribute in practice, through the explanation of how to build a research laboratory to explore the Mediterranean Diet´s area and how it aggregates value to the development of new ingredients, as well as, services that contribute to improving quality, sustainability and local heritage.

The aim of the HoSTLab Project[1] is to understand and study the feelings associated with the eno-gastronomic experiences that are based on the elements of the Mediterranean Diet, and the characteristics of flavour and the surrounding atmosphere, which can condition the sensory result of the experiences (in a real context, testing along the restaurants or hotels; and, in a laboratory context, in a simulated environment).

Innovating through traditional products and resources correspond to the proposal of HoSTLab, directing to be a reference for research and development applied to culinary tourism and eno-gastronomic experiences associated with the Mediterranean Diet (MD), which will study the determinants of experience and the atmosphere of places, promoting sensory evaluation with tourists and stakeholders, of new products and services (based on local resources and products, such as cereals, olive trees, vines, and other foods), enhancing the qualification and diversification of the Algarve as a sustainable tourist destination.

This chapter reports the literature framework and findings related to the first phase of the exploratory research developed by the University of the Algarve team that aimed at better understanding present local food systems and their relation to tourism in the regional community and the Mediterranean Diet recognition throughout the province of Algarve.

For the purposes of this initiative, it was necessary to develop a diagnosis of the existing research and define the case study theoretical framework. One of the main results of the ongoing empirical and still exploratory analysis is the enormous lack of studies and research about innovation and knowledge transfer in the food and tourism industry related to the Mediterranean Diet in the Algarve.

The chapter follows a structure of five sections, without considering the present introduction. The first section contextualizes the research problem, the relevance and expected contribution of the study and the research objectives. The second section is dedicated to the literature review about the surrounding topics involving Mediterranean Diet and innovation. The third section describes the methodology and justifies the aims of the design for the study. The fourth section discusses the results and the respective analysis. The fifth section presents the main conclusions of the study as well as its limitations and recommendations for future research.

2. THEORETICAL BACKGROUND

2.1. The Mediterranean Diet Concept

The Mediterranean Diet is a lifestyle, typical of those who live near the sea and in the middle of the land. The nomenclature comes from the Greek "δίαιτα" which means "way of living" and from the Latin "*mar mediterraneum*" which means "the sea in the middle of the lands" (Graça, 2014a: 19). For centuries, Phoenicians, Greeks, Romans, Arabs and all other peoples in the region of the Mediterranean Sea had a navigable inland sea, contributing to the sharing of knowledge, tools, genes, plants, and animals capable of influencing ways of thinking and living (Graça, 2014a).

In the 40´s/50´s, North American researchers from the Rockefeller Foundation, when arrived in Crete observed the health levels of the population of the Mediterranean island concluding that the longevity was much greater than in North America (Graça, 2014b).

The term "Mediterranean Diet" was created by Ancel Keys and his wife, in 1975, when they published the book "How to Eat Well and Stay Well. The Mediterranean way". The first representation of the Mediterranean Diet´s pyramid was in the "The Diets of the Mediterranean" congress, in Cambridge (Real and Graça, 2019a).

Just 50 years after the first approach to Mediterranean Diet and its characteristic, traditional food model, in 2005, the United Nations Educational, Scientific and Cultural Organization (UNESCO) recognized the Mediterranean diet as an intangible cultural heritage, generating the statement "The 2005 Rome Call for a Common Action on Food in the Mediterranean" and, in 2010, UNESCO approved the inscription of the Mediterranean Diet in the list of intangible heritage of humanity (Graça, 2014a; Saulle & La Torre, 2010). Among other countries already institutionalized as participants of the Mediterranean Diet, in 2013 Portugal, Cyprus and Croatia joined the group of countries that officially practice the Mediterranean Diet (Graça, 2014b).

UNESCO recognizes, values, and emphasizes the culinary practices around popular cultures in a broad way, focusing on the quality, simplicity, and wholesomeness of the food products native to the place (autochthonous) that, in addition, are involved with folkloric food practices, territoriality, and biodiversity, respecting seasonality (Saulle & La Torre, 2010; Valagão, 2015).

The Mediterranean Diet was first understood by Ancel Keys as food consumption that generates health benefits. The author focused his studies on the way of life of the people belonging to the places bathed by the Mediterranean Sea, based on the relationship between food consumption and disease/mortality (Brites, 2015). The health area remains in evidence regarding the development of studies. Publications in other areas began to be made in 2000, and topics such as lifestyle, culture, and sustainability were included in the discussion, linked to the concept of the Mediterranean Diet. The interpretation of the MD provides a new perspective on the subject, to highlight the multidimensionality of the Mediterranean Diet concept, reflecting the current principles of health, environmental sustainability, and lifestyle (Freitas *et al.*, 2015; Real & Graça, 2019; Saulle & Torre, 2010; Trichopoulou, 2021).

In Portugal, several events emerged that potentiated a broader view of the Mediterranean Diet: (i) the country's inclusion in the candidacy for UNESCO's distinction as Intangible Cultural Heritage of Humanity, approved in 2013; (ii) the creation of the *Monitoring Group for the Safeguarding and Promotion of the Mediterranean Diet*, which ended on December 31, 2017, (iv) development of the *Plan of Activities to Safeguard the Mediterranean Diet* in the Algarve from 2018 to 2021 and subsequent creation of a *Competence Center for the Mediterranean Diet* to integrate those involved in the research, preservation, dissemination, enhancement and promotion of the Mediterranean Diet in Portugal, with the mission of preserving and promoting it through the pillars of Portuguese cultural heritage, lifestyle, quality food pattern and influencing factor in the development of rural territories; (v) constitution of the *Dynamic Council for the Safeguarding and Promotion of the Mediterranean Diet,* to safeguarding and promoting the Mediterranean Diet (Real & Graça, 2019a; Real & Graça, 2019b; Trichopoulou, 2021).

Graça (2014b: 26), states the principles of the Mediterranean Diet in Portugal:

(i)   simple cooking based on preparations that protect nutrients such as soups and stews;

(ii) high consumption of vegetable products to the detriment of consumption of food products of animal origin, namely vegetables, fruit, quality bread and poorly refined cereals, dried and fresh legumes, nuts, and oilseeds;

(iii) consumption of plant products produced locally or nearby, fresh and in season;

(iv) consumption of olive oil as the main source of fat;

(v)   moderate consumption of dairy products;

(vi) the use of aromatic herbs to season instead of salt;

(vii) more frequent consumption of fish compared to low and less frequent consumption of red meat;

(viii)   consumption of wine low to moderate and only with main meals;

(ix) water as the main drink during the day; and,

(x)   sociability around the table.

In Portugal, Tavira is the representative community of the Mediterranean Diet, as a UNESCO Intangible Cultural Heritage. Located in Algarve (southern Portugal) and also the most Mediterranean due to its climate, production, and ways of life (Queiroz, 2015). As an initiative of the *Dynamic Council for the Safeguarding and Promotion of the Mediterranean Diet Plan,* was created the Mediterranean Diet Route with the purpose of qualify, organize and making available an extensive set of sites and heritage resources, where the identity elements of the Mediterranean Diet are clear and can be enjoyed with quality and depth (Dieta Mediterrânica, 2022). All over the Algarve, as identified in the Figure 1, were defined places to enjoy the gastronomy, meet artisanal producers, visit ancient heritage that reflects the deep connection between man and his territory and taste the local products that best represent the Mediterranean Diet, demonstrating the millennial lifestyle. The establishments included on the Algarve route, have a Stamp of the Mediterranean Diet Route. They are grocery stores with local products made with products from the land and the sea; craft shops that use materials and ancient techniques in the products; monuments and places with the natural and cultural heritage of the Mediterranean; and restaurants that offer Mediterranean Diet meals.

The Portuguese government established the *Competence Center for the Mediterranean Diet in Tavira* (CCDM), which has a national scope and serves as a forum for research, knowledge sharing, and articulation on the MD subject. It brings together agents involved in research and innovation, training, knowledge dissemination, and transfer, as well as economic agents and public administration bodies, to improve the respective cooperation at both the national and international levels.

Figure 1. Official Mediterranean diet Algarve route
Source: Dieta Mediterrânica (2022).[REMOVED HYPERLINK60 FIELD]

MD can be considered as a dietary pattern, considered the healthiest and most sustainable in the world that meets the SDG, which is fundamental in health and nutrition worldwide, but particularly in the Mediterranean area. MD, in addition to improving aspects of health and nutrition, is part of a cultural heritage that must be preserved and promoted in different areas: culture and tourism, health, agriculture, policy and economic development, and community well-being (Serra-Majem *et al.*, 2012).

2.1.1. Dimensions of the Mediterranean Diet

2.1.1.1. Social and Ritual Dimensions of the Mediterranean Food Model

Family can be defined as a group that meets to eat the same food at the table, sharing what comes from the same place, literally from the pan. It is the act of reuniting around the table that ritually reflects and strengthens vertical and horizontal relationships, whether with a domestic group or a larger collective. Behavioural changes also reflect changes on the social scale. Changes in the current rhythm of life are no longer allowing routine meetings at the table. However, there are meetings at least once a day/week or concrete celebrations where food and drink play a central role together (Moreno, 2015).

The culture of a society is reflected through its gastronomy, showing itself to be part of the people's lifestyle. An impact on traditions is generated, reflecting different cultural and unique legacies in each geographic location (Gálvez *et al.*, 2017).

The Mediterranean Diet is based on a food culture that adapts to food scarcity and availability, where coexistence and sociability around this culture translate into a way of life (Covas & Covas, 2014). Currently, cultures and societies present around the diversities of food systems have the ecosystem in common. Influenced by the conditions and resources of more inland territories: continental, Atlantic or Desert, the weather is similar in different countries and provides a complementarity of resources such as fishing or irrigated agriculture allowing extensive livestock farming; and the ability to adapt to the inclusion of food-producing plants from other parts of the world (Moreno, 2015).

2.1.1.2. Products and Concepts Associated With the Mediterranean Diet

Wine, olive oil, and bread are the products in common between countries, associated with the Mediterranean Diet. Portugal adds codfish consumption as part of its national food identity, being the biggest consumer country in the world; olive oil and lard are the main fats used to cook, and soups and broths, a dish with a Mediterranean base, use products from hunting, fishing, and agriculture. Mediterranean characteristics that identify Portugal's food practices with those of other Mediterranean countries (Valagão, 2015).

In Algarve, the relationship between Natural Heritage and the Cultural Heritage of the region is also reflected in food habits. Other signs are reflected in the Algarve landscape, such as raw materials linked to the arts of pottery, crafts and basketry (clay, palm, cane, rush, wicker and esparto). Pottery preserves food and supports cooking art. Bowls of different shapes and sizes date back to the Arab heritage and their flavours, which are characteristic of Algarve food. Other signs are food products resulting from agriculture and fishing. The farming results in olive trees, orange trees, fig trees, almond trees, carob trees; cereals, and legumes. The fishing results in fish, bivalve molluscs, and dried fish (octopus, litham, horse mackerel, garfish, tuna, with the well-known 'tuna moxama'). The mountain areas, together with the seafront, the estuary, and hunting resources, configure and reinforce the identity of Algarve food, which makes it distinctive from cuisines from different regions of the country (Queiroz, 2015).

2.2. Mediterranean Diet Innovation

Mediterranean Diet represents a cultural and symbolic expression between nature and human activities; both should be in syntony. Specifically, in Algarve, some agents of imbalance are being questioned: construction of dispersed buildings is causing disordered territory and degrading land capital; inexistent power to control real estate capital affects the conservation and multiple uses of natural capital; technical and technological intensification weakens the biophysical attributes of ecosystems and territory, reducing the provision of environmental services in the region; and the increasing speed of rotation of financial capital ends up on a collision course with the rhythms of regeneration typical of biophysical systems (Covas & Covas, 2015).

The impact of progress and technology results in the propagation of monocultures, biophysical monotony, and the reduction of social diversity, the different aspects of the same problem. Mediterranean Diet needs the development of a plan to preserve the concept and everything it encloses, from the risks of a policy of speed and replacement technologies. In different circumstances, the Mediterranean Diet ceases to represent what it consists of: a healthy and sustainable lifestyle and food consumption with local products from the land and sea. The Mediterranean diet may already be colliding with regional and international capitalism. The interpretation is a kind of counterculture and counter-rationality in an unequal dispute against an established rule that, in the meantime, publicly praise a new promise of regional and rural development (Covas & Covas, 2015).

For the future, the Mediterranean Diet as World Heritage by UNESCO, in Tavira, has the role of directing a social production of world quality, improving the material well-being of local populations, valuing material heritage, challenging a new territorial intelligence (Covas & Covas, 2014). Traditional rules and practices are insufficient to ensure quality and consumer exigence. For the implementation, the Mediterranean Diet must be expressed as a successful participatory production process, operating through a value chain that connects Humanity's intangible heritage to a regional material heritage, improving the economic, social, and business structure of a community or region significantly (Covas & Covas, 2015).

In the context of the Mediterranean Diet in the Algarve, quality production is held by the approach:

(i)   the market understands the existent client demand;
(ii)   the quality, as an attribute to be negotiated as a process of continuous rules with diversified strategies, from production to consumption.

In this assumption, quality results from a social consensus and a learning process with political and organizational involvement (Covas & Covas, 2015).

Production and consumption are carried out by specific producers and consumers in a specific place. Processes are established whenever quality is considered a common interest, shared and based on mutual trust. Quality social production must be directed to promote a rural development strategy, consisting of a plurality of agricultures based on traditional products of high biological, ecosystem and landscape value.

The Mediterranean Diet should be the reason for local innovation referred to the territorial intelligence through the instrument of the economy of conventions; a territorial pact that supports the development of a local agro-food system and a symbolic culture that respects and values the international prestige designation as an intangible heritage by UNESCO (Covas & Covas, 2015). Due to this, are identified opportunities for the innovation of the Mediterranean Diet, in the Algarve region, through:

(i)   value creation, using arts and culture as a central activity in the economy of the Algarve;
(ii)   development of new products and services, innovating and creating a new line of structured products and services, linking an own design to the product and impactful marketing campaigns, and;
(iii) collective territorial intelligence, promoting a new collective territorial intelligence through the creation of thematic networks or network territories.

In the Algarve, "the Mediterranean Diet is based on a vertical concept, which crosses the entire region, from intangible heritage as a symbolic representation to material heritage as a support for the Mediterranean diet" (Covas & Covas, 2015: 280). In the Algarve, the Mediterranean Diet is based on a vertical concept, which crosses the entire region, from intangible heritage as a symbolic representation to material heritage as a support for the Mediterranean diet. The two realities complement each other, and the conservation and development of the material heritage are a consequence of the preservation of the intangible heritage (Covas & Covas, 2014).

The concordance between local tradition and innovation results in the territorial sustainability concept. The verticalization of economic activity, intersected with a network of horizontal actions, results in new territorial intelligence, structuring new products and providing a convenient contemporary and cosmopolitan image of the region (Covas & Covas, 2015). The verticalization of the Algarve´s goat value chain and the adjacent tasks aggregate value to the local economy. The objective is to value systematically local economies and ecosystems, including the products around the Algarve goat concept as the keystone -honey, arbutus, wild fruits, traditional rainfed orchard, citrus, flowers, mushrooms, cork, hunting, among others- (Covas & Covas, 2015). According to Covas and Covas (2014), the intersection of activities would build a network territory and a network economy, and unique visitation. Some innovations in the Algarve, derived from the crossing could be listed as follows:

(i) a day in the vineyard and the cellar; get to know the products in the vineyard from the treading of the grapes to the winemaking process, wine tasting and activities involving wine tourism, the gastronomy adjacent to the associated Mediterranean diet, cultural sessions, scientific techniques related to the wine theme and its surroundings;

(ii) a day in the grazing; herding the indigenous goats and participating in activities involving milking, producing cheese, tasting the foods associated with the Mediterranean Diet, and attending cultural and recreational sessions associated with the agro-silvo pastoral system.

The Mediterranean Diet as an intangible heritage of humanity appears as a great opportunity to upgrade the local and regional economy of the Algarve, especially remote and mountain areas. The challenge is a long-term one, bringing benefits yet to be estimated at the national and regional levels. To concretize, a plan for thematic networks and territories-network must be established, through the implementation of an experimental collaborative economy trial, so that soon there will be a policy of regional certification of the Mediterranean diet based on solid foundations (Covas & Covas, 2014).

Research on complex systems explains that different ideas and perspectives are exchanged between individuals, resulting in innovative developments. It shows that connectivity and diversity among the engaging participants (people or organizations) that contribute to a food hub are equally important to the emergence of innovative insights. Trust and communication are elemental to promoting connections in human systems challenged by diversity (Moore & Westley, 2011; Stroink & Nelson, 2013). Given the above, MD is an educational tool food for the public in general, for the scientific community, and for the sustainable development of the regions (Manios *et al.*, 2006). In this context, the MD contributes to educating and promoting healthy lifestyles, where the consumption of food produced by local producers is privileged, with a view to adopting adequate lifestyles, beneficial nutritional behaviours and the promotion of well-being both in the community or with the human being himself, whether he is a resident or a tourist, so there is a need to create a space that promotes education and transmission of knowledge associated with MD, that is, the need to create a laboratory that scientific fulfil the aforementioned needs – The HoST Lab Project.

2.2.1. Mediterranean Diet Innovation: The HoST Lab Project

Due to the scarcity of studies and activities in the area of Mediterranean Diet innovation, the creation of HoST Lab is a necessity for the Algarve region and for Portugal, a laboratory that provides concrete answers to the challenges faced by local producers and tourist companies that will allow them to innovate in terms of products, services and experiences based on the Mediterranean Diet, Intangible Cultural Heritage of Humanity and, consequently, to strengthen the quality of its products and services and safeguard the social and economic dimensions of sustainability in the Algarve region. HoST Lab intends to present concrete solutions on the following levels:

(i) to research a set of products, services and experiences based on the Mediterranean Diet's main elements with the potential to be marketed by restaurants, accommodation units and local producers;

(ii) to test in a laboratory environment (phase 1) and in a real environment (phase 2), namely in the restaurants adhering to the *Rota do Petisco*, the respective products and experiences with foreign and Portuguese visitors. *Rota do Petisco* is a project managed by a third sector organization with an active role in the social sector and with experience in conceiving projects with a network of restaurants in the coastal and mountain areas of the Algarve;

(iii) to evaluate the eno-gastronomic experience and to develop a sentiment analysis index;

(iv) to assess their respective economic and financial viability;

(v) to create an economic nutrition label for each product, service, and experience to assess its financial and social sustainability.

The project will increase the knowledge in the development of innovative experiences using food, wine, and other elements associated with Mediterranean Diet, allowing the tourist companies to provide environmentally, socially, and economically sustainable food and beverage experiences. The research process and methodology include the design and testing of new culinary experiences, the development of data sheets and mapping of resources associated, tests, and the study of their respective viability and economic and financial sustainability. The project includes the production of a system for assessing and monitoring the sustainability of new experiences, products, and services associated with the Mediterranean Diet.

Sustainable tourism development guidelines and management practices are applicable to all forms of tourism in all types of destinations. Sustainability principles refer to the environmental, economic, and socio-cultural aspects of tourism development, considering that a suitable balance must be established between these dimensions to guarantee its long-term sustainability (United Nations Environmental Development Programme/World Tourism Organization, 2005).

2.3. Gastronomy, tourist experiences, Local Food, and Healthy and Sustainable Lifestyles

Gastronomy can contribute to the Sustainable Development Goals in (SDG) tourism destinations, holding an opportunity for the whole sector and for society in general (Pimental de Oliveira & Pitarch-Garrido, 2022). Gastronomy has the potential to provide visitors with new experiences and values regarding the culinary culture, ingredients, culture, and traditions of a place. The concept of gastronomy tourism has evolved to encompass the ethical and sustainable values of the territory, the landscape, the sea, local history and cultural heritage (Kiráľová & Malec, 2021). Safety, health, and food are central factors, which weigh in the decision when choosing the destination. Gastronomy tourism is part of regional culture and contributes to the tourist experiences, being a relevant part of regional development (Kiráľová & Malec, 2021). Food is a vital dimension of a place´s culture and human expression, contributing to a significant component of intangible heritage, and a persuasive attraction for tourists (Dupeyras, 2016).

Gastronomy is a demarcated cultural expression of the human being, distinguishing the traditions of communities. The food, drinks, and habits of a particular place, together with the food heritage, create an integral part of the tourist experience in the destination (Jong & Varley, 2018; Kiráľová & Hamarneh, 2017; Moreno, 2015). Tourists intentionally look for food experiences, authenticity, and distinction in the food of the local destiny. Local food is a foundation of a sustainable tourism experience and a topic that provides the development of tourist experiences in several ways (Tiganis & Tsakiridou, 2022). Authenticity and local dimensions allied to the food, are directly related to tourism and have a straight connection (Kiráľová & Hamarneh, 2017; Kiráľová & Malec, 2021). Observing history, tradition, place, and the distinctiveness of regional cuisine and culinary heritage, food tourism offers tourists an essential way to find authenticity (Zhang et al., 2019). When tourists look for authentic food experiences at the destination, they are seeking local and innovative gastronomy. Current culinary tourists feel excited to try unique and attractive food experiences. Therefore, food is a subject used to promote tourism experiences in several manners: linking culture and tourism, developing the meal experience, producing distinctive foods, developing the critical infrastructure for food production and consumption, and supporting local culture (Richards, 2012).

Gastronomy and food tourism promote local agricultural and economic evolution through the development of small and medium-sized enterprises (SMEs), contributing to the region´s development and enhancing a relationship with sustainability. Gastronomic tourism rises a relationship with sustainability, allowing investigation of the relationship between tourism and the socio-community system of the host society through culinary and how it is expressed by cultural heritage and cultural consumption, forms of product, forms of social relationships, and local trade (economic, social, and environmental benefits). Regional food and beverages contribute to the guests´ traffic growth, spending average, and enhances the local image (Carral *et al.*, 2020; Kiráľová & Hamarneh, 2017). This fact is an important tool with economic potential that aims the regional development, reinforces local individuality, increases the maintenance of the environment, and promotes the conservation of local heritage and its adjacent economies. The relationship between tourism and food production is undoubtedly a fundamental practice for the economic development of rural areas (Kiráľová & Malec, 2021).

Promoting local food supports regional development strengthens the sustainability and preserves regional character and authenticity. Local cuisine can enhance tourist experiences. Tourists perceived the food as more original when consumed in the local cultural environment, allowing visitors to understand cultural identity and local lifestyles (Santos et al., 2020).

Tourism is a driver to create new job positions as impulse foreign direct investment and foreign currency and boost economic growth. Economy growth occurs when exists correct support by the state and is controlled as required (Kiráľová & Malec, 2021). Tourism and food combine a relationship that promotes policies able to improve economic and social well-being (Richards, 2012).

Portugal, in particular the south regions of the River Tejo (Algarve), has an important influence on the Mediterranean culture, and gastronomy and wine play a central role in this identity. In the Algarve, gastronomy and wines, play a strategic role in the region's economy and the famous trilogy of bread, olive oil, and wine – representative of the Mediterranean lifestyle – are basic elements of culture and user-friendliness at the table (Serra, 2016). In this regard, several studies show how innovation can promote the quality and reputation of restaurants, and increase revenues and income, in addition to promoting competitive advantages in the long term. However, smaller restaurants with a local scope, have more difficulty in accessing high-quality resources in terms of products or professional chefs and can innovate differently, adopting partnerships. These companies can easily introduce innovation to their offers, processes, management structures or marketing techniques (Lee *et al.*, 2016). More recent studies show that the notion and perception of innovation are different when evaluated by companies or by customers since they have different evaluation perspectives. The concept of innovation in a company is centred on technical and functional aspects, while the customer focuses on creating and offering new experiences (Kim *et al.*, 2018).

Food nutrition labelling has created a revolution in the food industry, however, labelling usually focuses on nutritional composition. Economic nutrition labelling is a worldwide innovation and aims to show how money from a purchase is invested in the local community and the impacts it has. Shore fast, a registered Canadian charity with the mandate to secure a resilient future for Fogo Island developed the Economic Nutrition Label, which shows prospective purchasers where their money goes and how much is spent locally. The market trend towards the development of sustainable products leads us to believe that this type of labelling in the new products and services constitutes an important innovation that will be valued by consumers. On the other hand, considering the need for the visitors' evaluation of tourist services and experiences, the feelings expressed in social networks and associated with the gastronomy considered in Mediterranean Diet can be detected through the application of text mining techniques, which constitute sentiment analysis, contributing to the identification of the dishes that are mostly chosen and to the detection of aspects that need more attention (Yu & Zhang, 2018). The analysis of feelings helps to understand the expectations of visitors about the products and services, as well as to identify the potentialities associated with the novelty of certain products and services. Naturally, such products and services with similar characteristics associated with the Mediterranean Diet will motivate the advancement of well-being within the population and visitors as well as the sustainability of the region.

3. METHODOLOGY

3.1. The HoST Lab Project

HoSTLab- Hospitality, Sustainability and Tourism Experiences Innovation Laboratory is located at *Escola Superior de Gestão, Hotelaria e Turismo* (ESGHT), placed in the University of Algarve institution. The project is composed of a multidisciplinary team, from Tourism, Hospitality and Catering, Food Engineering, Management and Economics, that proposes to welcome and develop research applied to the innovation of new products and services related to basic foods integrated into the Mediterranean Diet, as well as to promote the main elements of this Common Heritage of Humanity, as the main value of the eno-gastronomic and cultural offer of the region.

The main objective of HoSTLab is: to develop, evaluate and test new products, services and experiences, for hotels, restaurants and tourism, associated with the basic elements integrated into the Mediterranean Diet – cereals, olive trees, vines, and other common foods in the South of Portugal -, which contribute: to the qualification and diversification of the Algarve as a sustainable tourist destination; to add emotional value to the eno-gastronomic experiences of those who visit us, as well as, to publicize the authenticity of the destination; and to enhance the set of knowledge and customs of the local population, aiming to improve the quality of life of the resident.

Specific aims include: (i) enhance the innovation activities of enterprises in the tourism area through the introduction of new products, services and differentiated experiences or differentiated products, services and experiences significantly improved in relation to their current characteristics; (ii) innovate in the use of local products, associated with Mediterranean roots in the development of tourist experiences that contribute to healthy and sustainable lifestyles; (iii) promote the structuring of the offer of gastronomic and enological experiences in relation to the Mediterranean Diet, which are economically sustainable for commercialization by the regional enterprises; (iv) to know the flavours, the traditional skills and the experiences, which provoke a greater sentimental relationship between the tourist visitors and the places they visit; (v) to study the determinants of emotional value associated with eno-gastronomic experiences based on the main elements of the Mediterranean Diet; (vi) identify feelings/emotions associated with the products and services of the Mediterranean Diet, through comments from tourists expressed on social networks; and (vii) research how tourist companies, related to the marketing of products and services related to the Mediterranean diet, can take greater advantage of technology.

The tourism experience is composed of multiple products and services (tangible and intangible) and combines a wide range of actors, with different motivations and needs. In the development of the research proposal contained, local resources and products will be used, related to the main foods and elements of the Mediterranean Diet, with the purpose of promoting local consumption, market sustainability and the rural economies of the Algarve based upon proposals of responsible consumption.

Figure 2. Sequence of HoST Lab's activities
Source: Own elaboration

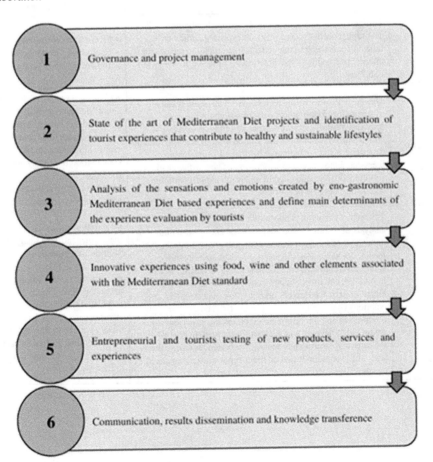

The activities related to the project consider six domains as presented in figure 2, and the corresponding objectives and investigation methods in table 1.

The *Competence Center for the Mediterranean Diet* (CCDM), which has a national focus and acts as a forum for research, knowledge sharing, and articulation on the MD topic, was created by the Portuguese government in Tavira. To enhance the corresponding collaboration at the national and international levels, it brings together agents engaged in research and innovation, training, knowledge dissemination, and transfer, as well as economic agents and public administration agencies. Different from HoSTLab's objective, which intends to promote the development of new ingredients for the MD, cooperating with research directly in the laboratory and enhancing the Mediterranean Diet to a symbol of innovation. In addition, its hub located in the heart of the UALG, acts as a hub that exalts and promotes scientific cooperation, testing in the laboratory to cooperate directly with businesses in the local tourist environment; transforming scientific information for application in business promoting sustainability, local food innovation and the well-being of communities and tourists. Additionally, it is located geographically in another location (Faro), facilitating access to researchers who are currently linked to research at UALG facilities.

Table 1. HOST Lab´s investigation methods

Activities	Objective	Research Methods
1. Governance and project management	This activity aims to guarantee that project outcomes will accomplish technical and research expected results on time and according to the budget. It is crucial to ensure the internal proceedings flow between all involved in the HOST Lab.	• Meetings to ensure the workload; • Using tools to support the tasks project administrative and financial management
2. State of the art of Mediterranean Diet projects and identification of tourist experiences that contribute to healthy and sustainable lifestyles	This activity aims to collect and systematize the results of research and best practice examples about the Mediterranean Diet and define local products associated with Mediterranean roots that can be used to develop tourist experiences and contribute to healthy and sustainable lifestyles. Innovation and structuring of new offers, and simultaneously create conditions for lab experimentation and testing of those new offers.	• Literature review to develop the state of art about Mediterranean diet; • Secondary data collection about best practices cases to develop case studies reports, involving tourist experiences and contribution to healthy and sustainable lifestyles; Qualitative data collection through focus group activity, to report innovation needs.
3. Analysis of the sensations and emotions created by eno-gastronomic MD based experiences and define main determinants of the experience evaluation by tourists	This activity intends to deepen the state of the art on the analysis of feelings associated with the eno-gastronomic dimension of Mediterranean cuisine, expressed in social media, not neglecting the perspectives of sustainability, and innovation in the offer of new products and services. In parallel, and to assess how much it adds value to the tourist experience through investigating the effect on emotions (basic emotions) and in the five senses (participant observation and interviews based on tasting experiences), we intend to develop a conceptual framework to evaluate the sensorial gastronomic touristic experiences (using the scales of evaluation tested on other types of experiences like events).	• Literature review to support the state of the art on sentiment analysis and the eno-gastronomic dimension of the MD, expressed in social networks; • Literature review to support the state of the art on the emotions and the five senses associated with gastronomic touristic experiences; • Data collection through survey; preprocessing, feature extraction, application of methods for sentiment analysis (according to literature review).

continues on following page

Table 1. Continued

Activities	Objective	Research Methods
4. Innovative experiences using food, wine and other elements associated with the Mediterranean Diet standard	The aim of the activity is to develop innovative experiences using foods associated with the Mediterranean Diet from Algarve. Ingredients of natural origin with emulsifying, gelling and stabilizing functions will be used to change the texture of traditional foods allowing different applications.	• Literature review to develop innovative experiences using foods associated with the MD from Algarve. • Quantitative data collection through survey in public events for tourists and national visitors; • Mixed data collection order to validate the acceptability of matching experiences, in gastronomy experts' workshop.
5. Entrepreneurial and tourists testing of new products, services and experiences	This activity aims to evaluate the sensory result of tourist testing of new products, services and experiences in a real context, doing it along the *Rota do Petisco* and to evaluate the economic and social improvements among the regional partners and local populations.	• Quantitative data collection through survey in *"Rota do Petisco"*, to assess the acceptability and feelings experienced by tourists in the context of restaurants; • Literature review and secondary data collection to develop a methodology and monitoring system to evaluate economic and social impact in partner restaurants and local populations by the introduction of innovations provided through new offers and Economic Nutrition Labels.
6. Communication, results dissemination and knowledge transference	With the main highlight for the eno-gastronomic dimension, the sharing of results will be carried out through a platform that will be developed to contribute to the transfer of knowledge between academia and the business sector, considering the agents related to gastronomy and the sustainability of tourists destinations. The platform will include information about the different communication public events. The dissemination of results will consist of several documents, reports and scientific publications and ensures open and online access for any user to all peer-reviewed scientific publications relating to the project results and documents produced.	All methods have the aim of disseminate the information collected along the project: • Creation of webpage and social networks; • Video/trailer with project timeline; • Final Brochures; • Video documentary based on Lab testing and experiences delivered on real • Conference and e-Book.

Source: own elaboration

4. RESULTS

As results expected from the HoSTLab creation, include the production of a system for assessing and monitoring the sustainability of new experiences, products and services associated with the Mediterranean Diet, including a set of reference indicators for assessing the tangible economic impact of these new offers (development of economic nutrition labels).

Sentiment analysis will aim to identify new opportunities to innovate in terms of products and services, as well as to find indicators that contribute to managing the reputation associated with the tourism sector. For new products or services, resulting from the project, it is intended to analyse consumer behaviour through the investigation of their feelings, expressed in the digital environment.

The research process includes: (i) the collection of information from tourism stakeholders (catering companies, hotels, local producers, visitors) and the HoST Lab foundation which will take place in the first phase of innovation and the structuring of new offerings.

The following phases of the research include: (ii) the design and testing of new products, services and culinary experiences as one of the dimensions of the global tourist experience; (iii) the development of data sheets and mapping of resources associated with product proposals, services, and experiences, as well as (iv) tests of developed products, and finally (v) the study of their respective viability and economic and financial sustainability and finally (vi) reporting and result in transference.

The methodology will include a triangulation of traditional methods (interviews, focus groups, workshops), but also creative labs, and developing a digital solution for sharing research results (webpage, digital survey, sentiment analysis). The research is on progress and results are in an early testing phase and the subject will be explored in other research paper.

5. CONCLUSION, LIMITATIONS, AND FUTURE RESEARCH

HoST Lab expects to be a joint forum for research, sharing, and learning between the academy, and the tourism and hotel sector, in which proposals for the innovation of new products, services, and experiences associated with the Mediterranean Diet are developed, and whose results will be transferred to the community, with the ambition to contribute to a more sustainable destination, and promoting the well-being of the local population and visitors.

For the moment, data is merely qualitative and limited to present situation analysis. For now, the materials to develop and the ones already developed are presented.

The project is in the initial phase. Due to the pandemic caused by Covid-19 and the uninterrupted lockdown, the HoSTLab building construction was finalised with a delay of around one year.

Some of the limitations of the present study come from the inherent nature of the object of study. Food and food innovation involves several different agents and stakeholders; experimentation and testing require specific availabilities not easy to congregate at the same time and space. The ingredients were developed for testing facing some obstacles. The next steps of the research will have a special focus on the evaluation of the first stage of the project implementation and the barriers detected. The next stage requires a negotiating process with the establishments' managers because they have different dimensions and quality patterns (restaurants and hospitality). For the testing to proceed successfully, in the different types of establishments, it was identified a need to transfer knowledge directly to the peers, since some difficulties were encountered regarding the application of the ingredients in the locations. The following adversities were identified: (i) difficulty in large-scale production due to high value-added base ingredients and no conditions for their financial acquisition; (ii) difficulty in transporting the products, to maintain the quality until the place and time of use; (iii) product storage capacity, maintaining its characteristics, until the moment of consumption. In addition to the testing issue, other difficulties were foreseen in the product development phase – the manufacture of some of the proposed products was not successful (texture problems, among others), and adaptations had to be made. The next steps include testing the ingredients in public events and integrate it in wine paring in the workshop for gastronomy experts (combinations of wine and ingredients).

Hotel restaurants have less pressure during the meals and display a bigger overture to experimentation than smaller restaurants. At this point, due to procedures inflexibility resulting from research practices at the University, and the lack of integrated experience with such transversal competencies among researchers, a collection of informal, administrative, and cultural barriers was documented identified a group of informal, administrative, and cultural barriers.

The enormous lack of studies and research about innovation and knowledge transference in the food, gastronomy and tourism industry related to the Mediterranean Diet in the Algarve do not allowed to deepen the topic at this moment. Focus group proposed on the activity " State of the art of Mediterranean Diet projects and identification of tourist experiences that contribute to healthy and sustainable lifestyles" will contribute to the scarcity of literature in the topic. The focus group, involving tourism stakeholders (restaurants, hotels, local producers) has the intention to understand the needs for innovation regarding MD in Algarve; and, contribute to the robustness of information regarding the use of MD in the business strategies of hotels, restaurants and tourism companies.

Sentiment analysis is not explored at this phase. Next research publication should include analysis applied to gastronomy, expressed in social media about MD (text mining techniques).

Development and introduction of the innovative economic nutrition labels in regional economics and evaluation of economic and financial viability and sustainability of each new product, service and experience are not explored at this stage. Is expected in future research to develop a scientific paper about methodology and monitoring system to evaluate economic and social impact in partners restaurants and local populations by the introduction of innovations provided through new offers and Economic Nutrition Labels.

A food innovation hub, with a special focus on the Mediterranean Diet, with the participation of the tourism and hospitality sector, will certainly help the knowledge transference and the discussion of the better ways to commercialize the results of the research developed where gastronomy is a key point for the Mediterranean Diet Concept.

ACKNOWLEDGMENT

This paper is financed by National Funds provided by FCT- Foundation for Science and Technology through project UIDB/04020/2020 and project HoST Lab nr. 72592 financed by SAICT (CRESC2020).

REFERENCES

Brites, C. (2015). Os cereais no contexto da dieta mediterrânica. In A. Freitas, J. P. Bernardes, M. P. Mateus, & N. Braz (Eds.), *Dimensões da Dieta Mediterrânica* (pp. 181–195). Universidade do Algarve.

Carral, E., del Río, M., & López, Z. (2020). Gastronomy and Tourism: Socioeconomic and Territorial Implications in Santiago de Compostela-Galiza (NW Spain). *International Journal of Environmental Research and Public Health*, 17(17), 6173. doi:10.3390/ijerph17176173 PMID:32854422

Covas, A., & Covas, M. M. (2014). *Os territórios-rede, a inteligência territorial da 2.ª ruralidade.* Editora Colibri.

Covas, A., & Covas, M. M. (2015). A Dieta Mediterrânica: entre a tradição e a inovação – Uma oportunidade para o rural tradicional algarvio. In A. Freitas, J.P. Bernardes, M.P. Mateus & N. Braz (Eds.). *Dimensões da Dieta Mediterrânica* (pp. 277-294). Faro: Universidade do Algarve Dieta Mediterrânica (2018).

Dupeyras, A. (2016). *Growth of Sharing Economy in Tourism: Developing a Balanced Policy Response.* OECD.

Freitas, A., Bernardes, J. P., Mateus, M. P., & Braz, N. (2015). *Dimensões da Dieta Mediterrânica.* Universidade do Algarve.

Freitas, A., Braz, N., Bernardes, J. P., Cruz, A. L., Quintas, C., Gonçalves, A., Romano, A., & Mateus, M. P. (2022). *Mediterranean Diet: a multidisciplinary approach to develop a new territorial strategy.*

Gálvez, J. C. P., Granda, M. J., & Guzmán-López, T. (2017). Local gastronomy, culture and tourism sustainable cities: The behavior of the American tourist. *Sustainable Cities and Society, 32,* 604–512. doi:10.1016/j.scs.2017.04.021

Graça, P. (2014a). Breve história do conceito de dieta Mediterrânica numa perspetiva de saúde. *Revista Fatores de Risco, 31,* 20–22.

Graça, P. (2014b). Dieta Mediterrânica: uma realidade multifacetada. In A. Freitas, J. P. Bernardes, M. P. Mateus, & N. Braz (Eds.), *Dimensões da Dieta Mediterrânica* (pp. 19–27). Universidade do Algarve.

Jong, A., & Varley, P. (2018). Food tourism and events as tools for social sustainability? *Journal of Place Management & Development, 11*(3), 277–295. doi:10.1108/JPMD-06-2017-0048

Kim, E., Tang, L., & Bosselman, R. (2018). Measuring customer perceptions of restaurant innovativeness: Developing and validating a scale. *International Journal of Hospitality Management, 74,* 85–98. doi:10.1016/j.ijhm.2018.02.018

Kiráľová, A., & Hamarneh, I. (2017). Local gastronomy as a prerequisite of food tourism development in the Czech Republic. *Marketing and Management of Innovations, 2*(2), 15–25. doi:10.21272/mmi.2017.2-01

Kiráľová, A., & Malec, L. (2021). Local Food as a Tool of Tourism Development in Regions. *International Journal of Tourism and Hospitality Management in the Digital Age, 5*(1), 54–68. doi:10.4018/IJTHMDA.20210101.oa1

Lee, C., Hallak, R., & Sardeshmukh, S. (2016). Innovation, entrepreneurship, and restaurant performance: A higher-order structural model. *Tourism Management, 53,* 215–228. doi:10.1016/j.tourman.2015.09.017

Manios, Y., Detopoulou, V., Visioli, F., & Galli, C. (2006). Mediterranean diet as a nutrition education and dietary guide: Misconceptions and the neglected role of locally consumed foods and wild green plants. *Forum of Nutrition, 59,* 154–170. doi:10.1159/000095212 PMID:16917178

Moore, M., & Westley, F. (2011). Surmountable Chasms: Networks and Social Innovation for Resilient Systems. *Ecology and Society, 16*(1), 5. doi:10.5751/ES-03812-160105

Moreno, I. (2015). Culturas mediterrânicas e sistemas alimentares: continuidades, imaginários e novos desafios. In A. Freitas, J. P. Bernardes, M. P. Mateus, & N. Braz (Eds.), *Dimensões da Dieta Mediterrânica* (pp. 51–79). Universidade do Algarve.

Ortega, C., Nogueira, C., & Pinto, H. (2013). Sea and Littoral Localities' Economy: Exploring Potentialities fora Maritime Cluster - An Integrated Analysis of Huelva, Spain andAlgarve, Portugal. *Journal for Maritime Research, 10*(2), 35–42.

Pimentel de Oliveira, D., & Pitarch-Garrido, M. D. (2022). Measuring the sustainability of tourist destinations based on the SDGs: the case of Algarve in Portugal: tourism agenda-2030. *Tourism Review*. Advance online publication. doi:10.1108/TR-05-2022-0233

Queiroz, J. (2015). A Dieta Mediterrânica e a UNESCO: memória breve de um reconhecimento mundial. In A. Freitas, J. P. Bernardes, M. P. Mateus, & N. Braz (Eds.), *Dimensões da Dieta Mediterrânica* (pp. 29–47). Universidade do Algarve.

Real, H. & Graça, P. (2019a). Marcos da história da Dieta Mediterrânica, desde Ancel Keys. *Acta Portuguesa de Nutrição*, *17*, 06-14. doi:10.21011/apn.2019.1702

Real, H. & Graça, P. (2019b). Perceções de utilização do conceito de Dieta Mediterrânica, potencial utilização indevida e perspetivas a explorar. *Revista española de comunicación en salud*, *10*(2), 147-159. doi:10.20318/recs.2019.4824

Richards, G. (2012). Food and the tourism experience: major findings and policy orientations. In D. Dodd (Ed.), *Food and the Tourism Experience* (pp. 13–46). OECD. doi:10.1787/9789264171923-3-en

Santos, J. A. C., Santos, M. C., Pereira, L. N., Richards, G., & Caiado, L. (2020). Local food and changes in tourist eating habits in a sun-and-sea destination: A segmentation approach. *International Journal of Contemporary Hospitality Management*, *35*(3), 3501–3521. doi:10.1108/IJCHM-04-2020-0302

Saulle, R., & la Torre, G. (2010). The Mediterranean Diet, recognized by UNESCO as a cultural heritage of humanity. *Italian Journal of Public Health*, *7*(4), 414–415. doi:10.2427/5700

Serra, M. (2016). Algarve - relação enogastronómica. Tese de mestrado. Faro: ESGHT-Universidade do Algarve.

Serra-Majem, L., Bach-Faig, A., & Raidó-Quintana, B. (2012). Nutritional and cultural aspects of the Mediterranean diet. *International Journal for Vitamin and Nutrition Research*, *82*(3), 157–162. doi:10.1024/0300-9831/a000106 PMID:23258395

Stroink, M. L., & Nelson, C. H. (2013). Complexity and food hubs: Five case studies from Northern Ontario. *Local Environment*, *18*(5), 620–635. doi:10.1080/13549839.2013.798635

Tiganis, A., & Tsakiridou, E. (2022). Local food consumption by foreign tourists in Greece. *International Journal of Tourism Policy*, *12*(2), 70–83. doi:10.1504/IJTP.2022.121898

Trichopoulou, A. (2021). Mediterranean diet as intangible heritage of humanity: 10 years on. *Nutrition, Metabolism, and Cardiovascular Diseases*, *31*(7), 1943–1948. doi:10.1016/j.numecd.2021.04.011 PMID:34059382

United Nations Environmental Development Programme/World Tourism Organization. (2005). Making tourism more sustainable. A guide for policymakers. United Nations Environment Programme, Division of Technology, Industry and Economics and World Tourism Organization. Paris, Madrid: UNEP, WTO.

Valagão, M. M. (2015). Identidade alimentar mediterrânica de Portugal e do Algarve. In A. Freitas, J. P. Bernardes, M. P. Mateus, & N. Braz (Eds.), *Dimensões da Dieta Mediterrânica* (pp. 155–179). Universidade do Algarve.

Yu, C.-E., & Zhang, X. (2020). The embedded feelings in local gastronomy: A sentiment analysis of online reviews. *Journal of Hospitality and Tourism Technology*, *11*(3), 461–478. doi:10.1108/JHTT-02-2019-0028

Zhang, T., Chen, J., & Hu, B. (2019). Authenticity, Quality, and Loyalty: Local Food andSustainable Tourism Experience. *Sustainability (Basel)*, *11*(2), 3437. doi:10.3390u11123437

ENDNOTE

[1]   HOST Lab is a project led by the University of the Algarve, related to basic foods integrated in the Mediterranean Diet, and to the promotion of the main elements of this Common Heritage of Humanity. The total investment is 644.291,94 euros and it is financed by Regional Operational European Program with an eligibility of: 599.273,30. The calendar includes a total of 36 months duration.

Chapter 10

Investigating the Implications of Virtual Reality and Augmented Reality in Tourism Marketing:
A Systematic Literature Review of Publications From 2010 to 2023

Nesenur Altinigne
Istanbul Bilgi University, Turkey

ABSTRACT

This literature review aims to investigate the developments of virtual reality (VR) and augmented reality (AR) research in tourism marketing. This chapter also highlights fruitful directions for tourism marketing research regarding VR and AR applications. A total of 31 full-length articles published between 2010 to 2023 were retrieved from the Web of Science database and reviewed. The theoretical backgrounds of the articles were thoroughly examined, and a detailed report on the research progress of the theories and research methodologies are presented. Finally, future research directions for the improvement of the existing literature are explained.

INTRODUCTION

The expansion of the internet and technological innovations have changed the form of the hospitality industry and influenced how consumers perceive and consume tourism destinations. Latest technological innovations (i.e., virtual reality, augmented reality, artificial intelligence) create opportunities for destination marketing companies to offer several options for their potential visitors to experience different destinations. Even though Virtual Reality (VR) and Augmented Reality (AR) provide numerous opportunities and challenges for companies to enrich consumer experience, practical applications have become prevalent in the last decade. The existing literature on the subject is unstructured. Therefore, this systematic literature review is conducted to synthesize the latest developments in the field and to propose avenues for future research.

DOI: 10.4018/978-1-6684-6985-9.ch010

AR and VR technologies establish various sources of value, such as creating entertainment value, supporting purchase decisions, and improving product usage (Kostin, 2018). AR differs from VR in that it enhances the perception of reality rather than replacing it. Whereas VR generates an artificial environment in which the user engages in a simulation of the real world or a totally fictitious world (Bretonès et al., 2010), AR applications allow users to merge computer-generated content and real-life and to communicate with virtual elements in the real world (Shabani et al., 2018). The approaches that AR and VR embrace inspired hospitality and tourism marketing researchers to further investigate the role of these technologies in promoting e-marketing (Shabani et al., 2018), generating impactful advertising strategies (Phua and Kim, 2018), and designing compelling customer experiences.

The COVID-19 pandemic put the tourism industry in an extremely difficult situation. According to the UN report, in January 2021, tourist arrivals worldwide decreased by 87% compared to January 2020 (UN, 2021). This fall forced the hospitality sector to change its business models to overcome the challenge. Thus, service providers in the tourism sector adopted new technologies such as virtual travel, virtual tours, and AR-enhanced options. Virtual travel helps tourists to explore the destinations before/ without physically visiting there. For instance, the Patagonia VR experience presents a lively experience that allows users to explore Monte Fitzroy and Laguna Sucia, where accessing is difficult in real life. Also, virtual tours of hotels provide potential visitors an opportunity to experience the atmosphere before they book their rooms. Various upscale hotels are now adopting these tours as a tool to present their locations and facilities online that can be reached by simply using a smartphone. After visitors make their reservations with the help of VR and start their dream vacation, then AR will be there with different applications to enhance real-life experiences (Marr, 2021).

VR / AR applications have become a necessity for the hospitality sector. In 2023 the global AR and VR market is expected to grow to $304.4 million (AR Report, 2020). Hence, this chapter mainly aims to investigate the recent advances in virtual reality (VR) and augmented reality AR) research in tourism marketing. It also aims to identify fruitful directions for tourism and hospitality marketing research on VR and AR applications. A literature review of full-length articles published between 2010 to 2023 (including the post-pandemic period) is conducted. Since the Web of Science (WOS) is the database that includes the main body of VR and AR research in diverse scientific disciplines in peer-reviewed journals (Antons and Breidbach, 2018), it is examined with the keywords of the chapter.

The scope of this systematic literature review comprises both empirical and conceptual studies of VR and AR usage in tourism and hospitality marketing. First, a non-systematic literature search with the aim of extracting additional keywords (Boolean phrases based) is conducted. Then, a more cautious approach is adopted. Articles studying VR and AR in the hospitality marketing context across disciplines and related articles in fields that are core or adjacent to hospitality and tourism (i.e., business, management, or psychology) are included. Then, screening processes are executed. At the beginning of January 2023, the completeness of the dataset was validated by cross-checking it with existing reviews and the literature review sections of published articles on VR and AR. And the final dataset is adopted (n 31).

This paper represents one of the first attempts to extensively review research improvements on VR and AR in tourism and hospitality marketing from 2010 to 2023. This chapter summarizes the distribution of the published articles. Then, it synthesizes the motivations, dimensions, and consequences of AR- and VR-related experiences and develops a theoretical framework. This chapter also presents a report on the progress of theories and research methodologies, providing valuable background to indicate the gaps in the literature. Finally, it explains promising directions for improving the current knowledge about VR and AR utilization in hospitality and tourism management. By doing so, this chapter shows various outcomes of using VR and AR technologies to encourage marketers in tourism and hospitality marketing to integrate recent technological developments into their future plans. Besides, this chapter indicates the research gap regarding the adverse effects of enhanced VR and AR applications in tourism marketing and the different customer groups' diverse needs and expectations (such as the different types of authenticity).

VR AND AR POTENTIAL FOR TOURISM MARKETING

The rapid changes in technology are altering all industries. Virtual Reality (VR) and Augmented Reality (AR), two of the latest offspring of technology, are widely adopted by different industries. VR and AR are two types of extended reality. But there are several key differences between the two. First, VR presents an engaging virtual environment, while AR enhances a real-world scene. Second, a headset is required to experience VR but not for AR. Third, VR helps users experience an entirely fictional world, whereas AR keeps users in contact with the real world. And fourth, VR users' presence is controlled by the system, while AR users' presence is under their control.

Especially after the COVID-19 pandemic, the importance and the speed of adoption of VR and AR increased in the hospitality sector, significantly altering the industry (Loureiro et al., 2020). The ability of VR to depict virtual landscapes allows potential visitors to experience and compare different destinations and enables them to gain information about several places and attenuate risk perceptions (Ying et al., 2022; Yung and Khoo-Lattimore, 2019). Such an opportunity (living the experience before you go) encourages potential visitors to try the firms that provide this option. It also helps marketers to promote places effectively (Yung and Khoo-Lattimore, 2019; McLean and Barhorst, 2022). In contrast with the traditional marketing channels that are less engaging–such as static images and texts, videos, and websites, VR and AR offer more enriched experiences even during the promotion process of touristic destinations (Liu and Huang, 2023).

There is growing literature about the role of VR / AR in tourism marketing. And the current knowledge is limited regarding the antecedents of VR and AR adoption and their effects on revisit intention (Guttentag, 2010; Huang et al., 2016; Hudson et al., 2019). Furthermore, the body of knowledge at hand is dispersed. Bringing together and integrating the present knowledge will help unveil the challenges and problems of using these technologies. This chapter aims to fill this gap in the literature by synthesizing the research studies on VR / AR in tourism marketing published between the years 2010 and 2023.

METHODOLOGY

Search Process

A systematic literature review presented in **Table 1** (Siddaway et al., 2019, Law et al., 2014) was conducted to integrate the existing literature on the role of VR and AR in tourism marketing. The literature review was carried out between November 2022 and March 2023. All the refereed journal publications from the Web of Science (WOS) database, including the main body of VR/AR research in diverse fields (Antons & Breidbach, 2018), were included as a first step. The extent of the literature review encompasses both empirical and conceptual studies.

Table 1. Explanation of the research process

Literature Search	Selection & Coding	Synthesis
Search in the Web of Science (WOS) database using the following Boolean Formula, TS=((virtual reality OR VR OR augmented reality OR AR) AND (tourism OR hospitality) AND (marketing))	**Title / Abstract Screening** Selection of publications according to WOS category and screening of 65 articles titles and abstracts. **Full-Text Screening** Review of 40 publications' full texts according to exclusion/inclusion criteria. **Coding** Coding of the remaining 34 based on the criteria. **Final Selection** Identification of the studies in the 31 articles according to hospitality and tourism marketing.	Qualitative content analysis. Development of a framework. Identification of areas for future research.

Screening and Data Extraction

At the beginning of the study, as a first step of screening, some of the WOS categories (e.g., chemistry, engineering) were left outside the research scope. A highly conservative approach is embraced, and all articles (book chapters and proceedings were removed) studying VR /AR in tourism marketing, destination marketing, or hospitality marketing across various disciplines between 2010 and 2023 were included. Then, the titles and abstracts of the remaining sixty-five were assessed according to their fit to this chapter's scope and research question (Calabro et al., 2019). Next, the full texts of all remaining articles (n=40) were examined to find out if they directly feed the chapter's main topic. In January 2023, the data was reevaluated, and its completeness was checked by investigating the existing reviews and literature review sections of the empirical articles published in the WOS database. As a result of the screening process final dataset (n=31 articles) was identified.

FINDINGS

Distribution of the Articles

Thirty-one papers about VR and AR technologies in tourism marketing were investigated. The papers were broken down by the research design as experimental design, survey, thematic analysis, and netnography. A comparison between 2010 and 2023, in terms of research progress of the VR and AR literature, is presented in Figure I. It can easily be seen that the last two years have witnessed a radical increase in published articles about VR / AR in tourism marketing. A total of 20 articles were published during the years 2020-2023, and 2022 is the peak year with 12 articles. VR / AR related publications were published in various journals such as Current Issues in Tourism, Journal of Travel & Tourism Marketing, and Journal of Vacation Marketing.

Figure 1. Progress of the VR/AR studies in the literature according to years

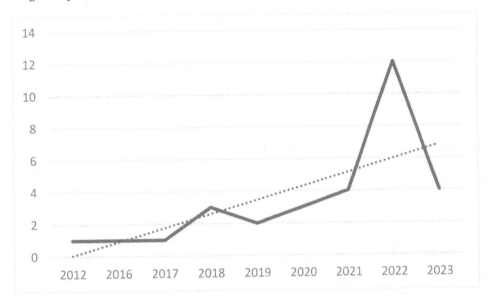

Research Cultural Contexts and Design

The findings of this literature review showed that most of the studies about the role of VR / AR in tourism marketing had been conducted in North America (n=9) (United States, Canada) and Asia Pacific (n=12) (China, Taiwan, Japan). On the other hand, eight studies were conducted in Europe, the U.K., Italy, Spain, Portugal, and two were in Australia.

Among the thirty-one studies, surveys (questionnaires) are identified as the most common method of data collection with nineteen studies (Marasco et al., 2018; Jung et al., 2018; Li & Chen, 2019; Lee et al., 2020; Pinto et al., 2022; Huang, 2023). The experimental design was the second favorite approach in terms of the research design with seven studies (Flavian et al., 2019; Huang et al., 2020; Slevitch et al., 2022; Bogicevic et al., 2021). In the studies adopting qualitative methods, thematic analysis, in-depth interviews, netnography, and content analysis have been utilized (tom Dieck et al., 2018; Subawa et al., 2021; Sancho-Esper et al., 2022; Dybsand, 2022; Zheng et al., 2022. In qualitative studies, respondents first try a VR / AR application and then attend an in-depth interview about their experience. For instance, in tom Dieck et al. (2018), a VR application is developed to promote the Lake District and attract potential visitors. The newly developed VR application incorporating natural sounds offers potential visitors to experience landscapes and bird views over the district in a peaceful environment. After they try the application, they then join a semi-structured interview and talk about its usability level and emotional, personal, and hedonic benefits.

Theoretical Background of the Studies

This study follows a widely used approach adopted by review papers (Nicholls, 2010; Yung & Khoo-Lattimore, 2017; Wei, 2019) to provide a useful perspective on theoretical progress regarding VR / AR related studies in tourism marketing. The present literature review synthesized the theories that prior

VR/AR research was utilized to clarify VR/AR-related behaviors/experiences in tourism marketing. This section of the paper presents a detailed categorization of the theories helping to comprehend VR / AR behaviors or experiences.

Antecedents-related theories. With the rapid progress of extended reality devices (VR / AR) for consumer use worldwide and the undeniable significance of their influence on potential visitors' decision-making processes, an increasing focus has been recently put on their acceptance process (tom Dieck et al., 2018; Lee et al., 2020). The theories in this category mainly focus on identifying the antecedents of VR / AR experience. Several papers reviewed for this study adopted the TAM (technology acceptance model) and UTATUT (Unified theory of acceptance and use of technology), such as Liu and Huang (2023), Sancho-Esper et al. (2022), Marasco et al. (2018), and Jung et al. (2018), tom Dieck et al. (2018). These studies utilized TAM and UTAUT to investigate the effects of perceived visual appeal, perceived usefulness and perceived ease of use, the emotional involvement of the user, aesthetics, perceived enjoyment, social influence, VR-related arousal, and technology anxiety on consumers' attitudes and behavioral intentions in different settings (e.g., cultural heritage and destinations). Furthermore, Huang (2023) integrated the unified theory of acceptance and use of technology (UTAUT) and the theory of planned behavior (TPB) to investigate users' VR tourism behavior intention. Huang (2023) concluded that the perceived benefits (perceived ease of use, perceived usefulness) mediated the relationship between UTAUT concepts and VR usage intention and the mechanisms of UTAUT, such as performance expectancy, effort expectancy, social influence, facilitating conditions, hedonic motivation, price value, habit have attracted increasing attention in the context of VR. Also, Pinto et al. (2022) employed UTAUT 3 to further comprehend the factor leading to Mobile Augmented Reality adoption in Tourism in Porto. The authors' primary purpose was to contribute to the literature on Mobile Augmented Reality; thus, they utilized an extended UTAUT model with greater predictive power and attempted to understand the moderating role of demographic (i.e., gender, age and experience). Pinto et al. (2022) concluded that habit, hedonic motivations, and facilitating conditions are the determinants of using Mobile Augmented Reality in Tourism.

Medai and Wu (2022) enriched the TAM with VAM (value acceptance model) and attempted to examine the effect of the value perceptions of the participants on the intention to participate in online tours and the mediating effects of telepresence and authenticity. Results of the analysis suggested that positive emotions and telepresence positively affect individuals' intention to participate in online tours. Furthermore, interpersonal limitations (an incompatibility of schedules with companions) were found to enhance the effect of telepresence but attenuate the impact of authenticity.

Sancho-Esper et al. (2022) developed a conceptual model that recognizes the role of perceived ease of use and usefulness in users' attitudes and intentions to embrace VR technology. Their paper analyzed older adults' acceptance of VR technology in a tourism marketing context. Hence, technology anxiety regarding using new technology was at the focal point of the study as a contextual and affective factor (Dogruel et al., 2015; Heerink et al., 2010). All the antecedents in their proposed model (perceived usefulness, perceived ease of use, technology anxiety, attitudes towards VR technology) seem to be essential factors among older adults. In other words, when VR technology devices are easy to use and more practical, older consumers have stronger attitudes toward using them. This finding is parallel with the previous conclusions on VR acceptance in tourism marketing (tom Dieck & Jung, 2018; Tussyadiah et al., 2018) and technology acceptance regarding older adults (Mendes-Filho et al., 2018; Tsai et al., 2020). Also, as emphasized by former research (e.g., Wei, 2019), the anxiety level in using new technology (VR) negatively affects both perceived ease of use and perceived usefulness.

Furthermore, Sancho-Esper et al. (2022) concluded that some older adults perceived themselves as incapable of using the VR device without assistance. Therefore, the initial support of an expert is required for these consumer groups.

The reviewed articles also employed the motivational theory to investigate the antecedents of VR / AR related behaviors of consumers in the tourism marketing context (Jung et al., 2018; Li & Chen, 2019). Jung et al. (2018) conducted a cross-cultural study based on TAM and motivational theory to uncover the antecedents of intention to adopt AR, which depends on cultural traits and differences. Their paper focuses on the aesthetic and hedonic characteristics of AR applications and the impact of social influence. Data was collected from two strongly contrasting countries (South Korea and Ireland) according to Hofstede's cultural dimensions. Their study supported the idea that high-power distance, collectivism, and high uncertainty avoidance culture (South Korea) perceive stronger dependence on social influence and the hedonic features of AR. Also, the aesthetics of AR have a strong influence on the perceived enjoyment of individuals.

Another theory embraced by the researchers to shed light on the antecedents of virtual tours in tourism marketing was sensory theory (Dybsand, 2022; Lin & Yeh, 2020). Dybsand (2020) utilized this theory to examine how individuals experienced virtual tours by analyzing 3661 TripAdvisor reviews. The study results indicated that virtual guided tours had a critical role for the participants during the pandemic. Individuals were entertained and socialized during the virtual tour experiences, and as a result, they had a positive mood. Therefore, the author concluded that virtual tours positively affect consumers' desire to visit the destination. This study also unveiled different benefits of virtual tours, such as escaping reality and feeling present.

Process-related theories. Another group of theories is categorized under the roof of process theories because they try to clarify the conceptualizations of the process through which VR/AR technology implications influence consumers. For instance, Zheng et al. (2022) followed the dual processing theory approach to develop a holistic framework and explain the underlying psychological mechanisms of virtual tourism. They facilitated both quantitative and qualitative methods (survey and in-depth interviews). Mainly, they attempted to examine how mental imagery processing of stimuli in virtual tourist attractions affects cognition (learning) and emotion. Furthermore, they aimed to contribute to the existing literature focusing on the positive effects of mental imagery processing. Two rounds of surveys conducted in China indicated that mental imagery processing impacts cognition and emotion, which in turn, together affect future visit intention. Even though prior studies implied a positive effect of mental imagery processing on consumption (Le, Scott, and Lohmann, 2019), Zheng et al. (2022) tried to shed light on the negative impact. If a VR stimulus triggers negative emotions, then the mental imagery processing of the unpleasant feeling may cause a decreased visit intention. The intensity of these negative emotions may hinder the interest in the destination and reduce the intention to visit.

Wang et al. (2022) aimed to explore the underlying mechanism behind the relationships among virtual reality tourism involvement (three dimensions: pleasure, sign, and centrality), place dependence, place identity, and behavioral intentions. The model of the study was built on the involvement theory and the results indicated that pleasure and sign had a positive effect on visit intention, whereas centrality had a negative one. Furthermore, they revealed the mediating role of pleasure, centrality, sign, and behavioral intentions. This study provides new insights into destination marketing by enriching the content with the usage of involvement theory.

The telepresence mediation model was also embraced by various research published between 2010 and 2022 (Lee et al., 2020; Choi, Ok & Choi, 2016; McLean & Barhorst, 2022) to examine the processes of VR /AR intentions. The concept of telepresence is specifically generated in virtual reality research. Basically, telepresence is the "feeling of being there" (Steuer, 1992), and different mediums can produce a certain level of telepresence. (Hyun and O'Keefe, 2012).

The level of telepresence depends on how closely a tool simulates the real world (Wei, 2019).

Choi, Ok and Choi (2016) proposed a model that adopts the telepresence mediation hypothesis model to test the process during which telepresence serves as a mediator for the relationship among informativeness, entertainment, utilitarian performance, hedonic performance, and navigation outcomes. The paper built its argument on the uses and gratifications theory and confirmed the mediating role of telepresence.

Stimulus – Organism – Response paradigm was another widely-used theory regarding process-related ones. The majority of the papers in this review adopted this paradigm (Flavian et al., 2019; Leung et al., 2022; Shin & Jeong, 2022; Talwar et al., 2022). For instance, Talwar et al. (2022) examined the impact of the environmental impact of travel and pandemic travel anxiety (stimulus) on willingness to forgo the pleasure of in-site tourism and post-pandemic VR continuance intention (response). Their study also investigated the mediating roles of attitude toward VR tourism and Eco-guilt (organism) on the beforementioned relationship; travel mode and COVID-19 vaccination status were elaborated as moderators. The findings of the paper indicated a positive relationship between the environmental impact of touristic travel and pandemic travel anxiety with eco-guilt. Pandemic travel anxiety was also associated with attitude towards VR tourism.

Subjective experience-related theories. The third group of theories helps to depict the subjective experiences about the usage of VR/AR from a user's perspective. Flow theory is the most widely used one regarding VR / AR applications in tourism marketing. Chang and Chian (2022) examined the impact of VR as a communication tool for the advertising of destination marketing. The impact of four components (friendliness, interactivity, telepresence, and realness) of the flow experience on destination image and attitude change was examined. Besides, the mediating role of destination image between the aforementioned four factors and attitude change was also investigated. The research participants first viewed a tourism destination marketing video through VR gear and then answered questions about their experience. The findings revealed that two factors (friendliness and telepresence) of flow experience have a positive impact on destination image, and destination image mediates the relationship between these two factors and attitude change.

One such research employing the flow theory was conducted by Huang, Wei and Leung (2020) to understand "how travelers achieve flow states in VR experiences". In their proposed model, the relationship between VR experience type (sightseeing, interactive) and flow experience (focused attention, time distortion, enjoyment), and the mediating roles of arousal (self-reported measure, heart rate measure) and perceived control (self-reported measure, respiratory measure) was scrutinized. Based on the flow theory, this paper combined the physiological measures with self-reported ones to comprehend the individuals' flow experiences. The findings of the study exhibited that arousal has a stronger role in mediating travelers' flow states compared to perceived control.

Outcomes-related theories/theoretical models and concepts. The last set of theories forms the fourth and final category. These theories present an insight into the notion of VR / AR applications in tourism marketing. The remaining theories are briefly explained in this part.

Hyun and O'Keefe (2012) investigated the formation of a virtual destination's image (cognitive, affective and conation) due to incorporating web-mediated virtual information. Bogicevic et al. (2021) utilized the brand personality concept and examined the mediating role of brand coolness in the relationship between VR applications and self-brand connection. The results of the study confirmed the mediating role of brand coolness and its positive impact on self-brand connection.

Subawa et al. (2021) employed three theories (the theory of social change, the theory of knowledge and power, theory of hegemony) to analyze the practices of virtual reality marketing by pursuing a netnographic approach. The findings indicated "hegemony in the practice of virtual reality marketing to tourists and potential consumers, using significant technology". They concluded that virtual reality marketing could be the starting point of a new pattern of marketing.

The details regarding the journal names, authors, and theories employed are all presented in **Table 2**.

CHALLENGES, FUTURE RESEARCH AND LIMITATIONS

In the post-COVID era, VR and AR are both extensively popular tools for tourism marketing. There is an increasing demand for the innovative use of these technologies to create more immersive experiences for potential customers. By employing VR / AR tools in tourism marketing, a firm can establish awareness, increase interest, and generate positive attitudes toward tourism attractions. Even though these tools seem enormously beneficial, there are several challenges that tourism marketers need to overcome.

According to the reviewed literature, features (aesthetic, hedonic characteristics) and quality (functional characteristics) of VR and AR practices are important to the outcome. Specifically, the vividness, the colors, and the telepresence are associated with the outcomes, such as the intention to visit (Jung et al., 2018). On the one hand, investing in AR / VR tools to enhance the quality of the experience may generate positive results, such as an increased number of customers in the short run. But on the other hand, closely simulating the real world might also hurt some customer groups' experiences. For instance, enhanced features might require more complex and detailed interfaces, and older people may feel threatened by this technology. Since the global 70+ population grew by 627 million, from 5% of the total population to 12% in 2022 (HBR, 2022), many of these older adults are potential customers for tourism firms; their abilities and needs should be taken into consideration.

Customers looking for authentic experiences might also be disturbed by the enhanced level of VR / AR applications. Although prior studies found a positive effect on mental imagery processing, if a VR / AR stimulus triggers negative emotions, then processing unpleasant feelings might hurt visit intentions (Zheng et al., 2022). Hence, the long-term effects of VR / AR applications should also be investigated.

The enhanced features of VR / AR will increase telepresence (Geng, 2022). The virtual tours with enhanced telepresence will exactly meet the needs of some visitors and, as a result, might cannibalize the demand for physical tours. Unfortunately, this cannibalization effect is widely observed in different products or services. According to Bakhshi and Throsby (2012), the majority of virtual visitors who previously preferred physical tours now visit the service provider for virtual tours, which signals the cannibalization effect.

The research gap regarding the adverse effects of enhanced VR and AR applications in tourism marketing might be a starting point to close this gap in the literature. Besides, the sensitivities (older adults) and the different needs and expectations (such as the different types of authenticity) of different customer groups should be investigated in detail.

Table 2. Reviewed articles of VR/AR usage in hospitality marketing

No	Authored work	Journal name	Nature of work	Research Design	Cultural Context	Theoretical Background
1	Hyun, M.Y. and O'Keefe, R.M. (2012)	Journal of Business Research	Empirical	Survey	Oceania	Cognition, affect and conation Telepresence
2	Choi, J; Ok, C. and Choi, S (2016)	Journal of Travel & Tourism Marketing	Empirical	Survey	North America	Uses and gratifications theory
3	Tom Dieck et al. (2018)	Leisure Studies	Empirical	Thematic Analysis	Europe	Uses and gratification theory (U&GT), Unified theory of acceptance and use of technology (UTAUT)
4	Marasco et al. (2018)	Journal of Destination Marketing & Management	Empirical	Survey	Europe	Technology acceptance model (TAM) Hedonic theory
5	Jung et al. (2018)	International Journal of Contemporary Hospitality Management	Empirical	Survey	Cross-continental (Asia Pacific and Europe)	Motivational theory Technology acceptance model (TAM)
6	Li and Chen (2019)	Journal of Destination Marketing & Management	Empirical	Survey	Asia Pacific	Motivational theory
7	Flavian et al. (2019)	Journal of Travel & Tourism Marketing	Empirical	Experimental Design	Europe	Stimulus – Organism - Response
8	Lee at al. (2020)	International Journal of Hospitality Management	Empirical	Survey	North America	Attitude-behavior theory
9	Huang, Wei and Leung (2020)	Asia Pacific Journal of Tourism Research	Empirical	Experimental Design	Asia Pacific	Flow theory
10	Slevitch, Chandrasekera and Sealy (2022)	Journal of Hospitality & Tourism Research	Empirical	Experimental Design	North America	Cognitive load theory
11	Bogicevic et al. (2021)	International Journal of Hospitality Management	Empirical	Experimental Design	North America	Brand personality
	Table 2. Continued					
No	**Authored work**	**Journal name**	**Nature of work**	**Research Design**	**Cultural Context**	**Theoretical Background**
12	Subawa et al. (2021)	Current Issues in Tourism	Empirical	Netnography	Asia Pacific	Theory of social change, Theory of knowledge and power, Theory of hegemony
13	An, Choi and Lee (2021)	Journal of Destination Marketing & Management	Empirical	Survey	Oceania	Flow theory
14	Le, Scott and Wang (2021)	Journal of Hospitality and Tourism Management	Empirical	Survey	Cross-continental (Australia and international)	Process theory Construal-level theory
15	Lin and Yeh (2022)	Current Issues in Tourism	Empirical	Zmet	Asia Pacific	Sensory theory

continues on following page

Table 2. Continued

No	Authored work	Journal name	Nature of work	Research Design	Cultural Context	Theoretical Background
16	Sancho-Esper et al. (2022)	Journal of Vacation Marketing	Empirical	Survey and Thematic Design	Europe	Technology acceptance model (TAM)
17	Chang and Chian (2022)	Journal of Hospitality and Tourism Technology	Empirical	Survey	Asia Pacific	Flow theory
18	Wang et al. (2022)	Asia Pacific Journal of Tourism Research		Survey	Asia Pacific	Involvement theory
19	Leung et al. (2022)	Journal of Hospitality and Tourism Technology	Empirical	Survey	Asia Pacific	Stimulus – Organism - Response
20	Dybsand (2022)	Current Issues in Tourism	Conceptual	Content Analysis	Europe	Sensory theory
21	Zheng et al. (2022)	Journal of Travel Research	Empirical	In-depth interview and Survey	Asia Pacific	Dual processing theory
22	Shin and Jeong (2022)	Journal of Travel & Tourism Marketing	Empirical	Survey	North America	Stimulus – Organism - Response

Table 2. Continued

No	Authored work	Journal name	Nature of work	Research Design	Cultural Context	Theoretical Background
23	Medai and Wu (2022)	Current Issues in Tourism	Empirical	Survey	Asia Pacific	Value-based adoption model (VAM)
24	McLean and Barhorst (2022)	Journal of Travel Research	Empirical	Experimental Design	Europe	Presence theory
25	Talwar et al. (2022)	Journal of Sustainable Tourism	Empirical	Survey	Europe	Stimulus – Organism - Response
26	Pinto et al. (2022)	Journal of Information Systems Engineering and Management	Empirical	Survey	Europe	Unified theory of acceptance and use of technology 3 (UTAUT 3)
27	Griffin et al. (2023)	Journal of Vacation Marketing	Empirical	Experimental Design	North America	Cognition, affect and conation
28	Liu and Huang (2023)	Journal of Hospitality and Tourism Management	Empirical	Survey	Asia Pacific	Unified theory of acceptance and use of technology (UTAUT), Technology acceptance model (TAM)
29	Geng (2023)	Current Issues in Tourism	Empirical	Mathematical Modelling	Asia Pacific	Presence theory
30	Guo et al. (2023)	Journal of Hospitality & Tourism Research	Empirical	Experimental Design	North America	Disruption theory
31	Huang (2023)	Journal of Retailing and Consumer Services	Empirical	Survey	Asia Pacific	Unified theory of acceptance and use of technology (UTAUT), Theory of planned behavior (TPB)

The articles for this literature review were retrieved from the Web of Science database; future research may consider other databases such as Scopus.

MANAGERIAL AND THEORETICAL CONTRIBUTIONS

This paper is one of the first attempts to review the research improvements on both VR / AR related tourism marketing applications. The previous literature review papers were generally conducted before 2020 (Wei, 2019). Since the pandemic altered the way of living and doing business, a literature review including the pandemic and post-pandemic period was essential.

One of the aims of this review paper is to present a detailed report on the theories, which helps understand the role of VR / AR in tourism marketing as studied in prior research, to offer a valuable background of theoretical progress. The present research extends the existing knowledge by integrating the theories utilized by other articles related to VR / AR in tourism marketing applications.

The existing literature review also provides implications for the practitioners. This study indicates that the tourism marketing industry utilizes the latest VR / AR Technologies according to different objectives. This paper implies that there is a rapid adoption regarding the usage of VR / AR Technologies. However, since their negative effects have not yet been visible, small organizations might hesitate to invest in these technologies (tom Dieck and Jung, 2017). This study integrates studies about the negative and positive effects of VR / AR technologies in the tourism marketing context. This synthesis might encourage small organizations to learn from the existing findings and assist them in their decision-making processes.

CONCLUSION

After the COVID-19 pandemic attenuated and halted the tourism market, tourism firms sought solutions to overcome the disruption. Focusing on the trend of VR / AR technologies in marketing efforts accelerated the industry and attracted customers by offering opportunities for socialization and entertainment. After the pandemic precautions were removed (e.g., lockdown, mask, distance education and working), many individuals sustained the lifestyle that the pandemic brought to them. This inheritance of the pandemic fastened the adoption process to further technological progress. Hence, VR and AR applications have become a regular part of many customers' daily life.

Even though VR / AR technologies are prevalent in many industries, including tourism, the reviewed articles in this study revealed their negative and positive effects. By understanding the double edge sword nature of VR and AR technologies, firms can make better decisions regarding their marketing efforts.

REFERENCES

An, S., Choi, Y., & Lee, C. K. (2021). Virtual travel experience and destination marketing: Effects of sense and information quality on flow and visit intention. *Journal of Destination Marketing & Management*, *19*, 100492. doi:10.1016/j.jdmm.2020.100492

Antons, D., & Breidbach, C. F. (2018). Big data, big insights? Advancing service innovation and design with machine learning. *Journal of Service Research*, *21*(1), 17–39. doi:10.1177/1094670517738373

Bakhshi, H., & Throsby, D. (2012). New technologies in cultural institutions: Theory, evidence and policy implications. *International Journal of Cultural Policy*, *18*(2), 205–222. doi:10.1080/10286632. 2011.587878

Bogicevic, V., Liu, S. Q., Seo, S., Kandampully, J., & Rudd, N. A. (2021). Virtual reality is so cool! How technology innovativeness shapes consumer responses to service preview modes. *International Journal of Hospitality Management*, *93*, 102806. doi:10.1016/j.ijhm.2020.102806

Bretonès, D. D., Quinio, B., & Réveillon, G. (2010). Bridging virtual and real worlds: Enhancing outlying clustered value creations. *Journal of Strategic Marketing*, *18*(7), 613–625. doi:10.1080/096525 4X.2010.529157

Calabro, A., Vecchiarini, M., Gast, J., Campopiano, G., Massis, A., & Kraus, S. (2019). Innovation in family firms: A systematic literature review and guidance for future research. *International Journal of Management Reviews*, *21*(3), 317–355. doi:10.1111/ijmr.12192

Chang, H. H., & Chiang, C. C. (2022). Is virtual reality technology an effective tool for tourism destination marketing? A flow perspective. *Journal of Hospitality and Tourism Technology*, *13*(3), 427–440. doi:10.1108/JHTT-03-2021-0076

Choi, J., Ok, C., & Choi, S. (2016). Outcomes of destination marketing organization website navigation: The role of telepresence. *Journal of Travel & Tourism Marketing*, *33*(1), 46–62. doi:10.1080/1054840 8.2015.1024913

Dogruel, L., Joeckel, S., & Bowman, N. D. (2015). The use and acceptance of new media entertainment technology by elderly users: Development of an expanded technology acceptance model. *Behaviour & Information Technology*, *34*(11), 1052–1063. doi:10.1080/0144929X.2015.1077890

Dybsand, H. N. H. (2022). 'The next best thing to being there'–participant perceptions of virtual guided tours offered during the COVID-19 pandemic. *Current Issues in Tourism*, 1–14. doi:10.1080/1368350 0.2022.2122417

Fong, S. C. (1998). Conceptualizing consumer experiences in cyberspace. *European Journal of Marketing*, *32*(7/8), 655–663. doi:10.1108/03090569810224056

Gartner, Inc. (2016). *Gartner says worldwide wearable devices sales to grow 18.4 percent in 2016.* Gartner, Inc. http://www.gartner.com/newsroom/id/ 3198018/

Geng, W. (2023). Whether and how free virtual tours can bring back visitors. *Current Issues in Tourism*, *26*(5), 823–834. doi:10.1080/13683500.2022.2043253

Griffin, T., Guttentag, D., Lee, S. H., Giberson, J., & Dimanche, F. (2023). Is VR always better for destination marketing? Comparing different media and styles. *Journal of Vacation Marketing*, *29*(1), 119–140. doi:10.1177/13567667221078252

Guo, K., Fan, A., Lehto, X., & Day, J. (2021). Immersive digital tourism: The role of multisensory cues in digital museum experiences. *Journal of Hospitality & Tourism Research (Washington, D.C.)*, 10963480211030319. doi:10.1177/10963480211030319

Guttentag, D. A. (2010). Virtual reality: Applications and implications for tourism. *Tourism Management*, *31*(5), 637–651. doi:10.1016/j.tourman.2009.07.003

Heerink, M., Kröse, B., Evers, V., & Wielinga, B. (2010). Assessing acceptance of assistive social agent technology by older adults: The Almere Model. *International Journal of Social Robotics*, *2*(4), 361–375. doi:10.100712369-010-0068-5

Huang, X. T., Wei, Z. D., & Leung, X. Y. (2020). What you feel may not be what you experience: A psychophysiological study on flow in VR travel experiences. *Asia Pacific Journal of Tourism Research*, *25*(7), 736–747. doi:10.1080/10941665.2019.1711141

Huang, Y. C. (2023). Integrated concepts of the UTAUT and TPB in virtual reality behavioral intention. *Journal of Retailing and Consumer Services*, *70*, 103127. doi:10.1016/j.jretconser.2022.103127

Huang, Y. C., Backman, K. F., Backman, S. J., & Chang, L. L. (2016). Exploring the implications of virtual reality technology in tourism marketing: An integrated research framework. *International Journal of Tourism Research*, *18*(2), 116–128. doi:10.1002/jtr.2038

Hudson, S., Matson-Barkat, S., Pallamin, N., & Jegou, G. (2019). With or without you? Interaction and immersion in a virtual reality experience. *Journal of Business Research*, *100*, 459–468. doi:10.1016/j.jbusres.2018.10.062

Hyun, M. Y., & O'Keefe, R. M. (2012). Virtual destination image: Testing a telepresence model. *Journal of Business Research*, *65*(1), 29–35. doi:10.1016/j.jbusres.2011.07.011

Jonathan, S. (1992). Defining virtual reality: Dimensions determining telepresence. *Journal of Communication*, *42*(4), 73–93. doi:10.1111/j.1460-2466.1992.tb00812.x

Jung, T. H., Lee, H., Chung, N., & tom Dieck, M. C. (2018). Cross-cultural differences in adopting mobile augmented reality at cultural heritage tourism sites. *International Journal of Contemporary Hospitality Management*, *30*(3), 1621–1645. doi:10.1108/IJCHM-02-2017-0084

Kostin, K. B. (2018). Foresight of the global digital trends. *Strategic Management-International Journal of Strategic Management and Decision Support Systems in Strategic Management*, *23*(1).

Law, R., Buhalis, D., & Cobanoglu, C. (2014). Progress on information and communication technologies in hospitality and tourism. *International Journal of Contemporary Hospitality Management*, *26*(5), 727–750. doi:10.1108/IJCHM-08-2013-0367

Le, D., Scott, N., & Lohmann, G. (2019). Applying Experiential Marketing in Selling Tourism Dreams. *Journal of Travel & Tourism Marketing*, *36*(2), 220–235. doi:10.1080/10548408.2018.1526158

Le, D., Scott, N., & Wang, Y. (2021). Impact of prior knowledge and psychological distance on tourist imagination of a promoted tourism event. *Journal of Hospitality and Tourism Management*, *49*, 101–111. doi:10.1016/j.jhtm.2021.09.001

Leung, W. K., Cheung, M. L., Chang, M. K., Shi, S., Tse, S. Y., & Yusrini, L. (2022). The role of virtual reality interactivity in building tourists' memorable experiences and post-adoption intentions in the COVID-19 era. *Journal of Hospitality and Tourism Technology*, *13*(3), 481–499. doi:10.1108/JHTT-03-2021-0088

Li, T., & Chen, Y. (2019). Will virtual reality be a double-edged sword? Exploring the moderation effects of the expected enjoyment of a destination on travel intention. *Journal of Destination Marketing & Management*, *12*, 15–26. doi:10.1016/j.jdmm.2019.02.003

Lin, L. Z., & Yeh, H. R. (2022). Using ZMET to explore consumers' cognitive model in virtual reality: Take the tourism experience as an example. *Current Issues in Tourism*, 1–15. doi:10.1080/13683500.2022.2147052

Liu, C., & Huang, X. (2023). Does the selection of virtual reality video matter? A laboratory experimental study of the influences of arousal. *Journal of Hospitality and Tourism Management*, *54*, 152–165. doi:10.1016/j.jhtm.2022.12.002

Loureiro, S. M. C., Guerreiro, J., & Ali, F. (2020). 20 years of research on virtual reality and augmented reality in tourism context: A text-mining approach. *Tourism Management*, *77*, 104028. doi:10.1016/j.tourman.2019.104028

Marasco, A., Buonincontri, P., Van Niekerk, M., Orlowski, M., & Okumus, F. (2018). Exploring the role of next-generation virtual technologies in destination marketing. *Journal of Destination Marketing & Management*, *9*, 138–148. doi:10.1016/j.jdmm.2017.12.002

Marr, B. (2021, June 11). Extended reality in tourism: 4 Ways VR and AR can enhance the travel experience. *Forbes*. https://www.forbes.com/sites/bernardmarr/2021/06/11/extended-reality-in-tourism-4-ways-vr-and-ar-can-enhance-the-travel-experience/?sh=50729fbd82ff

McLean, G., & Barhorst, J. B. (2022). Living the experience before you go... but did it meet expectations? The role of virtual reality during hotel bookings. *Journal of Travel Research*, *61*(6), 1233–1251. doi:10.1177/00472875211028313

Medai, N., & Wu, L. (2022). A study of determinants that affect the intention to participate in online tours and the role of constraints under COVID-19 pandemic. *Current Issues in Tourism*, 1–15.

Mendes-Filho, L., Mills, A. M., Tan, F. B., & Milne, S. (2018). Empowering the traveler: An examination of the impact of user-generated content on travel planning. *Journal of Travel & Tourism Marketing*, *35*(4), 425–436. doi:10.1080/10548408.2017.1358237

Nicholls, R. (2010). New directions for customer-to-customer interaction research. *Journal of Services Marketing*, *24*(1), 87–97. doi:10.1108/08876041011017916

Phua, J., & Kim, J. J. (2018). Starring in your own Snapchat advertisement: Influence of self-brand congruity, self-referencing and perceived humor on brand attitude and purchase intention of advertised brands. *Telematics and Informatics*, *35*(5), 1524–1533. doi:10.1016/j.tele.2018.03.020

Pinto, A. S., Abreu, A., Costa, E., & Paiva, J. (2022). Augmented reality for a new reality: Using UTAUT-3 to assess the adoption of mobile augmented reality in tourism (MART). *Journal of Information Systems Engineering & Management*, 7(2), 14550.

All The Research. (2020, February 21). *AR-VR in travel and tourism market*. All the Research. https://www.alltheresearch.com/press-release/ar-vr-in-travel-and-tourism-market-ecosystem-worth-304-4-million-by-2023

Sancho-Esper, F., Ostrovskaya, L., Rodriguez-Sanchez, C., & Campayo-Sanchez, F. (2022). Virtual reality in retirement communities: Technology acceptance and tourist destination recommendation. *Journal of Vacation Marketing*, 13567667221080567.

Shabani, N., Munir, A., & Hassan, A. (2018). E-Marketing via augmented reality: A case study in the tourism and hospitality industry. *IEEE Potentials*, 38(1), 43–47. doi:10.1109/MPOT.2018.2850598

Shin, H. H., & Jeong, M. (2022). Does a virtual trip evoke travelers' nostalgia and derive intentions to visit the destination, a similar destination, and share?: Nostalgia-motivated tourism. *Journal of Travel & Tourism Marketing*, 39(1), 1–17. doi:10.1080/10548408.2022.2044972

Siddaway, A. P., Wood, A. M., & Hedges, L. V. (2019). How to do a systematic review: A best practice guide for conducting and reporting narrative reviews, meta-analyses, and meta-syntheses. *Annual Review of Psychology*, 70(1), 747–770. doi:10.1146/annurev-psych-010418-102803 PMID:30089228

Slevitch, L., Chandrasekera, T., & Sealy, M. D. (2022). Comparison of virtual reality visualizations with traditional visualizations in hotel settings. *Journal of Hospitality & Tourism Research (Washington, D.C.)*, 46(1), 212–237. doi:10.1177/1096348020957067

Subawa, N. S., Widhiasthini, N. W., Astawa, I. P., Dwiatmadja, C., & Permatasari, N. P. I. (2021). The practices of virtual reality marketing in the tourism sector, a case study of Bali, Indonesia. *Current Issues in Tourism*, 24(23), 3284–3295. doi:10.1080/13683500.2020.1870940

Suh, K. S., & Chang, S. (2006). User interfaces and consumer perceptions of online stores: The role of telepresence. *Behaviour & Information Technology*, 25(2), 99–113. doi:10.1080/01449290500330398

Talwar, S., Kaur, P., Nunkoo, R., & Dhir, A. (2022). Digitalization and sustainability: Virtual reality tourism in a post pandemic world. *Journal of Sustainable Tourism*, 1–28. doi:10.1080/09669582.2022.2029870

Dieck, M. C., Jung, T., & Moorhouse, N. (2018). Tourists' virtual reality adoption: An exploratory study from Lake District National Park. *Leisure Studies*, 37(4), 371–383. doi:10.1080/02614367.2018.1466905

Dieck, M. C., & Jung, T. (2018). A theoretical model of mobile augmented reality acceptance in urban heritage tourism. *Current Issues in Tourism*, 21(2), 154–174. doi:10.1080/13683500.2015.1070801

Dieck, M. C., & Jung, T. H. (2017). Value of augmented reality at cultural heritage sites: A stakeholder approach. *Journal of Destination Marketing & Management*, 6(2), 110–117. doi:10.1016/j.jdmm.2017.03.002

Tsai, T.-H., Lin, W.-Y., Chang, Y.-S., Chang, P.-C., & Lee, M.-Y. (2020). Technology anxiety and resistance to change behavioral study of a wearable cardiac warming system using an extended TAM for older adults. *PLoS One*, 15(1), 1–24. doi:10.1371/journal.pone.0227270 PMID:31929560

Tussyadiah, I. P., Wang, D., Jung, T. H., & tom Dieck, M. C. (2018). Virtual reality, presence, and attitude change: Empirical evidence from tourism. *Tourism Management*, *66*, 140–154. doi:10.1016/j.tourman.2017.12.003

United Nations Report. (2021, April 12). *UN Tourism News #23*. UNWTO. https://www.unwto.org/un-tourism-news-23

Wang, F., Huang, S., Morrison, A. M., & Wu, B. (2022). The effects of virtual reality tourism involvement on place attachment and behavioral intentions: Virtual reality tourism of the Yellow Crane Tower in Wuhan. *Asia Pacific Journal of Tourism Research*, *27*(3), 274–289. doi:10.1080/10941665.2022.2061363

Wei, W. (2019). Research progress on virtual reality (VR) and augmented reality (AR) in tourism and hospitality: A critical review of publications from 2000 to 2018. *Journal of Hospitality and Tourism Technology*.

Ying, T., Tang, J., Ye, S., Tan, X., & Wei, W. (2022). Virtual reality in destination marketing: Telepresence, social presence, and tourists' visit intentions. *Journal of Travel Research*, *61*(8), 1738–1756. doi:10.1177/00472875211047273

Zheng, C., Chen, Z., Zhang, Y., & Guo, Y. (2022). Does vivid imagination deter visitation? The role of mental imagery processing in virtual tourism on tourists' behavior. *Journal of Travel Research*, *61*(7), 1528–1541. doi:10.1177/00472875211042671

KEY TERMS AND DEFINITIONS

Augmented Reality: The real-time virtual enhancements (graphics, audio, scent) integrated with real-world objects.

Authenticity: The quality of being honest and true to one's own values.

Technology Anxiety: An emotional response to using new/unfamiliar technologies (products or services).

Telepresence: A state of being at an imaginary/virtual place.

Virtual Reality: A simulation that provides an entirely fictional world by creating an engaging environment.

Visit intention: Thoughts about visiting a specific destination.

Vividness: The quality of being clear and bright.

Section 3

Development Dynamics, Governance, and Promotion of Tourism and Hospitality

Chapter 11
Tourism Entrepreneurship in Innovation Sustainability:
Challenges and Opportunities

Albérico Travassos Rosário
GOVCOPP, IADE, Universidade Europeia, Portugal

ABSTRACT

The increasing wealth and economic development have dramatically driven rapid growth in the tourism industry, making it one of the fastest growing and developing industries worldwide. Tourism entrepreneurship should involve adopting a green business model innovation that is more efficient and leads to low carbon production to balance the needs of the tourists with those of their desired destinations. Despite sustainability being a core concept in current policies and trends in the last decade, most company managers in the tourism industry are yet to incorporate it into the agenda. Based on this research gap, a systematic review of the bibliometric literature was conducted, and data was synthesized from 80 documents identified through the Scopus indexation using. This chapter aims to evaluate the challenges and opportunities of innovation sustainability in tourism entrepreneurship, thus building a clear image of what should be done to overcome the obstacles and increase awareness of the need for sustainable tourism.

INTRODUCTION

Innovation is vital for sustainable development in the tourism sector. It is a multifaceted concept involving generating new ideas, research and development, and successfully utilizing and commercializing them. In tourism entrepreneurship, innovation helps improve efficiency and productivity and increases customer loyalty (Alkier et al., 2017). However, recent years have seen a shift of focus from mere innovation to sustainable innovation, which involves generating and implementing ideas and technologies that foster sustainable performance and growth. Tourism entrepreneurship should be evaluated from the three primary sustainability dimensions; economic, environmental, and social. Tourism sustainability

DOI: 10.4018/978-1-6684-6985-9.ch011

innovation aims to lessen the adverse effects of tourist activities on the natural, historical, cultural, or social environment while maintaining economic and social advantages (Moscardo, 2008). In addition, sustainable innovation involves developing innovative technologies, services and products for people and organizations while respecting the environment, including natural resources and the earth's regenerative capacity (Elkhwesky et al., 2022). Therefore, sustainability innovation in tourism can help entrepreneurs and businesses deliver products and services characterized by improved economic, environmental, and social performance. As a result, adopting sustainable business models can help achieve higher innovation and value creation. This approach can ensure that organizations engage in business practices that improve their financial performance while protecting the environment and the communities they serve.

However, the tourism industry lags in adopting and implementing sustainable innovations. For instance, Garay et al. (2019) indicated that tourism firms do not prioritize sustainability in their innovation strategies and dedicate limited resources to sustainability learning. Similarly, Peeters et al. (2006) argue that tourism primarily focuses on three goals, satisfying customer needs, increasing competitiveness and gaining a more significant market share. As a result, the industry neglects its environmental impacts. Examples of the ecological effects of tourism include increased pressure on local land use, overconsumption of resources in areas where they are already scarce, increased pollution, loss of natural habitat, pressure on endangered species, and soil erosion (Elkhwesky et al., 2022). The diversity of these issues indicates the need for prioritizing sustainable innovation in the tourism sector to help create a balance between engaging in sustainable business practices that ensure environmental protection and promote social-cultural wellbeing while achieving the desired financial goals. Therefore, this systematic review of the bibliometric literature synthesizes data from 79 sources to identify opportunities that can be leveraged for sustainable tourism innovation and challenges that may hinder success. These insights help promote innovative, sustainable tourism entrepreneurship that cares about the business and respects people and the environment.

METHODOLOGICAL APPROACH

The methodology used was a systematic review of the bibliometric literature to identify and select relevant sources and synthesized data to create an evidence-based report implementable in tourism business practice. This methodology was selected based on Xiao and Watson's (2019) recognition of systematic reviews as a critical feature of knowledge advancement and academic research. The scholars explain that understanding the breadth and depth of existing literature helps identify gaps to explore, inconsistencies, and contradictions. In this regard, summarizing, analyzing, and synthesizing literature will help understand the sustainability innovations in the tourism sector, opportunities for growth, and challenges anticipated and hindering progress. Such insights can improve the planning and integration of sustainability goals and efforts into the overall tourism organizational processes and practices. Consequently, the methodology can be used to decipher and map cumulative scientific knowledge and emerging variations of a well-established topic (Rosário, 2021; Rosário & Dias, 2022; Rosário, et al., 2021)

Thus, the use of bibliometric analysis can help understand its development and adoption in businesses to identify potential challenges. The use of SRB review process is divided into 3 phases and 6 steps (Table 1), as proposed by Rosário (2021), Rosário and Dias (2022) and Rosário, et al. (2021).

Table 1. Process of systematic LRSB

Fase	Step	Description
Exploration	Step 1	formulating the research problem
	Step 2	searching for appropriate literature
	Step 3	critical appraisal of the selected studies
	Step 4	data synthesis from individual sources
Interpretation	Step 5	reporting findings and recommendations
Communication	Step 6	Presentation of the SRB report

Source: adapted from Rosário (2021),Rosário and Dias (2022) and Rosario, et al. (2021)

The database of indexed scientific and/or academic documents used was Scopus, the most important peer review in the scientific and academic world. The literature search includes peer-reviewed scientific and/or academic documents published up to December 2022.

The keyword "tourism" was used for the initial literature search, resulting in 134,007 document results. The researcher then added the keyword "entrepreneurship" and limited the search to the subject area "business, management and accounting", reducing the document results to 878. While the sources generated provided information about tourism entrepreneurship, they did not offer exact knowledge on sustainable innovation, which is the primary topic for this research. Therefore, the researcher added the exact keywords "innovation" and "sustainability" to identify more relevant documents, resulting in 79 sources (N=79), which were analyzed, summarized, and synthesized in this literature review (Table 2).

Table 2. Screening methodology

Database Scopus	Screening	Publications
Meta-search	keyword: *tourism*	134,007
First Inclusion Criterion	keyword: *tourism,* en*trepreneurship*	1,389
Second Inclusion Criterion	keyword: tourism, entrepreneurship subject area: business, management and accounting	878
Screening	keyword: tourism, entrepreneurship subject area: business, management and accounting Exactly word: innovation, sustainability Published until December 2022	79

Finally, content and theme analysis techniques were used to identify, analyze and report the various documents as proposed by Rosário (2021), Rosário and Dias (2022) and Rosário, et al. (2021). The 79 scientific and/or academic documents indexed in Scopus are later analyzed in a narrative and bibliometric way to deepen the content and possible derivation of common themes that directly respond to the research question (Rosário, 2021; Rosário & Dias, 2022; Rosário, et al., 2021). Of the 79 selected documents, 69 are Articles; 4 are Conference Paperare; 3 are Book Chapter; 2 Review; and 1 Book.

PUBLICATION DISTRIBUTION

Peer-reviewed articles on tourism entrepreneurship in innovation sustainability until December 2022. The year 2021 had the highest number of peer-reviewed publications on the subject, reaching 14.

Figure 1 summarizes the peer-reviewed literature published by year since 2003 to 2022.

The publications were sorted out as follows: Journal of Sustainable Tourism (12); Scandinavian Journal of Hospitality And Tourism (4); Tourism Management (4); Annals of Tourism Research (3); Current Issues In Tourism (2); Journal of Environmental Management And Tourism (2); Tourism Geographies (2); Tourism Review (2); the remaining publications with 1. Interest in the topic with a positive evolution from 2016.

Figure 1. Documents by year
Source: own elaboration

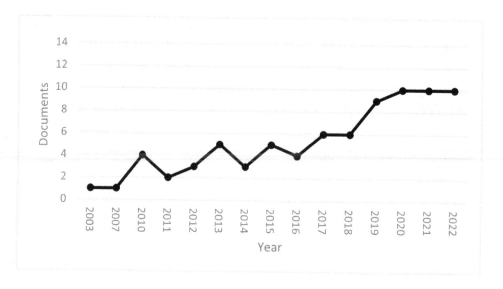

In Table 3 we analyze for the Scimago Journal & Country Rank (SJR), the best quartile and the H index by publication Journal Of Management Studies is the most quoted publication with 4,640 (SJR), Q1 and H index 194.

There is a total of 24 publications in Q1, 9 publications in Q2 and 11 publications, Q3 and 3 publications in Q4. Publications from best quartile Q1 represent 16% of the 56 publications titles; best quartile Q2 represents 20%, best quartile Q3 represents 5%, and best Q4 represents 3% of each of the titles of 56 publications, finally, 9 publications still do not have data available, representing 16% of publications.

As shown in Table 3, the significant majority of publications still do not have indexed data.

The subject areas covered by the 79 scientific and/or academic documents were: Business, Management and Accounting (79); Social Sciences (43); Economics, Econometrics and Finance (15); Environmental Science (6); Decision Sciences (4); Computer Science (3); Engineering (3); Energy (2); Mathematics (1); and Psychology (1).

Table 3. Scimago journal and country rank impact factor

Title	SJR	Best Quartile	H index
Journal Of Management Studies	4,640	Q1	194
Research Policy	3,530	Q1	255
Tourism Management	3,380	Q1	216
Annals Of Tourism Research	3,150	Q1	187
Journal Of Sustainable Tourism	2,480	Q1	114
Tourism Geographies	2,270	Q1	73
Business Strategy And The Environment	2,240	Q1	115
Journal Of Cleaner Production	1,920	Q1	232
Current Issues In Tourism	1,840	Q1	82
Service Industries Journal	1,800	Q1	70
Tourism Management Perspectives	1,760	Q1	54
Journal Of Hospitality And Tourism Management	1,610	Q1	45
Tourism Review	1,480	Q1	38
International Journal Of Entrepreneurial Behaviour And Research	1,210	Q1	75
Journal Of Hospitality And Tourism Technology	1,210	Q1	34
Scandinavian Journal Of Hospitality And Tourism	1,170	Q1	50
International Journal Of Tourism Research	1,140	Q1	67
Asia Pacific Journal Of Tourism Research	0,890	Q1	44
Tourism Planning And Development	0,840	Q1	36
Big Data And Cognitive Computing	0,830	Q1	18
Journal Of Tourism Futures	0,810	Q1	21
Tourism And Hospitality Research	0,780	Q1	41
Leisure Studies	0,670	Q1	69
International Journal Of Entrepreneurship And Innovation	0,630	Q2	21
Journal Of Asia Business Studies	0,580	Q1	20
Journal Of Hospitality Leisure Sport And Tourism Education	0,530	Q2	29
Journal Of Small Business Strategy	0,530	Q2	16
Journal Of Workplace Learning	0,500	Q2	52
Journal Of Enterprising Communities	0,480	Q2	31
European Journal Of Tourism Research	0,450	Q2	19
Tourism	0,380	Q2	25
Journal Of Teaching In Travel And Tourism	0,350	Q3	24
Rural Society	0,330	Q3	25
Journal For International Business And Entrepreneurship Development	0,300	Q2	6
Lecture Notes In Business Information Processing	0,300	Q3	52
International Journal Of Entrepreneurship And Small Business	0,290	Q3	38
International Journal Of Entrepreneurship And Innovation Management	0,270	Q3	26

continues on following page

Table 3. Continued

Title	SJR	Best Quartile	H index
Journal Of Environmental Management And Tourism	0,240	Q3	18
International Journal Of Innovation And Learning	0,240	Q3	27
Problems And Perspectives In Management	0,240	Q2	23
Deturope	0,230	Q3	8
Innovations	0,230	Q3	11
World Review Of Entrepreneurship Management And Sustainable Development	0,230	Q3	17
African Journal Of Hospitality Tourism And Leisure	0,210	Q3	14
International Journal Of Tourism Policy	0,200	Q4	15
Universidad Y Sociedad	0,130	Q4	4
Jordan Journal Of Business Administration	0,120	Q4	2
International Journal Of Scientific And Technology Research	0	_*	22
Iza Journal Of European Labor Studies	0	_*	16
Sinergie	0	_*	2
Social Sustainability In The Global Wine Industry Concepts And Cases	_*	_*	_*
Springer Proceedings In Business And Economics	_*	_*	_*
Strategies And Best Practices In Social Innovation An Institutional Perspective	_*	_*	_*
Studies On Entrepreneurship Structural Change And Industrial Dynamics	_*	_*	_*
Tourismos	_*	_*	_*
Wiley Blackwell Companion To Tourism	_*	_*	_*

Note: *data not available.

The most quoted article was "A review of innovation research in tourism" from Hjalager (2010) with 837 quotes published in Tourism Management 3,380 (SJR), the best quartile (Q1) and with H index (216), this paper "reveals that there is still only limited systematic and comparable empirical evidence of the level of innovative activities and their impacts and wider implications for destinations and national economies".

In Figure 2 we can analyze the evolution of citations of the documents published until December 2022. The number of citations shows a positive net growth with R2 of 14% for between 2013 and 2023, with 2023 reaching 3149 citations.

The h-index was used to ascertain the productivity and impact of the published work, based on the largest number of articles included that had at least the same number of citations. Of the documents considered for the h-index, 24 have been cited at least 13 times.

Citations of all scientific and/or academic documents from the period 2013 to December 2022, with a total of 3149 citations, of the 80 documents 8 were not cited. The self-citation of documents in the period between 2013 to December 2023 was self-cited 2919 times.

The bibliometric study was carried out to investigate and identify indicators of the dynamics and evolution of scientific and/or academic information in documents based on the main keywords (Figure 3). The results were extracted from the scientific software VOSviewer, which aims to identify the main search keywords "tourism" and "entrepreneurship".

Figure 2. Evolution of citations between 2013 and 2023

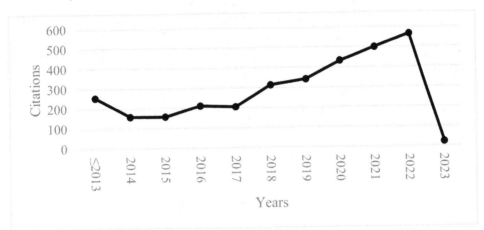

Figure 3. Network of all keywords

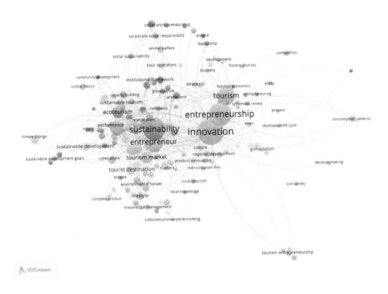

The research was based on scientific and/or academic documents on future challenges of tourism entrepreneurship in innovation sustainability. In Figure 4, we can examine the linked keywords, and thus, it is possible to highlight the network of keywords that appear together / linked in each scientific article, allowing to identify the topics studied by research and identify trends in future research. Finally, in Figure 5, a profusion of co-citation with a unit of analysis of the cited references is presented.

THEORETICAL PERSPECTIVES

Sustainable innovation encourages tourism entrepreneurs and their teams to integrate economic, environmental, and social factors when designing processes, products, and organizational structures. Embracing this approach improves production methods' sustainability, promotes sustainable consumption, and creates sustainable marketing structures (Baratta et al., 2022). As a result, sustainability is perceived as a driver of innovation and value creation. Organizations must regularly modify their thinking patterns and actions to improve effectiveness and facilitate innovation for the tourism industry to be sustainable. In this case, sustainable innovation in tourism entrepreneurship requires analyzing the existing situation from different perspectives to determine potential areas of growth and challenges (Dias & Silva, 2021). Therefore, this literature review section integrates data from multiple research to identify trends, opportunities and challenges for sustainable innovation in tourism to provide insights that can be implemented in practice.

Figure 4. Network of linked keywords

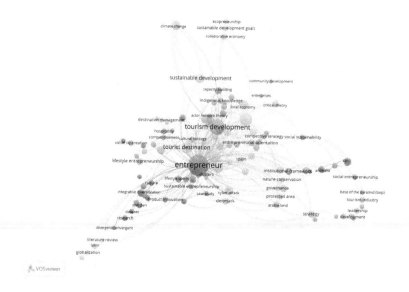

Figure 5. Network of co-citation

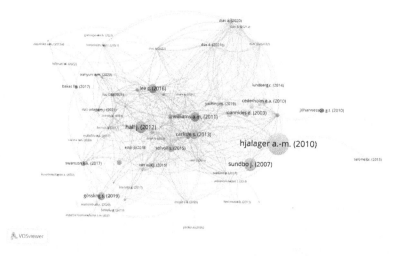

Defining the Main Concepts

Tourism is critical to the economies of many countries. In recent years, tourism has grown to be a significant source of revenue for many governments and major employers worldwide (Ferri & Aiello, 2017). The industry creates diverse career opportunities and facilitates local development by constructing roads leading to tourist sites, hotels, airports, and other facilities required for the tourism business to thrive (Bertella, 2017). As a result, each country is increasing its tourist attractions, while those with the most effective method of attracting tourists are becoming more successful. While governments play critical roles in building vital tourism industries worldwide, entrepreneurship is essential (Herrmann & Kritikos, 2013). For instance, tourism entrepreneurs identify and pursue opportunities and operate commercial businesses by funding needs and want through innovation. They create new ideas, innovatively use resources, and leverage opportunities based on their knowledge and risk acceptance (Forbes et al., 2019). Adopting sustainable innovation in the sector would require the support and leadership of these entrepreneurs. For example, they are needed to approve and fund the sustainable initiatives adopted by tourism organizations. These aspects make it essential to understand the concept of tourism entrepreneurship and its correlation with sustainable innovation.

Tourism Entrepreneurship

Entrepreneurship is a process in which individual entrepreneurs generate new ideas and identify great opportunities to mobilize resources to engage in professions and job creation. In addition, Alexandre-Leclair and Liu (2014) define it as the process of producing something valuable by committing the necessary time and effort, accepting the associated financial, emotional, and social risks, and reaping the rewards of monetary and personal fulfilment and independence. Entrepreneurship also involves building new enterprises and creative organizations (Bertella & Vidmar, 2019; Dias et al., 2022). As a result, this process facilitates value creation using resources that help take advantage of existing opportunities. For example, countries with rich histories have historical sites that function as tourist attractions sites (Williams, 2014). An entrepreneur can identify this opportunity and create a travel agency that provides transport for tourists visiting these sites (Cardow & Wiltshier, 2010). The entrepreneurship procedure necessitates high-risk acceptance, which results in introducing a new product or service to the community (Hillman & Radel, 2022). These entrepreneurship efforts result in higher job creation and stability in society, enabling people to meet their family and financial obligations, and achieve better social welfare, income distribution, social justice, and social anxiety reduction.

Tourism entrepreneurship is increasingly gaining momentum as the industry grows. For instance, tourism-related industries contributed approximately US$7,170.3 billion to the GDP in 2015, accounting for 9.8% of the total GDP (Sotiriadis, 2018). This market size is projected to grow by 4% to reach $10,986.5 billion by 2026, equivalent to 10.8% of total GDP, and support more than 370 million jobs. This growth is anticipated in various tourism-related sectors, including food and beverage, accommodation, transportation, retail trade, culture, sports, and entertainment (Solvoll et al., 2015). They create opportunities for tourism entrepreneurs to invest, thus resulting in overall growth and development. However, for entrepreneurship to take place, there are various activities that individual entrepreneurs must do (Valeri & Baiocco, 2012). These include: 1) identifying new, potential business opportunities,

2) obtaining financial resources, 3) marketing the idea, product or service, 4) remaining creative and innovative, and 5) organizing and controlling resources, monitoring performance, and ensuring organizational profitability and sustainability.

Other than entrepreneurs' roles, entrepreneurship is hugely affected by government policies and activities. For instance, the national government in any country establishes legal frameworks through specified agencies in which businesses are installed and operated (Hoppstadius & Möller, 2018). For example, using natural and artificial resources is subject to government control, meaning tourism companies must adhere to specific laws and policies. In addition, the education system for tourism courses must meet national and local standards established by the department of education and other agencies (Jóhannesson, 2012; Hjalager, 2010). These examples indicate that tourism entrepreneurship goes beyond having investors willing to fund innovative ideas and tourists ready to buy the products and services to include other stakeholder roles, including the government and society in general (Lundberg et al., 2014; Hjalager & Johansen, 2013). This holistic collaboration is critical in adopting sustainable innovation in tourism, which may require the entrepreneur to provide resources and the government to provide laws and regulations on sustainable tourism practices.

Sustainable Innovation

Production and consumption patterns have substantially changed in the last decades, causing environmental and societal transformations. In addition, these changes have created demands and constraints for companies, so organizational competitiveness is associated with innovation management in ways that include sustainability (Carlisle et al., 2013). As a result, companies have realized the importance of sustainable innovation practices in enhancing corporate performance and reducing social and environmental impacts from their business activities. Sustainable innovation refers to creating something new that improves performance from the three sustainability dimensions; social, economic and ecological (Dias et al., 2022). While innovation is often defined from the perspective of developing new technologies, it also includes changes in operational practices and processes, business models, systems and thinking.

While some businesses make conscious decisions to adopt sustainable innovation and integrate it within their core strategies, others are prompted b external forces. For instance, recent years have seen increased pressure from society and legislation demanding organizations to accompany their products, services, processes and business models innovation with responsibility for sustainable development (Cederholm & Hultman, 2010). Alonso-Gonzalez et al. (2018) observed an increasing awareness in society about the need for solutions addressing the ongoing socio-environmental crisis through sustainable innovation and lifestyles. As a result, consumers demand businesses embrace sustainability in their business practices, processes, and strategies to ensure sustainable development.

Sustainability-oriented innovation involves modifying mindset and organizational principles, as well as goods, processes, and practices, with the primary goal of producing and achieving social and environmental value beyond economic returns. From this perspective, sustainable innovation can be defined as a firm's strategic and systematic approach to economic, social, and ecological factors (Chou et al., 2020; Matsiliza, 2017). This indicates that the innovation should not only focus on creating new environmentally responsible processes and products but also ensuring that they are economically viable and enhance society's wellbeing. Therefore, sustainable innovation goes beyond eco-innovation, primarily focusing on business practices' environmental and economic impacts to include social objectives and ethical aspects of conducting business (Lindvert et al., 2022). In addition, sustainability-oriented innovation accommodates sustainability issues, and the acquisition of new customers and market segments increases a firm's market value and position in the global markets (Ahmad et al., 2022). Thus, sustainable innovation is a holistic and long-term process of achieving sustainable development. As a result, sustainable innovation practices are associated with improved organizational performance.

Sustainable Tourism

Tourism is a crucial aspect of the global culture since it allows people to travel to different parts of the world, interact with people from various backgrounds, and learn about new cultures and activities. Given the ongoing globalization and migration, it has become vital for people to develop cultural competence, which allows one to function in different regions and around people from varying cultures (Marx & Klotz, 2021). Tourism provides an excellent opportunity for people to develop these skills by traveling and interacting with different people. As a result, tourism is typically seen as a force for good since it provides several advantages to travelers and communities (Kusumowidagdo & Rembulan, 2022). The sector, however, is evolving. Global populations are becoming more conscious of the problem of climate change and their part in exacerbating it. This awareness encourages them to consider methods to reduce tourism practices' environmental and social effects without hindering the industry's growth. This desire has led to the development of the concept of sustainable tourism.

Sustainable tourism aims to strike a proper balance between tourist development's environmental, economic, and socio-cultural components and plays a vital role in biodiversity conservation. For instance, coastal areas are popular tourist destinations (Singh & Wagner, 2022). While traveling to these areas can benefit the business and local communities, tourism activities negatively impact marine ecosystems (Moyle et al., 2020). For instance, effluents and untreated sewage heavily pollute the coastal waters and beaches are left covered with trash. In addition, global tourism generates a high carbon footprint and causes large-scale environmental degradation as previously pristine natural areas are converted into busy travel destinations (Romão et al., 2018; Iskakova et al., 2021). Sustainable tourism is a solution to these problems as it makes an effort to reduce its influence on the environment and local culture so that it will be available for future generations while also contributing to revenue generation, job creation, and the protection of local ecosystems (Salome et al., 2013). Creating this balance enables sustainable tourism to maximize tourism's invaluable contributions to biodiversity protection and, as a result, reduce poverty and fulfil common goals toward sustainable development.

Sustainable tourism provides critical economic incentives for habitat preservation. Habitat loss, poaching, and lack of funding for protection are some problems threatening wildlife and biodiversity (Sofield & Lia, 2011). However, recent years have seen an increase in nature-based tourism, encouraging tourism-related businesses and tourists to engage in wildlife conservation and inject money into local communities to strengthen growth and protection measures. In addition, countries worldwide are recognizing this opportunity and providing financial resources to promote wildlife and biodiversity protection (Thompson et al., 2018; Van Wijk et al., 2015). In these perspectives, revenues generated by tourist expenditure are frequently reinvested in environmental conservation or capacity-building initiatives for local communities to administer protected areas (Jóhannesson et al., 2021). Furthermore, tourism plays a vital role in increasing awareness and encouraging positive behavior change for biodiversity conservation among the millions of individuals who traverse the world each year (Valeri & Fadlon, 2016). Thus, integrating sustainability in nature-based tourism provides a practical solution for conserving and protecting the environment, local communities, and the economy.

Opportunities for Sustainable Tourism

Sustainable tourism creates opportunities for long-term business success since it results in net benefits for the social, economic, natural and cultural environments it takes place. For instance, it supports local communities, lowers ecological impact, encourages travelers to be conscious of their choices, and protects wildlife and biodiversity (Wahyuni & Sara, 2020). However, adopting sustainable innovations that would strengthen the industry's sustainability and create these benefits lag behind other sectors. This section explores various opportunities that can be leveraged to make the global tourism industry more sustainable.

Technologies

The operationalization of sustainable tourism requires changing the parameters within which economic tourism activities are organized, including adopting new technologies. Over the past years, ICT technologies and increased access to the internet have led to a global transformation of the tourism sector (Sinno, 2019). For example, with user-generated content, tourists can access information about various destinations and reviews regarding businesses in the tourism industry, including hotels and travel agencies. As a result, technologies have played a critical role in the ongoing development and growth of the sector (Triantafillidou & Tsiaras, 2018). These technologies can be leveraged to promote sustainable tourism practices. Below is an overview of some technologies causing dramatic changes in the tourism sector that can be used to ensure sustainability.

Computer Simulation

Computer simulation modeling is a tool/application based on ICT that may be used to manage information needs and enhance the sustainability of tourism destinations. This innovation is a computer-aided simplification of reality (Gössling & Michael Hall, 2019). The simulation uses mathematical modeling to predict behavior or outcome of a real-world or physical system. Recent research efforts recognize the effectiveness of computer-based simulation modeling in developing and managing nature-based tourism and protected natural areas (Chang et al., 2015). Its energy is evident in the ability of simulation approaches to envision the intended condition of the physical environment. In this case, computer simulation can be designed to showcase how a system works over time, providing invaluable tools for testing and observing a system's different components and its interactions with other parts (Pailis et al., 2020). In the case of the tourism industry, computer simulation can be used to design a system representing how different tourism activities and stakeholders contribute to sustainable innovation. In addition, the simulation can be used to test new techniques before implementation to assess their effectiveness (Triantafillidou & Tsiaras, 2018). It can be used to mimic real-world tourism issues and design new practical solutions addressing the various components associated with sustainable tourism.

Additionally, computer simulation can be used to explain current visitor use, which is typically challenging to see in protected areas due to their size and multiple methods of access and exit. Computer simulation makes it easy to collect information regarding visitor usage patterns, such as data acquired at parking lots, trailheads, and visitor centres (Ali & Frew, 2013). Managers can engage their tech teams to leverage this data using a computer simulation model to generate exact estimations of how tourists use protected natural areas (Fernandes, 2021; Kordel, 2016). As a result, the managers can gather spatial and temporal information that helps identify hazards such as trouble spots and opportunities for growth, including areas for further improvement to an area's natural and cultural elements (Parasotskaya et al., 2021). Another use of computer simulation in sustainable tourism is the enhanced capability to monitor indicator variables that are difficult to measure. For example, land use in tourism is a major issue causing environmental degradation (Reid, 2019). Destination managers can use computer simulation to monitor heavily used areas and facilitate the transition to those with limited use to reduce environmental impacts.

Environment Management Information System (EMIS)

Environment management information system (EMIS) is another technology that can be leveraged to facilitate sustainable innovation in the tourism industry. Ali and Frew (2013) define EMIS as a computer-based technology, including computer hardware, software, and professional services used to support an environmental management system. Destination managers use this technology to collect, analyze, and report business information on environmental management in a systematic manner (Gössling, 2017). As a result, its adoption enables them to track, refine, and gradually improve an organization's environmental management practices. Integrating globally networked computer systems within a firm's systems and processes helps incorporate ecological health and safety functions into the company's information system, thus aiding efficient business operations (Ali & Frew, 2013). In the case of sustainable tourism, it can help coordinate activities such as monitoring emissions, ensuring accurate documentation of sustainability-related issues, tracking waste, scheduling tasks, coordinating permits, and providing information necessary to select alternative activities, processes, or systems. For instance, companies that have integrated EMIS into their operating systems can monitor their carbon footprint or waste management strategies and initiate appropriate measures to curb potential problems before they escalate (Kummitha, 2020). Through an EMIS system, tourism businesses obtain real-time insights that would have been impossible to collect due to large data volumes, incompatibility, and diversity.

Geographical Information Systems (GIS)

Geographical information systems (GIS) are information systems used to collect, store, analyze and retrieve geographical data. GIS captures data from multiple databases, thus functioning as a decision support system and allowing managers to solve problems (Ali & Frew, 2013). Examples of its application in supporting sustainable tourism include auditing tourism resources and conditions, identifying areas for potential development and modeling, and testing different scenarios to project possible outcomes (Gössling, 2017). Using GIS can help tourism companies understand the impacts of their business practices on the environment and local communities. For example, conducting an audit of resources can reveal critical information regarding environmental pollution, an area's socio-economic and ecological characteristics, and potential damage from natural disasters (Gössling & Michael Hall, 2019). This information can aid tourism planning and management and facilitate decision-making and transition towards sustainable practices.

Leveraging Trade Associations

Trade associations (TAs) are organizational fields characterized by connectedness and structural equivalence, thus enhancing field homogeneity. These organizations are founded by businesses within the same industry, making them reasonably resilient to disruptions. The aspect of connectedness reflects operations that connect actors and is evidenced by a myriad of opportunities provided by TAs to their members and their active engagement in them (Font et al., 2019). Although TA memberships are primarily voluntary and the organizations' primary focus is to represent the interests and needs of their members, the companies are connected through normative or cultural values, which include corporate social responsibility (CSR) and sustainability (Teruel-Sánchez et al., 2021). For instance, TAs that promote sustainable innovation are more likely to influence their members to embrace sustainable tourism to maintain the industry's homogeneity. Sustainable innovation in different sectors tends to emerge as an outcome of competing for emerging and existing values since trade organizations are normative entities (Teixeira & Ferreira, 2019). This means that for TAs to advocate for change, there must be competition between existing and new standards of practice. For instance, some members of the TA might recognize the necessity of integrating sustainability into their business processes. In contrast, others continue with the traditional business models that primarily focus on the economic impacts of their activities. These varying business practices are more likely to push for change as the TAs try to find common ground and promote practices that facilitate growth and compliance with internal and external demands, including laws and regulations.

Adopting changes within the TA can happen due to various factors. For instance, institutional entrepreneurs have the motive and means to disrupt an industry's institutional logic, making them the key players who initiate and actively participate in executing divergent reforms (Font et al., 2019). In addition, TAs form and implement divergent change by adopting strategies intended to shield the institution's conduct from being questioned, thus setting off a cascade of events that eventually twist the TA's thinking and become unintentionally divergent. For example, a TA can advocate for members to implement domestic and international policies on environmental protection to avoid legal hiccups. Such a decision can result in long-term reforms leading to the adoption of sustainable innovation by tourism entrepreneurs.

Changing Consumers' Attitudes Towards Sustainable Tourism

With the increasing attention devoted to sustainable development, climate change and environmental protection, more customers are becoming aware of the need to shift their consumption habits toward sustainability. This transition may require the tourism industry to move its sustainability initiatives from the operational to the strategic management level (Swanson & DeVereaux, 2017). For instance, customers are willing to pay more for sustainable or eco-friendly products and services. Thus, tourism-related companies can leverage sustainable tourism products by marketing them as distinct product categories competing with other product categories that follow either traditional or innovative business practices but do not particularly address sustainability (Sørensen & Grindsted, 2021; Ruiz-Ortega et al., 2021). As a result, most managers working in the tourism sector are beginning to raise concerns about the role of sustainability in travel possibilities throughout customer journeys. They are increasingly advocating for the integration of sustainable alternatives during consumer preparation and destination selection phases to ensure that the innovations developed match the ongoing customer trends, preferences, and demands.

Furthermore, customer reasons and intentions to pick a sustainable destination are intriguing. Intention refers to the probability of engaging in specific behaviours. Ensuring sustainable tourism requires businesses and their management to understand their customers' behavioral choices to implement strategies that build loyalty (Saviera et al., 2022; Dias et al., 2021). Research shows that customer perceptions of an organization's efforts to respect the environment influence their intention to select a sustainable tourism destination (Chan et al., 2016). In addition, factors such as individual commitment to the environment, conservation, and nature care determine the motivation behind a client's decision to travel or visit any tourism destination. With more tourists becoming pro-environment, tourism-related companies must offer sustainable products, which are considered more competitive in visitors' eyes. This shift in offerings may provide better satisfaction, which in turn may increase loyalty.

Multi-Stakeholder Partnerships

A multi-stakeholder approach to tourism development is one of the most effective techniques for increasing and creating sustainable tourism and improving the completeness of local and regional environments. According to Alhammad (2020), entrepreneurship and entrepreneurial skills can influence rural and agri-tourist microeconomics at the local community level when applied in a bottom-up development approach. In this regard, the multi-stakeholder partnerships focus on promoting sustainable tourism through the active engagement of community members. That is, it recognizes the role of consumers in achieving sustainable development (Bakas, 2017; Robbins & Devitt, 2017). In addition, this initiative requires agents, public and private companies, and society to collaborate for mutual goals towards building a sustainable tourism industry (Ndou et al., 2019). It allows the incorporation of diverse perspectives and ideas, skills and competencies, and increases access to resources required to implement sustainable innovations.

Challenges in Achieving Sustainable Tourism

Sustainable tourism development can facilitate national, regional, and local economic growth without compromising the ability of future generations to meet their needs. It encourages adopting sustainable innovation that incorporates the three main dimensions of sustainability, including environmental, social, and economic (Luu, 2021). Therefore, it creates a balance between economic growth, environmental protection, and societal wellbeing. However, multiple inadequate laws and regulations to enforce and monitor sustainability initiatives often hinder achieving sustainable tourism. This literature review section explores the various challenges undermining sustainable tourism development.

Inadequate Laws and Regulations

In recent decades, tourism has been identified as a significant cause of unsustainable development. The general conclusion has been that appropriate policies and their implementation offers a practical solution to make tourism sustainable. Despite the industry's importance in economic growth, tourism policy is relatively neglected compared to other sectors. Stoffelen et al. (2020, p.2) indicate that government policies that support sustainable tourism are "weak, too centralized and too bureaucratic in decision-making or without the instruments to facilitate local skills creation and empowerment." As a result, most sustainable innovations do not succeed due to weak institutional support and low private-sector engagement.

Without government support and regulations that instill pressure, most private companies are unwilling to invest in sustainability initiatives, mainly due to a lack of clear return on investment. Despite creating international sustainable tourism policies such as the 1998 WTO guide for local authorities, research shows tiny implementation due to lack of adequate enforcement (Estol et al., 2018; Sundbo et al., 2007). In addition, the implementation of sustainable tourism policies requires dealing with the overall scope of sustainability at a smaller scale. This means that putting into practice sustainable tourism policies requires implementation at the local level through the land use policies (Preko, 2020). However, most employees working at these local level agencies have limited understanding of the sustainability concept, which makes it challenging to identify a technical solution to the problem. Consequently, governments cannot reinforce and monitor sustainability initiatives within the tourism sector, thus slowing down the adoption of sustainable innovations.

Inability to Control Pollution

The increasing number of tourists and their excellent mobility have made it impossible to manage pollution in the tourism sector. The industry's operations involve technologies and systems that result in multiple types of pollution, including the release of sewage, air emissions, noise, solid waste and littering, and oil and chemicals (Yachin, 2019). The rising number of tourists translates to higher demand for air, road, and rail transport, which increases CO2 emissions and contributes to severe local air pollution. In addition, the tourism transport sector is a massive contributor to noise pollution resulting from airplanes, cars, buses, and recreational vehicles such as snowmobiles and jet skis (Lee et al., 2016). This challenge can have severe implications on human life, including causing hearing loss, stress, annoyance, and distress for wildlife.

While innovations such as electric cars may partly help reduce air pollution from cars and buses, they cannot adequately address the problem. Developing alternative solutions, for example, to reduce pollution from air travel, may take a long time before the goal is realized (Zapalska & Brozik, 2015). In addition, waste disposal is a major challenge in areas with high concentrations of tourists and tourism activities. The waste is often disposed of in rivers, oceans, roadsides, or other scenic locations, causing environmental problems (Williams & Shaw, 2011). In addition, improper waste disposal causes the death of marine animals. However, despite recognizing these problems, the tourism industry is struggling with developing alternative sustainable innovations that would address them.

Unsustainable Tourism Development Practice

As the tourism industry grows, the demand for facilities such as accommodations and restaurants has led to unsustainable development practices. For instance, the construction activities and infrastructure development involve sand mining and extensive paving, which can result in soil, sand and beach erosion (Dias et al., 2021). In addition, government efforts to enhance access to tourist destinations such as historical sites and beaches involve the construction of roads, airports, and railways, which results in environmental degradation, deterioration of scenery, and loss of wildlife habitats. Moreover, tourism development is often associated with deforestation and intensified and unsustainable use of land, which causes further environmental problems (Galdon et al., 2013; Giannopoulos et al., 2022). For companies to construct ski resort accommodations and facilities, they often have to clear large tracts of land or drain coastal wetlands to create space for infrastructural development. Overbuilding disrupts ecosystems and damages habitats. Sustainable tourism would require identifying and implementing alternative development strategies that do not involve massive destruction of the environment and disruption of the ecosystems (Zapalska et al., 2015). The primary challenge arises from attempts to meet the demand for tourism infrastructure while preserving the environment and ensuring the well-being of the local communities (Hall et al., 2012). For instance, the industry's key players, including entrepreneurs and government agencies, should collaborate to determine practical ways of pursuing development without damaging wildlife habitats. Therefore, while infrastructural development is critical for the survival and growth of the tourism sector, there is a need to innovate sustainable approaches to achieving the desired development levels.

CONCLUSION

Attaining sustainable tourism requires innovation, which involves generating new ideas, R&D, and successful use and commercialization of these ideas. Innovation goes beyond developing and integrating technologies, including products, services, and organizational processes and systems. In the case of sustainable tourism, innovation plays a critical role in providing solutions that help businesses strike a balance between the economic, environmental, and social factors associated with their business practices. Unlike in traditional markets where companies prioritized profitability, the increased global awareness of sustainable business practices requires them to include environmental protection and strategies ensuring the overall wellbeing of society. The literature review revealed that market trends and consumer behaviors are shifting toward sustainable practices. For instance, customers are more likely to purchase sustainable products and services or visit a tourism destination shown to respect the environment and surrounding communities. Thus, tourism-related businesses can leverage this opportunity to strengthen their competitive advantage, performance, and productivity.

Other opportunities for sustainable tourism include integrating ICT into tourism planning, management, and strategies. Emerging technologies such as computer simulation, environment management information systems (EMIS), and geographical information systems (GIS) provide tools and systems that improve operational efficiency and facilitate the adoption of sustainable innovation. In addition, trade associations (TAs) provide an opportunity to reinforce sustainability policies and standards. TAs are voluntary membership organizations within an industry formed to protect the interests and needs of the member companies. They help drive change and innovation by structuring the standards of practice based on existing principles and ongoing changes. TAs can help encourage businesses to comply with domestic and international sustainability regulations to avoid legal consequences. Other opportunities include the changing consumer attitudes towards sustainable tourism and multi-stakeholder partnerships, which increase access to markets, ideas, resources, skills and knowledge. Exploring these opportunities can help tourism-related businesses develop and implement sustainable business strategies, thus achieving early success.

However, multiple challenges often hinder the exploitation of these opportunities to achieve sustainable tourism. For instance, adequate laws and regulations do not encourage businesses to implement sustainable practices and monitor progress. The bureaucracy in decision-making slows progress. Another challenge is the inability to control pollution. The increasing number of tourists worldwide has significantly increased the demand for tourism services and facilities, including transport and accommodation. As a result, more airplanes, buses, and cars are transporting these travelers to their desired destinations, thus resulting in higher CO_2 emissions and destruction of the environment. Finally, unsustainable tourism development practices are another challenge hindering sustainable innovation. For example, the construction of accommodation and restaurant facilities and infrastructure such as roads, airports, and railways severely impacts the environment.

However, multiple challenges often hinder the exploitation of these opportunities to achieve sustainable tourism. For instance, adequate laws and regulations do not encourage businesses to implement sustainable practices and monitor progress. The bureaucracy in decision-making slows progress. Another challenge is the inability to control pollution. The increasing number of tourists worldwide has significantly increased the demand for tourism services and facilities, including transport and accommodation. As a result, more airplanes, buses, and cars are transporting these travelers to their desired destinations, thus resulting in higher CO_2 emissions and destruction of the environment. Finally, unsustainable tourism development practices are another challenge hindering sustainable innovation. For example, the construction of accommodation and restaurant facilities and infrastructure such as roads, airports, and railways severely impacts the environment.

Tourism entrepreneurship can play a crucial role in promoting sustainability and innovation within the tourism industry. By definition, entrepreneurship involves taking risks and pursuing new opportunities, which can lead to the development of new products, services, and business models that are more sustainable and innovative. One of the key ways in which tourism entrepreneurship can contribute to sustainability is by promoting eco-tourism and responsible travel practices. Entrepreneurs in this field can create new experiences that allow travellers to engage with the natural environment in a responsible and sustainable way. This can include activities such as hiking, wildlife watching, and conservation volunteering.

Another important way in which tourism entrepreneurship can promote sustainability is by developing new technologies and business models that reduce the environmental impact of tourism. For example, entrepreneurs can create new transportation systems that are more efficient and environmentally friendly, or develop new technologies that reduce waste and energy consumption in hotels and other tourist accommodations.

In addition to promoting sustainability, tourism entrepreneurship can also drive innovation within the industry. Entrepreneurs can create new products and services that enhance the tourist experience, such as virtual reality tours, personalized travel recommendations, and mobile apps that provide real-time information about local attractions and events.

However, it is important to note that not all tourism entrepreneurship is necessarily sustainable or innovative. Some entrepreneurs may prioritize short-term profits over long-term sustainability, or may simply replicate existing business models rather than developing truly new and innovative ideas. Therefore, it is important to encourage and support entrepreneurship that is aligned with sustainability goals and promotes genuine innovation within the tourism industry.

In conclusion, tourism entrepreneurship has the potential to play a critical role in promoting sustainability and innovation within the tourism industry. By supporting entrepreneurs who are focused on creating sustainable and innovative products and services, we can help to create a more responsible and enjoyable travel experience for all.

However, we consider that the study has the limitation of considering only the Scopus decomposition database, excluding other scientific and academic bases.

Tourism entrepreneurship is a constantly evolving field, and there are several relevant research lines that focus on its sustainability and innovation. Some possible research lines are: (i) Social and environmental impact of tourism entrepreneurship: This research line analyzes how tourism entrepreneurs can create a positive impact on the local community and the environment while generating profit. For example, how entrepreneurs can implement sustainable practices such as waste reduction, recycling, conservation of natural resources, and community engagement, to ensure that their businesses are sustainable in the long term; (ii) Innovation in tourism products and services: This research line explores how tourism entrepreneurs can develop new products and services that meet the needs of tourists and are sustainable and innovative. For example, how entrepreneurs can use emerging technologies such as virtual reality or artificial intelligence to create unique and personalized tourism experiences; (iii) Development of tourism entrepreneurship networks: This research line focuses on how tourism entrepreneurs can collaborate with each other to share knowledge, resources, and experiences, and increase the sustainability and innovation of the sector. For example, how entrepreneurs can create tourism clusters, promote collaborative tourism, or create tourism innovation ecosystems; (iv) Tourism entrepreneurship in rural and/or remote communities: This research line examines how tourism entrepreneurs can develop sustainable businesses in rural and/or remote areas that are often overlooked by traditional tourism. For example, how entrepreneurs can use communication technologies to promote their businesses to a wider audience or create authentic and immersive tourism experiences that value local culture and heritage.

These are just some possible research lines in tourism entrepreneurship. The important thing is that research is guided by relevant questions and the challenges faced by the sector, in order to provide practical and innovative solutions that promote the sustainability and success of tourism businesses.

ACKNOWLEDGMENT

I would like to express gratitude to the Editor and the Arbitrators. They offered extremely valuable suggestions or improvements. The author was supported by the GOVCOPP Research Center of the University of Aveiro.

REFERENCES

Ahmad, N., Youjin, L., & Hdia, M. (2022). The role of innovation and tourism in sustainability: Why is environment-friendly tourism necessary for entrepreneurship? *Journal of Cleaner Production, 379*, 134799. doi:10.1016/j.jclepro.2022.134799

Alexandre-Leclair, L., & Liu, Z. (2014). Innovation and entrepreneurship: The case of the tourism sector in Paris. *Innovations, 44*(2), 169–185. doi:10.3917/inno.044.0169

Alhammad, F. A. (2020). Trends in tourism entrepreneurship research: A systematic review. *Jordan Journal of Business Administration, 16*(1), 307–330. www.scopus.com

Ali, A., & Frew, A. (2013). *Information and communication technologies for sustainable tourism.* Routledge. doi:10.4324/9780203072592

Alkier, R., Milojica, V., & Roblek, V. (2017). Challenges of the social innovation in tourism. *ToSEE – Tourism in Southern and Eastern Europe, 4,* 1-13. doi:10.20867/tosee.04.24

Alonso-Gonzalez, A., Chacon, L. A. P., & Peris-Ortiz, M. (2018). Sustainable social innovations in smart cities: Exploratory analysis of the current global situation applicable to Colombia. Strategies and best practices in social innovation: An institutional perspective (pp. 65-87). Springer. doi:10.1007/978-3-319-89857-5_5

Bakas, F. E. (2017). Community resilience through entrepreneurship: The role of gender. *Journal of Enterprising Communities, 11*(1), 61–77. doi:10.1108/JEC-01-2015-0008

Baratta, R., Bonfanti, A., Cucci, M. G., & Simeoni, F. (2022). Enhancing cultural tourism through the development of memorable experiences: The "Food democracy museum" as a phygital project. *Sinergie, 40*(1), 213–236. doi:10.7433117.2022.10

Bertella, G. (2017). The emergence of Tuscany as a wedding destination: The role of local wedding planners. *Tourism Planning & Development, 14*(1), 1–14. doi:10.1080/21568316.2015.1133446

Bertella, G., & Vidmar, B. (2019). Learning to face global food challenges through tourism experiences. *Journal of Tourism Futures, 5*(2), 168–178. doi:10.1108/JTF-01-2019-0004

Cardow, A., & Wiltshier, P. (2010). Indigenous tourism operators: The vanguard of economic recovery in the Chatham islands. *International Journal of Entrepreneurship and Small Business, 10*(4), 484–498. doi:10.1504/IJESB.2010.034027

Carlisle, S., Kunc, M., Jones, E., & Tiffin, S. (2013). Supporting innovation for tourism development through multi-stakeholder approaches: Experiences from africa. *Tourism Management, 35,* 59–69. doi:10.1016/j.tourman.2012.05.010

Cederholm, E. A., & Hultman, J. (2010). The value of intimacy - negotiating commercial relationships in lifestyle entrepreneurship. *Scandinavian Journal of Hospitality and Tourism, 10*(1), 16–32. doi:10.1080/15022250903442096

Chan, J. H., Iankova, K., Zhang, Y., McDonald, T., & Qi, X. (2016). The role of self-gentrification in sustainable tourism: Indigenous entrepreneurship at Honghe Hani rice terraces world heritage site, china. *Journal of Sustainable Tourism, 24*(8-9), 1262–1279. doi:10.1080/09669582.2016.1189923

Chang, L., Chen, Y., & Liu, H. (2015). Explaining innovation in tourism-retailing contexts by applying Simon's sciences of the artificial. *Journal of Hospitality and Tourism Technology, 6*(1), 40–58. doi:10.1108/JHTT-02-2015-0012

Chou, S., Horng, J., Liu, C., Huang, Y., & Zhang, S. (2020). The critical criteria for innovation entrepreneurship of restaurants: Considering the interrelationship effect of human capital and competitive strategy a case study in Taiwan. *Journal of Hospitality and Tourism Management, 42,* 222–234. doi:10.1016/j.jhtm.2020.01.006

Dias, Á., Cascais, E., Pereira, L., Lopes da Costa, R., & Gonçalves, R. (2022). Lifestyle entrepreneurship innovation and self-efficacy: Exploring the direct and indirect effects of marshaling. *International Journal of Tourism Research, 24*(3), 443–455. doi:10.1002/jtr.2513

Dias, Á., & Silva, G. M. (2021). Lifestyle entrepreneurship and innovation in rural areas: The case of tourism entrepreneurs. *Journal of Small Business Strategy*, *31*(4), 40–49. doi:10.53703/001c.29474

Dias, Á., Silva, G. M., Patuleia, M., & González-Rodríguez, M. R. (2020). Developing sustainable business models: Local knowledge acquisition and tourism lifestyle entrepreneurship. *Journal of Sustainable Tourism*. doi:10.1080/09669582.2020.1835931

Dias, Á., Silva, G. M., Patuleia, M., & González-Rodríguez, M. R. (2021). Transforming local knowledge into lifestyle entrepreneur's innovativeness: Exploring the linear and quadratic relationships. *Current Issues in Tourism*, *24*(22), 3222–3238. doi:10.1080/13683500.2020.1865288

Dias, Á. L., Silva, R., Patuleia, M., Estêvão, J., & González-Rodríguez, M. R. (2022). Selecting lifestyle entrepreneurship recovery strategies: A response to the COVID-19 pandemic. *Tourism and Hospitality Research*, *22*(1), 115–121. doi:10.1177/1467358421990724

Elkhwesky, Z., El Manzani, Y., & Elbayoumi Salem, I. (2022). Driving hospitality and tourism to foster sustainable innovation: A systematic review of COVID-19-related studies and practical implications in the digital era. *Tourism and Hospitality Research*, *14673584221126792*. doi:10.1177/14673584221126792

Estol, J., Camilleri, M. A., & Font, X. (2018). European Union tourism policy: An institutional theory critical discourse analysis. *Tourism Review*, *73*(3), 421–431. doi:10.1108/TR-11-2017-0167

Fernandes, S. (2021). Which way to cope with covid-19 challenges? Contributions of the IoT for smart city projects. *Big Data and Cognitive Computing*, *5*(2), 26. doi:10.3390/bdcc5020026

Ferri, M. A., & Aiello, L. (2017). Tourism destination management in sustainability development perspective, the role of entrepreneurship and networking ability: Tourist kit. *World Review of Entrepreneurship, Management and Sustainable Development*, *13*(5-6), 647–664. doi:10.1504/WREMSD.2017.086334

Font, X., Bonilla-Priego, M. J., & Kantenbacher, J. (2019). Trade associations as corporate social responsibility actors: An institutional theory analysis of animal welfare in tourism. *Journal of Sustainable Tourism*, *27*(1), 118–138. doi:10.1080/09669582.2018.1538231

Forbes, S. L., De Silva, T., & Gilinsky, A. (2019). Social sustainability in the global wine industry: Concepts and cases, pp. 1-204. SCOPUS. www.scopus.com doi:10.1007/978-3-030-30413-3

Galdon, J. L., Garrigos, F., & Gil-Pechuan, I. (2013). Leakage, entrepreneurship, and satisfaction in hospitality. *Service Industries Journal*, *33*(7-8), 759–773. doi:10.1080/02642069.2013.740464

Garay, L., Font, X., & Corrons, A. (2019). Sustainability-oriented innovation in tourism: An analysis based on the decomposed theory of planned behavior. *Journal of Travel Research*, *58*(4), 622–636. doi:10.1177/0047287518771215

Giannopoulos, K., Tsartas, P., & Anagnostelos, K. (2022). A targeted multi-parameter approach of Greek start-ups, related to tourism, culture, and leisure. *Paper presented at the Springer Proceedings in Business and Economics*, (pp. 215-226). IEEE. 10.1007/978-3-030-92491-1_13

Gössling, S. (2017). Tourism, information technologies and sustainability: An exploratory review. *Journal of Sustainable Tourism*, *25*(7), 1024–1041. doi:10.1080/09669582.2015.1122017

Gössling, S., & Michael Hall, C. (2019). Sharing versus collaborative economy: How to align ICT developments and the SDGs in tourism? *Journal of Sustainable Tourism, 27*(1), 74–96. doi:10.1080/0 9669582.2018.1560455

Hall, J., Matos, S., Sheehan, L., & Silvestre, B. (2012). Entrepreneurship and innovation at the base of the pyramid: A recipe for inclusive growth or social exclusion? *Journal of Management Studies, 49*(4), 785–812. doi:10.1111/j.1467-6486.2012.01044.x

Herrmann, B., & Kritikos, A. S. (2013). Growing out of the crisis: Hidden assets to Greece's transition to an innovation economy. *IZA Journal of European Labor Studies, 2*(1), 14. doi:10.1186/2193-9012-2-14

Hillman, W., & Radel, K. (2022). The social, cultural, economic and political strategies extending women's territory by encroaching on patriarchal embeddedness in tourism in Nepal. *Journal of Sustainable Tourism, 30*(7), 1754–1775. doi:10.1080/09669582.2021.1894159

Hjalager, A. (2010). A review of innovation research in tourism. *Tourism Management, 31*(1), 1–12. doi:10.1016/j.tourman.2009.08.012

Hjalager, A., & Johansen, P. H. (2013). Food tourism in protected areas - sustainability for producers, the environment and tourism? *Journal of Sustainable Tourism, 21*(3), 417–433. doi:10.1080/0966958 2.2012.708041

Hoppstadius, F., & Möller, C. (2018). 'You have to try being a role model' – learning for sustainability among tourism entrepreneurs in a Swedish biosphere reserve. *European Journal of Tourism Research, 20*, 28–45. www.scopus.com. doi:10.54055/ejtr.v20i.338

Iskakova, M. S., Abenova, M. K., Dzhanmuldaeva, L. N., Zeinullina, A. Z., Tolysbaeva, M. S., Salzhanova, Z. A., & Zhansagimova, A. (2021). Methods of state support of innovative entrepreneurship. The example of rural tourism. *Journal of Environmental Management and Tourism, 12*(2), 466–472. doi:10.14505//jemt.12.2(50).14

Jóhannesson, G. T. (2012). "To get things done": A relational approach to entrepreneurship. *Scandinavian Journal of Hospitality and Tourism, 12*(2), 181–196. doi:10.1080/15022250.2012.695463

Jóhannesson, G. T., Huijbens, E. H., & Sharpley, R. (2010). Icelandic tourism: Past directions - future challenges. *Tourism Geographies, 12*(2), 278–301. doi:10.1080/14616680903493670

Kordel, S. (2016). Selling ruralities: How tourist entrepreneurs commodify traditional and alternative ways of conceiving the countryside. *Rural Society, 25*(3), 204–221. doi:10.1080/10371656.2016.1255475

Kummitha, H. R. (2020). Eco-entrepreneurs organizational attitude towards sustainable community ecotourism development. *DETUROPE, 12*(1), 85-101. www.scopus.com

Kusumowidagdo, A., & Rembulan, C. L. (2022). *The sense of place value and the actors involved: Indigenous entrepreneurship in Indonesia.* Springer. doi:10.1007/978-981-16-4795-6_7

Lee, C., Hallak, R., & Sardeshmukh, S. R. (2016). Innovation, entrepreneurship, and restaurant performance: A higher-order structural model. *Tourism Management, 53*, 215–228. doi:10.1016/j.tourman.2015.09.017

Lindvert, M., Laven, D., & Gelbman, A. (2022). Exploring the role of women entrepreneurs in revitalizing historic Nazareth. *Journal of Sustainable Tourism*, 1–19. doi:10.1080/09669582.2022.2145291

Lundberg, C., Fredman, P., & Wall-Reinius, S. (2014). Going for the green? The role of money among nature-based tourism entrepreneurs. *Current Issues in Tourism*, 17(4), 373–380. doi:10.1080/1368350 0.2012.746292

Luu, T. T. (2021). Green creative behavior in the tourism industry: The role of green entrepreneurial orientation and a dual-mediation mechanism. *Journal of Sustainable Tourism*, 29(8), 1290–1318. doi: 10.1080/09669582.2020.1834565

Marx, S., & Klotz, M. (2021). Entrepreneurship during crisis: Innovation practices of micro and small tour operators. *International Journal of Entrepreneurship and Innovation*. doi:10.1177/14657503211061025

Matsiliza, N. S. (2017). Seeking strategies for sustainability in tourism entrepreneurship in South Africa. *African Journal of Hospitality, Tourism and Leisure*, 6(4), 1–10. www.scopus.com

Moscardo, G. (2008). Sustainable tourism innovation: Challenging basic assumptions. *Tourism and Hospitality Research*, 8(1), 4–13. doi:10.1057/thr.2008.7

Moyle, C., Moyle, B., & Burgers, H. (2020). Entrepreneurial strategies and tourism industry growth. *Tourism Management Perspectives*, 35, 100708. doi:10.1016/j.tmp.2020.100708

Ndou, V., Mele, G., & Del Vecchio, P. (2019). Entrepreneurship education in tourism: An investigation among European universities. *Journal of Hospitality, Leisure, Sport and Tourism Education*, 25, 100175. Advance online publication. doi:10.1016/j.jhlste.2018.10.003

Pailis, E. A., Fatkhurahman, & Arif, A. (2020). Indigenous community approach through indigenous leaders social entrepreneurship in five Luhak in Rokan Hulu regency Riau Indonesia. *International Journal of Scientific and Technology Research*, 9(3), 5249-5255. www.scopus.com

Parasotskaya, N., Berezyuk, V., Prasolov, V., Nazarova, V., & Mezentseva, T. (2021). Comparative analysis of small and medium-sized businesses and its impact on the development of tourism. *Journal of Environmental Management and Tourism*, 12(6), 1586–1602. doi:10.14505//jemt.12.6(54).15

Peeters, P., Gossling, S., & Becken, S. (2006). Innovation towards tourism sustainability: Climate change and aviation. *International Journal of Innovation and Sustainable Development*, 1(3), 184–200. doi:10.1504/IJISD.2006.012421

Preko, A. (2020). Tourism development: National policies and tourism priorities in Ghana. *International Journal of Tourism Policy*, 10(4), 380–391. doi:10.1504/IJTP.2020.112644

Reid, S. (2019). Wonderment in tourism land: Three tales of innovation. *Journal of Teaching in Travel & Tourism*, 19(1), 79–92. doi:10.1080/15313220.2018.1560533

Robbins, P., & Devitt, F. (2017). Collaboration, creativity and entrepreneurship in tourism: A case study of how design thinking created a cultural cluster in Dublin. *International Journal of Entrepreneurship and Innovation Management*, 21(3), 185–211. doi:10.1504/IJEIM.2017.083454

Romão, J., Machino, K., & Nijkamp, P. (2018). Integrative diversification of wellness tourism services in rural areas–an operational framework model applied to east Hokkaido (Japan). *Asia Pacific Journal of Tourism Research, 23*(7), 734–746. doi:10.1080/10941665.2018.1488752

Rosário, A., Vilaça, F., Raimundo, R., & Cruz, R. (2021). Literature review on Health Knowledge Management in the last 10 years (2009-2019). *The Electronic Journal of Knowledge Management, 18*(3), 338-355. doi:10.34190/ejkm.18.3.2120

Rosário, A. T. (2021). The Background of articial intelligence applied to marketing. *Academy of Strategic Management Journal, 20*(6), 1–19. 1939-6104-20-S6-118

Rosário, A. T., & Dias, J. C. (2022). Sustainability and the Digital Transition: A Literature Review. *Sustainability (Basel), 14*(7), 4072. doi:10.3390u14074072

Ruiz-Ortega, M. J., Parra-Requena, G., & García-Villaverde, P. M. (2021). From entrepreneurial orientation to sustainability orientation: The role of cognitive proximity in companies in tourist destinations. *Tourism Management, 84*, 104265. doi:10.1016/j.tourman.2020.104265

Salome, L. R., van Bottenburg, M., & van den Heuvel, M. (2013). 'We are as green as possible': Environmental responsibility in commercial artificial settings for lifestyle sports. *Leisure Studies, 32*(2), 173–190. doi:10.1080/02614367.2011.645247

Saviera, T. M., Kusumastuti, R., & Hidayanto, A. N. (2022). Importance-performance analysis towards sustainable indigenous tourism (a lesson learned from Indonesia). *International Journal of Innovation and Learning, 31*(1), 91–116. doi:10.1504/IJIL.2022.119638

Singh, S., & Wagner, R. (2022). Indian wine tourism: New landscape of international spillovers. *Journal of Asia Business Studies*. doi:10.1108/JABS-01-2022-0004

Sinno, N. (2019). *The effect of digital transformation on innovation and entrepreneurship in the tourism sector: The case of Lebanese tourism services providers*. Springer. doi:10.1007/978-3-030-30874-2_3

Sofield, T., & Lia, S. (2011). Tourism governance and sustainable national development in China: A macro-level synthesis. *Journal of Sustainable Tourism, 19*(4-5), 501–534. doi:10.1080/09669582.2011.571693

Solvoll, S., Alsos, G. A., & Bulanova, O. (2015). Tourism entrepreneurship – review and future directions. *Scandinavian Journal of Hospitality and Tourism, 15*(sup1), 120–137. doi:10.1080/15022250.2015.1065592

Sørensen, F., & Grindsted, T. S. (2021). Sustainability approaches and nature tourism development. *Annals of Tourism Research, 91*, 103307. doi:10.1016/j.annals.2021.103307

Sotiriadis, M. (2018). Entrepreneurship and Entrepreneurs in Tourism. In *The Emerald Handbook of Entrepreneurship in Tourism, Travel and Hospitality* (pp. 3–17). Emerald Publishing Limited. doi:10.1108/978-1-78743-529-220181001

Stoffelen, A., Adiyia, B., Vanneste, D., & Kotze, N. (2020). Post-apartheid local sustainable development through tourism: An analysis of policy perceptions among 'responsible' tourism stakeholders around Pilanesberg national park, South Africa. *Journal of Sustainable Tourism*, *28*(3), 1–20. doi:10.1080/09 669582.2019.1679821

Sundbo, J., Orfila-Sintes, F., & Sørensen, F. (2007). The innovative behavior of tourism firms-comparative studies of Denmark and Spain. *Research Policy*, *36*(1), 88–106. doi:10.1016/j.respol.2006.08.004

Swanson, K. K., & DeVereaux, C. (2017). A theoretical framework for sustaining culture: Culturally sustainable entrepreneurship. *Annals of Tourism Research*, *62*, 78–88. doi:10.1016/j.annals.2016.12.003

Teixeira, S. J., & Ferreira, J. J. M. (2019). Entrepreneurial artisan products as regional tourism competitiveness. *International Journal of Entrepreneurial Behaviour & Research*, *25*(4), 652–673. doi:10.1108/ IJEBR-01-2018-0023

Teruel-Sánchez, R., Briones-Peñalver, A. J., Bernal-Conesa, J. A., & de Nieves-Nieto, C. (2021). Influence of the entrepreneur's capacity in business performance. *Business Strategy and the Environment*, *30*(5), 2453–2467. doi:10.1002/bse.2757

Thompson, B. S., Gillen, J., & Friess, D. A. (2018). Challenging the principles of ecotourism: Insights from entrepreneurs on environmental and economic sustainability in Langkawi, Malaysia. *Journal of Sustainable Tourism*, *26*(2), 257–276. doi:10.1080/09669582.2017.1343338

Triantafillidou, E., & Tsiaras, S. (2018). Exploring entrepreneurship, innovation and tourism development from a sustainable perspective: Evidence from greece. *Journal for International Business and Entrepreneurship Development*, *11*(1), 53–64. doi:10.1504/JIBED.2018.090020

Valeri, M., & Baiocco, S. (2012). The integration of a Swedish minority in the hotel business culture: The case of Riva del sole. *Tourism Review*, *67*(1), 51–60. doi:10.1108/16605371211216378

Valeri, M., & Fadlon, L. (2016). Sustainability in tourism: An originality and hospitality business in Italy. *Tourismos*, *11*(1), 1–18. www.scopus.com

Van Wijk, J., Van der Duim, R., Lamers, M., & Sumba, D. (2015). The emergence of institutional innovations in tourism: The evolution of the African wildlife foundation's tourism conservation enterprises. *Journal of Sustainable Tourism*, *23*(1), 104–125. doi:10.1080/09669582.2014.927878

Wahyuni, N. M., & Sara, I. M. (2020). The effect of entrepreneurial orientation variables on business performance in the SME industry context. *Journal of Workplace Learning*, *32*(1), 35–62. doi:10.1108/ JWL-03-2019-0033

Williams, A. M. (2014). Tourism innovation products, processes, and people. The Wiley Blackwell companion to tourism (pp. 168-178). Wiley. doi:10.1002/9781118474648.ch13

Williams, A. M., & Shaw, G. (2011). Internationalization and innovation in tourism. *Annals of Tourism Research*, *38*(1), 27–51. doi:10.1016/j.annals.2010.09.006

Xiao, Y., & Watson, M. (2019). Guidance on conducting a systematic literature review. *Journal of Planning Education and Research*, *39*(1), 93–112. doi:10.1177/0739456X17723971

Yachin, J. M. (2019). The entrepreneur–opportunity nexus: Discovering the forces that promote product innovations in rural micro-tourism firms. *Scandinavian Journal of Hospitality and Tourism, 19*(1), 47–65. doi:10.1080/15022250.2017.1383936

Zapalska, A. M., & Brozik, D. (2015). The life-cycle growth and development model and leadership model to analyzing tourism female businesses in poland. *Problems and Perspectives in Management, 13*(2), 82–90. www.scopus.com

Zapalska, A. M., Brozik, D., & Zieser, N. (2015). Factors affecting success of small business enterprises in the polish tourism industry. *Tourism (Zagreb), 63*(3), 365–381.

KEY TERMS AND DEFINITIONS

Entrepreneurship: the initiative process of implementing new businesses or changes in existing businesses.

Environment management information system (EMIS): defined as 'technical-organizational systems for obtaining, processing and systematically making available relevant environmental information in companies'.

GDP: gross domestic product represents the sum of all final goods and services produced in a given region during a given period.

Geographical information systems (GIS): is a system of hardware, software, spatial information, computational procedures and human resources that allows and facilitates the analysis, management or representation of geographic information.

Sustainable innovation: involves making intentional changes to a company's products, services, or processes to generate long-term social and environmental benefits while creating economic profits for the firm.

Sustainable tourism: it is a type of tourism that includes concern for sustainability (economic, social and environmental), attention to improving tourist experiences and meeting the needs of host communities.

World Trade Organization (WTO): It is an institution created with the objective of supervising and liberalizing international trade.

Chapter 12
Assessing the Accessibility of Tourist Destination Promotional Information:
The Case of Portugal

Gorete Dinis

GOVCOPP, CITUR, Polytechnic Institute of Portalegre, Portugal

Zélia Breda

iD https://orcid.org/0000-0002-5882-063X

DEGEIT, GOVCOPP, University of Aveiro, Portugal

ABSTRACT

Few studies have analysed the accessibility level of information sources used by persons with disabilities when making tourism-related purchases. Consequently, the main objective of this chapter is to gain insights into whether Destination Management Organisations are actively developing inclusive destination promotion and advertising materials. To accomplish this, an exploratory study was conducted, with a specific focus on the tourism of Portugal. Portugal was chosen as the subject of the study due to its recognition by the World Tourism Organisation as the world's first accessible tourism destination. This recognition encompasses various aspects, including the official promotional tourism website (visitportugal.com), which features digital brochures. It is important to note that the exploratory nature of this study limited the ability to make direct comparisons with previous research. However, for future investigations, it is recommended that the framework employed in this study be applied to assess the accessibility of promotional materials from other DMOs and tourism stakeholders.

DOI: 10.4018/978-1-6684-6985-9.ch012

INTRODUCTION

According to the World Health Organization (WHO), the global population of people with disabilities exceeds 1 billion, accounting for approximately 15% of the world's population (WHO, 2021). The percentage of accessible tourists, defined as individuals with disabilities, varies significantly depending on the type of disability and country. When considering a broader definition of accessible tourists to include older adults or those with temporary mobility limitations, the overall percentage of the population with accessibility needs while travelling is even higher. For instance, the baby boomer generation in many countries represents a substantial and growing demographic of older adults who may require specific accessibility provisions when travelling (UNWTO, 2019). This highlights the importance of ensuring that tourist destinations are accessible for persons with disabilities (PwDs), as they constitute a significant portion of the tourist market. Nevertheless, much work still needs to be done to prioritise accessibility in the tourism industry.

Travelling with a disability requires extensive planning. Even though PwDs frequently select locations closer to home when planning a vacation, they still require accurate information on accessible facilities and services (Eichhorn et al., 2008). The level of information needed increases with the degree of disability (Buhalis, n.a.), affecting PwDs' willingness to participate in the travel planning process and make reservations. Unfortunately, information specifically tailored to PwDs is often lacking. If PwDs had access to the necessary information, it is estimated that about 50% of this demographic would travel more frequently because they would have knowledge of available accessible facilities and services (Buhalis, n.a.; Lonely Planet, 2020a, 2020b). The Internet serves as an important and primary source of information for PwDs; therefore, it is imperative for Destination Management Organisations (DMOs) to develop adequate information channels and use online platforms to communicate destination accessibility effectively. This can be accomplished through the creation of user-friendly websites, the provision of detailed accessibility information, and the promotion of accessible tourism options through social media platforms. Moreover, DMOs can collaborate with organisations for PwDs to ensure the accuracy and suitability of the information provided for this segment.

Despite the significance of accessible tourism, only a few studies have analysed the accessibility level of the information sources utilised by PwDs when purchasing tourism-related products (Mills, A. et al., 2008; Mills, J. E. et al., 2008; Vila et al., 2018). Thus, the primary objective of this chapter is to examine whether DMOs develop inclusive destination promotion and advertising information. To achieve this, an exploratory study was conducted with Tourism of Portugal, the country's national DMO, analysing the accessibility of online brochures. Portugal was chosen as the subject of this study because the World Tourism Organisation (UNWTO) has distinguished it as the world's first accessible tourism destination (UNWTO, 2021), namely its official promotional tourism website (VisitPortugal. com). Although the exploratory nature of this study limits direct comparisons with previous research, the analytical framework can be applied to other DMOs and promotional materials within the tourism industry in future investigations.

The structure of this paper is as follows: firstly, a review of the literature on accessible tourism is provided, followed by an examination of the accessibility of tourism information, specifically focusing on the information provided by official tourism destination websites. Subsequently, the methodology employed in this study is described in more detail. Finally, the research findings are reported and discussed, along with their implications for future research in the field.

LITERATURE REVIEW

Accessible Tourism

In recent years, there has been a growing interest in accessible tourism, which has emerged as a field of academic research and industry practice. As a result, the concept of accessible tourism has evolved "from the idea of accommodation or adaptation so that PwDs can participate in tourism, towards a concept of quality tourism for all, understanding that accessibility is an important part of that quality" (UNWTO, 2016, p. 19). Although there is no universally accepted definition of accessible tourism, the definition proposed by Darcy and Dickson (2009) has been widely adopted in academic literature (UNWTO, 2016). According to their perspective, accessible tourism involves collaborative processes among stakeholders to deliver universally designed tourism products, services, and environments that allow individuals with various access requirements, including mobility, vision, hearing, and cognitive dimensions, to function independently and with equity and dignity (Darcy & Dickson, 2009). Accessible tourism, thus, refers to ensuring that all individuals, regardless of their physical or mental abilities, can enjoy travel and tourism experiences.

Accessible tourism encompasses two main groups of travellers with access needs: (i) persons with disabilities, mainly mobility-related but also cognitive, visual and hearing impairments; and (ii) senior citizens (aged 65 or older). While the specific disabilities may vary, the requirements for accessibility are often similar (CBI, 2022). Additionally, a significant portion of the population can benefit from accessibility measures and the development of accessible tourism at some point in their lives, such as pregnant women or individuals with temporary impairments (UNWTO, 2016). Therefore, the accessibility market is not homogenous but rather diverse and wide-ranging (OSSATE / University of Surrey, 2006).

According to the World Health Organization (WHO), approximately 15% of the world's population has a disability (UNWTO, 2020). In Europe, in 2011, the number of people with access needs was around 140 million in 2011, and it is projected to reach 160 million by 2025 (CBI, 2022). In Portugal, recent census data from 2021 indicate that about 10.9% of the population aged five years or more face difficulties in performing at least one of the six basic daily activities, such as bathing, dressing, eating, etc. (INE, 2022). These statistics underscore the need for increased support, resources, and accessibility measures to cater to individuals with disabilities.

The economic and social integration of PwDs has led to their increased participation in tourism activities (UNWTO, 2020). This trend is further amplified when considering senior citizens and travelling companions. According to estimates by the World Tourism Organization (UNWTO), 70% of Europeans with accessibility needs possess both the financial means and physical ability to travel, making accessible tourism a financially viable market (GFK, 2014). In the European Union (EU), accessible tourism contributed significantly to the economy, with a total economic output of 786 billion Euros, a gross value added of 356 billion Euros, and a GDP impact of 394 billion Euros (GFK, 2014). Therefore, improving accessibility not only enhances the lives of persons with disabilities and provides equal opportunities for all travellers but also stimulates economic growth within the tourism industry.

Despite the increasing recognition of accessibility as the norm, many tourist destinations and businesses still lack the necessary infrastructure and services to cater to individuals with access needs, thereby limiting market expansion and excluding a significant portion of potential customers. Travellers with disabilities often face a limited selection of tourism products, and numerous obstacles and barriers hinder their access to inclusive experiences (CBI, 2022, UNWTO, 2016). These barriers can be classified as intrinsic (e.g., overprotection of parents), environmental (e.g., unsuitable transport) and interactive (e.g., the inadequacy of the information's format in meeting the needs of those with hearing or visual impairments) (Smith, 1987, cited in Zajadacz & Lubarska, 2019). As a result, individuals with special needs tend to exhibit loyalty towards accessible destinations, more likely to revisit and recommend them (CBI, 2022). Thus, tourism businesses and destinations need to develop products to meet the requirements of travellers with disabilities, necessitating a comprehensive understanding of their travel patterns, motivations, needs, and expectations.

France, Germany and the United Kingdom are the largest European source markets for accessible tourism (CBI, 2022). Additionally, countries with high-income populations and a propensity for travel, such as Scandinavia and the Netherlands, present potential source markets. While most PwDs in Europe choose domestic destinations for their travels, increasing the length of stay or encouraging long-haul trips requires investment in the development and marketing of accessible products. This expansion aims to offer a wider range of activities for PwDs and ensure that relevant information reaches and remains accessible to them (CBI, 2022). PwDs typically travel in groups, and friends and relatives often participate in making reservations (Project Newcast, n.a.). With the Internet serving as the primary source of information for many Europeans, tourist destinations and products must be promoted through accessible websites (CBI, 2022).

Accessibility of Information in Tourism

In order to develop responsible and sustainable tourism policies and strategies, accessibility must be incorporated as an integral component (Buhalis & Darcy, 2020; UNWTO, 2020). Tourist destinations that consider the needs of travellers with special needs offer added-value products, enhancing the overall tourism experience and improving the standard of living for both locals and visitors alike (Buhalis & Darcy, 2020; UNWTO, 2016, 2021).

To embark on a journey, advance planning is necessary, and acquiring relevant information becomes essential. Travelling with a disability requires thorough preparation and significant organisation (CBI, 2022). PwDs must have comprehensive information about the destination's conditions and the accessibility of its facilities and services to ensure a smooth travel experience and mitigate potential issues. Consequently, they often need to conduct more extensive research than the average traveller (Buhalis & Darcy, 2020; Project Newcast, n.a.). One of the most significant challenges faced by PwDs when travelling is the lack of information, which severely restricts their options for choosing suitable products and destinations (ENAT, 2013), making their trips more challenging. The specific information needed varies for each individual and is unique to their circumstances (England Athletics, n.a.). However, wheelchair users typically have the highest need for detailed accessibility information (Buhalis & Darcy, 2020; Project Newcast, n.a.). It is, therefore, crucial to provide a range of easily accessible information tailored to the specific needs of each PwD (Buhalis & Darcy, 2010). This can include details about accessibility, transportation, accommodations, and local attractions, enabling PwDs to make informed decisions and enjoy a comfortable and stress-free travel experience.

Accessible information should be available to the traveller before, during, and after the trip (UNWTO, 2020). Regardless of the format in which it is provided, accessible information should meet specific criteria, such as accuracy, detail, relevance, reliability, ease of access, timeliness, and being up-to-date. It should also clearly indicate the organisation responsible for providing the information and the publication date (Project Newcast, n.a.). Furthermore, accessible information should (i) provide data on the accessibility of local infrastructure and services or provide references to where such information can be found; (ii) include contact details for further inquiries; (iii) maintain consistency across all communication channels; (iv) be provided by trained information managers and technical staff with the required knowledge to deliver accessible services; and (v) be regularly updated to ensure the safety and comfort of visitors (Buhalis & Darcy, 2020; CBI, 2022).

DMOs bear specific responsibility for collecting and organising this information, either independently or in collaboration with service providers. They can employ a variety of tools and methods to effectively communicate information to PwDs, but it is crucial that all stakeholders involved understand the relevance of the shared information and to whom it applies. Additionally, the images used in DMOs' marketing campaigns should depict diverse visitors of different ages and abilities (Turismo de Portugal, 2017). This approach not only ensures clarity and effectiveness in conveying the message, but also promotes inclusivity and diversity within the tourism industry. Creating an environment where all individuals feel welcome and represented is of paramount importance (Buhalis & Darcy, 2020).

Online Information Accessibility: Official Websites of Tourist Destinations

In recent years, the Internet has become a relevant tool for tourism businesses and destinations, providing access to a greater amount of information (UNWTO, 2016) and serving as an important booking platform (ENAT, 2013) and source of tourism information (Law et al., 2010). It is also considered a vital channel for communication and promotion of tourism destinations (Lee & Gretzel, 2012). Websites, in particular, play a crucial role as they offer detailed information and serve as a significant source of information for travellers in general, including PwDs (GFK, 2014).

DMOs have increasingly invested in their online presence, with almost all tourist destinations now having a website (Crouch, 2000). As mentioned earlier, DMOs are responsible for promoting and disseminating information about the destination, often managing the official destination website. Furthermore, official destination websites play an essential role in the travel cycle, as "tourists use them for trip planning and selecting destinations" (Morrison, 2013, p. 372), which justifies the need to research the accessibility of official tourism websites (Domínguez Vila et al., 2017).

DMOs or tourism service providers should evaluate the accessibility of their websites, ensuring accessibility for all. Web development must adhere to accessibility guidelines (Project Newcast, n.a.), guaranteeing access to all users regardless of their browsing context and technical means (UNWTO, 2016), thereby ensuring accessibility for PwDs (Pérez & Velasco, 2003).

To ensure website accessibility, guidelines exist for web managers and designers to assess the website's level of accessibility (Pérez & Velasco, 2003). One commonly used document for evaluating and measuring website accessibility is the Web Content Accessibility Guidelines (WCAG 2.1). According to its recommendations and criteria, websites can be categorised into three levels of accessibility: A, AA, and AAA. Level AA is the minimum standard that all public institutions should meet, while AAA signifies full compliance with all requirements.

Oertel's study (2004, cited in Buhalis & Darcy, 2010) has shown that national tourism web pages in the European Union did not present information in an accessible manner and failed to meet the minimum conformance level (A). GFK (2014) analysed ten tourism board websites and discovered that half of them had even lower accessibility levels when considering technical accessibility. Another study by ENAT (2013) examined the websites of 41 European national tourism organisations (NTOs) and found that only 19 provided some accessibility information for PwDs or individuals with special access needs.

None of the websites fully complied with the WCAG, even at the basic level (level A). Recent studies have also highlighted the importance of accessible tourism and the challenges it faces (Laws et al., 2020; Sigala & Gkritzali, 2020).

In recent years, digital platform providers and operating systems have made efforts to improve the accessibility of their platforms. Portable Document Format (PDF) has become more accessible for PwDs, both with and without assistive technology software and devices, thanks to accessibility guidelines and standards for electronic documents (PDF/UA (ISO14289-1)) and accessibility features in Adobe Acrobat, Adobe Reader, and PDF (Adobe, n.a.). However, it is generally recommended that document accessibility begins in the original format rather than relying solely on PDF as the destination file format. While the Department of Health (2010) suggests providing online documents in PDF format, GOV.UK (n.a.) notes that information should ideally be published in HTML, whenever possible, because PDF files make content harder to find, use and maintain, and are less compatible with assistive technologies.

Despite the prevalence of online information, some individuals still prefer printed materials such as leaflets, flyers, and brochures, which can also be obtained online (GFK, 2014). Different target groups, especially older people who may lack access to or prefer not to use the Internet, find brochures appealing (British Tourist Authority, 2019; GFK, 2014). In a survey by the British Tourist Authority in 2019, 86% of travellers with disabilities found brochures and leaflets to be very or somewhat helpful in making travel arrangements. However, it is important to note that travellers with disabilities have expressed a need for brochures with specific information catering to their needs, which is often lacking (GFK, 2014). This information must be presented in an accessible manner across different formats and provide alternative formats when necessary. Table 1 provides recommendations for improving the overall structure, appearance, and content of promotional print materials.

METHODOLOGY

Objectives and Case Study Selection Rationale

The main objective of this study is to assess the accessibility for PwDs of promotional materials, specifically the brochures available on Portugal's official tourism website, www.VisitPortugal.com, which serves as the primary platform for promoting tourism in Portugal. To achieve this, and considering the study's exploratory nature, the initial step was to select the accessibility parameters used in data collection. In situations where there is limited prior knowledge of the phenomenon being studied, which is the case here, the case study methodology is the most appropriate approach (Yin, 2009). Therefore, this study focuses on Portugal as a tourist destination and examines the actions taken by the national DMO, Tourism of Portugal, in providing information to potential accessible tourists.

Portugal was selected for this study because, in 2019, it was distinguished by the UNWTO, in partnership with Fundación ONCE, as the world's first accessible tourist destination. This prestigious award positioned the country as a global leader in accessibility and recognised Portugal's commendable efforts to promote accessibility in tourism. In line with this recognition, Portugal has implemented legislation to ensure accessibility for PwDs. The Portuguese Law No. 83/2017 establishes the legal framework for promoting accessibility, ensuring equal opportunities, and improving the quality of life for people with disabilities.

Table 1. Accessibility standards for print materials

Publication design guidelines
Good contrast between the paper and the text, using a different colour or bold type. Backgrounds should be as pale as possible, and printing ink as dark as possible. Black on yellow and black on white work the best.
Do not surround or cross-reference images with text.
Avoid highly stylised or ornate fonts. Instead, use typefaces such as Sans Serif, Arial, Univers or Verdana.
All text should be left aligned.
Avoid blocks of text in capitals and italics.
Publication structure and content
The publication must have a maximum of 24 pages.
Content must be organised in a logical sequence.
New chapters or topics should start on a new page.
Headings and page numbers should be placed consistently throughout each page.
The text should be written clearly and concisely.
Use large fonts, at least 14 or 16.
Use short sentences of 10 to 15 words.
Each sentence should have just one idea and one verb.
Use short headlines.
Include a glossary explaining abbreviations and jargon.
Add an index at the end of the document.
Use complete words, not abbreviations.
Make active, not passive, sentences.
Use even word spacing.
Avoid over-textually filling the page by not adding text until the very end.
Visual elements
Use pictograms or symbols to help readers navigate text; these are useful for people from other nationalities. Using Quick Response (QR) codes is advised.
Use images to clarify the text and support its meaning. Pictures must be positioned to the left of the text.
Illustrations should depict the population and recognise diversity – including age, disability, ethnicity, faith and gender.
Use testimonials and opinions from visitors with special needs.
Use any grading or awards logos.
Relevant information
Put the organisation's full address.
Provide information about the conditions at the destination (e.g., accessible routes, parking availability), transportation, accommodation, addresses for local tourist information centres and other accessible services in the area.

Source: Based on British Tourist Authority (2019), UNWTO (2016), Gov.UK (n.a.), Department of Health (2010)

As the national tourism authority, Tourism of Portugal was established with the goals of promoting Portugal as a tourist destination and supporting the growth of the industry. Since the launch of the "All for All" programme, which includes the "Accessible Festivals" programme and the "Manual on Accessible Tourism Destination Management," Tourism of Portugal has been actively promoting, stimulating, and supporting tourist destinations and stakeholders to foster the growth of accessible tourism and enhance services for this market segment. Furthermore, Tourism of Portugal collaborates as a partner with the Tur4all platform (www.tur4all.pt), which provides comprehensive information on the accessibility of tourist resources in Portugal, further supporting the dissemination of accessibility-related information.

By conducting this study on Portugal's official tourism website and examining the actions taken by Tourism of Portugal, we aim to evaluate the accessibility of promotional materials for PwDs and contribute to the ongoing development of accessible tourism practices in the country, aligning with the legislative framework and the nation's commitment to inclusivity and equal opportunities.

Data Collection and Analysis

In order to comprehensively assess the accessibility of tourism information for PwDs, an analysis was conducted on a range of promotional brochures. These brochures encompassed both printed versions and their digital counterparts, which are accessible through Portugal's official tourism website (Figure 1).

Figure 1. Brochures collected for analysis

To ensure a representative sample, a total of 12 brochures were diligently identified and collected for detailed examination (Table 2). Among these, nine brochures were available in Portuguese, two in English, and one in Japanese, showcasing the commitment to multilingual accessibility. Each brochure provided unique insights into specific regions within Portugal, with the exception of the central region, which was extensively covered in three brochures. To showcase the diverse facets of Portugal, the remaining seven regional tourism areas each had their dedicated brochure, highlighting the distinctive attractions and experiences they offer. Additionally, to cater to the comprehensive needs of potential visitors, three brochures provided an overview of the entire country, granting an all-encompassing glimpse into the attractions of Portugal.

Table 2. Brochures collected for analysis

Nº	Title	Language	Region
1	*Açores – Certificado pela Natureza*	Portuguese	Azores
2	*Alentejo – Tempo para ser Feliz*	Portuguese	Alentejo
3	*Algarve – Ideias e Inspirações*	Portuguese	Algarve
4	*Best of Center of Portugal 2020*	English	Centre of Portugal
5	*Center of Portugal – Don't Miss it!*	Japanese	Centre of Portugal
6	*Centro de Portugal Roundtrip*	English	Centre of Portugal
7	*Escolha Portugal*	Portuguese	Portugal
8	*Guia Madeira e Porto Santo*	Portuguese	Madeira
9	*Lisboa City Breaks*	Portuguese	Lisbon
10	*Porto e Norte*	Portuguese	Porto and North of Portugal
11	*Sentir Portugal*	Portuguese	Portugal
12	*Surf*	Portuguese	Portugal

In order to provide a comprehensive assessment of the VisitPortugal.com website, a thorough evaluation of its technical aspects was conducted. The webgrader tool (https://website.grader.com) was used to analyse the website's technical performance, while its accessibility was assessed using Achecker (https://achecker.ca/), following the guidelines outlined in the Web Content Accessibility Guidelines.

Following the initial technical evaluation, a direct data collection approach was employed to further examine the website's accessibility. The data collection took place in July 2022 and was performed manually. To begin, a quick review of the website was conducted to determine if any accessibility-related information was present. The location, ease of access, multilingual availability, content, and level of detail of such information were carefully examined.

To facilitate the observation process, an observation grid consisting of indicators that defined the specific behaviours to be observed was used, as suggested by Quivy and Van Campenhoudt (2008). The selected parameters for analysis were based on the principles outlined in WCAG 2.1 and the UNWTO recommendations for online communication in accessible tourism. Additionally, the defined criteria identified in the existing literature (Table 1) were taken into consideration.

To ensure the objectivity of the study, all authors strictly adhered to the established selection criteria and parameters. Descriptive statistics were employed to analyse the obtained data, with each parameter evaluated based on its presence and the observed situation within each brochure. For example, the "Centro de Portugal Roundtrip" brochure solely consisted of images without any text blocks. On the other hand, due to the brochure "Center of Portugal – Don't Miss it!" being in Japanese, certain parameters within the observation grid, specifically those related to the brochure's text, could not be observed.

RESULTS

Using the webgrader tool for the analysis of VisitPortugal.com yielded valuable insights. The website obtained a performance rating of 53%, indicating the presence of certain issues with page requests and speed, potentially resulting in a slower browsing experience that may lead visitors to leave the website. Furthermore, there are concerns regarding the time taken to download pages, which could be addressed through the compression of JavaScript. On a positive note, VisitPortugal.com demonstrates exceptional performance in terms of content optimisation for search engines, with a 100% optimisation rate, highlighting the deliberate efforts made to achieve favourable positioning in search engine results. Additionally, the website boasts a 100% responsive design, ensuring ease of readability and usability on mobile devices. Overall, there were relatively few technical issues identified with the website.

Subsequently, the accessibility of the website was assessed to determine its compliance with WCAG 2.1 standards. The findings revealed the presence of 33 accessibility problems, including issues related to text on images, potentially meaningless link text, and an inaccessible user interface. There may be additional problems requiring manual identification. It is worth noting that, contrary to the Tourism of Portugal website (www.turismodeportugal.pt), the analysed website does not mention its level of accessibility or its adherence to the regulations governing access for people with special needs to government websites, services, and public institutions (Resolução do Conselho de Ministros nº 155/2007, de 2 de outubro de 2007).

VisitPortugal.com stands as the primary and most significant official source of information pertaining to travel to Portugal. Alongside Portuguese, the website is available in nine other languages, including English, Spanish, Russian, French, German, Italian, Chinese, Dutch, and Japanese. The main menu offers a comprehensive range of links to valuable information, photos, videos, and brochures. Navigating through the main menu enables users to access details about Portugal's tourist regions, tourist services, key tourist resources, and things to do. By selecting the "what to do" option, users can explore information about accessible travel, encompassing accessible highways, beaches, modes of transportation, parking facilities, and parks. The information provided for PwDs is accessible, extensive, and highly detailed.

Regarding the brochures, all are published by national or regional official tourism organisations responsible for promoting tourism in Portugal (Table 3). Nine brochures are available in both print and digital formats and produced by the regional official tourism bodies. Each brochure follows a logical structure. Brochures encompassing the mainland regions are organised by regions, while those focusing on the archipelagos of Madeira and the Azores are organized by islands. Regional destination brochures follow a thematic organisation, except for "Lisboa City Breaks", which combines specific themes, such as gastronomy and electric cars, with distinct locations (e.g., Bairro Alto, Sintra).

Among the brochures analyzed, the majority consisted of extensive content, with only three brochures ("Alentejo – Time to be Happy", "Lisboa City Breaks" and "Porto e Norte") having fewer than 24 pages. Conversely, there were three large brochures: "Best of Center of Portugal 2020" (236 pages), "Guia Madeira e Porto Santo" (83 pages) and "Azores – Certificado pela Natureza" (60 pages). The typefaces used in the brochures featured straight lines, with minimal styles or ornate elements. However, it was observed that almost all brochures employed capitalisation for titles and included words or text blocks in italics.

The font size used in the brochures was generally small, except for the titles, which were set at 18 points. The remaining text predominantly had a font size of 10 or smaller, as observed in nine brochures. Additionally, approximately half of the text in five brochures was left-aligned. The colour scheme primarily consisted of grey (six brochures) and black (five brochures), with only one brochure employing blue text. The background colour tended to be pale, typically white. The highest contrast between the paper and the text was achieved when the text was black. Two brochures incorporated other colours, such as orange and red, mainly in titles, while the use of white text against pictures was commonly observed.

Table 3. Partial results of the observed parameters

Brochure	Entity	Letter fonts	Text alignment	Number of pages	Font size
Açores - Certificado pela Natureza	Azores Promotion Board	Myriad	Justified	60	12
Alentejo - Tempo para ser Feliz	Tourism of Alentejo / Regional Tourism Promotion Agency of Alentejo	Myriad	Left-aligned	14	10.5
Algarve - Ideias e Inspirações	Algarve Tourism Board	Myriad	Left-aligned	38	9
Best of Center of Portugal 2020	Centre of Portugal / Regional Tourism Promotion Agency of the Centre of Portugal	NoahGrotesc	Left-aligned	236	10
Center of Portugal - Don't Miss it!	Centre of Portugal	Mplus	Justified	44	10
Centro de Portugal Roundtrip	Centre of Portugal	Libertad –light (Titles)	Not applied	31	18
Escolha Portugal	Tourism of Portugal	Gotham	Justified	32	9
Guia Madeira e Porto Santo	Madeira Tourism Board	Helvetica	Left-aligned	83	8
Lisboa City Breaks	Tourism of Lisbon	MyriadPro	Justified	11	10
Porto e Norte	Porto and North Tourism Association	Calibri	Left-aligned	21	7
Sentir Portugal	Tourism of Portugal	Gotham	Justified	40	9
Surf	Tourism of Portugal	Calibri	Justified	48	9

Regarding images, every brochure contained images with overlaid text. While some images were positioned to the left of the text, others had text placed around, above, or below them, suggesting a lack of consistent pattern or structure in image placement. The brochures extensively used images to illustrate and enhance the accompanying text. Pictograms or symbols were employed in four brochures to aid users in navigating the text, while QR codes were included in two brochures. Only the "Porto e Norte" brochure displayed an award logo, although other brochures, such as "Guia Madeira e Porto Santo", "Porto e Norte", and "Surf", made textual references to distinctions or awards. People were occasionally depicted in the images, with men, women, or children being the most frequent subjects. Senior citizens were the primary subject in only seven brochures. Notably, no images representing PwDs were observed, and there was no integration of testimonials or opinions from visitors with special needs.

Seven brochures included an index, typically located at the beginning rather than the end. Each page generally covered a single topic or theme, occasionally extending to multiple pages without breaks in the content. The text, however, was not consistently written in a clear and concise manner, as passive sentences and scientific terms, including the scientific names of flora and fauna in Madeira, as well as descriptive adjectives, were occasionally employed.

The sentence structure varied, with sentences often containing multiple ideas and verbs and not being consistently short (between ten to fifteen words). The "Azores – Certified by Nature" brochure posed challenges in text comprehension due to the excessive use of hyphens to break words at the end of lines. Except for the "Center of Portugal – Don't Miss it!" brochure, where the language barrier hindered the analysis of these criteria, only the "Surf" brochure included a glossary explaining jargon. Additionally, seven brochures incorporated acronyms without accompanying full words. Headings across brochures were generally brief and frequently located in the same position on each page, as were the page numbers. Some blank space was observed at the end of pages due to the text not completely filling the allocated area, while the spacing between words and phrases remained consistent.

In terms of content analysis, three brochures provided general information about accommodation, four brochures addressed transportation, and two brochures focused on local tourist information. Notably, only two brochures included information specifically related to the conditions and accessibility for PwDs. The "Algarve – Ideias e Inspirações" brochure mentioned accessible beaches, while the "Best of Center of Portugal 2020" brochure featured accessible accommodations.

CONCLUSION

PwDs require access to comprehensive and inclusive information about tourist destinations before their visit. This is crucial for them to make informed decisions and ensure their safety, thus minimising the risk of disappointment or even discouragement from travelling. Regardless of the information source used during trip planning and organisation, PwDs need clear, up-to-date, and accessible information that caters to their specific needs. While PwDs are increasingly relying on the Internet as a primary source of information for selecting destinations and services, it is important to note that some individuals, particularly older generations, still prefer consulting printed materials like brochures. Therefore, the accessibility of these materials becomes a critical issue, despite being rarely discussed in the existing literature.

The significance of accessibility extends beyond the PwD community alone; it applies to everyone. Disabilities can be temporary or permanent, and it is essential to ensure equal opportunities for people of all ages and disabilities to travel to tourist destinations and enjoy the services available, just like any other traveller. International organizations, such as the UNWTO, are actively engaged in recognising and promoting accessible tourist destinations. In fact, Portugal achieved the distinction of being the first country awarded as an Accessible Tourist Destination, showcasing its commitment to inclusive tourism.

The official promotional website of Portugal's national tourism board plays a central role in marketing the country's attractions and services. However, an analysis of the website reveals certain accessibility issues when evaluated against international standards, such as WCAG 2.1. While the website performs well in terms of search engine optimisation and security, there is room for improvement in terms of accessibility. On a positive note, the information on the website, including promotional brochures, can be easily accessed through the main menu.

The study's findings highlight two distinct scenarios. While the website provides PwDs with relevant and adequate information, none of the promotional brochures caters explicitly to this market or considers their unique requirements. Although some parameters and criteria suggested in the literature for creating accessible print materials were followed, certain aspects were overlooked, such as font size, italicised text, brochure size, and the choice of images that fail to represent the PwD segment. Additionally, there is a lack of comprehensive glossaries, and the use of acronyms without full-word explanations makes the text difficult to comprehend. Regarding the brochure content, only two brochures mention accessible resources or services, but the references remain superficial. It is worth noting that while the Tourism of Portugal is responsible for the website, only three brochures have been produced with accessibility considerations in mind.

Although there are international guidelines available for developing accessible web content and companies dedicated to creating accessible software (e.g., Acrobat), it is crucial for official tourist organisations responsible for promoting and communicating Portugal and its regions as tourist destinations to prioritise destination accessibility in their development strategies. This requires working in a coordinated manner, establishing standards, and adopting common procedures for creating promotional materials, such as brochures, leaflets, and maps. Furthermore, involving PwDs in producing information in accessible formats should be considered, as their insights and perspectives can greatly contribute to ensuring inclusivity in the dissemination of tourism-related information.

To the best of our knowledge, this study makes a significant theoretical contribution by pioneering the systematisation of criteria that should be adopted in the construction and dissemination of tourist promotional material. By identifying specific parameters, the study establishes guidelines for creating accessible print materials. This systematic approach provides a foundation for improving the accessibility of tourist promotional brochures and extends to other print materials and contexts beyond Portugal.

The study's theoretical contribution lies in its ability to fill a gap in the existing literature. While accessibility in tourism is a topic of growing importance, there is limited research that systematically addresses the criteria for developing accessible promotional materials. By consolidating and organising these criteria, the study offers valuable insights for both academia and practitioners involved in tourism marketing and destination management.

Moreover, this research goes beyond theoretical implications by highlighting the practical applications of the identified criteria. By applying the criteria to the analysis of tourist promotional brochures developed in Portugal, the study uncovers shortcomings and areas for improvement. This hands-on approach demonstrates how the identified criteria can be effectively used to assess and enhance the accessibility of promotional materials. The practical implications extend to the broader context of print materials in the tourism industry, enabling destinations worldwide to incorporate accessibility considerations into their promotional strategies. This inclusive approach not only benefits individuals with disabilities but also enhances the overall visitor experience and promotes a positive image of the destination.

Furthermore, the applicability of the criteria to other print materials and contexts reinforces the broader impact of this research. Beyond brochures, the identified criteria can be extended to leaflets, maps, guidebooks, and other printed resources used in tourism promotion. By adhering to these criteria, destinations can create a cohesive and comprehensive accessibility strategy that encompasses various types of print materials. This adaptability ensures that the benefits of the study extend beyond the specific case of tourist brochures in Portugal.

Although this study makes several contributions to the understanding of accessibility in the tourism industry, it has certain limitations that should be acknowledged. It focuses specifically on the brochures available on Portugal's official tourism website, VisitPortugal.com, as the sole source of promotional materials. It does not encompass other forms of promotional content or platforms, which may limit the comprehensive assessment of accessibility in the overall tourism marketing strategies. While brochures are a common promotional tool, it is essential to recognise that accessibility extends beyond printed materials and also includes digital content, websites, and other communication channels. Additionally, the manual collection and observation of data from numerous brochure pages present challenges in terms of quantification and parameter identification.

Future research should provide more up-to-date insights into the current status of website accessibility in the tourism industry and the advancements made in this regard. More specificlly, it should consider the following areas to enhance the understanding and implementation of accessibility in tourist destinations:

- Expand the application of accessibility guidelines to other promotional brochures, leaflets, and maps, in addition to website content. This would ensure consistency in providing accessible information across various platforms.
- Develop technological tools to automate the analysis and evaluation of promotional materials. This would save time and effort while improving the accuracy and reliability of assessments.
- Involve PwDs in the production of information to ensure that their perspectives and needs are adequately represented.
- Conduct more studies on the current status of website accessibility in the tourism industry and examine the advancements made in improving accessibility across different sectors.
- Investigate the impact of accessible information on the travel experiences and decision-making processes of PwDs, as well as its effects on the tourism industry as a whole.

By addressing these research areas, scholars, researchers, and tourist destinations can contribute to creating more inclusive and accessible environments for all individuals, ultimately benefiting both tourists and the tourism industry.

REFERENCES

Adobe (n.a.). *PDF accessibility overview.* Adobe. https://www.adobe.com/accessibility/pdf/pdf-accessibility-overview.html (accessed: 23.12.2020)

British Tourist Authority. (2019). *Speak Up! guide: A guide to marketing your accessibility.* BTA. https://www.gov.uk/government/publications/inclusive-communication/accessible-communication-formats

Buhalis, D. (n.a.) *Accessible tourism marketing strategies and social media.* Europe Without Barriers. https://www.europewithoutbarriers.eu/download/21_Dimitrios-Buhalis.pdf

Buhalis, D., & Darcy, S. (2010). *Accessible tourism: Concepts and issues.* Channel View Publications. doi:10.21832/9781845411626

Buhalis, D., & Darcy, S. (2020). The need for accessible tourism. *Annals of Tourism Research, 83,* 102944.

CBI. (2022). *The European market potential for accessible tourism.* CBI. https://www.cbi.eu/market-information/tourism/accessible-tourism-europe/

Crouch, G. I. (2000). Services research in destination marketing: A retrospective and prospective appraisal. *International Journal of Hospitality & Tourism Administration, 1*(2), 65–86. doi:10.1300/J149v01n02_04

Darcy, S., & Dickson, T. (2009). A whole-of-life approach to tourism: The case for accessible tourism experiences. *Journal of Hospitality and Tourism Management, 16*(1), 32–44. doi:10.1375/jhtm.16.1.32

Department of Health. (2010). *Making written information easier to understand for people with learning disabilities: Guidance for people who commission or produce easy read information.* Department of Health.

Domínguez Vila, T., Alén González, E., & Darcy, S. (2017). Website accessibility in the tourism industry: An analysis of official national tourism organisation websites around the world. *Disability and Rehabilitation.* doi:10.1080/09638288.2017.1362709 PMID:28793789

Eichhorn, V., Miller, G., Michopoulou, E., & Buhalis, D. (2008). Enabling disabled tourists? Accessibility tourism information schemes. *Annals of Tourism Research, 35*(1), 189–210. doi:10.1016/j.annals.2007.07.005

ENAT. (2013). *Accessibility review of European national tourist boards' websites 2012.* ENAT. https://www.accessibletourism.org/resources/enat-nto-websites-study-2012_public.pdf

England Athletics (n.a.). *Providing accessible information formats.* England Athletics. https://d192th-1lqal2xm.cloudfront.net/2018/11/Providing-accessible-information-guidance-PDF-189kB.pdf (accessed:03.08.2020)

GFK. (2014). *Economic impact and travel patterns of accessible tourism in Europe.* GFK. https://ec.europa.eu/docsroom/documents/5567/attachments/1/translations/en/renditions/native

GOV.UK. (n.a.). *Why GOV.UK content should be published in HTML and not PDF.* GDS. https://gds.blog.gov.uk/2018/07/16/why-gov-uk-content-should-be-published-in-html-and-not-pdf/ (accessed: 23.04.2022)

INE. (2022). *O que nos dizem os censos sobre as dificuldades sentidas pelas pessoas com incapacidade.* INE. https://www.ine.pt (accessed: 02.01.2023)

Law, R., Qi, S., & Buhalis, D. (2010). A review of website evaluation in tourism research. *Tourism Management, 31*(3), 297–313. doi:10.1016/j.tourman.2009.11.007

Laws, E., Scott, N., & Brouder, P. (2020). *Accessibility, disability and inclusive tourism: Concepts, issues and global perspectives.* Channel View Publications.

Lee, W., & Gretzel, U. (2012). Designing persuasive destination websites: A mental imagery processing perspective. *Tourism Management, 33*(5), 1270–1280. doi:10.1016/j.tourman.2011.10.012

Lonely Planet. (2020a). *Travel for all: Join Lonely Planet's accessible travel project.* Lonely Planet. https://www.lonelyplanet.com/articles/travel-for-all-join-lonely-planets-accessible-travel-project

Lonely Planet. (2020b). *Travel for all: Accessible travel solutions.* Lonely Planet. https://www.lonelyplanet.com/travel-tips-and-articles/travel-for-all-accessible-travel-solutions

Mills, A., Wood, R., & Darcy, S. (2008). The Internet and accessible tourism: Representing disability online. *Journal of Vacation Marketing, 14*(3), 207–217.

Mills, J. E., Han, J.-H., & Clay, J. M. (2008). Accessibility of hospitality and tourism websites. *Cornell Hospitality Quarterly, 49*(1), 28–41. doi:10.1177/1938965507311499

Morrison, A. (2013). *Marketing and managing tourism destinations*. Routledge. doi:10.4324/9780203081976

OSSATE/University of Surrey. (2006). *Accessibility market and stakeholder analysis: One-stop-shop for accessible tourism in Europe*. OSSATE.

Pérez, D., & Velasco, D. (2003). *Turismo accesible: Hacia un turismo para todos*. Madrid: CERMI.

Quivy, R., & Campenhoudt, L. V. (2008). *Manual de investigação em ciências sociais*. Gradiva.

Sigala, M., & Gkritzali, A. (2020). Accessible tourism at the crossroads: Lessons learnt, challenges and future directions. *Inaugural International Conference on Smart Tourism, Smart Cities and Enabling Technologies (STATE 2020)* (pp. 56-65). Springer.

Turismo de Portugal. (2017). *Accessible tourism destination handbook*. Turismo de Portugal.

Turismo de Portugal. (2020) *Portugal é o destino turísstico acessível*. Turismo de Portugal. https://travelbi.turismodeportugal.pt/en-us/Pages/Portugal-%C3%A9-o-Destino-Tur%C3%ADstico-Acess%C3%ADvel-2019.aspx

UNWTO. (2016). *Manual on accessible tourism for all: Principles, tools and best practices – Module I: Accessible Tourism – Definition and context*. UNWTO. https://cf.cdn.unwto.org/sites/all/files/docpdf/moduleieng13022017.pdf

UNWTO. (2019). *Accessible tourism for all: An opportunity within our reach*. UNWTO. https://www.e-unwto.org/doi/abs/10.18111/9789284421147

UNWTO. (2020). *Accessible tourism*. UNTWO. https://www.unwto.org/fr/accessibility

UNWTO. (2021). *Portugal takes first step to becoming a more accessible destination*. UNWTO. https://www.unwto.org/news/portugal-takes-first-step-to-becoming-a-more-accessible-destination

Vila, T. D., González, E. A., & Darcy, S. (2019). Accessible tourism online resources: A Northern European perspective. *Scandinavian Journal of Hospitality and Tourism, 19*(2), 140–156. doi:10.1080/15022250.2018.1478325

WHO. (2021). *Disability*. WHO. https://www.who.int/news-room/fact-sheets/detail/disability

Yin, R. K. (2009). Case study research. Design and methods (4. ed.). Thousand Oaks.

Zajadacz, A., & Lubarska, A. (2019). Development of a catalogue of criteria for assessing the accessibility of cultural heritage sites. *Studia Periegetica, 2*(26). doi:10.26349t.per.0026.06

ADDITIONAL READING

Buhalis, D., & Darcy, S. (Eds.). (2011). *Accessible tourism: Concepts and issues*. Channel View Publications.

Figueiredo, E., Eusébio, C., & Kastenholz, E. (2012). How diverse are tourists with disabilities? A pilot study on accessible leisure tourism experiences in Portugal. *International Journal of Tourism Research, 14*(6), 531–550. doi:10.1002/jtr.1913

Michopoulos, E., & Buhalis, D. (2013). Information provision for challenging markets: The case of the accessibility requiring market in the context of tourism. *Information & Management, 50*(5), 229–239. doi:10.1016/j.im.2013.04.001

UNWTO. (2013). *Manual on accessible tourism for all: Public-private alliances and good practices.* World Tourism Organization.

KEY TERMS AND DEFINITIONS

Accessibility: This refers to the design and provision of products, services, environments, and information that can be accessed and used by all people, including those with disabilities, without the need for adaptation or specialised accommodations.

Assistive technologies: Assistive, adaptive, and rehabilitative devices for people with disabilities (e.g. Walkers and wheelchairs)

Brochure: Promotional document, consisting of more than one page, used to advertise a tourist destination and / or the tourist services or products to the potential visitors.

Promotional material: Any type of communication or media used to promote tourist destinations, attractions, or services, including brochures, leaflets, maps, websites, and digital content.

PwDs: Abbreviation for people with disabilities, referring to individuals with physical, sensory, cognitive, or other impairments that may affect their participation in various activities.

Tourism marketing strategies: Plans and actions employed by tourist organisations to promote and communicate tourist destinations, attractions, and services to target audiences, with the goal of attracting visitors and generating tourism revenue.

WCAG 2.1: The abbreviation for Web Content Accessibility Guidelines 2.1, which are internationally recognized standards developed by the World Wide Web Consortium (W3C) for creating accessible web content.

Website accessibility: The degree to which a website is designed, developed, and maintained in a way that allows users, including those with disabilities, to perceive, navigate, and interact with its content effectively.

Chapter 13
Tourism Differentiation Through Social Media Branding:
A Qualitative Exploration of the Moroccan Case

Yassine El Bouchikhi

University of Al Akhawayn, Morocco

ABSTRACT

With a focus on Morocco, this chapter reviews the literature on nation branding and its connection to social media in the context of tourism. The first part gives a foundation for understanding the many tactics used to distinguish one country from another by reviewing the literature on nation branding, social media, and tourism. The use of social media to market travel and to promote cultural and natural assets is examined in the second section. Then, an examination of the significant developments and trends in the Moroccan tourism industry are addressed. A Netnographic study is conducted on six influential media accounts to explore the strategies and tactics utilized to promote Morocco's image internationally.

INTRODUCTION

The phenomenon of nation branding has garnered increased scholarly and practical interest in recent times(Theodoropoulou, 2020). According to Adler, nation branding is not a new phenomenon, since world fairs and international exhibitions were organized in the late nineteenth century and used to attract vast crowds eager to see the presentation of national culture, and heritage, and these events are thought to be the places where nation branding first emerged (Adler-Nissen, 2014). Nation branding has emerged as a key component of international competition as nations work to improve their standing abroad and draw in investment, tourism, and skilled labor. Some authors refer also to it as a form of "Soft Power" (Nye Jr, 2004). Soft power involves influencing people's attitudes and behaviors via culture and economy rather than using physical force or coercive sanctions.

DOI: 10.4018/978-1-6684-6985-9.ch013

The tourism industry is constantly looking for innovative ways to grow, and nation branding techniques are often at the heart of national strategies for promoting destinations, their heritage, and their culture. Marketing tactics can be realized through classic advertising campaigns called push strategies, or indirect strategies, like creating marketing content through pull strategies on digital platforms (Kotler & Armstrong, 2020). The latter, innovative approach is particularly used in the field of tourism, where online audiences and prospective tourists find such content more authentic and exciting especially if it looks like genuine user-generated content -UGC- instead of classic advertising campaigns instigated by governments. One of the interesting cases to study is the Kingdom of Morocco, a famous tourist destination, but for which there are still few academic works that explore the role of social networks in building a narrative about this country and influencing the perception of future tourists (especially through digital accounts and platforms that seem to be private or semi-official).

This chapter starts by reviewing the nation branding literature, including key definitions, important theories, and tactics used for differentiation purposes. This will provide a strong basis for understanding nation branding and the diverse strategies used to set one country apart from another. The second section of the literature review will focus on social media's function in tourism branding and how it affects the perception of tourists. The third part presents a qualitative online study that was carried out to learn more about how Morocco is promoting itself on the Internet directly through official social networks and indirectly through semi-official or private accounts that allow the spreading of continuous information, photos, and videos concerning Moroccan heritage. Morocco is an interesting case since it has been promoting itself as a tourist destination by utilizing its wealth of cultural heritage and natural beauty while extensively mobilizing social media platforms. Finally, the author examines major trends and future opportunities for the Moroccan tourism sector.

The utilization of Nation Branding theories, the qualitative Netnographic study conducted, and the resulting findings provide a novel theoretical, methodological, and managerial outlook on the online management of Moroccan tourism marketing practices.

LITERATURE REVIEW

Nation Branding: Definitions, Theories, and Some Differentiation Strategies

The origins of branding theory can be traced back to the 1950s, primarily within the context of consumer products and later extended to countries. However, it was not until the 1990s that branding began to receive attention as a concept within tourism destination marketing, capturing the interest of practitioners in the field (Hankinson, 2015). Place-based branding became increasingly common as governments and nonprofits needed to draw more residents, visitors, customers, and businesses to the places they serve. According to Sevin, a place brand can be defined as a network of associations in the minds of individuals about a given place (Sevin, 2021; Zenker et al., 2017). These associations 'are constructed from various contributory elements including what a place has to offer (e.g., its landscape, architecture, goods, services), what it communicates with the outside world (e.g. its promotional campaigns), and what others communicate about the place (e.g. word of mouth on social media, reviews on websites and conversations among friends) (Sevin, 2021).

Sevin posits that place branding practice includes two major functions: a communication aspect and a policy side. Regarding communication, places have used available marketing and advertising tools to disseminate their messages as well as to monitor the chatter about themselves (Sevin, 2021). On the policy side, practitioners have supported 'policies that aimed at improving the place to the benefit of residents, businesses, and visitors' (Sevin, 2021).

Theodoropoulou prefers a larger perspective and talks about "Nation branding". She defines it as the process that aims to turn a nation into a commodity that can be marketed internationally and domestically (Theodoropoulou, 2020). Attributes such as stability, diversity, or exoticness can be projected onto the commodity that is the nation (Theodoropoulou, 2020).

According to Ojo, nation branding refers to the use of branding and marketing communications techniques to alter the perceptions of an international audience, this definition focuses on the positioning strategy of a nation (Ojo, 2020). On the other hand, Hao explored nation branding research throughout the past 20 years and defined it as branding and marketing communication strategies to advance a nation's reputation (Hao et al., 2021).

Pike indicates that great care and selectivity should be exercised by brands and branding actors in constructing geographical associations in branded goods and services commodities for particular spatial and temporal markets (Pike, 2015). Pike identifies 10 different territorial layers that can be nested and combined according to the desired strategy, for example, the supra-national level (Europe, Latin America, etc.), or national (Morocco, France, etc.), down to the level of granularity of the boulevard (Madison Avenue, Champs Elysées, etc.) (Pike, 2015).

Figure 1. The nation brand hexagon
Source: Adapted from Simon Anholt (Anholt, 2005)

On the other hand, Simon Anholt, one of the most renowned authorities on national branding has identified six essential aspects to consider when assessing a country. Anholt developed a tool that he called the Nation Brands Index, where he polled 10,000 consumers in 10 nations about their opinions of America's cultural, political, economic, and human resources as well as its investment potential and tourist attraction. These six dimensions are represented below in Figure 1.

The six elements identified by Anholt are:

1. **Governance:** it refers to the respect for human rights and the rule of law, as well as the efficiency and openness of a nation's government. It is essentially related to the legal and institutional aspects of a nation.

2. **Culture**: this component describes a nation's historical, artistic, literary, and traditional contributions as well as its current cultural offers. This part is related to heritage, lifestyle, artisanal products, history, and all the elements that make a culture unique.

3. **People:** this factor considers a nation's population's knowledge, friendliness, and talents as well as how welcoming visitors and foreigners are.

4. **The tourism component** includes a nation's architectural and natural landmarks, as well as its hospitality sector and infrastructure proposed to the visitors.

5. **Investment and Immigration** describe the economic opportunities available in a nation as well as the methods it uses to draw in and keep immigrants and investments from elsewhere. This is essentially related to how nations brand themselves to the external economic world.

6. **Export** considers a country's commercial relations with other countries as well as the standard and appeal of its goods and services. For example, France is known for several luxury brands like Channel, Hermes, Louis Vuitton, etc. Germany presents itself as the tenant of the intellectual technical capital of Europe as the land of ideas and advanced technology, especially in the automotive industry (Tovar, 2020).

Anholt asserts that by mixing these six components, countries can create powerful and admirable national brands that can aid them in attracting and retaining talent, investment, and tourism as well as in promoting their values and interests worldwide (Anholt, 2005).

The closest author to Anholt's philosophy when it comes to nation branding is Joseph Nye, even if his framework of analysis is more suitable for geopolitics and international relations. Nye's book "Soft Power: The Strategies to Succeed in International Politics" hints at some aspects of nation branding that Simon Anholt described. However, Nye focuses more on the idea of building soft power, which he describes as a nation's capacity to influence others without using force or coercion (Nye Jr, 2004). According to Nye, one of the most effective methods for a nation to develop such soft power is through promoting its institutions, values, and culture. This is consistent with Anholt's "Governance" and "Culture" dimensions, which are crucial components of a nation's soft power. In addition, Nye makes the case that a nation's capacity to draw in creative individuals and promote innovation is essential to enhance its soft power, which is comparable to Anholt's "People" and "Investment" dimensions. Despite this similarity, the "Tourism" and "Export" components of nation branding, for example, are not addressed by Nye. These dimensions may play a significant role in a nation's total reputation and brand, but they seem to have marginal importance when it comes to building soft power.

Hecht defines nation branding as a helpful subfield of global history. The author states that country branding entails boosting national development, retaining talent, attracting foreign investment, boosting political influence abroad, and reversing any unfavorable perceptions that might arise in the wake of international credit rating downgrades (Gienow-Hecht, 2019). This definition seems quite similar to the approach of Anholt when it comes to some dimensions involved like foreign investment and talent retention (Anholt, 2005).

Recently, some authors have begun to use the concept of "cultural diplomacy". It refers to using culturally unique elements to brand a country, whereas nation branding is more generic and designates the creating, organizing, and disseminating of a nation's name and identity to enhance its international profile and manage or build its reputation (Hurn, 2016). Some recent work explored the Moroccan case and the tools used by the Moroccan Ministry of Foreign Affairs to implement such cultural diplomacy. The results showed how Morocco manages to market its image of cultural openness and tolerance in particular – arising from religion, historicity, and tradition – to advance its soft power and geopolitical interests (Wüst & Nicolai, 2022).

Balakrishnan asserts that destinations and places have distinctive qualities and should not be handled as goods or services (Stephens Balakrishnan, 2009). The author identifies 4 distinctive qualities represented in Figure 2.

Figure 2. The distinctive qualities of destinations and countries
Source: Strategic branding of destinations (Stephens Balakrishnan, 2009)

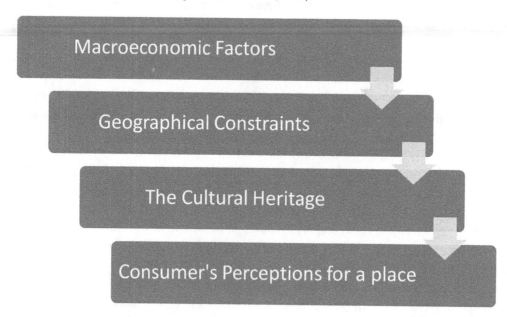

1. The tourism industry depends on **macroeconomic variables**, such as weather, security, political stability, and monetary fluctuations.
2. **Geographical constraints** have an impact on access to natural resources, infrastructure, characteristics, and lifestyles of residents of the community of destination.

3. **The cultural heritage**, history, and culture of a place cannot be changed and are an integral part of the identity.

4. The idea of a destination's brand is complicated and dynamic since it depends on many variables that affect how **customers and tourists perceive a location**, such as image and media promotion.

Place branding is now a real topic of interest for decision-makers and studies regularly measure the attractiveness of nations. Every year, the top 100 national brands in the world are ranked by their worth and strength in the annual Brand Finance Nation Brands report. This annual index has offered over 20 years of important benchmarks for diplomats, tourism boards, trade organizations, geographical indication brands, nation brand managers, and consultants (*Brand Finance Nation Brands Forum 2022 in Association with DPAAL*, 2022). The results of Brand Finance's research on nation brands in 2022 are shown in Figure 3.

Figure 3. Top ten most valuable nation brands 2022
Source: (Brand Finance Nation Brands Forum 2022 in Association with DPAAL, 2022)

The United States continues to have a powerful ability to attract. The powerful American film industry, the Westernized lifestyle with the "Mcdonaldisation of the world" to mimic Ritzer's theory, and the U.S. military might be cornerstones at the heart of American global influence (Ritzer, 2021). Nevertheless, new serious competitors are emerging from the BRICS alliance such as China as well as India.

At the end of this section, nation branding can be defined as the collection of tools used to build national identity through marketing and branding strategies. Nation branding is employed to sway public opinion and can be used to foster a sense of nationalism and patriotism (Theodoropoulou, 2020). Such strategies intend to establish a national shared identity and brand it as a commodity to promote (Graan, 2016). A nation brand is an endeavor that gives a country a positive reputation and is part of a nation's soft power. As noted at the beginning of this chapter with Joesph Nye's approach to soft power, Nation branding has begun to play a vital role in international diplomacy, replacing the classic soft power tools such as propaganda and public diplomacy that were used over the last century (Dinnie, 2015).

Social Media in Tourism Branding: Uses, Benefits, and Risks

The culture of digital, online exchange, and social networks is nowadays fully integrated into our lives. Social media platforms allow us to discover new perspectives, ideas, and experiences, and have social interaction enriched by a combination of language, images, symbols, and videos. This section addresses the use of social media as a means of promoting countries and tourism. Social media platforms offer a new medium for any brand to advertise its products through enhanced exposure, a greater capacity to interest the customer, and a sophisticated way to influence perceptions and create value (Shareef et al., 2019).

Social media use can promote a country's culture, increase soft power, and help it establish a good reputation. There are several positive benefits of branding a nation, among them, we find, for example, increased tourism revenue. In Morocco, tourism represented approximately $ 16 billion in 2019 (before the COVID-19 pandemic), since then this number has been divided by two and it represented $9 billion in 2021 (*Tourism in Morocco | Statista*, 2021). Nation branding on social media allows a deeper sense of national identification and pride, which can serve to promote a nation's distinctive culture and heritage. Morocco for example, often promotes the idea of being a country that has twelve history centuries and a Moorish heritage (Wüst & Nicolai, 2022). A Strong Nation Brand and successful presence in the digital realm can aid in fostering international relations, cooperation, and understanding among nations. Morocco often uses the concept of cultural diplomacy to advance causes that are important to it such as the Sahara case or its foreign policy (Wüst & Nicolai, 2022).

There is a rich academic body of knowledge regarding the use of social media in national and tourism branding. For example, Rita has shown that social media can be used effectively and yield interesting outcomes in hospitality and tourism branding (Moro & Rita, 2018). Moro conducted an automated literature analysis that unveiled the key factors that encourage the adoption of social media in tourism and hospitality branding strategies. The results showed that out of 213 articles analyzed, there was a strong academic focus on brand-building stages (Moro & Rita, 2018).

Karjaluoto mobilized social media and the social identity theory to explore group dynamics and foster strong feelings and a sense of identity in a community. The Social Identity Theory is used in psychology and explains how a person's sense of self is influenced by their affiliation with particular groups (Hogg, 2016; Karjaluoto, 2017). Antonín Pavlíčeki has explored the key social media characteristics that have an impact on destination communication strategies (Kiráľová & Pavlíčeka, 2015). According to the latter, more than 11 types of goals were identified when it comes to using social media in campaigns:

1. Creating/increasing awareness of the destination
2. Reaching global publicity
3. Encouraging visitors to plan their journey.
4. Strengthening the destination image as a favorite destination
5. Targeting new/specific market
6. Increasing the number of visitors
7. Creating a buzz around the destination
8. Increasing the number of email subscribers

9.   Increasing the number of Facebook fanbase
10. Changing the position of destination in the mind of visitors
11. Bringing back the destination as a favorite one for visitors

Thellefsen's research mobilized semiotic interpretation and defined the brand as "an artifact or a sign whose meaning is based on a continuous reinforcement of habits" and branding as "a process sign that goes beyond the brand" (Thellefsen et al., 2006). Another interesting example of the application of semiotic analysis to the case of places/nations is the one carried out for the island of Mykonos in Greece by mobilizing social networks to design the brand image of this place through a process of symbolization, which transfers unique qualities and properties to be associated with the place to be promoted to make it "a tangible artifact" that symbolizes wealth and success (Thellefsen et al., 2006; Theodoropoulou, 2020).

Nation branding researchers have wisely not overlooked the potential negative societal effects or non-ethnic approaches observed when branding countries via social media, especially developing nations like romanticizing poverty. The danger of romanticizing or exoticizing poverty while marketing a country may lead to locals having a skewed or false perception of reality. Due to their perceived genuineness, tourists may be drawn to poor or suffering communities, but this may cause them to overlook or ignore the complicated social and political contexts that lead to poverty and inequality (Karnani, 2007). The good intention of wanting to promote a destination online can quickly become a political whitewashing of authoritarian regimes or minimization of the challenging living conditions of the population. Examples of these practices can be the fact of disseminating photos of children playing barefooted football and talking about passion, street vendors smiling despite their torn clothes, or showing obvious signs of poverty.

Poverty in Morocco remains a real challenge, and some promotional campaigns might present Aboriginals in precarious situations and portray them as a form of authentic existence giving the impression that it was a chosen lifestyle (Karnani, 2007). Another possible risk concerns reinforcing economic inequalities. In places where poverty and inequality are pervasive, tourism has the potential to aggravate inequalities between visitors and local communities. For instance, foreign or affluent investors may negatively impact local standards of living driving up house prices and food, while locals are forced to work in low-paying service positions. Instead of enabling local communities to enhance their own economic and social conditions, mass tourism may maintain systems of exploitation and dependency (Lee & Syah, 2018).

Finally, cultural appropriation is also one of the potential risks of country branding online. Occasionally it leads to the appropriation of cultural components for commercial ends, which the original culture may view as offensive. Regular controversies break out in the press between Morocco and Algeria on this theme, one of the most recent concerns Zellige which is a kind of artisanal faience that is part of the Moorish architecture that Algeria and Morocco have in common (*Adidas Admits It Used Morocco's Zellige Design*, 2022). Other controversies emerged about the origin of the caftan (a traditional female habit decorated with embroidery), or gastronomy (such as the traditional couscous food present in both Morocco and Algeria) are regularly a source of grievances from both sides (*Is Cultural Heritage the New Battleground?* 2021).

In the following section, the Moroccan heritage is explored, its distinctive characteristics are highlighted, and examples of online campaigns launched to promote the image of the country and foster tourism are discussed.

The Moroccan Unique Heritage: Key Characteristics and Prospects

Morocco is one of the most popular destinations for international tourists traveling to Africa. According to the World Bank, the number of tourists arriving in the country has increased in the last decade and reached 13 million in 2019 (*World Bank, Number of Arrivals - Morocco*, 2022). Morocco has a large diaspora living abroad residing particularly in European countries. This diaspora represents almost half of the yearly international tourist arrivals.

Tourism in Morocco is one of the major employers and contributes significantly to economic and social growth. The COVID-19 pandemic has significantly impacted the world and Moroccan tourism industry. The government implemented initiatives to lessen the financial impact on individuals and businesses, but the industry still needs to adjust to the "new normal" created by the pandemic (*Tourism in Morocco | Statista*, 2021). Since 2020, numbers were significantly affected by the coronavirus (COVID-19) pandemic, and the country deployed intensive promotional efforts to recover its previous performances. A compound annual growth rate of 5 percent is expected in the coming years according to experts (*Tourism in Morocco | Statista*, 2021).

Experts are also noticing a shift in the Moroccan strategy since Morocco is moving toward adopting sustainable tourism practices as it realizes the value of ethical and environmentally friendly travel.

Morocco has a variety of attractions, including snow-capped mountains, arid deserts, and labyrinthine medinas. Morocco offers a varied cultural experience because of its unique environment and architecture, which have been influenced by France, Spain, and Africa. Morocco provides a variety of cultural experiences in addition to its varied landscapes, including touring the nation's energetic towns, going to traditional markets, testing local food, and taking part in cultural festivals. Moroccan historical sites include the Koutoubia Mosque in Marrakesh and the Hassan II Mosque in Casablanca. The nation also provides a variety of outdoor pursuits, including surfing, camel trekking, and hiking.

Only a few scholars explored the Moroccan branding case, and this movement is still burgeoning since it started during the last decade. The explored themes mainly revolved around topics such as Jewish heritage in Morocco (Boum, 2020), the recent branding of the city of Casablanca (Belkadi, 2020; Sedra & El Bayed, 2022), the role of the Moroccan diaspora in the promotion of Morocco (Cascón-Pereira & Hernández-Lara, 2013, 2014; El Aouni, 2015), or more recently through the prism of cultural diplomacy and soft power to advance its political causes internationally (Wüst & Nicolai, 2022).

The Moroccan economic model is like the French one. It is based on a developed agri-food industry, the tourism industry as well as rich and sophisticated craftsmanship. On this last point, even if Morocco doesn't own yet global luxury brands like France, Morocco is widely renowned for its beautiful handmade crafts and leather products that visitors can find in imperial cities like Fes, Marrakech, and Meknes. Tourists can also discover beautiful pottery, stunning medieval architecture, and Moroccan faience called "Zellige", which are handcrafted clay tiles known for their variations in tone, shine, and flatness and historically used in geometric and floral shapes in the most beautiful palaces both in Spain and North Africa. This rich heritage has roots in the European, African, Berber, and Arab interactions and influences resulting from Morocco's strategic geographical location and its role as a civilizational hub and junction point between sub-Saharan Africa, Western Europe, the Mediterranean influence, and finally the Arab-Moorish culture.

Morocco proudly exhibits its lifestyle, gastronomy, and architecture to seduce tourists. Among the craft and terroir products that are part of the Morocco signature, we find:

- Traditional Moroccan clothing like the Djellaba. It is a long, loose dress worn by both men and women and the Kaftan which is a more formal piece typically adorned with embroidery worn during special ceremonials and weddings.

- Argan Oil, is a cosmetic and culinary oil extracted from the Argan tree, and often used in Moroccan Food and Beauty Products. Argan trees, or Argania Spinosa, are indigenous to Morocco's sub-Saharan region, which lies in the southwest of the country. They thrive in arid and semiarid environments. Argan is one of the most expensive oils in the world, it is revered as the "liquid gold" of Morocco. Scientific evidence supports its benefits for the skin as well as its capacity to prevent cardiovascular diseases (United Nations, 2023).

- Moroccan tea is a sweetened green tea beverage with a mint flavor combined with other local plants. Tea is another local signature, presented in tiny cups as a gesture of friendliness and hospitality.

- Moroccan rugs made by Berber women have intriguing geometric patterns and vibrant colors. Moroccan rugs are well-known throughout the world for their heavy wool. For certain carpets, every process—from spinning the yarn to dying the wool to weaving—is carried out by hand which makes them expensive.

- Moroccan ceramics include colorful plates and bowls with traditional designs and ornamentation and "Tagine" dishes. With its conical shape and clay composition, the Tajine creates a particularly wet, heated atmosphere for the food being cooked.

- Morocco is also known for accessories of high-quality leather, such as bags, shoes, and belts, usually with intriguing embroidery and design.

Morocco as a destination and culture has a lot of diversity and richness. Conscious of the importance of its image, Morocco has launched several campaigns to promote its heritage. After Covid19 and its strong negative impact on the tourism industry, Morocco has launched several campaigns like the 'Aji' and 'We Are Open' Campaigns to Revive Tourism (Anouar, 2022). Moreover, by eschewing cliches and emphasizing Morocco's modern cultural attractions, such as street art and music, "Kingdom of Light" (2021) sought to redefine Morocco as a contemporary, imaginative, and adventurous travel destination (Bhutia, 2022). The efforts of the national office of Tourism are constant to ensure an appropriate promotion that corresponds to the ambitious standing that Morocco wants to offer itself on the world tourism scene. To better understand the narrative around Morocco's destination, a qualitative study was conducted to understand how the image of Morocco is built and disseminated online. The key research question was to explore how does the innovative utilization of social media platforms contribute to the branding of Morocco as a tourist destination?

METHODOLOGY

The Netnographic Approach

To study how the Moroccan digital landscape builds narratives that invoke the imagination of tourists and uses symbolism on social media to promote the heritage of the Kingdom, it seemed judicious to proceed via a qualitative observational approach, namely a Netnography to observe the Moroccan digital presence online. Netnography is defined as a qualitative investigation that utilizes secondary data available online on social networks, forums, and web pages to better understand the discourses, stories, narratives conveyed online on a given subject (Kozinets, 2002; Malhotra, 2008). Netnography is ethnography adapted to the study of online materials and communities. This method is faster, and simpler compared to traditional ethnography, it has also the advantage of being unobtrusive than focus groups or interviews.

To carry out the research, the author collected secondary data via a web scrapping method for three Facebook pages that promote Morocco's tourism and culture, and which total more than 1 million visitors. These Facebook pages are public and have been selected because they are the most influential and important showcases on social media platforms for the promotion of Morocco. To cross-fertilize data sources, data was also collected from Twitter, where 3 major accounts were identified with between 10k to 20k followers each, and more than 6000 tweets were extracted and analyzed in detail. Table 1 below details these data sources.

Table 1. Data sources used in the ethnographic study as of February 18th, 2023

Account/Page Name	Social Media Platform	Description	Number of Followers
@simplymorocco	Facebook	"Simply Morocco is one of the top inspiring travel pages on social media. Our main mission is to promote the culture and the fascinating beauty of our beloved kingdom of Morocco"	619146
@heartof.morocco	Facebook	"Morocco pictures music food traditions and everything Morocco has to offer"	389000
@VisitMoroccoEN	Facebook	"The official Facebook page of the Moroccan National Tourism Office (MNTO)"	162658
@MoorishMovement	Twitter	A Twitter account sharing content about "History, Culture, News, Pictures, Conservatism & Nationalism"	46636
@Visit_Morocco_	Twitter	"The official Twitter account of the Moroccan National Tourism Office (MNTO)"	24783
@WealthofMorocco	Twitter	"Everything you need to know about Moroccan culture and heritage through time."	21440

Source: Facebook and Twitter official accounts

For the 3 Facebook accounts, the period covered for Netnography is December 2022, January, and February 2023 (until February 18th 2023). Concerning the 3 Twitter accounts selected, the Tweets analyzed respectively are the *MoorishMovement* with 3000 tweets published between 2021 and 2023, the official *VisitMorocco* account which includes 1952 tweets from 2015 to 2023, and finally, the Wealth of Morocco account with 1676 tweets shared between 2020 and 2023.

Only the *VisitMorocco* account represents an official initiative of the Moroccan state, all the other accounts are private or semi-private initiatives on the internet to enhance the image of Morocco internationally, mainly in English and French with few posts in Arabic.

The Analytical Approach

The first approach used a lexicometry approach (word counts, collocates, keyword in Context Analysis -KWIC-) and analyzed the data collected while following the approach used by Tovar on the study of the image of the German republic and its evolution (Tovar, 2020). Tovar examined the most used words that are associated with Germany using a corpus from the branding newsletters to determine whether any deliberate attempts to evoke favorable associations were made.

A preliminary work of cleaning and preparing the data was carried out to remove from the database the incomplete observations, the special characters, and the internet links to keep only the text of the verbatims to be considered. Also, depending on the language used (English or French), a procedure of lemmatization and tagging has been carried out.

For English, the Sandford POS tagger was used, and for the French language, the "Snowball Stemmer" method was used to prepare the extracted words for the analysis. These methods allow words to be brought back to their common root before analysis. The qualitative data analysis used the Japanese software KhCoder and the online website Voyant-Tools.org.

The second analytical phase mobilized content analysis. Bardin defines it as a group of methods for studying communications that employ methodical and objective procedures to describe the content of messages. This analytical method is used to qualitatively determine the presence of certain words, themes, or concepts within the verbatim.

DATA ANALYSIS AND DISCUSSION

Results of the Lexicometry Analysis

Before analyzing the textual corpus collected, it is appropriate to present certain characteristics of the accounts selected, table 2 below synthetizes this information.

Table 2. Characteristics of selected Facebook accounts

Facebook Page	Likes	Comments	Shares	Posts	Engagement Ratio	Virality Ratio
Heart of Morocco	301795	12659	33372	176	1787	190
Simply Morocco	123144	4557	20332	93	1373	219
Visit Morocco	7319	344	736	48	160	15
	432258	**17560**	**54440**	**317**		

Source: Created by the Author

The author started by computing some metrics to measure the virality and engagement of the different accounts studied. The engagement indicator was calculated by dividing the number of interactions (likes and comments) by the number of publications for each of the pages.

The virality ratio was obtained by dividing the number of shares by the number of publications. On average the Facebook pages of these three accounts publish between 1 to 2 posts per day, in December there are 109 publications, 90 in January, and 118 in February.

First, out of the three Facebook pages, "Heart of Morocco" has the most likes, comments, shares, and posts. This means that it is the most visited page and that its audience is the most engaged. It has a high engagement ratio of 190, which indicates that its followers actively participate in each post on average. Its virality ratio is also rather high, suggesting that the platform users spread its material widely.

Although "Simple Morocco" has fewer likes, comments, shares, and posts than "Heart of Morocco," its engagement ratio of 219 is still rather strong. This implies that its followers are likewise very involved. Finally, "Visit Morocco" has the lowest engagement and virality rates, as well as the fewest likes, comments, shares, and posts. As a reminder, the latter is the only official page among the three selected Facebook accounts.

For the Twitter platform, table 3 below synthesizes the data. The *MoorishMovement* and Wealthof-Morocco accounts were created during the same period in 2020. On the other hand, the official "Visit Morocco" account was created in December 2015.

An analysis of 6628 Tweets from the three accounts indicates that most shares are made at 6 p.m. (1280 tweets). Some accounts (*MoorishMovement*) seem to be less active during certain periods of the year, notably June and September in general.

Table 3. Characteristics of selected Twitter accounts

Twitter Account	Number of Tweets	Favorited Tweets	Retweeted	Engagement Ratio	Virality Ratio
MoorishMovement	6840	1261305	781987	299	114
*Visit_Morocco_*	1954	77616	20959	50	11
WealthofMorocco	1695	322102	63332	227	37
	1695	1661023	866278		

Source: Created by the Author

Overall, the *MoorishMovement* account has the most tweeted, favorited, and retweeted content. It also has the highest engagement ratio and virality ratio among the three accounts, indicating a more active and engaged fan base and a higher likelihood that its followers will spread its material.

The Visit Morocco_ account has the lowest rates of engagement and virality, as well as the smallest number of tweets, favorited tweets, and retweets. This means that it has a less active audience or less interesting content compared to the other accounts.

Hashtags act as keywords that drive visitors and traffic to content. A study of the top 10 hashtags used was conducted. More than 150 different hashtags were identified in the corpus of tweets analyzed. Table 4 above presents these results. Regarding Facebook, the use of Hashtags seemed to be less recurrent, the 3 most frequent hashtags were #morocco, #simplymorocco, and #moroccanarchitecture.

The Results of the Content Analysis

The results showed a strong presence and activity of accounts that promote the image of Morocco on the cultural level: Gastronomy, Architecture, Craftsmanship, History, etc. There is a great diversity of content that is essentially visual and generates strong interaction depending on the type of content (nationalist and historical content was particularly popular on Twitter, while visual content appeared to be more effective on Facebook).

Table 4. List of the most used hashtags by the selected Twitter accounts

Hashtag	Count
#visitmorocco	750
#morocco	595
#Marrakech	171
#Maroc	129
#MarocSolutionsCop21	124
#ONMT	89
#COP22	44
#Fez	42
#Chefchaouen	34
#Casablanca	28

Source: Created by the Author

The conceptual framework developed by Anholt in 2006 seemed to offer an interesting theoretical prism to use for exploring the digital presence of Morocco on social networks. Indeed, even if the Facebook pages and Twitter accounts mobilized within the framework of this analysis were mainly private or individual initiatives (apart from the Facebook page of *VisitMorocco*), it is, therefore, possible to find the 6 dimensions presented by the Anholt to strengthen the nations online presence.

- The **first dimension**: concerns people, it is about the reputation of the local population and its perceived qualities. Indeed, the openness, warm welcome, generosity, the friendly aspect of the native populations are all elements that the countries mobilize to seduce internationally. Thus, it is possible to identify a strong mobilization of this dimension on all the accounts analyzed within the framework of this study.
- The **second dimension** concerns the exports for which a country is known. Germany is reputed to be the land of ideas and engineering (Tovar, 2020), while Mykonos is renowned for luxury tourism (Theodoropoulou, 2020). It is therefore the public image of the products and services exported by the country. In the case of Morocco, an analysis of the data confirms the existence of several publications, visuals, and videos, highlighting local craft products such as leather products, pottery such as tajines (traditional utensils used in the kitchen for cooking), other handmade decorative products such as floor lamps, cushions, products made from argan oil (a unique tree that only grows in Morocco). The accounts studied on Twitter and Facebook valued and promoted local products in several posts.
- The **third dimension** indicates the degree of interest that foreigners may have in visiting the country or the power of attraction to a geographical area. According to Anholt, this could entail actions like selecting brand ambassadors and performing research to serve as the foundation for promotional initiatives. Anholt recommends enticing weather and landscapes, high standards of customer service, safety, law and order. An analysis of the results of the corpus collected as part of this Netnography showed that these elements were strongly mobilized by these online accounts explaining that Morocco was a safe destination and showing pictures of foreign visitors and tourists to reinforce this image.

- The **fourth dimension** according to Anholt's conceptual framework, relates to cultural heritage. All the accounts mobilized a set of visuals, publications, videos, and reels (short video format on Facebook) to promote Moorish architecture, the beauty of Moroccan riads, the history of the country, the Moroccan lifestyle, gastronomy as well as the harmony that exists between the different ethnic (Berber, Arab, Sahrawi, etc.) and religious groups by presenting Morocco as a thousand-year-old land of coexistence and dialogue between religions and cultures.

Finally, the last two dimensions identified by Anholt seemed to be less obvious in the Moroccan case. These are the governance and public opinion dimension (1), and the investment and immigration (2). These dimensions are not non-existent, but they are rare to find in the analyzed content.

- **The fifth dimension** identified by Anholt concerns governance and public opinion about local government. The period covered in the analysis was characterized by the World Cup and the sporting achievement achieved by Morocco which finished 4th in the Qatar FIFA 2022 World Cup ranking, all the groups analyzed shared photos of Moroccan flags and photos of the national team as a form of patriotic pride and national communion around the symbol embodied by this team. In the analyzed corpus, one account in particular shared content about governance, it is the *MoorishMovement* account, which seemed to indirectly act as the semi-official spokesperson for the authorities to defend the government's political orientations, stir up feelings of patriotism and nationalism, and responds to certain online criticisms against Morocco, particularly regarding the difficulties brought to light regarding free speech and human rights. To the knowledge of the author, the owner of the *MoorishMovement* account is unknown to the public, however, since the initiative is presented as a proactive action by a patriot and not backed by the state, the information shared gave the impression of greater authenticity and credibility, even if the doubt remains on who propels such accounts. In 2021, for example, Facebook announced that it had removed more than 400 accounts, pages, and groups linked to propaganda campaigns to which certain Moroccan media outlets known for their proximity with the government were linked (*Facebook supprime près de 400 faux comptes*, 2021).
- The **sixth and last dimension** of Anholt concerns investment and immigration: it is about the power of attraction of a country internationally to attract people to live, work and study there. Morocco has traditionally acted as a bridge between the West and sub-Saharan Africa, it has always been a land of passage and continues to be so, for migrants heading for Europe. These migration flows do not intend to settle on Moroccan territory but see it more as a springboard to Europe, which often creates challenges for local authorities. In 2022, a large number of migrants died in clashes against the Moroccan authorities trying to force entry into Melilla, a city under Spanish control in the north of Morocco, which caused an international outcry and embarrassment for Morocco and Spain ("Governance & Politics of Morocco," 2023). International economic opportunities, as well as the lack of development in certain sectors such as health, education, or human rights, might partially explain why Morocco has a large diaspora with approximately 5 million Moroccans and their immediate descendants living abroad. Despite these challenges, Morocco is effective in attracting foreign investments and ranks third in Africa in terms of foreign direct investment after South Africa and Egypt. The political stability, the excellent geographical location, as well as the efforts of the authorities make it attractive to several international investors. One account among the 6 studied, seemed to heavily promote Morocco's assets internationally on this dimension: the *MoorishMovement* account.

Table 5. Indicative illustration of the application of the Anholt 6-dimensional model

Dimension	Illustration example	Translation (if applies)	Source
People	The experience of eating in an Amazigh house with locals was beyond incredible. The hospitality, culture, and food were all so unique and rich. I left feeling a greater appreciation for the warmth and kindness of the Amazigh people		Heart of Morocco on Facebook
Exports	*L'huile d'argan.* *Après tous ces soins nettoyants, les marocaines utilisent pour hydrater la peau et les cheveux l'huile d'argan. Originaire de la région de Souss au Maroc MA, cette huile est en effet riche en vitamine A et en vitamine E. https://t.co/HRMgj6BcuO*	*Argan oil* *After all these cleansing treatments, Moroccan women use argan oil to moisturize their skin and hair. Originally from the Souss region of Morocco MA, this oil is indeed rich in vitamin A and vitamin E. https://t.co/ HRMgj6BcuO*	The Wealth of Morocco on Twitter
Tourism	Big dreams, big emotions, passion, sharing, come to live it in Morocco!		VisitMorocco on Facebook
Culture	Architecture fans, this one is for you! Tinmel Mosque is one of the most mesmerizing sights in Morocco. Located in the High Atlas mountains, this is a place you need to see to understand just how extraordinary it is.		VisitMorocco on Facebook
Governance	Ceci dit, nous devons nous poser les questions suivantes: qu'avons-nous fait pour renforcer le sentiment patriotique de nos immigrés ? Les procédures sont-elles adaptées à leurs attentes du moment? Leur avons-nous assuré l'encadrement religieux et éducatif nécessaire ?	That said, we need to ask ourselves the following questions: What have we done to strengthen the patriotic feeling of our immigrants? Are the procedures adapted to their current expectations? Have we provided them with the necessary religious and educational support?	MoorishMovement on Twitter
Investment and Immigration	« Nous exhortons les jeunes et les porteurs de projets marocains, résidant à l'étranger, à profiter des multiples opportunités d'investissement offertes par la mère-patrie »	"We urge young people and Moroccan project leaders residing abroad to take advantage of the multiple investment opportunities offered by the motherland"	MoorishMovement on Twitter

Source: Created by the Author

Table 5 below shows the 6 dimensions of Anholt discussed as a theoretical framework for understanding the Moroccan national branding story and illustrates each of the dimensions by an example extracted from the analyzed corpus.

On the managerial level, this work helps to guide reflection on the discourse held by the Moroccan authorities to promote the country and its culture to foreign audiences. The analysis of unofficial accounts makes it possible to deliver content that seems more credible with very good results in terms of virality and engagement compared to traditional accounts like *VisitMorocco*. This may be due to the diversity of content shared by private accounts, as they have fewer constraints than official accounts. A visual analysis of the extracted content showed the following results:

The Heart of Morocco Facebook page was the richest page in terms of visual content used. It included the use of more than 350 types of emojis (smileys, flags, etc.). The page mainly shared photos of architecture, landscapes, and local landmarks.

SimplyMorocco's Facebook account used more than 43 types of different emojis (Moroccan flags, smileys, etc.). The account shares photos mainly around nature and the Moroccan lifestyle (gastronomy, Moroccan living rooms, etc.) with more than 116 photos over the period covered.

Finally, *VisitMorocco*'s Facebook account had more than 63 different types of emojis. This page contained 70 images and was published less frequently but given the official nature of the page as part of the Moroccan tourist office, the photos were more professional and better in quality.

On the managerial level, the research found that Morocco's branding seemed to be mainly aimed at foreign populations in English and French, and it is, therefore, important for local decision-makers to fully understand how the narrative around the "Morocco" brand is built by these digital actors who are influential and visible on the web, and which are therefore de facto a digital showcase of the country that seems more authentic and generates more engagement.

On the theoretical level, the mobilization of Anholt's framework was interesting to study the Moroccan case. Indeed, the analysis showed that a large majority of the six dimensions were covered by the pages studied, in particular: people, tourism, exports, and culture. The remaining two dimensions were mainly covered by the *MoorishMovement* Twitter account which embodies a patriotic neo-nationalist movement focusing on governance and investment and immigration.

On the instrumental level, the mobilization of a Netnography, the crossing of several data sources (Facebook vs Twitter), the diversity of accounts selected (Private accounts versus Official accounts), and finally the crossing of thematic analysis vs the lexicometry analysis allowed a rich cross-fertilization of data and approaches (Bardin, 2001; Malhotra, 2008). The results show that there is a similarity in the content shared on certain themes, but complementarities between these accounts studied. Some pages are more visual, others are known for sharing more content, other accounts are presenting playful content, while others focus on informational arguments.

As part of this work, some content has been observed that criticizes the romanticization of poverty. In some photos, poverty is portrayed romantically or nostalgically, sometimes as a real, or even desirable way of life, that practice is known as "romanticizing poverty." Presenting precarious water sellers, or villagers living in misery as a form of authenticity is sometimes frowned upon by Internet users, and this issue should proactively be addressed (Crossley, 2012; Karnani, 2007). Another issue concerns cultural appropriation. It consists of the adoption or application of components from one culture by individuals from another, frequently without adequate knowledge of or respect for the original culture (*Is Cultural Heritage the New Battleground?*, 2021). In addition to perpetuating power disparities and prejudice, this can lead to the erasure or misrepresentation of the culture that is being appropriated. such as the very recent controversy surrounding the sports equipment brand Adidas accused by the Kingdom of Morocco of cultural appropriation by using Moroccan tiles as patterns on some of its products (*Adidas Admits It Used Morocco's Zellige Design*, 2022).

This work contributed to an embryonic body of research where very few works exist to date about the Moroccan destination. On the methodological level, the mobilization of a netnographic approach was an appropriate way to have an original, non-invasive, and reliable method that makes it possible to better explore the terminology used, the type of visuals, the campaigns used, and the profile of people interested in such content. On the managerial level,

FUTURE RESEARCH DIRECTIONS

Like all studies, this research has also some limitations that are opportunities for future research. First, future research might examine the use of additional techniques like surveys and interviews to gain insights from target audiences or decision-makers about their perception and their assessment regarding the promotional content shared online. These techniques might offer a deeper comprehension of the perspectives, preferences, and motivations of various tourist market segments and might also shed insight into the strategies and regulations of Morocco's tourism sector.

Second, future research could expand the studied corpus by including additional official (i.e., government-owned) social network accounts. This will enable a more thorough understanding of the social media environment and the various viewpoints and voices that influence the online discussion around Moroccan tourism.

Finally, the comparison of the results may be hampered by the differing time windows used in the current study for the data gathered from Facebook and Twitter. Future studies could compare the frequency of publication on Twitter and Facebook using the same time window.

Additionally, incorporating data from other social media sites like Instagram, TikTok, or YouTube may offer a more complete picture of Morocco's online tourism presence and the various ways in which consumers engage with the content, especially among younger audiences known to use such platforms.

CONCLUSION

National branding in the modern day entails more than just logos and catchphrases; it also entails campaigning, raising awareness, and fostering meaningful interaction to engage the appropriate target consumers and key markets. It provides a strong and clear foundation for the country's communication strategies with a multisectoral development strategy to promote a favorable view of the nation.

This research work was intended to deduce how Morocco proceeds directly and indirectly to project a positive image of its heritage, politics, and assets. The results demonstrated that there is an ecosystem of influential accounts that enhance the image of Morocco internationally, each account brings something different with a different style, but with a certain complementarity between their actions.

Private initiatives (vs. official accounts) seem to receive better engagement rates, perhaps because they are less constrained and freer in spreading their messages. Nevertheless, the kingdom suffers regarding communication on the aspect of governance, local quality of life, and immigration, because very few accounts seem to cover this part, and the Moroccan regime is sometimes singled out by international NGOs for its administration of human rights or the immigration challenges which can create a certain imbalance in the way of promoting the image of the country.

REFERENCES

Adidas admits it used Morocco's zellige design. (2022). Middle East Eye. https://www.middleeasteye.net/news/qatar-world-cup-2022-adidas-morocco-algeria-zellige-top-admits

Adler-Nissen, R. (2014). Stigma management in international relations: Transgressive identities, norms, and order in international society. *International Organization*, *68*(1), 143–176. doi:10.1017/S0020818313000337

Anholt, S. (2005). Anholt Nation Brands Index: How Does the World See America? *Journal of Advertising Research*, *45*(3), 296–304. doi:10.1017/S0021849905050336

Anouar, S. (2022). Morocco Launches "Aji" and 'We Are Open' Campaigns to Revive Tourism. *Morocco World News.* Https://Www.Moroccoworldnews.Com/. https://www.moroccoworldnews.com/2022/02/347121/morocco-launches-aji-and-we-are-open-campaigns-to-revive-tourism

Bardin, L. (2001). L'analyse de contenu (10e éd.). Paris: Presses Universitaires de France.

Belkadi, E. (2020). City Branding of Casablanca in Morocco. In *Strategic Innovative Marketing and Tourism* (pp. 129–138). Springer. doi:10.1007/978-3-030-36126-6_15

Bhutia, P. D. (2022). *Creative New Tourism Pitch.* Skift. https://skift.com/2022/05/13/moroccos-covid-policies-work-against-creative-new-tourism-pitch/

Boum, A. (2020). Branding Convivencia: Jewish Museums and the Reinvention of a Moroccan Andalus in Essaouira. *Exhibiting Minority Narratives: Cultural Representation in Museums in the Middle East and North Africa*, 205–223. Academia.

Brand Finance. (2022). *Nation Brands Forum 2022 in Association with DPAAL.* Brand Finance. https://brandfinance.com/events/brand-finance-nation-brands-forum-2022

Cascón-Pereira, R., & Hernández-Lara, A. (2013). Town and city management papers Building the 'Morocco'brand as a tourist destination: The role of emigrants and institutional websites. *Journal of Urban Regeneration and Renewal*, *6*(3), 252–263.

Cascón-Pereira, R., & Hernández-Lara, A. B. (2014). The Morocco brand from the Moroccan emigrants' perspective. *Place Branding and Public Diplomacy*, *10*(1), 55–69. doi:10.1057/pb.2013.27

Crossley, É. (2012). Poor but happy: Volunteer tourists' encounters with poverty. *Tourism Geographies*, *14*(2), 235–253. doi:10.1080/14616688.2011.611165

Dinnie, K. (2015). *Nation branding: Concepts, issues, practice.* Routledge. doi:10.4324/9781315773612

El Aouni, F. (2015). *Destination branding and the role of emigrants: The case of morocco.* [PhD Thesis, Universitat Rovira i Virgili].

Facebook supprime près de 400 faux comptes. (2021, March 4). Hespress Français. https://fr.hespress.com/192812-facebook-supprime-pres-de-400-faux-comptes-lies-a-une-propagande-dun-media-marocain.html

Gienow-Hecht, J. (2019). Nation branding: A useful category for international history. *Diplomacy and Statecraft*, *30*(4), 755–779. doi:10.1080/09592296.2019.1671000

Fanack. (2023). Governance & Politics of Morocco. *Fanack.Com*. https://fanack.com/morocco/politics-of-morocco/

Graan, A. (2016). The nation brand regime: Nation branding and the semiotic regimentation of public communication in contemporary Macedonia. *Signs and Society (Chicago, Ill.)*, *4*(S1), S70–S105. doi:10.1086/684613

Hankinson, G. (2015). Rethinking the place branding construct. In *Rethinking place branding* (pp. 13–31). Springer. doi:10.1007/978-3-319-12424-7_2

Hao, A. W., Paul, J., Trott, S., Guo, C., & Wu, H.-H. (2021). Two decades of research on nation branding: A review and future research agenda. *International Marketing Review*, *38*(1), 46–69. doi:10.1108/IMR-01-2019-0028

Hines, K. (2022, January 14). Social Media Usage Statistics For Digital Marketers In 2022. *Search Engine Journal*. https://www.searchenginejournal.com/top-social-media-statistics/418826/

Hogg, M. A. (2016). *Social identity theory*. Springer.

Hurn, B. J. (2016). The role of cultural diplomacy in nation branding. *Industrial and Commercial Training*, *48*(2), 80–85. doi:10.1108/ICT-06-2015-0043

Is cultural heritage the new battleground? (2021). The Africa Report. https://www.theafricareport.com/146636/algeria-morocco-is-cultural-heritage-the-new-battleground/

Karjaluoto, H. (2017). *Influence of Social Media on Corporate Heritage Tourism Brand*. Springer. doi:10.1007/978-3-319-51168-9_50

Karnani, A. (2007). Romanticizing the poor harms the poor. *Metamorphosis*, *6*(2), 151–162. doi:10.1177/0972622520070206

Kiráľová, A., & Pavlíčeka, A. (2015). Development of Social Media Strategies in Tourism Destination. *Procedia: Social and Behavioral Sciences*, *175*, 358–366. doi:10.1016/j.sbspro.2015.01.1211

Kotler, P., & Armstrong, G. (2020). *Principles of Marketing (18th Globa)*. Pearson Educatio n Limited.

Kozinets, R. V. (2002). The field behind the screen: Using netnography for marketing research in online communities. *JMR, Journal of Marketing Research*, *39*(1), 61–72. doi:10.1509/jmkr.39.1.61.18935

Lee, J. W., & Syah, A. M. (2018). *Economic and Environmental Impacts of Mass Tourism on Regional Tourism Destinations in Indonesia*. SSRN Scholarly Paper No. 3250133. https://papers.ssrn.com/abstract=3250133 doi:10.13106/jafeb.2018.vol5.no3.31

Malhotra, N. K. (2008). *Marketing research: An applied orientation, 5/e*. Pearson Education India. doi:10.1108/S1548-6435(2008)4

Moro, S., & Rita, P. (2018). Brand strategies in social media in hospitality and tourism. *International Journal of Contemporary Hospitality Management, 30*(1), 343–364. doi:10.1108/IJCHM-07-2016-0340

Nye Jr, J. S. (2004). *Soft power: The means to success in world politics*. Public affairs.

Ojo, S. (2020). Interrogating place brand–a case of two cities. *Qualitative Market Research, 23*(4), 907–932. doi:10.1108/QMR-11-2017-0151

Pike, A. (2015). *Origination: The geographies of brands and branding*. John Wiley & Sons. doi:10.1002/9781118556313

Ritzer, G. (2021). The McDonaldization of society. In *In the Mind's Eye* (pp. 143–152). Routledge. doi:10.4324/9781003235750-15

Sedra, D., & El Bayed, H. (2022). Branding the city: The case of Casablanca-Morocco. *Place Branding and Public Diplomacy, 18*(2), 181–189. doi:10.105741254-020-00195-y

Sevin, E. (2021). Computational approaches to place branding: A call for a theory-driven research agenda. In *A Research Agenda for Place Branding* (pp. 33–45). Edward Elgar Publishing. doi:10.4337/9781839102851.00010

Shareef, M. A., Mukerji, B., Dwivedi, Y. K., Rana, N. P., & Islam, R. (2019). Social media marketing: Comparative effect of advertisement sources. *Journal of Retailing and Consumer Services, 46*, 58–69. doi:10.1016/j.jretconser.2017.11.001

Stephens Balakrishnan, M. (2009). Strategic branding of destinations: A framework. *European Journal of Marketing, 43*(5/6), 611–629. doi:10.1108/03090560910946954

Thellefsen, T., Sørensen, B., Vetner, M., & Andersen, C. (2006). *Negotiating the meaning of artefacts: Branding in a semeiotic perspective*. De Gruyter.

Theodoropoulou, I. (2020). The case of Mykonos, Greece on Facebook. *Research Companion to Language and Country Branding*, 313.

Statista. (2021). *Tourism in Morocco*. Statistia. https://www.statista.com/topics/8256/tourism-industry-in-morocco/

Tovar, J. (2020). Reimaging and branding a post-reunification Germany. *Research Companion to Language and Country Branding*, 183.

United Nations. (2023). *International Day of Argania*. UN. https://www.un.org/en/observances/argania-day

World Bank. (2022). *Number of arrivals—Morocco*. World Bank. https://data.worldbank.org/indicator/ST.INT.ARVL?locations=MA

Wüst, A., & Nicolai, K. (2022). Cultural diplomacy and the reconfiguration of soft power: Evidence from Morocco. *Mediterranean Politics*, 1–26. doi:10.1080/13629395.2022.2033513

Zenker, S., Braun, E., & Petersen, S. (2017). Branding the destination versus the place: The effects of brand complexity and identification for residents and visitors. *Tourism Management, 58*, 15–27. doi:10.1016/j.tourman.2016.10.008

KEY TERMS AND DEFINITIONS

Cultural appropriation: The term used to describe the adoption or usage of components from one culture by members of another culture, frequently without the originating culture's proper knowledge or respect.

Nation Branding: A nation state's attempts to use appropriate methods and strategies to market its country abroad in order to maximize the advantages that can be gained.

Romanticizing poverty: This refers to the representation of poverty in a romantic or nostalgic light, frequently presenting it as a more straightforward, real, or even desirable way of life.

Chapter 14
A Bibliometric Investigation of Electronic Word–of–Mouth in Tourism and Hospitality Research

Mahmut Bakır
Samsun University, Turkey

Ali Emre Sarılgan
Eskisehir Technical University, Turkey

ABSTRACT

Today, electronic word-of-mouth (eWOM) has a substantial impact on consumers' decisions to purchase tourism and travel services. Therefore, it is essential to have a comprehensive understanding of eWOM in tourism and hospitality research. The chapter aims to conduct a comprehensive examination of the existing body of knowledge pertaining to eWOM communication within the domain of tourism and hospitality. To this end, bibliometric data was obtained from the Web of Science database, utilizing the Bibliometrix R package. The PRISMA flowchart was employed to ensure a systematic examination, which included a total of 515 scholarly documents published between 2008 and 2021. The utilization of a word cloud facilitated the identification of the most prevalent terms within the field, while a thematic map was employed to reveal the themes that guide the existing body of knowledge. Furthermore, co-occurrence analysis was utilized to discern four distinct research themes that have emerged as particularly prominent within the field.

INTRODUCTION

Internet-based technologies and platforms have had a significant influence on marketing. Using digital platforms, organizations can engage customers directly and receive quick feedback through likes, clicks, and comments. On online portals such as e-commerce websites, social media platforms, weblogs, and

DOI: 10.4018/978-1-6684-6985-9.ch014

peer-to-peer networking sites, users can voice their opinions regarding acquired items and services. The customer's feedback and opinion about a product or service are called as electronic word-of-mouth (eWOM; Donthu et al., 2021). eWOM can be represented in many different forms, such as likes, comments, ratings, reviews, video testimonials, and tweets (Nam et al., 2020). eWOM is regarded as more reliable and convincing than traditional media, given that consumers are exposed to many advertising stimuli daily. Customers increasingly rely on online reviews day by day, a form of eWOM, before making their final purchasing decisions (Cheung & Thadani, 2012).

The importance of eWOM on consumer behavior has also been established in the context of tourism and hospitality services (Litvin et al., 2008; Sparks & Browning, 2011; Vermeulen & Seegers, 2009). Customers are making booking or purchase selections largely after reading reviews on famous sites like TripAdvisor, Twitter, and Google (Donthu, Kumar, Pandey, et al., 2021). One of the main reasons for this is the risk aversion effort by using the past experiences of others, as travel and accommodation services are high-risk purchasing decisions (Huang et al., 2010).

Numerous research (Jalilvand & Samiei, 2012; Ran et al., 2021; Sparks & Browning, 2011) have addressed the role of eWOM in the tourist and hospitality industries. Furthermore, retrospective analyses have also been carried out to delineate the boundaries and structure of the body of knowledge and illustrate its evolution (Chen & Law, 2016; Hlee et al., 2018; Litvin et al., 2018; Pourfakhimi et al., 2020; Sotiriadis, 2017). Retrospective literature analyses can be performed through systematic literature review, meta-analysis, and bibliometric analysis. However, a systematic literature review can handle few studies on a particular topic and lacks quantitative metrics, making it prone to researcher bias (Gölgeci et al., 2022). On the other hand, although meta-analysis is a powerful technique, it is limited to a certain field and is susceptible to publication bias (Donthu et al., 2021). In light of these challenges, tracing the growth of the existing literature requires a rise in bibliometric research that can evaluate a vast number of works and have analytical rigor (Gölgeci et al., 2022). Nonetheless, there is a paucity of investigations that examines the theoretical and conceptual structure of eWOM research in the tourism and hospitality fields (Mukhopadhyay et al., 2022). To bridge this gap, this chapter aims to examine the evolution of the eWOM literature in the field of tourism and hospitality and to increase understanding of its theoretical and conceptual structure. In doing so, it conducts bibliometric analysis using the R package Bibliometrix.

The organization of the remainder of the chapter is as follows. Section 2 presents a comprehensive overview of the concept of eWOM in general and in the field of tourism and hospitality. The research methodology, inclusive of data collection procedures, is outlined in Section 3. In Section 4, bibliometric analysis findings are presented and discussed in detail. Section 5 outlines potential directions for future research. The final section concludes the chapter by highlighting the managerial implications, theoretical contributions, and limitations of the research.

BACKGROUND

Word-of-mouth (WOM) is one of the oldest forms of information sharing between individuals (Huete-Alcocer, 2017). Although marketing research on WOM began in the 1960s, the definition of WOM has evolved over time (Litvin et al., 2008). Initially, WOM was defined as face-to-face communication based on sharing product and service evaluations between private parties but was later used as all kinds of informal communication (Chen & Law, 2016). Very early, Katz and Lazarsfeld (1966) defined WOM as the exchange of marketing information with a pivotal role in shaping consumer behavior and changing attitudes towards products. Conversely, Westbrook (1987, p. 261) defined WOM as "all informal communications directed at other consumers about the ownership, usage, or characteristics of particular goods and services or their sellers."

WOM communication is interpersonal communication between consumers regarding their experiences with a company or product that is not influenced by commercial sources (Gretzel & Yoo, 2008). Litvin et al. (2008) stated that the independence of the message source is the defining characteristic of WOM. Consumers find WOM communication to be more credible and compelling than traditional media channels (e.g., newspapers, radio, and broadcast television) (Bhaiswar et al., 2021; Cheung & Thadani, 2012). Therefore, WOM is regarded as one of the most effective sources of information after product and service consumption. Customers trust other consumers and their information more because they have no different benefits for firms. This is particularly crucial in industries that supply intangible items, such as tourism, hospitality, and travel, because the products cannot be tested prior to purchase (Huete-Alcocer, 2017).

The emergence and widespread use of the Internet have paved the way for a new form of WOM, referred to as electronic WOM (eWOM) (Huete-Alcocer, 2017). As a result, consumers are increasingly utilizing Web 2.0 platforms (e.g., online discussion forums, consumer review sites, social network sites, microblogs, etc.) to voice their opinions and share their product experiences. Hennig-Thurau et al. (2004) defined eWOM communication as "any positive or negative statement made by potential, actual, or former customers about a product or company, which is made available to a multitude of people and institutions via the Internet" (Hennig-Thurau et al., 2004, p. 39). Similarly, Litvin et al. (2008, p. 461) define eWOM as "all informal communications directed at consumers through Internet-based technology related to the usage or characteristics of particular goods and services, or their sellers."

The information shared among consumers can be found on various platforms such as online forums (e.g., Reddit, Quora.com), social networking sites (e.g., Facebook, Instagram), microblogs (e.g., Twitter), online review platforms (e.g., Tripadvisor, Yelp), and even on the sellers' own websites (e.g., Amazon, Aliexpress, ebay.com) (Cheung & Thadani, 2012; Huete-Alcocer, 2017). Likes, comments, reviews, video testimonials, images, and blog posts are all examples of how customers can express their ideas and recommendations (Bhaiswar et al., 2021; Donthu et al., 2021). One advantage of eWOM is that it provides a convenient platform for consumers to share their opinions and product reviews with other users on online platforms (Huete-Alcocer, 2017). Consumers now turn to online comments (eWOM) for information about products and services instead of relying on WOM, which was previously reflected by friends and family (Nieto et al., 2014).

Two dimensions comprise the eWOM typology: communication scope and level of interaction. The communication scope can be classified as one-to-one (emails), one-to-many (review sites), or many-to-many (virtual communities), and the level of interaction can be categorized as asynchronous (emails, review sites, blogs) or synchronous (chatrooms, newsgroups, instant messaging) (Litvin et al., 2008; Serra Cantallops & Salvi, 2014).

Compared to traditional media channels, eWOM is considered a more reliable and helpful information source for assisting consumers in avoiding risks and making better purchasing decisions, particularly in industries where the intangible attribute is prevalent. The intangible attribute is one of the distinguishing characteristics of tourism and hospitality products. A traveler staying in a hotel for two days, a tourist participating in a cultural tour, or a traveler traveling by airline are unable to evaluate these services before purchasing, making it difficult to evaluate by using senses prior to consumption. Thus, interpersonal influence is particularly important in purchasing tourism and hospitality products (Chen & Law, 2016).

On the other hand, the use of technology by customers to share their opinions about products or services can pose a significant risk, as it may become an uncontrollable factor for companies. Companies try to mitigate this effect by controlling customers' online reviews through comment boxes and chatbots on their own websites, where consumers can leave comments and share their opinions (Huete-Alcocer, 2017).

eWOM refers to interpersonal communication just like WOM. It also contains positive, negative, and neutral information about companies and their products and services (Bhaiswar et al., 2021). However, there are some fundamental differences between these two forms of communication. Table 1 summarizes the differences between WOM and eWOM.

Table 1. Differences between WOM and eWOM communication

	WOM	eWOM
Credibility	The recipient of the information possesses prior familiarity with the sender.	A condition of anonymity exists between the sender and the recipient of the information.
Privacy	The mode of communication employed is characterized as being private, interactive, and occurring in real-time.	The content shared is not confidential and, as it is recorded in written form, may be subject to perusal by any individual.
Diffusion speed	Information moves slowly. The sharing of information requires the presence of users.	Users can communicate more quickly and at any time with one another using the Internet.
Accessibility	Less accessible	Easily accessible

Source: Huete-Alcocer (2017)

The credibility of the information source is the first major difference between WOM and eWOM. Accordingly, the anonymity of online reviews might have a negative effect on the credibility of the messages (Huete-Alcocer, 2017). Nonetheless, Sotiriadis and Van Zyl (2013) discovered that eWOM is regarded as more reliable than WOM in circumstances where customers have previously used it. Another characteristic is the privacy of the message, referring to the fact that WOM communication consists of private and face-to-face dialogues, while the information shared in eWOM is not private and can be shared online and viewed by everyone (Cheung & Thadani, 2012). The third difference is the diffusion speed of the message. eWOM spreads rapidly in a very short period of time because it is spread online (Huete-Alcocer, 2017). Lastly, whereas WOM is less accessible because it is conducted among real people in a restricted space, eWOM information is readily available online.

In addition, eWOM communication is more lasting and accessible than traditional WOM. The majority of text-based material published on the Internet is archived and may therefore be made accessible indefinitely (Cheung & Thadani, 2012). On the other hand, Serra Cantallops and Salvi (2014) stated that the primary difference between WOM and eWOM is the review impact (number of individuals influenced) and interaction speed. Moreover, eWOM communication is more measurable because the information shared on online platforms leaves permanent evidence. Additionally, due to the high volume of information available from many users on online platforms, eWOM communication has a large impact on consumers, with no geographical restriction (Bhaiswar et al., 2021).

One of the most frequently used fields for eWOM is apparently tourism and hospitality. The development of the internet has created the possibility of forming a form of eWOM known as user-generated content (UGC) in virtual communities. Therefore, in the tourism industry, customers can easily communicate through eWOM on various social media platforms (Chen & Law, 2016). According to Litvin et al. (2018), the growing use of the internet has created a substantial space for consumers to voice their opinions about brands and their hospitality-related products. Customers provide innumerable reviews for tourist and hospitality services on platforms such as TripAdvisor, Yelp, Urbanspoon, Expedia, and Hotels.com (Litvin et al., 2018). Research shows that most tourists engage in online communication not only when planning their travels but also while on holiday (Chen & Law, 2016). Consumers tend to rely more on consumer reviews when purchasing high-involvement products such as travel and accommodation. As a result, customers commonly consult online reviews in these industries (Gretzel & Yoo, 2008).

The motivation for sharing customer experiences in the tourism and hospitality industry has been studied in various studies. The majority of tourists believe it is vital to share their experiences to help other travelers in making informed decisions (Bronner & de Hoog, 2011). In addition, tourists may be inclined to share information online to protect others from purchasing inaccurate or substandard goods and services (Munar & Jacobsen, 2014). In a review of the literature, Sotiriadis (2017) identified the influence of several factors, including service quality and customer satisfaction, pre-purchase service expectations, a sense of community belonging, social identity/social support, and personality traits, on sharing tourism experiences on social media. On the other hand, Kim et al. (2011) concluded the existence of three factors, convenience and quality, risk reduction, and social reassurance, that are effective in tourists' search for eWOM information. Serra Cantallops and Salvi (2014) listed the reasons for posting online reviews in the hospitality industry as service quality, customer satisfaction, customer dissatisfaction, sense of community belonging, social identity, pre-purchase expectations, helping other vacationers, helping companies, and failure and recovery.Consumer reviews serve two distinct purposes, namely providing information and giving advice about products and services. Consumer reviews are considered effective because they are written from the consumer's perspective and are therefore considered indirect and credible experiences. It is also considered more reliable than traditional marketing efforts (Gretzel & Yoo, 2008).

Numerous studies have been conducted on eWOM in the tourism and hospitality sectors, as evidenced by the existing literature (Chen & Law, 2016; Hlee et al., 2018; Litvin et al., 2018; Pourfakhimi et al., 2020; Sotiriadis, 2017). Sotiriadis (2017) analyzed publications on sharing tourism experiences on social media from 2009 to 2016 using content analysis. Mukhopadhyay et al. (2022) summarized existing eWOM research in leading hospitality and tourism journals using bibliometric analysis, examining 398 selected tourism publications. Chen and Law (2016) conducted a review of electronic word-of-mouth research in the hospitality and tourism fields, while Hlee et al. (2018) presented a systematic review of online review research in the hospitality and tourism sector using 55 research articles. Upon reviewing these review studies, it was observed that network analysis, which addresses complex relationships, was not conducted (Chen & Law, 2016; Hlee et al., 2018; Sotiriadis, 2017) or included in fewer publications (Sotiriadis, 2017). Therefore, this research is designed to address the current gap in the literature.

METHODOLOGY AND DATA COLLECTION

Bibliometric Analysis

Bibliometrics analysis is a quantitative methodology that utilizes metadata from published studies in a discipline or journal to provide insights into the performance of research constituents, such as authors, institutions, countries, studies, journals, etc. (Hashemi et al., 2022). Bibliometric analysis employs visual and quantitative tools to understand the trends and evolution of a particular research field or journal (Palácios et al., 2021; Tanrıverdi & Durak, 2022). The initial discussions about bibliometrics began in the 1950s, making this methodology well-grounded (Donthu et al., 2021). The popularity of bibliometrics in recent years is due to the advancement and availability of bibliometric analysis software, as well as a shift towards retrospective studies in fields beyond library science, such as business and tourism (Donthu et al., 2021). The ability to easily retrieve large volumes of bibliometric material from databases such as Scopus and Web of Science, and the ability of software such as VOSviewer, CiteSpace, and SciMat to effectively analyze this large amount of data has recently gained popularity in this area of research (Donthu et al., 2021).

Rather than a straightforward literature review, the bibliometric analysis offers much deeper insights, such as identifying past and current trends. Another benefit of bibliometric studies is that they inform editors and potential authors about future research avenues in a particular field (Derudder et al., 2019). Two distinct components comprise bibliometric analysis: performance analysis and science mapping (Donthu et al., 2021). Performance analysis is a descriptive analysis that conveys a general picture of scientific production from different perspectives. On the other hand, science mapping graphically depicts the conceptual, social, and intellectual structure of the scientific research discipline and offers deeper insights (Cobo et al., 2015; Donthu, Kumar, Mukherjee, et al., 2021).

This chapter conducts bibliometric analysis in three steps in line with past literature (Bakır et al., 2022; Donthu et al., 2021; Palácios et al., 2021). In the first step, bibliometric material was retrieved from the Web of Science based on certain inclusion criteria, as explained in the material section of this present chapter. Next, a performance analysis that quantitatively evaluates the contributions of research constituents (authors, institutions, countries, and journals) was carried out. In the third and final step, science mapping techniques were employed (Hashemi et al., 2022). This was done by identifying research themes using the author's keywords and identifying collaboration networks among research constituents (authors, institutions, countries, and journals) (Bakır et al., 2022). The Bibliometrix R package was used to perform the performance analysis and science mapping analysis (Aria & Cuccurullo, 2017). Although there are different bibliometric analysis tools available, the Bibliometrix package was chosen as the most suitable tool due to its advanced visualization features and ability to conduct network analysis (Palácios et al., 2021; Raza et al., 2020).

In performance analysis, there are numerous measures available. However, there is no consensus on which ones perform better. The most prominent measures are the number of publications and the number of citations. The number of publications serves as a proxy for productivity, while the number of citations is a measure of influence (Donthu et al., 2021). On the other hand, the Hirsch index (h-index), the impact factor, and the total number of citations per year are other influence metrics. The Hirsch index measures the adjusted impact of research constituents and shows the number of x studies that a research constituent has received x citations (Bakır et al., 2022). The impact factor represents the influence of academic journals and is obtained by dividing the number of citations received by the number of studies published in the last x years. The total number of citations per year is another measure of influence.

Science mapping examines the networks among research constituents. It also discovers research themes and the future research agenda using the authors' keywords. Co-occurrence analysis often derives from the authors' keywords, and similar to co-citation analysis, keyword co-occurrence analysis assumes that keywords that frequently appear together have thematic relationships (Donthu et al., 2021). A thematic map, based on the keywords in the analyzed documents, prepares a thematic map consisting of four quadrants. This map is a simplified diagram that groups the current research themes (Raza et al., 2020).

Word cloud also visualizes the keywords present in the bibliometric material in the form of a cloud. Lastly, the co-authorship analysis examines the collaborations among research constituents (e.g., authors, institutions, countries, etc.) and how the constituents are connected to each other (Bakır et al., 2022).

Material

In this present chapter, the selection process of the bibliometric material is depicted in Figure 1. The author developed a PRISMA flow chart to achieve a systematic and rigorous analysis process. The PRISMA flow diagram is a useful tool for outlining the boundaries of the process of searching, selecting, and including documents in retrospective studies such as bibliometric analysis, systematic literature reviews, and meta-analysis (Moher et al., 2009). PRISMA contributes to transparency and completeness in the selection of documents in the analysis and improves the reproducibility of the scientific process. Furthermore, it facilitates the determination and sharing of any biases and inclusion criteria that may be reported in the final analysis of the studies included in the analysis (Kim & So, 2022).

In this research, although there are different databases such as Scopus, Google Scholar, and ProQuest, the Web of Science (WoS) database, which covers the most prestigious journals in their fields and enjoys great scholarly attention, despite being less inclusive, was used (Mulet-Forteza et al., 2018; Zhao et al., 2020). Web of Science is a well-known and widely respected database in the academic community, especially in the field of bibliometrics (Tanrıverdi & Durak, 2022). It provides a comprehensive collection of scientific works in numerous fields, including tourism and hospitality (Li et al., 2017; Palácios et al., 2021). In addition, it features a powerful search engine and advanced filtering tools, which contribute to a high level of precision and accuracy in bibliometric analysis (Bhaiswar et al., 2021; Kim & So, 2022). In addition, it provides access to a variety of citation measures, such as the Journal Impact Factor, which is frequently used in academic research to assess the impact and influence of a journal or article.

As of June 21, 2022, the author reached 682 records by searching the term *("electronic word of mouth" OR "electronic word-of-mouth" OR "eWOM" OR "electronic WOM" OR "online word of mouth" OR "online WOM" OR "e-Word of Mouth" OR "eWord of Mouth") AND (aviation OR airline OR airlines OR airport OR airports OR "air transport" OR "air transportation" OR hotel OR tourism OR hospitality OR travel)* as target title, abstract, and keywords on the WoS database. While the first publication appeared in 2008 within the scope of the search, the research was restricted until 2021, the last year of investigation. The following flowchart resulted in the inclusion of 515 studies in the main analysis. Donthu et al. (2021) considered the research area wide enough to be analyzed using bibliometric analysis if there are hundreds (e.g., 500 or more) publications in a research area.

The Mendeley software was utilized to eliminate duplicated publications and manage bibliometric material. As a part of the evaluation process, the selected articles were manually controlled to ensure they were relevant to the topic. Irrelevant topics were excluded (Kim & So, 2022). To determine the eligibility of the studies, the author first analyzed the abstract and keywords of each publication (Moher et al., 2009). In conclusion, the bibliometric investigation was finalized with 515 studies.

Figure 1. The PRISMA flow diagram
Source: Author's elaboration.

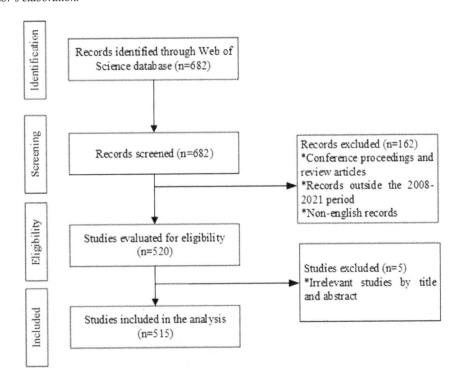

FINDINGS AND DISCUSSION

Performance Analysis

This section presents the findings of a bibliometric analysis of scholarly articles on the topic of eWOM in tourism and hospitality research. As seen in Table 2, the analysis reveals that the first study on this issue published in 2008 and that a total of 515 papers were analyzed. These articles appeared in 184 scientific journals, and the yearly growth rate of publications on this topic from 2008 to 2021 is 43.15%.

In addition, the analysis reveals that the average number of citations per study is 40,1 and that 872 keywords plus (ID) and 1659 author keywords (DE) are utilized in the articles. There are a total of 1123 authors for the publications, with 49 articles being single-authored and 466 being multi-authored. The average number of co-authors per study is 2.78. The findings of this bibliometric analysis shed light on the current status of research on eWOM in tourism and hospitality. The high annual growth rate of 43.15% indicates that this is a rapidly growing area of research and that there is a growing interest among scholars in studying the impact of eWOM on the tourism and hospitality industry.

Additionally, the high average number of citations per study (40.1) indicates that the articles in this field are highly influential and have a significant impact on the academic community. A large number of keywords and authors also implies a diverse and active research community. However, the large number of multi-authored papers compared to the small number of single-authored articles demonstrates that collaboration is an integral component of the research process and also implies that papers are more likely to get published in high-impact journals when they have multiple authors.

Table 2. Basic information about bibliometric materials

Description	Results
Timespan	2008-2021
Sources (Journals, Books, etc.)	184
Documents	515
Annual Growth Rate %	43.15%
Average citations per doc	40.1
Keywords Plus (ID)	872
Author's Keywords (DE)	1659
Authors	1123
Single-authored docs	49
Multi-authored docs	466
Co-Authors per Doc	2.78

Source: Author's elaboration.

Figure 2 represents the annual scientific production by year. Based on the data provided, it appears that the amount of tourism and hospitality-related articles published on eWOM has increased steadily over time. The number of articles published in 2008 is 1, and this number increases to 2 in 2009, 5 in 2010, and 9 in 2011. This trend continues, with the number of articles produced each year reaching 13, 19, and 14 in 2012, 2013, and 2014, respectively. A large increase in the number of articles was noticed in 2015, with 29 articles, and this number continued to increase in the following years, reaching 38 articles in 2016, 45 in 2017, 59 in 2018, 79 in 2019, 96 in 2020, and 106 in 2021. Figure 2 demonstrates that eWOM is a rapidly growing area of research in the field of tourism and hospitality and that there is a significant interest in the topic among researchers and practitioners in the field. Additionally, the significant number of articles published in 2020 and 2021 suggests that the topic remains relevant and timely. These findings may suggest that eWOM has become a critical aspect of the tourism and hospitality industry and that more research is needed to fully understand its impact on the industry.

In the next stage of bibliometric analysis, the most productive authors have been identified. Identifying the most productive authors in the field of eWOM can help us understand researchers who have made significant contributions to the field and define the boundaries of the field (Bhaiswar et al., 2021). Therefore, this step is an important step in performance analysis. Table 3 lists the authors who have made the most contributions to the area of eWOM in tourism and hospitality. This table provides metrics of productivity in terms of total studies, influence in terms of total citations, and h-index. According to this, the most productive authors in the eWOM literature for tourism and hospitality are Law R ($n=10$), Xu X ($n=8$), and Zhang Z ($n=7$). Furthermore, it is observable that Law R and Ye Q are the most influential authors in terms of total citations. In terms of the h-index, Law R ranks first. In conclusion, Law R is the most prolific author, both in terms of productivity and influence in the extant literature.

Table 4 shows the top journals that have publications on eWOM in the field of tourism and hospitality. Accordingly, 8 of the top 10 journals where studies appear the most are from tourism, travel, and hospitality journals. Apart from that, one business and one sustainability science journal also contributed to the list, but surprisingly no scholarly journal existed on transportation. The most productive journal in the literature was found to be the International Journal of Hospitality Management ($n=32$), followed by the International Journal of Contemporary Hospitality Management (*n*=29) and Sustainability (*n*=29). Furthermore, the first ten journals contributed to 37% of the body of knowledge. In terms of total citations, Tourism Management, with 5586 citations, is the most influential academic outlet, followed by the International Journal of Hospitality Management, with 3597 citations. When the impact factors of the top 10 journals in 2020 are examined, it is observed that eWOM publications in the field of tourism and hospitality have been published in journals with an impact factor of at least 3, which are quite influential.

Figure 2. Annual scientific production by year
Source: Author's elaboration.

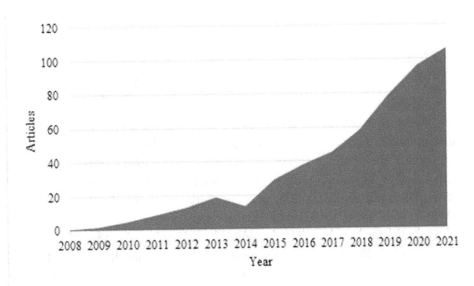

Table 3. Most productive authors

Author	Total Studies	Total Citations	*h*-Index
Law R	10	2060	10
Xu X	8	452	7
Zhang Z	7	650	7
Borghi M	6	160	6
Filieri R	6	994	6
Ilkan M	6	267	5
Kim J	6	60	5
Mariani MM	6	167	5
Ye Q	6	1709	5
Zhang Y	6	305	3

Source: Author's elaboration.

Table 4. Most contributing journals

Source	Total Studies	Total Citations	Impact Factor
International Journal of Hospitality Management	32	3597	9.237
International Journal of Contemporary Hospitality Management	29	1288	8.646
Sustainability	29	327	3.251
Tourism Management	25	5586	10.967
Journal of Hospitality and Tourism Technology	19	332	4.260
Current Issues in Tourism	12	383	7.430
Journal of Hospitality Marketing & Management	12	514	7.022
Journal of Travel & Tourism Marketing	12	677	7.564
Journal of Travel Research	12	1044	10.982
Journal of Business Research	9	184	7.550

Source: Author's elaboration.

The next step of performance analysis addresses the most prominent institutions and countries in the eWOM literature in the field of tourism and hospitality. Table 5 lists the most prominent institutions and countries in terms of their productivity based on the number of total studies. The University of Valencia leads with a total of 35 study contributions, followed by the Hong Kong Polytechnic University (25 contributions). Then, universities such as the University of Reading (*n*=23) and the University of Málaga (*n*=21) appeared in the list. In terms of countries, the USA ranks first with 334 studies, followed by China with 257 studies, and Spain with 189 studies. Table 5 also shows that eWOM literature in the field of tourism and hospitality has contributed from developing countries such as China, India, and Turkey, besides developed countries like the USA, Spain, United Kingdom, Australia, Portugal, and Italy.

Table 5. Most productive institutions and countries

Institution	Total Studies	Country	Total Studies
University of Valencia	35	USA	334
The Hong Kong Polytechnic University	25	China	257
University of Reading	23	Spain	189
University of Málaga	21	UK	130
Griffith University	20	Australia	71
University of Central Florida	19	India	66
Eastern Mediterranean University	18	Portugal	63
Bournemouth University	17	Turkey	53
National Pingtung University of Science and Technology	17	Italy	51

Source: Author's elaboration.

In the final step of performance analysis, Table 6 presents the most cited articles on eWOM in the field of tourism and hospitality. Identifying the most frequently cited studies in the eWOM literature helps to shape the research field and understand the key studies that have influenced the area (Bhaiswar et al., 2021). According to the table, the most cited article is "Electronic word-of-mouth in hospitality and tourism management" by Litvin et al. (2008), which has a total of 1308 citations, with an average of 81.75 citations per year. This article explains the online interpersonal influence or eWOM as a cost-effective means in hospitality and tourism management. The article is followed by Sparks and Browning (2011) and Vermeulen and Seegers (2009), which have been cited 781 and 701 times, respectively. The first study focuses on four core factors that influence trust perceptions and consumer choice in relation to online reviews of hotels, while the second study examines the effect of review valence, hotel familiarity, and reviewer expertise on consumer choice. Both of these studies are empirical, unlike Litvin et al. (2008), and have received an average of 60.08 and 46.73 citations per year, respectively.

Other studies in Table 6 also focused on the impact of online user reviews on various aspects of the hospitality and tourism industry, such as hotel room sales and online bookings, trust and reputation in the sharing economy, and the influence of user-generated content on traveler behavior. Additionally, the studies explore the motivations for sharing tourism experiences through social media and the effects of eWOM on the online popularity of restaurants. On the other hand, based on the average total citation count per year, Litvin et al. (2008), Ert et al. (2016), and Sparks and Browning (2011) lead the list. The publication year of the studies in Table 6 ranged from 2008 to 2016. In this regard, it is observed that the most influential studies were published in the first half of the publication period (2008-2021). Of course, more recent studies will appear at the top of the most cited publications in the future years (Bakır et al., 2022).

Table 6. Most cited publications

Author(s)	Study	Total Citations	TC per Year
Litvin et al. (2008)	Electronic word-of-mouth in hospitality and tourism management	1308	81.75
Sparks and Browning (2011)	The impact of online reviews on hotel booking intentions and perception of trust	781	60.08
Vermeulen and Seegers (2009)	Tried and tested: The impact of online hotel reviews on consumer consideration	701	46.73
Ye et al. (2009)	The impact of online user reviews on hotel room sales	669	44.60
Ert et al. (2016)	Trust and reputation in the sharing economy: The role of personal photos in Airbnb	616	77
Ye et al. (2011)	The influence of user-generated content on traveler behavior: An empirical investigation on the effects of e-word-of-mouth to hotel online bookings	597	45.92
Munar and Jacobsen (2014)	Motivations for sharing tourism experiences through social media	458	45.80
Serra Cantallops and Salvi (2014)	New consumer behavior: A review of research on eWOM and hotels	413	41.30
Filieri and McLeay (2014)	E-WOM and Accommodation: An Analysis of the Factors That Influence Travelers' Adoption of Information from Online Reviews	388	38.80
Zhang et al. (2010)	The impact of e-word-of-mouth on the online popularity of restaurants: A comparison of consumer reviews and editor reviews	370	26.43

Source: Author's elaboration.

Keyword Analysis

Following the presentation of the descriptive statistics of the bibliometric units in the phase of performance analysis, this section proceeds to conduct analyses utilizing the author's keywords as the basis. In this context, word cloud, thematic map, and co-occurrence analysis were performed. In doing so, text preprocessing steps were applied to standardize the text data for analysis (Bakır et al., 2022). First, the word cloud relying on the author's keywords is displayed in Figure 3. The word cloud generated from the author's keywords in the bibliometric analysis indicates several key themes in the research. With 85 occurrences, internet reviews are the most prevalent theme in the analysis. This suggests that the research is heavily focused on the use and impact of online reviews in the hotel and tourism industry. Another prominent theme is social media, with 57 occurrences. This suggests that the research investigates the role of social media in the hotel and tourism industries. In addition, the analysis reveals that customer satisfaction is a significant topic, occurring 42 times. This highlights the importance of understanding how online reviews and social media affect customer satisfaction in the hotel and tourism industry. The themes of hotel and tourism frequently appear, with 40 and 29 occurrences, respectively, indicating that the research is specific to these sectors. In addition, the analysis reveals that TripAdvisor and the concept of user-generated content are significant topics in the research, appearing 29 and 23 times, respectively.

Overall, these findings suggest that the research is primarily focused on the use of online reviews and social media in the hotel and tourism industry and how they impact customer satisfaction.

Figure 3. Word cloud for keywords
Source: Author's elaboration.

After generating a word cloud from the keywords, a thematic map was also created from the keywords. The thematic map provides a diagram with four quadrants consisting of main themes, niche themes, emerging or declining themes, and basic themes (Raza et al., 2020). The thematic map is shown in Figure 4. Based on your thematic analysis conducted through Bibliometrix, the following themes were identified:

- ***Motor themes:*** "Sharing economy" and "Satisfaction" are considered motor themes. These are themes that are important to the subject of study and are of great significance and relevance. They are well-established topics that have been extensively researched and have a substantial impact on the field. Sharing economy, for example, is a rapidly growing trend in various industries and has been the focus of extensive research in recent years. Similarly, satisfaction is a vital idea in many fields, including tourism and hospitality, and has been extensively examined in terms of customer experience and loyalty.

- ***Niche themes:*** "Mobile apps," "Destination image," and "Emotions" are considered niche themes. These are themes that are unique to a certain subject or area of study within the field and are of moderate importance and relevance. They may not be as well-established as motor themes, but they are still important and have a significant impact on the subfield or area of study they belong to. For example, mobile apps have become an increasingly important topic in the tourism industry as the use of smartphones and mobile devices has become more widespread among tourists. Similarly, destination image and emotions have emerged as major topics in destination marketing and management.

- ***Emerging themes:*** "Sustainable tourism" and "Booking intention" are regarded as emerging themes. These are topics that have recently garnered attention but have not yet been thoroughly investigated or established as motor or specialty topics. They are believed to be gaining traction and are anticipated to grow more significant and relevant in the future. For instance, sustainable tourism is gaining importance as a growing number of tourists want to lessen their environmental effects and make more sustainable travel decisions. Similarly, booking intention is a significant topic of research in the realm of online travel as it can provide valuable insights into the factors that influence travelers' decisions to book a trip.

- ***Basic themes:*** "Electronic word of mouth" and "Online reviews" are basic themes. These are themes that are fundamental to the field of study and have been extensively researched for many years, but they are no longer considered to be at the forefront of research. They are still significant, but they have been widely researched. Thus their applicability may be confined to particular subfields or areas of study. Electronic word-of-mouth and online reviews, for instance, are important themes in the disciplines of e-commerce and digital marketing, although their applicability may be limited to subfields such as online travel.

In the third part of the keyword investigation, keyword co-occurrence analysis was conducted. As shown in Figure 5, the keyword co-occurrence analysis with 50 nodes revealed four different clusters. The first cluster (*#green*) includes the keywords social media, destination image, and content analysis. This cluster is related to the use of social media in building and managing the image of a destination, as well as the use of content analysis to understand and analyze social media data. The second cluster (*#red*) contains TripAdvisor and hotel-related keywords. This suggests that this cluster of keywords is related to the use of TripAdvisor and other online review platforms in the hotel industry. The third cluster (*#blue*) consists of keywords associated with customer satisfaction, big data, and sentiment analysis. This suggests that this cluster is related to the use of big data and sentiment analysis in understanding and improving customer satisfaction in the hotel industry. The fourth cluster (*#purple*) contains keywords related to Airbnb and the sharing economy. This group of keywords relates to the hotel industry's usage of the sharing economy, notably Airbnb and similar sites.

As a result of the keyword occurrence analysis, it can be concluded that eWOM in the tourist and hospitality industry has changed in tandem with the information created through many channels, such as social media, user-generated content, big data, and the sharing economy.

Figure 4. Thematic map
Source: Author's elaboration.

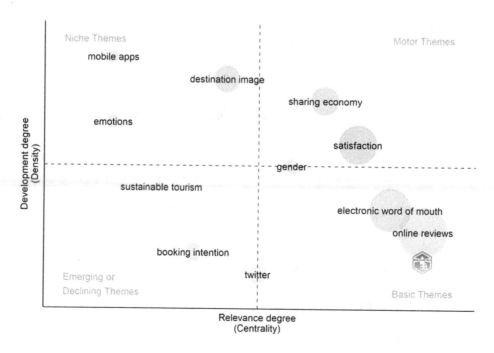

Co-Authorship Analysis

In the bibliometric analysis, the social structure of eWOM literature is also analyzed with the aid of collaboration networks. The analysis of the social structure shows how authors, institutions, and countries are related to each other and identifies the most influential authors, countries, and institutions. One of the most important social structures can be examined through the collaboration network of authors, which represents a collaboration among different authors. The collaboration network of authors is shown in Figure 6 (Palácios et al., 2021).

Figure 5. Keyword co-occurrence analysis
Source: Author's elaboration.

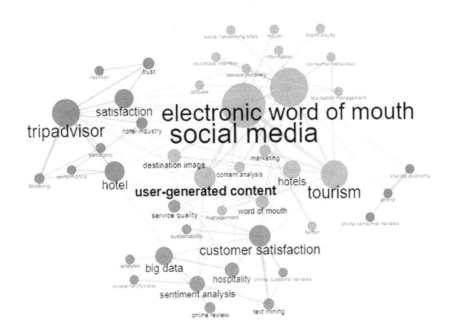

According to this, contributions to the eWOM literature in the field of tourism and hospitality have been provided by research groups consisting of between 2 and 5 researchers. For example, Figure 6 shows a five-author collaboration consisting of Law R, Zhang Z, Ye Q, Li H, and Yang Y. Furthermore, a four-person research team consisting of Stamolampros P, Buhalis D, Korfiatis N, and Viglia G also contributed to the field. In addition, three-person research groups consisting of Borghi M-Mariani MM-Mariani M, Sanz-Blas S-Buzova D-Cervera-Taulet A, and Andreu I-Ruiz C-Bigne E have also been present. Figure 6 also indicates a large number of 2-person author collaboration networks. Table 2 already shows that the co-authors per document were 2.78. The prevalence of 2 and 3-person research groups in this figure confirms this number.

Following the collaboration network of authors, the collaboration network of institutions was also obtained, as seen in Figure 7. According to this, one of the most extensive collaboration networks among institutions was observed between the College of Charleston, Florida State University, Temple University, and Pennsylvania State University. While there is also a significant collaboration between Hong Kong Polytechnic University, Sejong University, and the University of Denver, it should be noted that the "Sch management" node in this cluster is a misindexed term. Another collaboration network consisting of four institutions links Universidad Rey Juan Carlos, University of Malaga, University of Almería, and University of Algarve each other. Similarly, scientific collaboration has also been established between the institutions of University of Nevada, University of Florida, University of Mississippi and University of North Texas. Additionally, a large number of collaboration networks consisting of 2-3 institutions have contributed to the eWOM literature in the field of tourism and hospitality. However, it is also a finding that should be noted that collaboration networks are generally established at the national level and that the collaborating institutions are often from the same country.

Figure 6. Collaboration network of authors
Source: Author's elaboration.

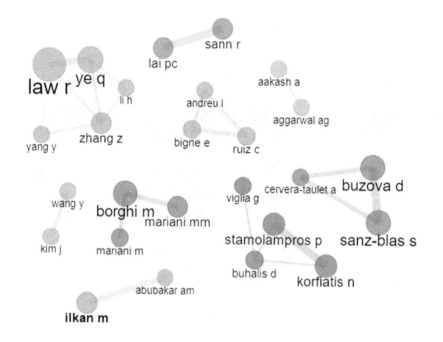

Figure 7. Collaboration network of institutions
Source: Author's elaboration.

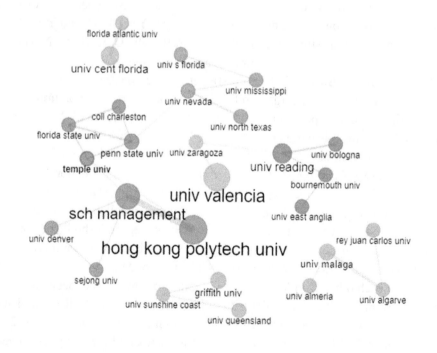

Finally, the network of country collaborations is determined based on the number of publications in co-authorship. As seen in Figure 8, researchers from the most influential countries, the USA and China, are involved in a strong scientific collaboration. Researchers from these countries frequently collaborate with those from South Korea, Denmark, Pakistan, Switzerland, and the United Arab Emirates. Additionally, the author has identified three additional collaboration clusters, represented by the blue, green, and red colors in Figure 8, comprising of 11, 10, and 4 countries, respectively. The collaborative countries within each cluster are from different geographical regions and continents. This fact highlights the internationality of the field, with researchers from various countries working together to advance knowledge in this area.

Figure 8. Collaboration network of countries
Source: Author's elaboration.

CONCLUSION

This chapter aimed to examine the evolution of the eWOM literature in the field of tourism and hospitality and to increase understanding of the theoretical and conceptual structure of the body of knowledge. To achieve this goal, a bibliometric analysis was used to systematically review the literature and identify key themes and trends using the R package Bibliometrix. Considering the drawbacks of meta-analysis and systematic literature reviews, bibliometric analysis was deemed appropriate for the research problem. Data were obtained from the Web of Science database. A total of 515 scholarly documents published between 2008 and 2021 were analyzed using the PRISMA flowchart. The results were presented through a word cloud, a thematic map, and a co-occurrence analysis, as well as a performance analysis. The findings revealed the most prevalent terms within the field, the themes that guide the existing body of knowledge, and the four distinct research themes that have emerged as particularly prominent within the field of eWOM in tourism and hospitality. The results have shown that eWOM is a dynamic and growing area of research, with a strong focus on consumer behavior, the impact of eWOM on destination image and reputation, and the role of social media and user-generated content in eWOM communication. Furthermore, the research has revealed that there is a need for more research in the field of eWOM in tourism and hospitality, which can provide a more comprehensive understanding of the theoretical and conceptual structure of eWOM research in the industry.

The current research presents the findings of an investigation into the field of eWOM within the context of the tourism and hospitality industry. According to the analysis, the first study on this issue was published in 2008. Additionally, the data shows that the yearly growth rate of publications on this topic from 2008 to 2021 is 43.15%. This high annual growth rate suggests that eWOM is a rapidly growing area of research and that there is a growing interest among scholars in studying the impact of eWOM on the tourism and hospitality industry. The examination of the literature also reveals that the most productive authors in this field are Law R, Xu X, and Zhang Z, with Law R being the most influential author in terms of total citations and h-index. In terms of institutions, the University of Valencia is the most prolific contributor, with 35 study contributions, followed by the Hong Kong Polytechnic University. In terms of countries, the USA ranks first with 334 studies, followed by China with 257 studies, and Spain with 189 studies. The most highly cited article on eWOM in the field of tourism and hospitality is "Electronic word-of-mouth in hospitality and tourism management" by Litvin et al. (2008), which has received a total of 1308 citations. A word cloud generated from the author's keywords in the bibliometric analysis indicates several key themes in the research. The most prevalent theme is online reviews, with 85 occurrences, suggesting that the research is heavily focused on the use and impact of online reviews in the hotel and tourism industry. Other prominent themes include social media and customer satisfaction. A keyword co-occurrence analysis revealed four distinct clusters, with the first cluster including keywords related to social media, destination image, and content analysis, the second cluster containing keywords related to TripAdvisor and hotels, the third cluster including keywords associated with customer satisfaction, big data, and sentiment analysis, and the fourth cluster containing keywords related to Airbnb and the sharing economy.

The theoretical contributions of this chapter can be enumerated as follows:

- This research examines the evolution of the eWOM literature in the field of tourism and hospitality,
- It contributes to a deeper comprehension of the theoretical and conceptual structure of eWOM research in the tourism and hospitality literature,
- This research applies a thorough PRISMA protocol, providing an approach for evaluating a vast number of works with analytical rigor,
- Discerns the research themes that emerged as prominent in the existing body of knowledge by using the keywords in the field.

This research is distinctive and unique due to the limited bibliometric research that covers the entire course of the available literature. Moreover, we believe that recognizing emergent themes will give scholars in the research topic a framework for future investigations.

The managerial implications of this chapter are that understanding the evolution and structure of eWOM research in the tourism and hospitality industries is essential for making strategic decisions about marketing and customer engagement strategies. The use of bibliometric analysis can provide valuable insights into the most prevalent themes and terms within the existing literature. Additionally, identifying key research themes through keywords' co-occurrence analysis can help tourism managers to better understand the current state of eWOM research and identify areas for future implementations. This information also helps managers identify promising and gaping areas to develop eWOM strategies to better engage with customers and increase customer satisfaction. The results of this research suggest that eWOM has a significant impact on consumer decisions in the tourism and hospitality industry. Therefore, professionals in this industry should pay close attention to managing and monitoring online reviews and social media presence. They should also prioritize creating a positive online reputation by encouraging satisfied customers to share their positive experiences online. Additionally, it is essential to engage with customers online promptly and effectively, addressing any negative feedback and complaints in a professional and empathetic manner. Professionals can also utilize the identified research themes to guide their efforts in researching and improving their eWOM strategies. By focusing on eWOM, hospitality industry professionals can improve their online reputation, attract more customers, and ultimately improve their business performance.

FUTURE RESEARCH DIRECTIONS

This bibliometric research identifies various avenues for future research on eWOM communication in the field of tourism and hospitality. Firstly, there is a need for further research to examine the impact of eWOM on different sections of the tourism and accommodation industry, such as airlines, hotels, and destinations. This can be achieved by analyzing the content of eWOM messages and comparing them in different segments to determine the main themes and trends. Secondly, research is also needed to investigate the impact of eWOM on consumer behavior in the tourism and hospitality industry. This can be done through surveys or experiments to measure the impact of eWOM on consumer attitudes and purchasing decisions.

Thirdly, future research can employ advanced analytical methods such as text mining and sentiment analysis to better understand the content and sensitivity of eWOM messages. Fourthly, it is important to research the role of social media platforms and online review sites in eWOM communication in the tourism and hospitality industry. This may involve analyzing the types of platforms used, the frequency and content of messages, and their impact on consumer behavior. Finally, it is crucial to conduct a comparative analysis between different databases in future studies. This analysis can be done by using different databases such as Scopus, Google Scholar, and ProQuest to compare the results of bibliometric analysis of eWOM in the tourism and hospitality industry.

REFERENCES

Aria, M., & Cuccurullo, C. (2017). bibliometrix: An R-tool for comprehensive science mapping analysis. *Journal of Informetrics, 11*(4), 959–975. doi:10.1016/j.joi.2017.08.007

Bakır, M., Özdemir, E., Akan, Ş., & Atalık, Ö. (2022). A bibliometric analysis of airport service quality. *Journal of Air Transport Management, 104*, 102273. doi:10.1016/j.jairtraman.2022.102273

Bhaiswar, R., Meenakshi, N., & Chawla, D. (2021). Evolution of Electronic Word of Mouth: A Systematic Literature Review Using Bibliometric Analysis of 20 Years (2000–2020). *FIIB Business Review*, *10*(3), 215–231. doi:10.1177/23197145211032408

Bronner, F., & de Hoog, R. (2011). Vacationers and eWOM: Who posts, and why, where, and what? *Journal of Travel Research*, *50*(1), 15–26. doi:10.1177/0047287509355324

Chen, Y. F., & Law, R. (2016). A Review of Research on Electronic Word-of-Mouth in Hospitality and Tourism Management. *International Journal of Hospitality & Tourism Administration*, *17*(4), 347–372. doi:10.1080/15256480.2016.1226150

Cheung, C. M. K., & Thadani, D. R. (2012). The impact of electronic word-of-mouth communication: A literature analysis and integrative model. *Decision Support Systems*, *54*(1), 461–470. doi:10.1016/j.dss.2012.06.008

Cobo, M. J., Martínez, M. A., Gutiérrez-Salcedo, M., Fujita, H., & Herrera-Viedma, E. (2015). 25 years at Knowledge-Based Systems: A bibliometric analysis. *Knowledge-Based Systems*, *80*, 3–13. doi:10.1016/j.knosys.2014.12.035

Derudder, B., Liu, X., Hong, S., Ruan, S., Wang, Y., & Witlox, F. (2019). The shifting position of the Journal of Transport Geography in 'transport geography research': A bibliometric analysis. *Journal of Transport Geography*, *81*, 1–9. doi:10.1016/j.jtrangeo.2019.102538

Donthu, N., Kumar, S., Mukherjee, D., Pandey, N., & Lim, W. M. (2021). How to conduct a bibliometric analysis: An overview and guidelines. *Journal of Business Research*, *133*, 285–296. doi:10.1016/j.jbusres.2021.04.070

Donthu, N., Kumar, S., Pandey, N., Pandey, N., & Mishra, A. (2021). Mapping the electronic word-of-mouth (eWOM) research: A systematic review and bibliometric analysis. *Journal of Business Research*, *135*, 758–773. doi:10.1016/j.jbusres.2021.07.015

Ert, E., Fleischer, A., & Magen, N. (2016). Trust and reputation in the sharing economy: The role of personal photos in Airbnb. *Tourism Management*, *55*, 62–73. doi:10.1016/j.tourman.2016.01.013

Filieri, R., & McLeay, F. (2014). E-WOM and Accommodation: An Analysis of the Factors That Influence Travelers' Adoption of Information from Online Reviews. *Journal of Travel Research*, *53*(1), 44–57. doi:10.1177/0047287513481274

Gölgeci, I., Ali, I., Ritala, P., & Arslan, A. (2022). A bibliometric review of service ecosystems research: Current status and future directions. *Journal of Business and Industrial Marketing*, *37*(4), 841–858. doi:10.1108/JBIM-07-2020-0335

Gretzel, U., & Yoo, K. H. (2008). Use and Impact of Online Travel Reviews. In P. O'Connor, W. Höpken, & U. Gretzel (Eds.), *Information and Communication Technologies in Tourism 2008: Proceedings of the International Conference in Innsbruck,* (pp. 35–46). Springer. 10.1007/978-3-211-77280-5_4

Hashemi, H., Rajabi, R., & Brashear-Alejandro, T. G. (2022). COVID-19 research in management: An updated bibliometric analysis. *Journal of Business Research*, *149*, 795–810. doi:10.1016/j.jbusres.2022.05.082 PMID:35669095

Hennig-Thurau, T., Gwinner, K. P., Walsh, G., & Gremler, D. D. (2004). Electronic word-of-mouth via consumer-opinion platforms: What motivates consumers to articulate themselves on the Internet? *Journal of Interactive Marketing, 18*(1), 38–52. doi:10.1002/dir.10073

Hlee, S., Lee, H., & Koo, C. (2018). Hospitality and tourism online review research: A systematic analysis and heuristic-systematic model. *Sustainability (Basel), 10*(4), 1–27. doi:10.3390u10041141

Huang, C. Y., Chou, C. J., & Lin, P. C. (2010). Involvement theory in constructing bloggers' intention to purchase travel products. *Tourism Management, 31*(4), 513–526. doi:10.1016/j.tourman.2009.06.003

Huete-Alcocer, N. (2017). A literature review of word of mouth and electronic word of mouth: Implications for consumer behavior. *Frontiers in Physiology, 8*, 1–4. doi:10.3389/fpsyg.2017.01256 PMID:28790950

Jalilvand, M. R., & Samiei, N. (2012). The impact of electronic word of mouth on a tourism destination choice: Testing the theory of planned behavior (TPB). *Internet Research, 22*(5), 591–612. doi:10.1108/10662241211271563

Katz, E., & Lazarsfeld, P. F. (1966). *Personal Influence: The Part Played by People in the Flow of Mass Communications*. Transaction Publishers.

Kim, E. E. K., Mattila, A. S., & Baloglu, S. (2011). Effects of gender and expertise on consumers' motivation to read online hotel reviews. *Cornell Hospitality Quarterly, 52*(4), 399–406. doi:10.1177/1938965510394357

Kim, H., & So, K. K. F. (2022). Two decades of customer experience research in hospitality and tourism: A bibliometric analysis and thematic content analysis. *International Journal of Hospitality Management, 100*, 103082. doi:10.1016/j.ijhm.2021.103082

Li, X., Ma, E., & Qu, H. (2017). Knowledge mapping of hospitality research – A visual analysis using CiteSpace. *International Journal of Hospitality Management, 60*, 77–93. doi:10.1016/j.ijhm.2016.10.006

Litvin, S. W., Goldsmith, R. E., & Pan, B. (2008). Electronic word-of-mouth in hospitality and tourism management. *Tourism Management, 29*(3), 458–468. doi:10.1016/j.tourman.2007.05.011

Litvin, S. W., Goldsmith, R. E., & Pan, B. (2018). A retrospective view of electronic word-of-mouth in hospitality and tourism management. *International Journal of Contemporary Hospitality Management, 30*(1), 313–325. doi:10.1108/IJCHM-08-2016-0461

Moher, D., Liberati, A., Tetzlaff, J., & Altman, D. G. (2009). Preferred Reporting Items for Systematic Reviews and Meta-Analyses: The PRISMA Statement. *PLoS Medicine, 6*(7), e1000097. doi:10.1371/journal.pmed.1000097 PMID:19621072

Mukhopadhyay, S., Pandey, R., & Rishi, B. (2022). Electronic word of mouth (eWOM) research – a comparative bibliometric analysis and future research insight. *Journal of Hospitality and Tourism Insights*. doi:10.1108/JHTI-07-2021-0174

Mulet-Forteza, C., Martorell-Cunill, O., Merigó, J. M., Genovart-Balaguer, J., & Mauleon-Mendez, E. (2018). Twenty five years of the Journal of Travel & Tourism Marketing: A bibliometric ranking. *Journal of Travel & Tourism Marketing, 35*(9), 1201–1221. doi:10.1080/10548408.2018.1487368

Munar, A. M., & Jacobsen, J. K. S. (2014). Motivations for sharing tourism experiences through social media. *Tourism Management, 43*, 46–54. doi:10.1016/j.tourman.2014.01.012

Nam, K., Baker, J., Ahmad, N., & Goo, J. (2020). Determinants of writing positive and negative electronic word-of-mouth: Empirical evidence for two types of expectation confirmation. *Decision Support Systems, 129*, 113168. doi:10.1016/j.dss.2019.113168

Nieto, J., Hernández-Maestro, R. M., & Muñoz-Gallego, P. A. (2014). Marketing decisions, customer reviews, and business performance: The use of the Toprural website by Spanish rural lodging establishments. *Tourism Management, 45*, 115–123. doi:10.1016/j.tourman.2014.03.009

Palácios, H., de Almeida, M. H., & Sousa, M. J. (2021). A bibliometric analysis of trust in the field of hospitality and tourism. *International Journal of Hospitality Management, 95*, 102944. doi:10.1016/j.ijhm.2021.102944

Pourfakhimi, S., Duncan, T., & Coetzee, W. J. L. (2020). Electronic word of mouth in tourism and hospitality consumer behaviour: State of the art. *Tourism Review, 75*(4), 637–661. doi:10.1108/TR-01-2019-0019

Ran, L., Zhenpeng, L., Bilgihan, A., & Okumus, F. (2021). Marketing China to U.S. travelers through electronic word-of-mouth and destination image: Taking Beijing as an example. *Journal of Vacation Marketing, 27*(3), 267–286. doi:10.1177/1356766720987869

Raza, S. A., Ashrafi, R., & Akgunduz, A. (2020). A bibliometric analysis of revenue management in airline industry. *Journal of Revenue and Pricing Management, 19*(6), 436–465. doi:10.105741272-020-00247-1

Serra Cantallops, A., & Salvi, F. (2014). New consumer behavior: A review of research on eWOM and hotels. *International Journal of Hospitality Management, 36*, 41–51. doi:10.1016/j.ijhm.2013.08.007

Sotiriadis, M. D. (2017). Sharing tourism experiences in social media: A literature review and a set of suggested business strategies. *International Journal of Contemporary Hospitality Management, 29*(1), 179–225. doi:10.1108/IJCHM-05-2016-0300

Sotiriadis, M. D., & van Zyl, C. (2013). Electronic word-of-mouth and online reviews in tourism services: The use of twitter by tourists. *Electronic Commerce Research, 13*(1), 103–124. doi:10.100710660-013-9108-1

Sparks, B. A., & Browning, V. (2011). The impact of online reviews on hotel booking intentions and perception of trust. *Tourism Management, 32*(6), 1310–1323. doi:10.1016/j.tourman.2010.12.011

Tanrıverdi, G., & Durak, M. Ş. (2022). A visualized bibliometric analysis of mapping research trends of airline business models (ABMs) from 1985 to 2021. *Journal of Aviation, 6*(3), 387–403. doi:10.30518/jav.1172121

Vermeulen, I. E., & Seegers, D. (2009). Tried and tested: The impact of online hotel reviews on consumer consideration. *Tourism Management, 30*(1), 123–127. doi:10.1016/j.tourman.2008.04.008

Westbrook, R. A. (1987). Product/Consumption-Based Affective Responses and Postpurchase Processes. *JMR, Journal of Marketing Research, 24*(3), 258–270. doi:10.1177/002224378702400302

Ye, Q., Law, R., & Gu, B. (2009). The impact of online user reviews on hotel room sales. *International Journal of Hospitality Management, 28*(1), 180–182. doi:10.1016/j.ijhm.2008.06.011

Ye, Q., Law, R., Gu, B., & Chen, W. (2011). The influence of user-generated content on traveler behavior: An empirical investigation on the effects of e-word-of-mouth to hotel online bookings. *Computers in Human Behavior, 27*(2), 634–639. doi:10.1016/j.chb.2010.04.014

Zhang, Z., Ye, Q., Law, R., & Li, Y. (2010). The impact of e-word-of-mouth on the online popularity of restaurants: A comparison of consumer reviews and editor reviews. *International Journal of Hospitality Management, 29*(4), 694–700. doi:10.1016/j.ijhm.2010.02.002

Zhao, X., Ke, Y., Zuo, J., Xiong, W., & Wu, P. (2020). Evaluation of sustainable transport research in 2000–2019. *Journal of Cleaner Production, 256*, 120404. doi:10.1016/j.jclepro.2020.120404

KEY TERMS AND DEFINITIONS

Bibliometric Analysis: Bibliometrics is the quantitative evaluation of published bibliographic material.

Collaboration Index: It is a metric that expresses the ratio of the total number of authors in multi-authored articles to the total number of multi-authored articles.

Co-Occurrence Network: A co-occurrence network is a graphical representation of how frequently units such as authors, institutions, and countries appear together.

Impact Factor: It refers to a metric that expresses the frequency of citation of an article published in a specific period.

Performance Analysis: It is an analysis that descriptively shows the contribution of bibliometric research units.

PRISMA: A systematic framework that allows for the synthesis of literature in retrospective analyses such as systematic literature reviews, meta-analyses, and bibliometric analyses.

Science Mapping: Science mapping is a technique that reveals deep relationships between keywords and bibliometric research units.

Thematic Map: It uses keywords to show current research themes on a diagram with four quadrants.

Chapter 15
Model for Action
and Cooperation on
Accessible Tourism:
ACT–MODEL

Borbála Gondos
Edutus University, Hungary

Márta Nárai
Széchenyi István University, Hungary

ABSTRACT

The study highlights the importance of accessible tourism, presents the actors of accessible tourism, including people with reduced mobility, and discusses their opportunities in tourism. The aim of the authors is to describe the ACT model and demonstrate its use in tourism. Creating an accessible environment provides comfort not only for people with reduced mobility but for almost everyone, so it has greater potential even from an economic point of view. The ACT model was born from the information obtained during the research in Hungary and summarizes and illustrates well what actors are involved in the topic of accessible tourism and what kind of relationship exists between them. Each actor and factor of the three levels of the theoretical model are important in order to ensure that accessibility is achieved, and that disabled people can participate in tourism in the same way as non-disabled people.

INTRODUCTION

In Hungary, the examination of the travel habits, roles and possibilities of people with disabilities, including those with reduced mobility, as well as the inherent potential in tourism, is a less researched area recently. In Hungary, until the change of regime, the existence and social presence of people with disabilities was a taboo subject. A real change of attitude took place from the second half of the nineties and the 2000s, (Nárai, 2013), so we cannot be surprised seeing the way most people relate to them even today, many are afraid of them, they avoid them because they are "different".

DOI: 10.4018/978-1-6684-6985-9.ch015

One of the goals of the doctoral dissertation (Gondos, 2020)[1], which forms the basis of the study, is to draw attention to this target group being equally valuable members of society with the same needs and rights as anyone else, such as for meaningful leisure time, gaining experiences, the right to travel should also be fundamental in their case. The segment of people with disabilities is very broad including people with reduced mobility, the blind and partially sighted, the deaf, the hard of hearing ones, the speech-impaired, and the mentally disabled. We did our research in 2018 and 2019 which was narrowed down to one target group, so a questionnaire survey was conducted among people over the age of 18 with reduced mobility because in our opinion this is one of the target groups easier to be reached and can be asked in independently used questionnaires. Appropriate infrastructure for people with reduced mobility such as ramps, elevators, and low-floor means of transportation, is essential for them to leave their homes and travel. However, this infrastructure can also be useful for a wider target group of people, such as elderly people and families with small children using strollers, because they require the same facilities, so the target group of accessible tourism can be much larger (UNWTO, 2016b). According to the data sources of the UN health organization, the World Health Organization (WHO, 2011), about 15% of the world's population, which means more than 1 billion people living with some kind of disability. In the forthcoming years, their number will be continuously increasing, also thanks to the growing average life expectancy (as we will see later, the proportion of people over 60, shows a constantly rising trend, they are sometimes included in the group of people with disabilities due to their deteriorating health or limited mobility).

In the recent decades it has become clear that handling the case of people with disabilities at both European and national levels, is not only a social or health issue but a human rights issue, too.[2] The essence of the change in attitude and paradigm is the fact that people, living with various disabilities, must be provided with the same chances and rights as non-disabled people. The opportunity to participate in everyday life and various services must be made available to them, including those that contribute to meaningful leisure time thus improving their quality of life (UN, 2006). Tourism is an activity through which visitors can gain experiences and become active participants in their experiences. This kind of experience can increase during multiple trips it can contribute to life satisfaction thus according to Veenhoven (2003) travel can lead to an awareness of happiness. Many people undertake to break away from their usual environment in order to be able to spend their free time properly, recharge mentally and physically and gain experiences that can become a source of their happiness and contribute to favorable indicators of the quality of life. Travel therefore has significant physiological effects, both in connection with reflection and consumption of various tourist products (Michalkó, 2010). Thanks to the aforementioned change in the attitude of people with disabilities have been receiving more and more attention in the field of tourism in recent years. The World Tourism Organization (UNWTO) announced themes in 2014 (community building) and 2016 (tourism for all) that focused on people with disabilities in connection with World Tourism Day and numerous conferences.

We can see and feel from above that the issue of accessible tourism is current interesting and relevant topic. It is important from several points of view to draw attention to the topic and to present research results that advance the matter and provide solutions to the problems raised. Regarding it, in this chapter the conceptual frameworks of people with disabilities/persons with reduced mobility will be presented, it will also cover the accessibility and accessible tourism, the theoretical model developed as a result of the information obtained from the authors' previous research results, the measure and cooperation model of accessible tourism, the ACT-model.

The theoretical model outlines the connection points and cooperation possibilities of the stakeholders (demand and supply side, as well as decision-makers). The development of the model was preceded by primary research during which a questionnaire was conducted among tourism service providers and people with reduced mobility as well as interviews with leaders of associations, organizations and federations, as well as specialists who are relevant to the topic.

During the investigation of the tourism supply side the respondents were the hotels and restaurants of Hungarian Hotel and Restaurant Association (MSZÉSZ), while in the case of the tourism demand side, the research was primarily conducted among the members of National Federation of Disabled Person's Associations (MEOSZ).

CONCEPTUAL FRAMEWORKS

Definition of World Health Organization in 1980 [WHO (1980)] defines disability as "an altered, reduced ability to perform certain activities of man: transportation, eating, drinking, washing, working, writing and so on" (Kálmán & Könczei, 2002, p. 81).

In 1997, the WHO (WHO, 2001) made a new definition: "disability is a comprehensive concept that includes disability, activity limitations and participation limitations, pointing to the individual and (social and personal) context" (WHO, 2001).

The UN Convention on the Rights of People with disabilities, entered into force on 2nd May 2008, defines the following: "A person with a disability is any person who has a long-standing physical, mental, mental or sensory impairment which may limit a person's full, effective and equal participation in society" (ENSZ, 2006).

The definition of disability is partially different in some member states of the European Union, Denmark does not even have one. According to the EU professional policy (EC, 2004), disabled persons include all those

Who need support, care or specific treatment in order to develop or regain their professional abilities, and receive help. This includes physical, mental, emotional and social disabilities. In the Finnish interpretation (Act on the Disabled, 1987), disabled persons are those who, due to their disability or illness, are unable to perform normal daily functions for a long time. For social security institutions, a person is disabled whose ability to work and earn has been decisively reduced. According to the Spanish, disabled are all those who, due to a congenital or acquired permanent limitation of their physical, mental or sensory abilities, can only participate to a limited extent in training, professional life and the life of society in general. (Laki, 2013, pp. 80-81)

Currently in Hungary, in accordance with the provisions of the Convention on the Rights of Persons with Disabilities (UN 2006), Act XXVI of 1998 on the Rights of Persons with Disabilities and the Commissioner for their Equal Opportunities. Act (Section 4 a)), the following definition applies to definitions of persons with disabilities:

Disabled person: a person who permanently or irreversibly has a sensory, communication, physical, intellectual, psychosocial impairment - or any combination of these - lives which, in interaction with environmental, social and other significant obstacles, limits or prevents effective and equal social participation with others.

Among the disabled, one group and it is also the target group of our research in 2018 and 2019, which helps to create the theoretical model. They are the disabled; the definition of mobility limitation is as follows:

Permanent damage to the active (muscle and nervous system) and passive (joint and bone system) organ systems of movement can lead to limitations in the ability to move and manipulate, and to the development of mobility limitations. (Bujdosó & Kemény, 2009, p. 36).

According to another source,

In the case of mobility impairment, congenital or acquired damage to the locomotor system permanently hinders the functional abilities of the body and the activity of the individual; it can cause a disadvantageous situation for the individual and a limited way of life. In the sense of special pedagogy, those persons are considered to be mobility-restricted who, due to congenital or acquired damage and/or dysfunction of the locomotor system (suspension, bone, joint system and/or motor-muscle-nerve system), have such a significant and permanent mobility impairment that changes the acquisition of movement experience, and socialization is only possible under difficult circumstances.[3]

For the integration of people with disabilities and their participation in tourism it is necessary to make various places, locations and services accessible. The argument against accessibility is usually that it affects only a few people, so it is not worth dealing with this issue. As we will see later, it is not only people with different disabilities that need to be considered, but also the elderly and families with small children pushing strollers because they need the same infrastructure. Due to illnesses or accidents, it may be necessary for us from time to time to be able to use the transport network unhindered, for example or in the case of administration we ask for the help of an escort.

The concepts of accessibility and accessible are related and primarily refer to the built environment. In Hungary, the requirement of accessibility access is defined by "Law LXXVIII of 1997 on the shaping and protection of the built environment law ", on the basis of which "Accessible: the built environment if its comfortable, safe, independent use is ensured for all people, including those individuals or groups of people who need special devices or technical solutions for this." (2.§ 1.)[4] The most important of these possibilities is that people with disabilities can use the built environment comfortably, safely and independently.

Regarding physical accessibility, the needs of people with reduced mobility are different from their peers with other disabilities. A blind and partially sighted person needs to have signs on the sidewalk (e.g. a guide lane), a signal light to emit some kind of sound, a menu to be read in Braille, for example, or make a map or attraction tangible. In the case of deaf and hard-of-hearing people, the visual information transfer can draw their attention with decisive, strong, flashing lights, or transfer the necessary information with the help of an induction loop/amplifier. People with disabilities, including those with reduced mobility, need an environment that they can use according to their condition, so accessibility is essential for them. Accessibility, "accessible, accessible environment, in particular is a basic condition for the usability of the environment and the chance of rehabilitation" (Bujdosó & Kemény, 2009, p. 90). "Without it, the person with limited mobility is in an insoluble and vulnerable situation in work, study, and in all areas of everyday life, dependent on others." (Bujdosó & Kemény, 2009, p. 90) In addition to physical accessibility, there is also the need for accessibility in terms of access to information and communication, which makes communication easier. The use of the concept of complex accessibility is widespread in Hungarian practice (Vid, 2017), which means accessibility according to several conditions and better meets the requirement of equal access. The term "well indicates that accessibility should be carried out according to the needs of several aspects and categories of disabilities" (Vid, 2017, p. 12).

Ensuring equal access already indicates and represents a new approach, as it already shows on a conceptual level that instead of eliminating obstacles, "the goal is for everyone to have the same chance to access what they need" (Vid, 2017, p. 12), be it a built environment element, service, website, form, whatever. In order to achieve this, the concept of universal design is more and more prominent these days, instead of subsequent accessibility.

By universal design we mean the design of products, the environment, programs and services in such a way that they are accessible to all people to the greatest extent possible, without the need for adaptation or special design. Universal planning cannot exclude the use of support and assistance tools and technologies necessary for groups of people with disabilities in justified cases.[5]

The great innovation of the approach is that the facilities, products, and services must be designed in the way that everyone can use them as much as possible.

At the same time, emphasizing overcoming the obstacles in 'head' in connection with accessible access and equal access, i.e. forming attitudes, sensitizing and changing attitudes, which is necessary in order to be fundamental both at the level of society as a whole and with regard to the agents providing the various services acceptance that people with disabilities are members of society with the same rights and similar demands, needs and desires as their non-disabled peers. "Accessibility in the mind" is necessary for all members of the society, but for this contact and communication with people, with disabilities, must be made part of childhood socialization, for example, so-called within the framework of chance classes, the kindergarten/school age group can/should be given the opportunity to meet people living with different disabilities, where they can see their everyday lives. In summary, providing accessibility and equal access means the following:

- Means more comfort, better quality, greater security and flexibility when using a product, building or service,
- Increases the range of users, ensures growth and employment,
- Creates a better quality of life for all of us,
- Industries dealing with leisure and tourism can take advantage of the opportunities if accessibility is interpreted in a way that covers everything and ensures quality for all guests, and develops and sells corresponding products (Polgár, n.d.).

CHANCES AND OPPORTUNITIES FOR PEOPLE WITH DISABILITIES IN THE TOURISM MARKET

Examining the supply and demand side is essential in order to create a supply and demand meeting that is suitable for both parties. People with disabilities have to deal with a lot of things, whether it be tasks related to their everyday life, but this is even more true if they want to travel to another settlement, region, or country. In recent years, progress has been going on in Hungary in terms of accessibility, development, and ensuring equal access, but the country is still far behind at the EU level. In order to have an attitude at the level of Western European or Northern European countries, in terms of social responsibility and infrastructure, it will take quite a few years, this was confirmed by in interviewees and partly by the results of questionnaire survey (Gondos, 2020).

In 2020 there was a research (Medarić et al., 2021) in connection with Lake Balaton to view their accessibility from the point of stakeholders' perspectives. 39 stakeholders were interviewed, including local governments, local destination management organizations and tourism service providers (accommodation, hospitality or attractions/sightseeing). In the Lake Balaton destination, tourism providers tend to target groups with low accessibility needs. The results show that public and private tourism providers do not use accessibility, and the very limited financial resources spent on accessibility issues. There are still negative attitudes in the tourism market that hinder the accessibility development path of the destination (Medarić et al., 2021). We had the same result in the former research (Gondos, 2020).

The disabled person as a tourist has only recently been on the agenda in the tourism industry. There are several reasons why the tourism industry should pay more attention to this growing group of potential tourists. The proportion of people with disabilities continues to increase and spend more money than is generally assumed (Ray & Ryder, 2003). In addition, guests with disabilities are loyal, spend longer than average in one place and spend a lot of money (Wagner 2010) if they find an accommodation or destination where they can enjoy the joys and experiences of travel despite their special needs.

People with disabilities have the same tourism needs and desires as everyone else, they also want to travel, learn new things, recharge, and gain experiences. However, since tourist attractions and activities are primarily designed for non-disabled people, this presents unique challenges (Yau et al., 2004). Tourism is an activity in which participation is not given to many people with disabilities, as it requires the coordinated cooperation of physical, mental and social domains, which are often adversely affected or threatened by disability (Yau et al., 2004). In addition, the main barriers to travel for people with reduced mobility include lack of access to physical infrastructure, accessible accommodations, and access to destinations and attractions. Another problem is the lack of reliable and accurate information on accessible conditions and service providers; being up-to-date and the availability of information is decisive during the trip and its planning (Gondos, 2020). In the absence of reliable information about accessibility, a significant number of disabled people do not travel due to the fear of disappointment (Wagner, 2010). The consequence of these limitations and shortcomings is the loss of tourist experiences, not experiencing them (Darcy, 1998).

Several authors note that providing tourism experiences for people with disabilities is more than an accessibility issue (Shelton & Tucker, 2005; Stumbo & Pegg, 2005; Yau et al., 2004), it is also an equal opportunity issue. At the same time, an accessible destination (Israeli, 2002) and suitable accommodation are the basis of tourism experiences for people with disabilities, including those with reduced mobility. In order to spend even one night away from their permanent place of residence, in order to be able to use the offer of tourism service providers, it is particularly important for them to have the proper design of the accommodation, without which, for example, they cannot use the bedroom and the bathroom (Darcy, 2007). According to the research, carried out by the Hungarian Tourist Agency in 2018 on the topic of senior and accessible tourism, the following should be taken into account in order to achieve physical accessible access: when designing parking spaces, it is ideal if they are as close as possible to the entrance of the accommodation, and within a hotel it is important that the different areas (restaurant, wellness, etc.) can be easily approached. In the wellness department, non-slip flooring, railings, handrails are required, and there must be a lifting structure into the pool. Wider doorways, more spacious spaces, rooms, equipment or the location of the switches are also important aspects, as well as the design of the special bathroom, where comfortable access and spaciousness are important, the door should open outwards, and have a handrail, a shower with a seat, and a special washbasin. Inside the rooms, it is also important that the bed is raised, that the floor is non-slip, and that there are no thresholds (MTÜ, 2019).

The following figure clearly illustrates the obstacles that people with disabilities encounter in everyday life and during their travels (Figure 1). These obstacles can be not only physical obstacles (architectural, environmental), but also very diverse, such as internal obstacles, interactive obstacles (social attitude, lack of accessibility of services, culture), lack of information and marketing, contradictory definitions of accessibility, regulations (standards, regulations), economic obstacles (lack of adequate income), social exclusion, discrimination (Figure 1). These are the factors that contribute to experiencing a deviation from the daily life of a non-injured person. Among these, there are and can be differences between the needs in several respects, because certain elements are affected in the same way by exclusion, discrimination, and economic obstacles. In order for a trip to take place, the supply side must, have suitable attractions, infrastructure[6] and superstructure[7], so the road network, the existence of means of transport, accessible parking, the creation of transport services and the creation of wide transport spaces contribute to the accessibility of a given destination area (Fekete, 2006), but they are not sufficient. The existence and accessibility of accessible information, the creation of accessible hiking trails (leisure, sports opportunities), the creation of accessible offers and products (entertainment, culture opportunities), ensuring equal access to individual facilities, services, and programs is essential and serves to complete the visitor experience.

In order for these to be realized, tourism providers also need help, information and support, for example different trainings, guidelines and recommendations can help them to be informed about tourists with special needs (Suárez-Henríquez et al, 2022). Suárez-Henríquez and his co-authors contribute to this with their study published in 2022, in which they examined the progress of scientific research related to accessible tourism by means of a bibliometric analysis of scientific articles published in the Scopus database between 1997 and 2021. The processed 254 articles provide information on the experiences of 52 countries, the bibliometric analysis can shed light on trends related to accessible tourism, future development opportunities, and can also provide insight into the mindset of the players in this segment of the tourism market (visitors with special needs, service providers, planning agencies, etc.).

There are several best practice in the tourism market. In Spain, the first travel agency was established, which specializes in people with disabilities, and Eurotaxi, a PPP initiative, undertakes the door-to-door transport of wheelchair users. In France, the beaches were redesigned to be accessible to everyone, just like in Malaga, where the "Enjoy the beach" project aimed to make the beaches accessible (UNWTO, 2016a). Since 2008, this latter project has won several awards in the field of accessibility (UNWTO, 2015). The publication "The Finished City" was produced in Geneva, which is not only aimed at people with reduced mobility, but also at elderly people and the visually impaired (thus treating several target groups together, who are also the target groups of accessible tourism), and its railway station is completely accessible (as is Zurich's) and Switzerland most of its museums as well (UNWTO, 2015). Scandic Hotels are also a good example because they operate according to accessible regulations, which is the standard of the hotel chain's own concept and Scandic is the first hotel to create allergy-friendly room standards (Scandic Press, 2019). In Spain, Ilunion Hotels provide their facilities and equipment with a high level of accessibility and provide specific training and qualifications for their employees, many of whom are some degree of disability themselves (NTG, 2020).

Figure 1. Categories of barriers
Source: European Commission (2014, p. 308)

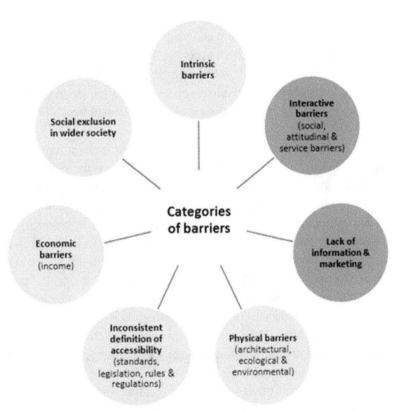

Characteristics, Conditions, and Target Groups of Accessible Tourism

The analysis of accessible tourism is essential as a result of accessibility and obstacles arising during travel, its concept was defined as follows:

Accessible tourism is a continuous effort aimed at making all destinations, tourism products and services accessible to all people, regardless of any physical limitation, disability or age. This applies to both privately and publicly owned tourist sites. (Polgár, n.d.).

Buhalis and Darcy (2011) define accessible tourism as a form of tourism that includes collaborative processes between stakeholders that enable people with disabilities to use tourism products and services, and which are taken into account during their design the universal design. This definition takes a life-long approach where people can enjoy accessible tourism services throughout their lives. These include people with permanent and temporary disabilities, the elderly, the obese, families with small children, and people with disabilities working in sheltered workplaces (Buhalis & Darcy, 2011). As an extension of accessible tourism, some authors use the concept of inclusive tourism, which means tourism in which vulnerable and/or marginalized groups participate (Scheyvens & Biddulph, 2018).

According to ENAT[8], accessible tourism includes:

- The barrier-free destination: infrastructure and opportunities
- Transport: by air, land, sea, suitable for all users
- High-quality services, including qualified and trained personnel
- Activities, exhibitions, attractions: it enables everyone to participate in tourism
- Marketing, reservation systems, websites and services: information available to all (Souca 2010).

Nowadays, the conceptual boundaries of European accessible tourism have expanded, and society as a whole is affected by it. The number of people with disabilities is increasing, every fifth European citizen is affected at some level (Zsarnóczky, 2017a). Accessible tourism is a complex service area (and the target group of accessible tourism is a niche market), the operation and operation of which must meet high standards and serious expectations of those participating in it, requiring great expertise and routine. According to the recommendations published by the European Union in 2015, the demand for accessible tourism is not only and exclusively made up of people with disabilities, but also the senior age group, families with young children and those looking for safe tourism, i.e. target groups that are particularly interested in sustainable and quality services (Zsarnóczky, 2017a). Thus, disabled tourists include many more people. An important question for tourism is whether someone has a congenital or acquired disability, because the forms of disability can be very different and may not be immediately apparent to everyone. Since the risk of developing a disability is higher among elderly people, this should also be taken into account in the future (Zsarnóczky, 2017b). The elderly, so-called the factor that most strongly generates the participation of the senior age group in tourism is the improvement and preservation of their health and the feminization characteristic of aging is also an important trend (MTÜ, 2019).

Accessibility is an essential part of the integration of people with disabilities. An accessible environment is the first step towards realizing the right to participate in all areas of life for persons with disabilities. Accessibility is a very broad concept that includes ensuring that people with disabilities can participate in activities related to their daily life, but also in travel, tourism, and have the same choices as people without disabilities. Many environments, spaces, products, etc. they develop without taking into account people with disabilities (including older people) and often assume that such groups are either irrelevant or unimportant to planning. As a result, the environment and products become inaccessible to a special population (but sometimes to a large proportion of the population as a whole).

Nowadays, tourists are looking for authentic experiences from the places they visit. In countries where the historical environment is decisive for tourism, maintaining authenticity is essential in order to encourage potential visitors. Historical buildings, such as castles, were built at a time when it was not even thought that they would later be used by people with disabilities. Although all people may desire to participate in tourism, and similarly, there is a growing interest in authentic experiences, historic sites can only be enjoyed by all interested visitors if they are accessible to all (Barlow, 2012). The issue of accessibility is perhaps most easily identified with the needs and abilities of people with disabilities. This group receives relatively little attention in the literature regarding their unique needs and abilities as consumers (Kaufman-Scarborough, 2001). Accessible infrastructure creates value for owners, as a building that meets accessibility requirements will easily be able to adapt to changing needs, including the aging or emerging disabilities of occupants. The proper design of accommodation and services can be more expensive, however, the offer designed for people with disabilities, such as the rooms, "are also gladly used by the elderly and those with less good health, as they are much more comfortable than average, so there is no need to worry about them remaining unused" (Wagner, 2010). But we can also agree on the issue that since many people do not consider disability-related issues to be important, it is rather difficult to get financial support due to this attitude of society. According to Imrie and Hall (2001), the built environment of many countries remains largely inaccessible, which can be attributed to the general reason that statutory and legal provisions for the construction of accessible environments are weak or absent in most countries. Tourism for persons with disabilities is not only about removing physical barriers (Yau et al., 2004), but must provide an experience to ensure and improve their quality of life. Accessible tourism promotes human rights and equality by paying more attention to the needs, wants and requests of tourists with disabilities and recognizing that disabled people have no different tourism needs and desires than able-bodied people, and this is accessible tourism leads to the concept (Yau et al., 2004). The inclusion of persons with disabilities in tourism activities not only means income, but also a legal obligation (Takeda & Card, 2002).

Accessible tourism is a specific activity in which anyone can participate, ensuring full social accessibility of the tourist area (Darcy & Dickson, 2009). During the processes of accessible tourism, the goal for all participants, regardless of their form of disability is to be able to reach any tourist destination and use any tourist service there. During accessible tourism, it is advisable to implement the principle of complete accessibility by applying the concept of universal planning (Buhalis et al., 2012). Specialists dealing with the topic distinguish between congenital disability (caused by genes, created due to changes occurring during prenatal development or at the moment of birth), and acquired disability, i.e. a condition that occurs during life after birth as a result of some external cause. It is important to separate the two types because there may be differences in demand that greatly affect the consumption processes, and rehabilitation may become a prominent area of connectivity in accessible tourism (Packer et al., 2009). In order to satisfy the special needs of people with disabilities, it is often necessary for staff working in tourism to acquire skills that require almost professional knowledge, sufficient expertise, which is less common in Hungary, for example (MTÜ, 2019). In environmental accessibility, ergonomic design plays a prominent role, it adheres to a rehabilitation approach, harmony between people and the technical environment must be ensured, and efforts must be made to achieve efficient and safe use. In accessible tourism, ergonomically designed tools can fit well into everyday activities, but certain obstacles may arise during their use, such as differences in body size and different physiological characteristics (Dulházi & Zsarnóczky, 2018). Tourist attractions are visited by many people, so buildings must be able to meet the needs of all types of people, especially people with disabilities, as an accessible environment can attract more tourists of different ages and backgrounds to the country, thereby increasing the country's income.

The target groups of accessible tourism are not only people with disabilities, but also all people with temporary or permanent disabilities. Whether this is a difficulty due to age characteristics, such as changing places with a cane, wearing strong dioptric glasses resulting from deterioration of the eyes or even traveling with a stroller - which is illustrated in Figure 2 below - it is necessary to create equal opportunities for everyone in an accessible environment and by developing services.

As can be seen in Figure 2, accessible accessibility provides an opportunity to improve the quality of life for several target groups at the same time. It is essential for one-tenth of people (e.g. people with disabilities), for some it is necessary (e.g. the elderly, people pushing baby carriages), but at the same time it provides comfort and makes everyday life easier for the whole society (Polgár, n.d.). Among those enjoying the benefits of accessible tourism, people with disabilities are therefore a dominant, but not exclusive actor. The World Tourism Organization's publication on accessible tourism (UNWTO, 2016b) mentions a number of other groups and actors compared to what was previously written, in addition to the obvious old people, people using strollers, or people living with temporary disabilities, there are also people such as obese, short or tall people, people transporting large objects, or people with various sensitivities and allergies, which represent an increasingly large proportion of the population these days.

Figure 3 clearly shows that anyone can be considered a potential target market.

As we have seen above and previously, the target group of acessible tourism can be anyone except the disabled, either temporarily or for life. Since people with disabilities are a potentially significant customer base (more than 1 billion people), the development of accessible services could be a significant economic factor, so the concept of tourism accessible to everyone was accordingly defined based on the following: "Every person, regardless of their disability, can make countries you can travel between or within the country, and you can also get to attractions, events, or anywhere you want" (In "Accessible tourism for all", 2012, p. 2).

Figure 2. Beneficiaries of accessible tourism
Source: Based on Polgár (n.d.)

In recent decades, the number of people living with disabilities has been increasing worldwide, and this increase is mainly attributed to aging and other health factors (Darcy, 2002; Turco et al., 1998; Yau et al., 2004). Research has shown that there is a very strong correlation between aging and disability (WHO, 2007). Although this is true, Mann (2005) argues that older people with some form of disability do not necessarily become disabled if they find ways to adapt to that impairment. However, in most cases, most of these elderly people are unable to compensate for their disability and thus become disabled (Chikuta, 2016). The link between disability and aging is unquestionable and a challenge for the global tourism industry (WHO, 2007). Recognizing this in Europe and the United States, the tourism industry is looking for ways to ensure that its infrastructure and products are accessible to older people. However, as can be seen in Figures 2 and 3, the majority of people - and not only the elderly - will enjoy these developments, as they mean a design that meets not only comfort but also a wide range of health and safety requirements (Darcy, 1998; Preiser & Ostroff, 2001). Many developed countries have building codes and standards for accessible design that incorporate these ideas (Darcy & Dickson, 2009).

Figure 3. Beneficiaries of accessible tourism
Source: UNWTO (2016b, p. 36)

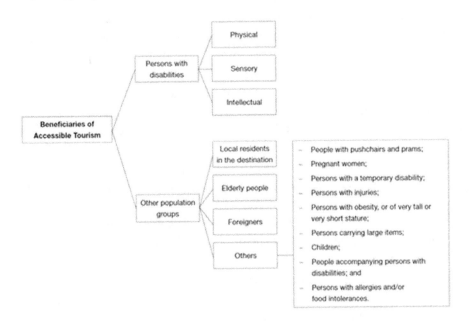

According to the general approach, people older than 55 who are active in tourism are classified as senior tourists, and we can separate the so-called young seniors (between 55 and 64 years old) and the group of elderly people over 64 years old (MTÜ, 2019). From a tourism point of view, the tourist offer should not be adapted to them, rather the existing offer should be customized. Age in itself is not decisive, because the same museum can be interesting for a 25-year-old as well as a 60-year-old. So it is not primarily the nature of the demand that distinguishes this age group from other demographic groups, but where and how, i.e. through which sales and promotion channels it is worth reaching and addressing them through. At the same time, an important change of the 21st century is that the digital competences of this age group have improved, and their internet usage habits have become diverse (MTÜ, 2019). It is more characteristic of senior travelers that they have significantly more free time, their trips are less seasonal, therefore, by creating a suitable offer, they can be motivated to travel in periods outside of the main season, outside of school holidays, thus reducing the problem of seasonality in destinations that are sensitive to this (Polgár, n.d.). Senior tourists are usually experienced travelers, in their case travel is often a lifestyle-maintaining priority, and an important motivating factor is the improvement and preservation of health (MTÜ, 2019).

THE ACT-MODEL

The most ideal case would be if people with disabilities should not be considered a special segment, but until society is at a level of development and preparedness to integrate and treat them in the same way as the able-bodied, we should still only talk about a special segment, so the authors it was considered necessary to create a theoretical model in which the cooperation of individual actors can result in a situation that is ideal for people with disabilities, including those with reduced mobility.

To create a theoretical model based on the results of research conducted in 2018 and 2019 among people with reduced mobility and tourism service providers, as well as organizations dealing with tourism. The authors used the model of Márton Lengyel's tourism system as a basis (Lengyel, 2004). Figure 4. illustrates the tourism system, which consists of tourism demand and supply. The two subsystems are related to each other in such a way that if one of the basic elements in demand, motivation, meets the attraction in the basic element of supply, then a trip can take place if the other basic elements exist together, because this alone does not mean the existence of the trip, it requires the intermediary sector (e.g. accommodation, transport), as well as the appropriate marketing activities on the part of service providers and destinations. Tourism is an open system, it is related to several areas, we can see this in Figure 4. This means areas such as society, politics, culture, nature, economy or technology, which have an impact on tourism and vice versa, so it means a mutual relationship with different factors (Lengyel, 2004).

Taking the above as a basis, the theoretical model was set up in relation to the following questions, which we basically created for Hungary, but it can also be applied internationally.

- How can people with reduced mobility be helped to participate in tourism in the same way as their able-bodied counterparts?
- Whose responsibility is this and what measures are required?

During the development of the model, we relied on the results of the interviews and the demand and supply questionnaires. We believe that these questions are decisive for future developments, so that the target group can more easily join the tourism market. The created model is shown in Figure 5.

Before explaining the theoretical model it is important to clarify that disability is interpreted on the basis of the social model and not the medical model. In the case of the latter disability is defined as a result of physical, sensory and intellectual impairment, while modern European disability policies are already based on the "social model of disability", which emphasizes the equal rights of people with disabilities, so this kind of new approach already respects and recognizes that people with disabilities have the same rights as people without disabilities. The social model recognizes that it is not the disability, but the environment that hinders, that social practices, social, physical and attitudinal barriers turn the individual's impairment into a disability, for example in education, transport, healthcare or even employment. As a result, the new approach has shifted from the person's disability to society's ability to ensure equal rights in these areas and elsewhere (NFÜ, 2012). Accordingly, the interpretation of disability in the theoretical model we created is based on the social model. It is also important to clarify in which welfare system we handle the issue of responsibility. We distinguish three types of the welfare state: liberal, conservative and social democratic (Esping-Andersen, 1991) (a detailed explanation of these is not the subject of the study). The individual systems differ fundamentally in their impact on social stratification, and the different institutional characteristics and financing techniques of the welfare state lie behind these differences. Approached from the point of view of people with disabilities (those with reduced mobility), in our opinion, the approach of the social democratic welfare state is the closest to dealing with the issue, because it provides a high level of care for everyone as a citizen, redistributes a significant amount of resources among citizens (poverty reduction), thus reducing social inequalities (Hicks & Esping-Andersen, 2005).

Figure 4. System of tourism
Source: Based on Lengyel (2004, p. 103)

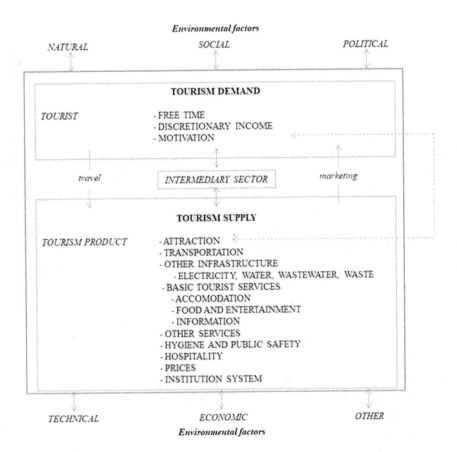

Figure 5. Model for action and cooperation on accessible tourism (ACT-model)
Source: Own editing

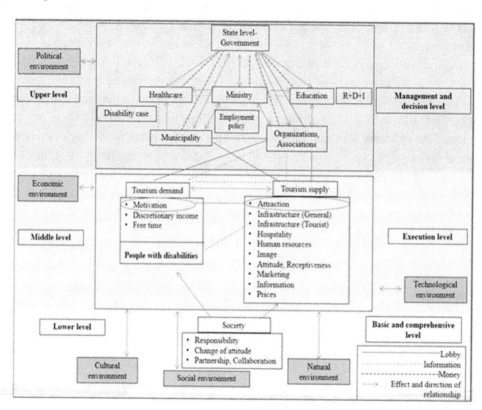

There are several international literatures in which models related to accessible tourism were created (Buhalis 2003; Eichhorn & Buhalis 2011; Leiper 1995; Makuyana & Saayman 2018; Polat & Hermans 2016; Trauer 2006), but none of them approached the issue from such a perspective (or only partially) as the action and cooperation model (ACT-model) of accessible tourism outlined by us (Figure 5). The name and appearance of the model was born from the totality of the information obtained during the research, which, in our opinion, summarizes and illustrates well what actors are involved in the topic of accessible tourism and what kind of relationship exists between them. The name ACT-model summarizes this system of relationships and mutual influence in accordance with international conditions (Figure 5). The word act by itself also means to act (it also means a law, but that is not important from the point of view of our topic), and if we read the initial letters of the individual words from the English name of the model, we still get the word ACT (for example, model for Action and Cooperation on accessible Tourism or model for action and cooperation on ACcessible Tourism), so the created theoretical model can be interpreted internationally.

The theoretical model consists of three levels, the lower, middle and upper levels, with different actors, spheres of influence and decision-making at each level. The lower, at the same time, basic level is a comprehensive level where society is included, the actors and factors in the middle level are the implementation level, while the upper level is the management and decision level. At all three levels there are important actors and factors whose task is to accept, support, integrate the segment with disabilities, and facilitate this. Each of the actors and factors of the three levels is important in order to ensure that accessibility is realized and that people with reduced mobility, the visually or hearing impaired, or those with other disabilities can participate in tourism in the same way as the able-bodied. All of them are needed in terms of responsibility and action, because cooperation is a key issue in this topic.

At the lower, basic level, there is society,[9] with most participants, but they have no decision-making authority. This is the basis of everything, and it is important in terms of attitude, responsibility, cooperation, and planning, because these can be implemented by individuals, groups, and organizations at the higher levels, so if the attitude of the members of society is appropriate on the subject, the role of the participants/representatives at the other levels is also unquestionable. The higher we go, the fewer there are, but the more responsibility and authority (power) they have in making decisions.

In the middle level, there is tourism demand and tourism supply with the basic elements related to them. As already mentioned in connection with the Lengyel model, the two subsystems (demand and supply) are connected by the meeting of motivation and attraction, then the trip can take place if the other conditions are met. On the demand side, in addition to motivation, free time and discretionary income, i.e. income that can be used freely, are also necessary, if any of the three factors are missing, we cannot participate in tourism. Gathering information and mapping their needs is also important for the disabled and other disability groups, because only then can we create a service or program package for them (the demand and supply sides are still connected, if there is sufficient information on both sides) that they need and we also need to know their special needs in order to be able to satisfy them. Thus, on the one hand, in the case of injured people, there must be the motivation to want to travel, for which, on the other hand, the existence of the appropriate infrastructure[10] and discretionary income are essential. On the supply side, in addition to the need for an attractiveness that makes us travel, the existence of the appropriate infrastructure is also necessary (this is of particular importance in the case of people with disabilities due to their special needs), accommodation, catering units, transport, as well as appropriate professionals, image, hospitality, information and other services. For supply-side tourism providers such as accommodation and restaurants, attitude, hospitality and adequate infrastructure are the most important factors. Here we can talk about the fact that if, for example, a visitor with reduced mobility has sufficient motivation and the transport infrastructure is also available, he would like to use the tourist infrastructure in the given destination and see the attractions. If a target area or service provider can be considered accessible, it can be reflected in their marketing activities and they can forge an advantage from it. Therefore, the existence of the general infrastructure can facilitate the travel of people with reduced mobility, thereby also the use of the tourist infrastructure.

Members of society obviously embody both sides, i.e. demand and supply, which depends on whether we participate in tourism as visitors on the demand side or represent service providers on the tourism supply side, or make up the population of the host location.

At the upper level, several connection points are possible between the individual actors, and between the upper and middle levels. At this top level, there is the state (government), ministry(s), health, education, actors of local political power (local governments) and disability and other organizations and associations, which are on the same level, but are also divided into sub-levels, as shown in Figure 5. At this level, the actors of the employment policy, as well as research, development and innovation policy, as well as the disability issue can be displayed. This is the management and decision-making level, because here we can find actors who have a decisive opinion on certain issues, have the authority to give opinions and/or make decisions, even regulate or provide support.

The municipality of the given settlement can use its revenues (e.g. rent, profit, dividends, local taxes) and the resources received from the state (normative central contribution, targeted support) for the development of the settlement and support for businesses. The attitude of local governments is also important in this matter, in the case of individual infrastructural developments, to what extent they take into account the fact that persons with disabilities can also use them, or to what extent is the application of the concept of universal planning at the local or national level, for example, since as we saw in the previous section of the study In the first part, the focus should not only be on the disabled, but on everyone, because the seniors can also have problems with stairs, for example, as well as families with small children. Some of the local governments employ a tourist/tourist officer, who is responsible for local events, communication with tourism service providers, and can forward any problems or requests that arise on the part of the service providers to the decision-makers within the local government. Local governments often consult with local organizations and associations (and vice versa) on professional issues, so the relationship between the two is important in order to represent a common position towards higher government bodies/levels.

With their lobbying and advocacy activities, professional associations, organizations and associations can contribute to the change of various laws, regulations, thus promoting the development of accessibility, for example among tourism service providers. If we talk about tourism organizations and associations, such as the Hungarian Hotel and Restaurant Association (MSZÉSZ) in Hungary, lobbying activities appear in addition to information between the supply and professional organizations, because the interest protection organizations unite the tourism service providers and their interests represented at higher levels. In this way, service providers and professional organizations mutually share information, while lobbying means protecting the interests of service providers on the part of the organizations towards higher levels. If we look at the demand side, the needs and opinions of the disabled are represented by the relevant professional associations and organizations (such as the National Federation of Disabled Person's Associations, the Hungarian Federation of the Blind and Partially Sighted, the Hungarian Association of the Deaf and Hard of Hearing, etc.) in possession of the appropriate information and needs, thus the organizations representing the demand and supply side must have adequate lobbying power to protect and enforce the interests of their segment.

At the level above local governments and interest protection organizations, government bodies, e.g. the ministry(s) responsible for the given area[11], health, education, R+D+I activity, and disability are included, which are important policy areas and social subsystems in themselves, and in the case of people with disabilities, they are amplified. In Hungary, the Ministry of Human Resources (EMMI) was responsible, among other things, for the operation of the social and health care system, the development of school education, the preservation of our cultural values, employment rehabilitation, the promotion of equal opportunities for disabled and disabled persons, and the coordination of science policy, with health, education, employment policy, R+D+I activities, and interest protection organizations. EMMI was able to support the work and cultural activities of local governments in connection with various grants, of which the support of tourism is also a part. Interest protection organizations could indicate new educational needs to EMMI, such as training for tourism professionals focusing on people with disabilities, so the ministry directly influences the development of education, and this is how interest protection organizations are connected to education. In the context of education, they could significantly contribute to shaping society's view of people with disabilities, for example sensitization programs in educational institutions, information about people with disabilities. In connection with integrated education, it would be important to treat the disabled together with the able-bodied in kindergartens and schools, thus making it natural that some people are different, but still the same people as the able-bodied. The relationship between EMMI and health and employment policy means, in addition to operation, that it supports people with disabilities properly, promotes their rehabilitation and protected employment at the highest possible level. The relationship between the local government and healthcare also depends on the financial situation of the given local government, because the level of social and health care coverage differs in different areas of the country.

The state/government stands at the very top of the model imagined by the authors, as it can create new laws through amendments to the law, make binding regulations, which the affected parties must definitely comply with, which is usually a long process. The state is made up of actors operating at different territorial levels and the relationships between them, which are designated accordingly in the theoretical model. If the given government is a committed supporter of accessible accessibility, it can do so with binding regulations, for which it can provide adequate support from various sources for the participants (e.g. municipalities, service providers). The task of the government is to support and operate the appropriate health care system and the education system, which is how these areas are connected to the state. In addition, it is important to encourage economic operators to employ people with disabilities/people with altered work ability, or even through the establishment and operation of a sheltered employment system, so that they can have an adequate income. The theoretical model outlined by the authors includes political, economic, cultural, social, natural and technological factors, all of which are illustrated at the corresponding levels and actors in Figure 5, but at the same time these areas act as a comprehensive level/system.

The political environment is placed at the top level of the model, because the task and role of the state (government) is to make a country politically stable and to take measures or provide subsidies that also promote tourism development. The economic environment is placed in the middle level next to the demand side, which also belongs to the supply side. In the case of the demand side, due to discretionary income, the economic environment is important by creating jobs and thereby obtaining income opportunities. On the supply side, the economic environment is important because the support of service providers is necessary even through tenders for development, and if there is demand for the given offer, the service providers generate income, which also contributes to a stable economic environment. The technological environment was shown on the supply side between the middle and lower levels. In the case of technical developments, such as transport, accommodation or catering units, it also helps to raise the level of services, thus achieving a higher number of arrivals and income. These technical developments can also promote and facilitate the travel of people with reduced mobility. The development of ICT technologies can add a lot to the access and delivery of information, on the one hand, and also to experiences, on the other hand with the use of digital technologies an attraction can be made enjoyable, for example, for the visually and hearing impaired.

At the lower, basic level, the cultural, social and natural environment was marked. The first is indicated by the model on the demand side, because the culture, tradition, and customs of society are decisive in the life of a destination, which can be an attraction for visitors (thus the cultural environment also belongs to the supply side), and this also requires that the people living in the given area actors of society to recognize the potential inherent in this. The natural environment is included on the supply side because it is the dominant type of attractions alongside the built attractions/environment. Since attractiveness plays a decisive role in the selection of a specific destination, it was placed lower in the model than the technological environment, also because the natural factors cannot be changed very much, or more precisely, they can be changed in order to remain in the same form and provide attraction as before.

The social environment is in the center at the bottom of the figure, which is important because, according to the model, the social environment is the most decisive factor in the life of a country, it is the basis of everything. It determines society's outlook, attitude, activity, and thinking on individual issues, so this type of opinion shapes all the other environments and factors already listed.

The listed factors and actors are those who/who play an important role in making a destination accessible so that each target group can use it.

The participants in the research mentioned the state and its example as the most important actor, and that professional organizations and associations should initiate the appropriate regulation towards the state, and that local governments should also take a role in this, because it is also their responsibility. In terms of mandatory regulations and their enforcement, they also attribute a decisive role to the state. Social responsibility is equally essential, the most important thing is that it is a shared responsibility. The activity of local governments could also be manifested if they were to advertise themselves as places with limited mobility/disability friendly, in which the role of service providers is also significant, in order to truly ensure equal access. In the case of the state, government, and local government, it is important to secure budgetary resources and tenders, while in the private sector, a change in the attitude of the ownership circle is necessary, just as a thorough change in attitude is also necessary among the injured and healthy people. We believe that the actors and factors mentioned above are important in order to achieve accessibility and to find responsible people to lead the individual measures and plans. At the social level, we must act together to ensure that everyone can participate in tourism, which is also the objective of the World Tourism Organization (UNWTO) (Jandala, 2015).

CONCLUSION

In the case of people with disabilities, participation in tourism, even the possibility of travel, can be considered a push factor that improves the quality of life (Yau et al., 2004). Since the 2000s, the international trend has clearly emerged, according to which the importance of accessibility in leisure time and tourism is increasing to a non-negligible extent, and the participation of people with disabilities in tourism is becoming more and more typical. The service providers are also starting to recognize the opportunities inherent in the segment, so more and more tourism decision-makers and service providers are turning their attention to accessibility in several countries. Destinations that play a pioneering role today and recognize the potential of accessibility also benefit from the tourism market[12] (Gondos, 2020). More and more hotels have accessible rooms, and in the case of more and more attractions they allow tourists with special needs to visit them, thereby getting the opportunity to join the trips. That is why it is worth separating the presence of people with disabilities in tourism, on the one hand, as the demand side, i.e. potential travelers, and the service providers on the supply side, who create the possibility of using their services (for example, in terms of design, attitude) or provide them with job opportunities.

According to the authors' research results and the opinion of both the service providers and the disabled people, Hungary is significantly behind the Northern and Western European countries by several decades (at least 10-20 years) in the realization of accessible tourism for everyone. With this study, the authors also aim to draw attention to the needs of people with disabilities, especially those with reduced mobility, so that the individual service providers who see potential in this can properly prepare to serve this special segment, which also contributes to the fact that the other target groups, such as the elderly or families with small children can also use their services more easily or can move around the given location more easily. The ACT model created using the research results, outlines the action and cooperation system of accessible tourism, defining the location, role, and decisiveness of the various actors and factors, as well as the relationships and effects between them, as well as indicating who/ what's tasks and have/may have responsibilities in the implementation of accessible tourism and equal access for everyone. The model can provide a useful framework for practical implementation for both decision-makers and service providers.

REFERENCES

Barlow, H. (2012). Gaining Access at Historic Tourism Sites: A Narrative Case Study of Physical Accessibility at Glamis Castle. Academic Press.

Bizottság. (2010). *Európai fogyatékosságügyi stratégia 2010–2020: megújított elkötelezettség az akadály-mentes Európa megvalósítása iránt* [European Disability Strategy 2010-2020: a renewed commitment to achieving an accessible Europe]. Európai Bizottság, 636.

Buhalis, D. (2003). eTourism: Information Technology for Strategic Tourism Management. Prentice Hall.

Buhalis, D., & Darcy, S. (2011). *Accessible tourism: Concepts and issues*. Channel View Publication.

Buhalis, D., Darcy, S., & Ambrose, I. (2012). *Best Practice in Accessible Tourism – Inclusion, Disability, Ageing Population and Tourism*. Channel View Publications. doi:10.21832/9781845412548

Bujdosó, B., & Kemény, F. (2009). *Fogyatékosság és rehabilitáció* [Disability and rehabilitation]. Budapest: Nemzeti Szakképzési és Felnőttképzési Intézet.

Chikuta, O. (2016). Is There Room in the Inn? Towards Incorporating People with Disability in Tourism Planning. *Review of Disability Studies: An International Journal, 11*(3).

CKH. (2012). *Az "Akadálymentes turizmus mindenkinek" ökonómiai impulzusai* [The economic impulses of "Accessible tourism for all"]. A Gazdasági és Technológiai Szövetségi Minisztérium megbízásából készült vizsgálat. Retrieved February 15, 2023, from http://www.ckh.hu/sites/default/files/fajlok_projekt/2012/2012-05-22- okonomiaiimpulzusokegyakadalymentesturizmusertm.pdf

Darcy, S. (1998). *Anxiety to access: Tourism patterns and experiences of New South Wales people with a physical disability*. Sydney, Australia: Tourism New South Wales.

Darcy, S. (2002). Marginalised participation: Physical disability, high support needs and tourism. *Journal of Hospitality and Tourism Management, 9*(1), 61–72.

Darcy, S. (2007). A methodology for testing accessible accomodation information provision formats. In I. McDonnell, S. Grabowski, & R. March (Eds.), *CAUTHE 2007: Tourism - Past Achievements, Future Challenges* (pp. 1–18). University of Technology Sydney.

Darcy, S., & Dickson, T. (2009). A Whole-of-Life Approach to Tourism: The Case for Accessible Tourism Experiences. *Journal of Hospitality and Tourism Management, 16*(1), 32–44. doi:10.1375/jhtm.16.1.32

Dulházi, F., & Zsarnóczky, M. (2018). Az akadálymentes turizmus, mint rehabilitációs "eszköz" [Accessible tourism as a rehabilitation "tool"]. *Conference Paper. LX. Georgikon napok, 60th Georgikon Scientific Conference*, 56-61.

Eichhorn, V., & Buhalis, D. (2011). Accessibility – A Key Objective for the Tourism Industry. In D. Buhalis & S. Darcy (Eds.), *Accessible Tourism: Concepts and Issues* (pp. 46–61). Channel View Publications.

EK. (2004). *Gleichstellung, Vielfalt und Erweiterung. Luxemburg: Amt für amtliche Veröffentlichungen der Europäischen Gemeinschaften*. Europäische Kommission.

ENSZ.(2006). *A Fogyatékossággal élő személyek jogairól szóló egyezmény [Convention on the Rights of Persons with Disabilities]*. ENSZ. file:///C:/Users/kk/Downloads/A%20fogyat%C3%A9koss%C3%A1ggal%20%C3%A9l%C5%91k%20jogair%C3%B3l%20sz%C3%B3l%C3%B3%20ENSZ%20egyezm%C3%A9ny.pdf

Esping-Andersen, G. (1991). Mi a jóléti állam? In Zs. Ferge & K. Lévai (Eds.), *A jóléti állam [The welfare state]*. (pp. 116–134). T-Twins.

European Commission (2014). *Economic impact and travel patterns of accessible tourism in Europe – Final report*. EC. file:///C:/Users/kk/Downloads/study%20A%20Economic%20Impact%20and%20travel%20patterns%20of%20accessible%20tourism_final

Fekete, M. (2006). *Hétköznapi turizmus. A turizmuselmélettől a gyakorlatig [Casual tourism. From tourism theory to practice]* [Unpublished doctoral dissertation, Nyugat-magyarországi Egyetem, Sopron, Hungary].

Giddens, A. (2008). *Szociológia [Sociology]* (2nd ed.). Osiris.

Gondos, B. (2020). *Speciális igények a turizmusban – A mozgáskorlátozottak helye, szerepe és lehetősége a turisztikai szektorban [Special needs in tourism – Role and opportunity of people with reduced mobility in tourism sector]* [Unpublished doctoral dissertation, Széchenyi István University, Győr, Hungary].

Hicks, A., & Esping-Andersen, G. (2005). Comparative and Historical Studies of Public Policy and the Welfare State. In T. Janoski, R. R. Alford, A. M. Hicks, & M. A. Schwartz (Eds.), *The Handbook of Political Sociology. States, Civil Societies, and Globalization* (pp. 509–525). Cambridge University Press.

Imrie, R., & Hall, P. (2001). *Inclusive Design. Designing and Developing Accessible Environments*. Spoon Press.

Israeli, A. A. (2002). A preliminary investigation of the importance of site accessibility factor for disabled tourists. *Journal of Travel Research*, *41*(1), 101–104. doi:10.1177/004728750204100114

Jandala, Cs. (2015). *2. modul: Turizmus-ipar* [Module 2: Tourism industry]. UNI NKE. https://www.uni-nke.hu/document/uni-nke-hu/jandala_2modul.pdf

Kálmán, Zs., & Könczei, Gy. (2002). *A Taigetosztól az esélyegyenlőségig [From Taygetos to equal opportunities]*. Osiris.

Kaufman-Scarborough, C. (2001). Accessible advertising for visually-disabled persons: The case of color-deficient consumers. *Journal of Consumer Marketing*, *18*(4), 303–318. doi:10.1108/07363760110392985

Laki, I. (2013). A fogyatékossággal élő emberekről [About people with disabilities]. *Neveléstudomány Tanulmányok, 1*(3), 79–84.

Leiper, N. (1995). *Tourism Management.* RMIT Press.

Lengyel, M. (2004). *A turizmus általános elmélete [General theory of tourism].* Budapest: Heller Farkas Gazdasági és Turisztikai Szolgáltatások Főiskolája.

Makuyana, T., & Saayman, M. (2018). The postulate for the systematic mainstreaming of impairments in Tourism Education in South Africa: A literature 204 synthesis. *African Journal of Hospitality, Tourism and Leisure, 7*(5), 1–28.

Medarić, Z., Sulyok, J., Kardos, S., & Gabruć, J. (2021). Lake Balaton as an accessible tourism destination – the stakeholders' perspectives. *Hungarian Geographical Bulletin, 70*(3), 233–247. doi:10.15201/hungeobull.70.3.3

Michalkó, G. (2010). *Boldogító utazás – a turizmus és az életminőség kapcsolatának magyarországi vonatkozásai [A happy trip - aspects of the relationship between tourism and quality of life in Hungary].* MTA Földrajztudományi Kutatóintézet.

MTÜ (2019, October). *Trendek és gyakorlatok a szenior, az akadálymentes és az orvosi turizmusban* [Trends and practices in senior, accessible and medical tourism]. Paper presented Tourism Academy at a series of events. Budapest: Magyar Turisztikai Ügynökség.

Nárai, M. (2013). *Fogyatékossággal élők szociális* ellátása [Social care for people with disabilities]. MTA KRTK RKI NYUTO, Győr, Hungary.

NFÜ. (2012). *Fogyatékos személyek akadálymentes hozzáférésének és megkülönböztetés-mentességének biztosítása* [Ensuring accessible access and non-discrimination of disabled persons]. Nemzeti Fejlesztési Ügynökség. https://www.kozbeszerzes.hu/data/filer_public/fa/fb/fafbe76a-2678-47d1-bc6d9737d614765d/fogyatekos_szemelyek_akadalymentes_hozzaferesenek_es_megkulonbo ztetes-mentessegenek_biztositasa.pdf

NTG. (2020). *Accessibility in Tourism: challenges and opportunities.* Next Tourism Generation. NTG. https://nexttourismgeneration.eu/accessibility-in-tourism-challenges-and-opportunities/?fbclid=IwAR2thEAC_SRvZaXTGO3zXzlcskcN

Packer, T. L., Mckercher, B., & Yau, M. K. (2009). Understanding the complex interplay between tourism, disability and environmental contexts. *Disability and Rehabilitation, 29*(4), 281–292. doi:10.1080/09638280600756331 PMID:17364778

Polat, N., & Hermans, E. (2016). A model proposed for sustainable accessible tourism (SAT). *TÉKHNE - Review of Applied Management Studies, 14*(2), 125-133.

Polgár, J. (n.d.). *Szakmai jegyzetek az akadálymentességről, akadálymentes turizmusról [Professional notes on accessibility and accessible tourism].* Unpublished professional notes.

Preiser, W., & Ostroff, E. (2001). *Universal design handbook.* McGraw-Hill.

Ray, N. M., & Ryder, M. E. (2003). "Ebilities" tourism: An exploratory discussion of the travel needs and motivations of the mobility-disabled. *Tourism Management*, *24*(1), 57–72. doi:10.1016/S0261-5177(02)00037-7

Scandic Press Release. (2019). *Scandic Hotels introduces standard for allergy-friendly rooms*. Scandic.

Scheyvens, R., & Biddulph, R. (2018). Inclusive tourism development. *Tourism Geographies*, *20*(4), 589–609. doi:10.1080/14616688.2017.1381985

Shelton, E., & Tucker, H. (2005). Tourism and disability: Issues beyond access. *Tourism Review International*, *8*(3), 211–219. doi:10.3727/154427205774791528

Souca, M. L. (2010). Accessible Tourism – The Ignored Opportunity. *Annals of Faculty of Economics*, *1*(2), 1154–1157.

Stumbo, N. J., & Pegg, S. (2005). Travelers and Tourists with Disabilities: A Matter of Priorities and Loyalties. *Tourism Review International*, *8*(3), 195–209. doi:10.3727/154427205774791537

Suárez-Henríquez, C., Ricoy-Cano, A. J., Hernández-Galan, J., & de la Fuente-Robles, Y. M. (2022). The past, present, and future of accessible tourism research: A bibliometric analysis using the Scopus database. *Journal of Accessibility and Design for All*, *7*(2), 26–60.

Takeda, K., & Card, J. A. (2002). US tour operators and travel agencies: Barriers encountered when providing package tours to people who have difficulty walking. *Journal of Travel & Tourism Marketing*, *12*(1), 47–61. doi:10.1300/J073v12n01_03

Trauer, B. (2006). Conceptualizing special interest tourism-frameworks for analyses. *Tourism Management*, *27*(2), 183–200. doi:10.1016/j.tourman.2004.10.004

Turco, D. M., Stumbo, N. J., & Garncarz, J. (1998). Tourism Constraints for People with Disabilities. *Parks & Recreation*, *33*(9), 78–84.

UNWTO. (2015). *Manual on Accessible Tourism for All Public-private Partnerships and Good Practices*. UNWTO. https://cf.cdn.unwto.org/sites/all/files/docpdf/aamanualturismoaccesibleomt-facseng.pdf

UNWTO. (2016a). *Accessible Tourism for All: An Opportunity within Our Reach*. UNWTO. https://cf.cdn.unwto.org/sites/all/files/docpdf/turismoaccesiblewebenok.pdf

UNWTO. (2016b). Manual on Accessible Tourism for All: Principles, Tools and Best Practices. *Module I: Accessible Tourism – Definition and Context. World Tourism Organization*. https://www.e-unwto.org/doi/pdf/10.18111/ 9789284418077

Veenhoven, R. (2003). Hedonism and happiness. *Journal of Happiness Studies*, *4*(4), 437–457. doi:10.1023/B:JOHS.0000005719.56211.fd

Vid, G. (Ed.). (2017). *Ne parázz! Információk az egyenlő esélyű hozzáférésről [Do not be afraid! Information on equal access]*. Camelot Mozgássérült Fiatalok Győri Egyesülete.

Wagner, Zs. (2010). *Fogyatékos fogadókészség [Impaired reception skills]*. Turizmus. https://turizmus.com/migracio/fogyatekos-fogadokeszseg-1092077

WHO. (1980). *International classification of impairments, disabilities and handicap.* World Health Organization. https://apps.who.int/iris/bitstream/handle/10665/41003/9241541261_eng.pdf?sequence=1

WHO. (2001). *International classification of functioning, disability and health (ICF).* WHO. https://apps.who.int/iris/bitstream/handle/10665/42407/9241545429.pdf?sequence=1

WHO. (2007). *Global age-friendly cities guide.* World Health Organization.

WHO. (2011). *World report on disability.* World Health Organization.

Yau, M. K., McKercher, B., & Packer, T. L. (2004). Traveling with a Disability: More than an Acces Issue. *Annals of Tourism Research, 31*(4), 946–960. doi:10.1016/j.annals.2004.03.007

Zsarnóczky, M. (2017a). Accessible Accessible tourism in the European Union. In K. Borsekova, A. Vanova, & K. Vitálisowa (Eds.), *6th Central European Conference in Regional Science "Engines of Urban and Regional Development": Conference Proceedings.* (pp. 30-39). Banská Bystrica: Faculty of Economics, Matej Bel University.

Zsarnóczky, M. (2017b). Developing Senior Tourism in Europe. *Pannon Management Review, 6*(3-4), 201–214.

KEY TERMS AND DEFINITIONS

Accessible Tourism: A continuous effort aimed at making all destinations, tourism products and services accessible to all people, regardless of any physical limitation, disability, or age.

ACT-Model (Model for Action and Cooperation on Accessible Tourism): This theoretical model summarizes and illustrates well what actors are involved in the topic of accessible tourism and what kind of relationship exists between them. The ACT-model consists of three levels: The lower level is a comprehensive level where society is included, the actors and factors are in the middle level, while the upper level is the management and decision level. Each of the actors and factors of the three levels are important in order to ensure that accessibility is realized and that people with reduced mobility can participate in tourism in the same way as non-disabled people.

Equal Access: The belief that everyone to have the same chance to access what they need, programs, services, environment, buildings, institutes etc.

People With Disabilities: A person who permanently or irreversibly has a sensory, communication, physical, intellectual, psychosocial impairment or any combination of these lives which, in interaction with environmental, social and other significant obstacles, limits or prevents effective and equal social participation with others.

People With Reduced Mobility: A person whose congenital or acquired damage to the locomotor system permanently hinders the functional abilities of the body and the activity of the individual; it can cause a disadvantageous situation for the individual and limited way of life.

Universal Design: Designing products, programs, services, and the built and designed environment in such a way that they are accessible to all people to the greatest extent possible, without the need for adaptation or special design.

ENDNOTES

1.   Another author of this paper helped prepare the doctoral dissertation, conduct the research and process the results as a supervisor. The study is not a reprint of the dissertation it is a revised, original publication that uses some of its parts.

2.   See e.g. UN Convention on the Rights of Persons with Disabilities (UN, 2006); Hungary: XXVI of 1998 Act on the rights of disabled persons and ensuring their equal opportunities.

3.   Unified Disability Information Portal: https://www.efiportal.hu/egeszsegugy/fogyatekossagi-tipusok/

4.   LXXVIII of 1997 Act on the Design and Protection of the Built Environment, § 2.1 (amended by: Act CLVII of 2012 § 66.1)

5.   Council of the European Communities: Council Decision (26 November 2009) on the conclusion by the European Community of the UN Convention on the Rights of Persons with Disabilities (2010/48/EC) Annex I Convention on the Rights of Persons with Disabilities, Article 2, Definitions.

6.   The tourist infrastructure includes narrower infrastructure dependent on tourism, supply and public cleanliness infrastructure that exceeds the needs of the residents, such as special means of transport, sports facilities, parks, spas, spas, entertainment venues, congress centers.

7.   Accommodation, places providing meals, program service facilities, locations.

8.   ENAT (European Network for Accessible Tourism) is a non-profit organization that aims to share information, studies and practices related to accessible tourism; available at https://www.accessibletourism.org/.

9.   *"Society is a system of structured social relations that connects those who live in it through a common culture."* (Giddens, 2008, p. 794)

10.   In the course of the authors' research among people with reduced mobility, it was revealed that the primary condition that must be present is the accessibility of transport.

11.   The current ministry(s) responsible for the given area is the Ministry of Human Resources (EMMI) in Hungary until 2022, but from the point of view of the topic, for example, construction regulations, which are important in terms of accessibility, are also included.

12.   According to an American research, the turnover of accessible hotels in the USA increased by 6% in one year (Gondos, 2020).

Compilation of References

Abbas, J., Mubeen, R., Iorember, P. T., Raza, S., & Mamirkulova, G. (2021). Exploring the impact of COVID-19 on tourism: Transformational potential and implications for a sustainable recovery of the travel and leisure industry. *Current Research in Behavioral Sciences*, 2, 1–11. doi:10.1016/j.crbeha.2021.100033

Abdoullaev, A. (2011). Keynote: A smart world: A development model for intelligent cities. In *Proceedings of the 11th IEEE International Conference on Computer and Information Technology (CIT)*. IEEE.

Abudaqa, A., Alzahmi, R. A., Almujaini, H., & Ahmed, G. (2022). Does innovation moderate the relationship between digital facilitators, digital transformation strategies and overall performance of SMEs of UAE? *International Journal of Entrepreneurial Venturing*, *14*(3), 330–350. doi:10.1504/IJEV.2022.124964

Adamczewski, P. (2018). Intelligent organizations in digital age. *MEST Journal, 6*(2), 1-11. https://doi.org/. doi:1-11

Adam, M., Wessel, M., & Benlian, A. (2021). AI-based chatbots in customer service and their effects on user compliance. *Electronic Markets*, *31*(2), 427–445. doi:10.100712525-020-00414-7

Adidas admits it used Morocco's zellige design. (2022). Middle East Eye. https://www.middleeasteye.net/news/qatar-world-cup-2022-adidas-morocco-algeria-zellige-top-admits

Adler-Nissen, R. (2014). Stigma management in international relations: Transgressive identities, norms, and order in international society. *International Organization*, *68*(1), 143–176. doi:10.1017/S0020818313000337

Adobe (n.a.). *PDF accessibility overview.* Adobe. https://www.adobe.com/accessibility/pdf/pdf-accessibility-overview.html (accessed: 23.12.2020)

Aggarwal, R., & Lal Das, M. (2012) RFID Security in the Context of Internet of Things. *First International Conference on Security of Internet of Things*, (pp. 51-56). ACM. 10.1145/2490428.2490435

Ahmad, A., Alshurideh, M. T., Al Kurdi, B. H., & Salloum, S. A. (2021). Factors impact organization digital transformation and organization decision making during COVID-19 pandemic. In: Alshurideh M., Hassanien A.E., Masa'deh R. (Eds), The Effect of Coronavirus Disease (COVID-19) on Business Intelligence. Springer, Cham. https://doi.org/ doi:10.1007/978-3-030-67151-8_6

Ahmad, N., Youjin, L., & Hdia, M. (2022). The role of innovation and tourism in sustainability: Why is environment-friendly tourism necessary for entrepreneurship? *Journal of Cleaner Production*, *379*, 134799. doi:10.1016/j.jclepro.2022.134799

Akca, H., Sayili, M., & Esengun, K. (2007). Challenge of rural people to reduce digital divide in the globalized world: Theory and practice. *Government Information Quarterly*, *24*(2), 404–413. doi:10.1016/j.giq.2006.04.012

AlBar, A. M., & Hoque, M. R. (2017). Factors affecting the adoption of information and communication technology in small and medium enterprises: a perspective from rural Saudi Arabia online: 21 Oct 2017. *Information Technology for Development*, 25(4), 715–738. doi:10.1080/02681102.2017.1390437

Albayrak, T., Caber, M., & Sigala, M. (2021). A quality measurement proposal for corporate social network sites: The case of hotel Facebook page. *Current Issues in Tourism*, 24(20), 2955–2970. doi:10.1080/13683500.2020.1854199

Alexandre-Leclair, L., & Liu, Z. (2014). Innovation and entrepreneurship: The case of the tourism sector in Paris. *Innovations*, 44(2), 169–185. doi:10.3917/inno.044.0169

Alexis, P. (2017). R-Tourism: Introducing the Potential Impact of Robotics and Service Automation in Tourism. *Ovidius University Annals, Series Economic Sciences, 17*(1)

Alhammad, F. A. (2020). Trends in tourism entrepreneurship research: A systematic review. *Jordan Journal of Business Administration*, 16(1), 307–330. www.scopus.com

Ali, A., & Frew, A. (2013). *Information and communication technologies for sustainable tourism*. Routledge. doi:10.4324/9780203072592

Alkier, R., Milojica, V., & Roblek, V. (2017). Challenges of the social innovation in tourism. *ToSEE – Tourism in Southern and Eastern Europe, 4*, 1-13. doi:10.20867/tosee.04.24

All The Research. (2020, February 21). *AR-VR in travel and tourism market*. All the Research. https://www.alltheresearch.com/press-release/ar-vr-in-travel-and-tourism-market-ecosystem-worth-304-4-million-by-2023

Aloini, D., Latronico, L., & Pellegrini, L. (2022). The impact of digital technologies on business models. Insights from the space industry. *Measuring Business Excellence*, 26(1), 64–80. doi:10.1108/MBE-12-2020-0161

Alonso-Gonzalez, A., Chacon, L. A. P., & Peris-Ortiz, M. (2018). Sustainable social innovations in smart cities: Exploratory analysis of the current global situation applicable to Colombia. Strategies and best practices in social innovation: An institutional perspective (pp. 65-87). Springer. doi:10.1007/978-3-319-89857-5_5

Alrawadieh, Z., Alrawadieh, Z., & Cetin, G. (2021). Digital transformation and revenue management: Evidence from the hotel industry. *Tourism Economics*, 27(2), 328–345. doi:10.1177/1354816620901928

Alsos, G. A., Eide, D., & Madsen, E. L. (2014). Introduction: innovation in tourism industries. In Handbook of Research on Innovation in Tourism Industries (pp. 1–24). Edward Elgar Publishing. doi:10.4337/9781782548416.00006

Ammirato, S., Felicetti, A. M., Linzalone, R., & Carlucci, D. (2021). Digital business models in cultural tourism. *International Journal of Entrepreneurial Behaviour & Research*, 28(8), 1940–1961. doi:10.1108/IJEBR-01-2021-0070

Anholt, S. (2005). Anholt Nation Brands Index: How Does the World See America? *Journal of Advertising Research*, 45(3), 296–304. doi:10.1017/S0021849905050336

Anouar, S. (2022). Morocco Launches "Aji" and 'We Are Open' Campaigns to Revive Tourism. *Morocco World News*. Https://Www.Moroccoworldnews.Com/. https://www.moroccoworldnews.com/2022/02/347121/morocco-launches-aji-and-we-are-open-campaigns-to-revive-tourism

An, S., Choi, Y., & Lee, C. K. (2021). Virtual travel experience and destination marketing: Effects of sense and information quality on flow and visit intention. *Journal of Destination Marketing & Management, 19*, 100492. doi:10.1016/j.jdmm.2020.100492

Antlova, K. (2014). Main Factors for ICT Adoption in the Czech SMEs. Springer. doi:10.1007/978-3-642-38244-4_7

Antlová, K., Popelínský, L., & Tandler, J. (2011). Long Term Growth of SME from the View of ICT Competencies and Web Presentations. *E+M. Ekonomie a Management, 14,* 125–139.

Antons, D., & Breidbach, C. F. (2018). Big data, big insights? Advancing service innovation and design with machine learning. *Journal of Service Research, 21*(1), 17–39. doi:10.1177/1094670517738373

Ara, J., Bhuiyan, H., Bhuiyan, Y. A., Bhyan, S. B., & Bhuiyan, M. I. (2021). Comprehensive analysis of augmented reality technology in modern healthcare system. *International Journal of Advanced Computer Science and Applications, 12*(6), 840–849. doi:10.14569/IJACSA.2021.0120698

Aral, S., & Weill, P. D. (2007). IT assets, organizational capabilities, and firm performance: How resource allocations and organizational differences explain performance variation. *Organization Science, 18*(5), 763–780.

Arbulú, I., Razumova, M., Rey-Maquieira, J., & Sastre, F. (2021). Measuring risks and vulnerability of tourism to the COVID-19 crisis in the context of extreme uncertainty: The case of the Balearic Islands". *Tourism Management Perspectives, 39,* 100857. doi:10.1016/j.tmp.2021.100857 PMID:34580625

Aria, M., & Cuccurullo, C. (2017). bibliometrix: An R-tool for comprehensive science mapping analysis. *Journal of Informetrics, 11*(4), 959–975. doi:10.1016/j.joi.2017.08.007

Arokiaraj, D., Ramyar, R. A., Ganeshkumar, C., & Gomathi Sankar, J. (2020b). An empirical analysis of consumer behaviour towards organic food products purchase in India. *Calitatea Qual Access Success, 21.*

Arokiaraj, D., Ganeshkumar, C., & Paul, P. V. (2020a). Innovative management system for environmental sustainability practices among Indian auto-component manufacturers. *International Journal of Business Innovation and Research, 23*(2), 168–182. doi:10.1504/IJBIR.2020.110095

Atkinson, J. (2021). The times they are a-changin': But how fundamentally and how rapidly? Academic library services post-pandemic. In *Libraries, digital information, and COVID* (pp. 303–315). Chandos Publishing. doi:10.1016/B978-0-323-88493-8.00019-7

Automation Anywhere University. (n.d.). *About.* Automation Anywhere University. https://university.automationanywhere.com/

Awad Alhaddad, A. (2015). The Effect of Advertising Awareness on Brand Equity in Social Media. *International Journal of E-Education, e-Business, e-. Management Learning, 5*(2), 73–84. doi:10.17706/ijeeee.2015.5.2.73-84

Aydin, A., Darici, B., & Taşçi, H. M. (2015). Uluslararası turizm talebini etkileyen ekonomik faktörler: Türkiye üzerine bir uygulama. *Erciyes Üniversitesi İktisadi ve İdari Bilimler Fakültesi Dergisi, 0*(45), 143–177. doi:10.18070/euiibfd.85938

Ayyagari, R., Grover, V., & Purvis, R. (2011). Technostress: Technological antecedents and implications. *Management Information Systems Quarterly, 35*(4), 831–858. doi:10.2307/41409963

Baines, A. (1998). Technology and tourism. *Work Study, 47*(5), 160–163. doi:10.1108/00438029810370492

Baird, K., Su, S., & Munir, R. (2018). The relationship between the enabling use of controls, employee empowerment, and performance. *Personnel Review, 47*(1), 257–274. doi:10.1108/PR-12-2016-0324

Bakas, F. E. (2017). Community resilience through entrepreneurship: The role of gender. *Journal of Enterprising Communities, 11*(1), 61–77. doi:10.1108/JEC-01-2015-0008

Bakhshi, H., & Throsby, D. (2012). New technologies in cultural institutions: Theory, evidence and policy implications. *International Journal of Cultural Policy, 18*(2), 205–222. doi:10.1080/10286632.2011.587878

Bakır, M., Özdemir, E., Akan, Ş., & Atalık, Ö. (2022). A bibliometric analysis of airport service quality. *Journal of Air Transport Management, 104*, 102273. doi:10.1016/j.jairtraman.2022.102273

Bakker, A. B., & Demerouti, E. (2007). The job demands-resources model: State of the art. *Journal of Managerial Psychology, 22*(3), 309–328. doi:10.1108/02683940710733115

Bannini, T. (2017). Proposing a theoretical framework for local territorial identities: concepts, questions and pitfalls. *Territorial Identities and development, 2*. http://doi.org/ doi:10.23740/TID2201722

Bannini, T., & Pollice, F. (2015). Territorial identity as a strategic resource for the development of rural areas. *Semestrale di Studi e Ricerche di Geografia Roma – 27*. https://www.academia.edu/en/16373583/

Barakazi, M. (2022). The use of Robotics in the Kitchens of the Future: The example of 'Moley Robotics'. *Journal of Tourism and Gastronomy Studies, 10*(2), 895–905. doi:10.21325/jotags.2022.1021

Baratta, R., Bonfanti, A., Cucci, M. G., & Simeoni, F. (2022). Enhancing cultural tourism through the development of memorable experiences: The "Food democracy museum" as a phygital project. *Sinergie, 40*(1), 213–236. doi:10.7433117.2022.10

Bardin, L. (2001). *L'analyse de contenu* (10e éd.). Paris: Presses Universitaires de France.

Barlow, H. (2012). *Gaining Access at Historic Tourism Sites: A Narrative Case Study of Physical Accessibility at Glamis Castle*. Academic Press.

Baum, T., & Hai, N. T. T. (2020). Hospitality, tourism, human rights and the impact of COVID-19. *International Journal of Contemporary Hospitality Management, 32*(7), 2397–2407. doi:10.1108/IJCHM-03-2020-0242

Bayyurt, Y., & Seggie, F. N. (2017). *Nitel araştırma yöntem, teknik, analiz ve yaklaşımları*. Anı Yayıncılık.

Beethem, H. (2016, February 13). *From digital capability to digital wellbeing: Thriving in the network*. Open Research Online. http://oro.open.ac.uk/72433/1/A%20From%20digital%20capability%20to%20digital%20wellbeing%20thriving%20in%20the%20network%20Helen%20Bee.pdf

Belanche, D., Casaló, L. V., Flavián, C., & Schepers, J. (2020). Service robot implementation: A theoretical framework and research agenda. *Service Industries Journal, 40*(3-4), 203–225. doi:10.1080/02642069.2019.1672666

Belkadi, E. (2020). City Branding of Casablanca in Morocco. In *Strategic Innovative Marketing and Tourism* (pp. 129–138). Springer. doi:10.1007/978-3-030-36126-6_15

Benckendorff, P. J., Xiang, Z., & Sheldon, P. J. (2019). *Tourism information technology*. Cabi international Australia.

Berezina, K., Ciftci, O., & Cobanoglu, C. (2019). Robots, artificial intelligence, and service automation in restaurants. In *Robots, artificial intelligence, and service automation in travel, tourism and hospitality*. Emerald Publishing Limited. doi:10.1108/978-1-78756-687-320191010

Berger, R. (2020). *Digital workplace in the era of COVID-19*. Roland Berger. https://www.rolandberger.com/en/Point-of-View/Digitalworkplace-in-the-era-of-Covid-19.html (Accessed: May 5, 2020).

Bertella, G. (2017). The emergence of Tuscany as a wedding destination: The role of local wedding planners. *Tourism Planning & Development, 14*(1), 1–14. doi:10.1080/21568316.2015.1133446

Bertella, G., & Vidmar, B. (2019). Learning to face global food challenges through tourism experiences. *Journal of Tourism Futures, 5*(2), 168–178. doi:10.1108/JTF-01-2019-0004

Bessière, J. (1998). *Local Development and Heritage: Traditional Food and Cuisine as Tourist Attractions in Rural Areas*. Blackwell Publishers.

Bhaiswar, R., Meenakshi, N., & Chawla, D. (2021). Evolution of Electronic Word of Mouth: A Systematic Literature Review Using Bibliometric Analysis of 20 Years (2000–2020). *FIIB Business Review*, *10*(3), 215–231. doi:10.1177/23197145211032408

Bhargava, V. (2017). Are social chatbots the Future of hassle-free travel? *The chatbot magazine*.

Bhimasta, R. A., & Kuo, P. Y. (2019, September). What causes the adoption failure of service robots? A Case of Henn-na Hotel in Japan. In Adjunct proceedings of the 2019 ACM international joint conference on pervasive and ubiquitous computing and proceedings of the 2019 ACM international symposium on wearable computers (pp. 1107-1112). ACM.

Bhutia, P. D. (2022). *Creative New Tourism Pitch*. Skift. https://skift.com/2022/05/13/moroccos-covid-policies-work-against-creative-new-tourism-pitch/

Bizottság. (2010). *Európai fogyatékosságügyi stratégia 2010–2020: megújított elkötelezettség az akadálymentes Európa megvalósítása iránt* [European Disability Strategy 2010-2020: a renewed commitment to achieving an accessible Europe]. Európai Bizottság, 636.

BluePrism University. (n.d.). *Certification*. BluePrism University. https://university.blueprism.com/certification

Bock, D. E., Wolter, J. S., & Ferrell, O. C. (2020). Artificial intelligence: Disrupting what we know about services. *Journal of Services Marketing*, *34*(3), 317–334. doi:10.1108/JSM-01-2019-0047

Boden, M. A. (2017). *Inteligencia artificial*. Turner.

Bogicevic, V., Bujisic, M., Bilgihan, A., Yang, W., & Cobanoglu, C. (2017). The impact of traveler-focused airport technology on traveler satisfaction. *Technological Forecasting and Social Change*, *123*, 351–361. doi:10.1016/j.techfore.2017.03.038

Bogicevic, V., Liu, S. Q., Seo, S., Kandampully, J., & Rudd, N. A. (2021). Virtual reality is so cool! How technology innovativeness shapes consumer responses to service preview modes. *International Journal of Hospitality Management*, *93*, 102806. doi:10.1016/j.ijhm.2020.102806

Bonanomi, M. M., Hall, D. M., Staub-French, S., Tucker, A., & Talamo, C. M. L. (2020). The impact of digital transformation on formal and informal organizational structures of large architecture and engineering firms. *Engineering, Construction, and Architectural Management*, *27*(4), 872–892. doi:10.1108/ECAM-03-2019-0119

Bonn, M. A., & Harrington, J. (2008). A comparison of three economic impact models for applied hospitality and tourism research. *Tourism Economics*, *14*(4), 769–789. doi:10.5367/000000008786440148

Bouabdellah, M. (2023). Digital Innovation in Healthcare. In A. Bouarar, K. Mouloudj, & D. Martínez Asanza (Eds.), *Integrating Digital Health Strategies for Effective Administration* (pp. 1–19). IGI Global., doi:10.4018/978-1-6684-8337-4.ch001

Bouarar, A. C., Mouloudj, S., & Mouloudj, K. (2021). Extending the theory of planned behavior to explain intention to use online food delivery services in the context of COVID -19 pandemic. In C. Cobanoglu, & V. Della Corte (Eds.), Advances in global services and retail management (pp. 1–16). USF M3 Publishing.

Bouarar, A. C., Mouloudj, K., & Mouloudj, S. (2020). The impact of coronavirus on tourism sector - an analytical study. *Journal of Economics and Management*, *20*(1), 323–335.

Bouarar, A. C., Mouloudj, S., & Mouloudj, K. (2022). Digital transformation: Opportunities and challenges. In N. Mansour & S. Ben Salem (Eds.), *COVID-19's Impact on the Cryptocurrency Market and the Digital Economy* (pp. 33–52). IGI Global. doi:10.4018/978-1-7998-9117-8.ch003

Boum, A. (2020). Branding Convivencia: Jewish Museums and the Reinvention of a Moroccan Andalus in Essaouira. *Exhibiting Minority Narratives: Cultural Representation in Museums in the Middle East and North Africa*, 205–223. Academia.

Bovsh, L. A., Hopkalo, L. M., & Rasulova, A. M. (2023). Digital Relationship Marketing Strategies of Medical Tourism Entities. In A. Bouarar, K. Mouloudj, & D. Martínez Asanza (Eds.), *Integrating Digital Health Strategies for Effective Administration* (pp. 133–150). IGI Global. doi:10.4018/978-1-6684-8337-4.ch008

Bowen, J., & Morosan, C. (2018). Beware hospitality industry: The robots are coming. *Worldwide Hospitality and Tourism Themes*, *10*(6), 726–733. doi:10.1108/WHATT-07-2018-0045

Bradić-Martinović, A. (2022). *Digitalne veštine građana Srbije*. Research Gate.

Brand Finance. (2022). *Nation Brands Forum 2022 in Association with DPAAL*. Brand Finance. https://brandfinance.com/events/brand-finance-nation-brands-forum-2022

Brandtzaeg, P. B., & Følstad, A. (2017). Why people use chatbots. In *Internet Science: 4th International Conference, INSCI 2017, Thessaloniki, Greece, November 22-24, 2017* [Springer International Publishing.]. *Proceedings*, *4*, 377–392.

Bredetzky, S. (1807). *Digitale - Sammlungen*. BSBD (Bayerische Staatsbibliothek Digital). https://reader.digitale-sammlungen.de/de/fs1/object/display/bsb10009138_00204.html

Breier, M., Kallmuenzer, A., Clauss, T., Gast, J., Kraus, S., & Tiberius, V. (2020). The role of business model innovation in the hospitality industry during the COVID-19 crisis. *International Journal of Hospitality Management*, *92*, 1–10. doi:10.1016/j.ijhm.2020.102723 PMID:36919038

Bretonès, D. D., Quinio, B., & Réveillon, G. (2010). Bridging virtual and real worlds: Enhancing outlying clustered value creations. *Journal of Strategic Marketing*, *18*(7), 613–625. doi:10.1080/0965254X.2010.529157

Brites, C. (2015). Os cereais no contexto da dieta mediterrânica. In A. Freitas, J. P. Bernardes, M. P. Mateus, & N. Braz (Eds.), *Dimensões da Dieta Mediterrânica* (pp. 181–195). Universidade do Algarve.

British Tourist Authority. (2019). *Speak Up! guide: A guide to marketing your accessibility*. BTA. https://www.gov.uk/government/publications/inclusive-communication/accessible-communication-formats

Bronner, F., & de Hoog, R. (2011). Vacationers and eWOM: Who posts, and why, where, and what? *Journal of Travel Research*, *50*(1), 15–26. doi:10.1177/0047287509355324

Büchi, M. (2021). Digital wellbeing theory and research. *New Media & Society*, 1–18.

Buckman, A. H., Mayfield, M., & Beck, S. B. (2014). What is a smart building? *Smart and Sustainable Built Environment*, *3*(2), 92–109. doi:10.1108/SASBE-01-2014-0003

Buhalis, D. (2003). eTourism: Information Technology for Strategic Tourism Management. Prentice Hall.

Buhalis, D. (n.a.) *Accessible tourism marketing strategies and social media*. Europe Without Barriers. https://www.europewithoutbarriers.eu/download/21_Dimitrios-Buhalis.pdf

Buhalis, D., & Moldavska, I. (2021). In-room voice-based AI digital assistants transforming on-site hotel services and guests' experiences. In *Information and Communication Technologies in Tourism 2021: Proceedings of the ENTER 2021 eTourism Conference, January 19–22, 2021* (30-44). Springer International Publishing.

Buhalis, D. (1998). Strategic use of information technologies in the tourism industry. *Tourism Management*, *19*(5), 409–421. doi:10.1016/S0261-5177(98)00038-7

Buhalis, D. (2020). Technology in tourism-from information communication technologies to eTourism and smart tourism towards ambient intelligence tourism: A perspective article. *Tourism Review*, *75*(1), 267–272. doi:10.1108/TR-06-2019-0258

Buhalis, D., & Darcy, S. (2010). *Accessible tourism: Concepts and issues*. Channel View Publications. doi:10.21832/9781845411626

Buhalis, D., & Darcy, S. (2020). The need for accessible tourism. *Annals of Tourism Research*, *83*, 102944.

Buhalis, D., Darcy, S., & Ambrose, I. (2012). *Best Practice in Accessible Tourism – Inclusion, Disability, Ageing Population and Tourism*. Channel View Publications. doi:10.21832/9781845412548

Buhalis, D., & Law, R. (2008). Progress in information technology and tourism management: 20 years on and 10 years after the Internet—The state of eTourism research. *Tourism Management*, *29*(4), 609–623. doi:10.1016/j.tourman.2008.01.005

Bujdosó, B., & Kemény, F. (2009). *Fogyatékosság és rehabilitáció* [Disability and rehabilitation]. Budapest: Nemzeti Szakképzési és Felnőttképzési Intézet.

Bulchand-Gidumal, J. (2022). Impact of artificial intelligence in travel, tourism, and hospitality. In *Handbook of e-Tourism*. Springer International Publishing. doi:10.1007/978-3-030-48652-5_110

Bulmer, S., Elms, J., & Moore, S. (2018). Exploring the adoption of self-service checkouts and the associated social obligations of shopping practices. *Journal of Retailing and Consumer Services*, *42*, 107–116. doi:10.1016/j.jretconser.2018.01.016

Burgers, A., de Ruyter, K., Keen, C., & Streukens, S. (2000). Customer expectation dimensions of voice-to-voice service encounters: A scale-development study. *International Journal of Service Industry Management*, *11*(2), 142–161. doi:10.1108/09564230010323642

Butler, D. (2020). Computing: Everything, Everywhere. *Nature*, *28*(3), 402–440. PMID:16554773

Büyükuysal, M. Ç. ve Öz, İ. İ. (2016). Çoklu doğrusal bağıntı varlığında en küçük karelere alternatif yaklaşım: Ridge regresyon. *Düzce Üniversitesi Sağlık Bilimleri Enstitüsü Dergisi*, *6*(2), 110–114.

Calabro, A., Vecchiarini, M., Gast, J., Campopiano, G., Massis, A., & Kraus, S. (2019). Innovation in family firms: A systematic literature review and guidance for future research. *International Journal of Management Reviews*, *21*(3), 317–355. doi:10.1111/ijmr.12192

Camilleri, M. A., & Camilleri, M. A. (2018). *Understanding customer needs and wants*. Springer International Publishing. doi:10.1007/978-3-319-49849-2_2

Candello, H., Pinhanez, C., & Figueiredo, F. (2017, May). Typefaces and the perception of humanness in natural language chatbots. In *Proceedings of the 2017 chi conference on human factors in computing systems* (pp. 3476-3487). IEEE. 10.1145/3025453.3025919

Cao, S. (2016). Virtual reality applications in rehabilitation, human-computer interaction. *Theory, Design. Development in Practice*, 3–10.

Cardow, A., & Wiltshier, P. (2010). Indigenous tourism operators: The vanguard of economic recovery in the Chatham islands. *International Journal of Entrepreneurship and Small Business*, *10*(4), 484–498. doi:10.1504/IJESB.2010.034027

Carlisle, S., Kunc, M., Jones, E., & Tiffin, S. (2013). Supporting innovation for tourism development through multi-stakeholder approaches: Experiences from africa. *Tourism Management*, *35*, 59–69. doi:10.1016/j.tourman.2012.05.010

Carral, E., del Río, M., & López, Z. (2020). Gastronomy and Tourism: Socioeconomic and Territorial Implications in Santiago de Compostela-Galiza (NW Spain). *International Journal of Environmental Research and Public Health*, *17*(17), 6173. doi:10.3390/ijerph17176173 PMID:32854422

Carvalho, L., & Costa, T. (2011). Tourism innovation–a literature review complemented by case study research 23-33. *Tourism & Management Studies*, 23–33.

Casaló, L. V., Flavián, C., Guinalíu, M., & Ekinci, Y. (2015). Avoiding the dark side of positive online consumer reviews: Enhancing reviews' usefulness for high risk-averse travelers. *Journal of Business Research*, *68*(9), 1829–1835. doi:10.1016/j.jbusres.2015.01.010

Cascón-Pereira, R., & Hernández-Lara, A. (2013). Town and city management papers Building the 'Morocco'brand as a tourist destination: The role of emigrants and institutional websites. *Journal of Urban Regeneration and Renewal*, *6*(3), 252–263.

Cascón-Pereira, R., & Hernández-Lara, A. B. (2014). The Morocco brand from the Moroccan emigrants' perspective. *Place Branding and Public Diplomacy*, *10*(1), 55–69. doi:10.1057/pb.2013.27

CBI. (2022). *The European market potential for accessible tourism*. CBI. https://www.cbi.eu/market-information/tourism/accessible-tourism-europe/

Cederholm, E. A., & Hultman, J. (2010). The value of intimacy - negotiating commercial relationships in lifestyle entrepreneurship. *Scandinavian Journal of Hospitality and Tourism*, *10*(1), 16–32. doi:10.1080/15022250903442096

Chae, H.-C., Koh, C. E., & Park, K. O. (2018). Information technology capability and firm performance: Role of industry. *Information & Management*, *55*(5), 525–546. doi:10.1016/j.im.2017.10.001

Chang, H. H., & Chiang, C. C. (2022). Is virtual reality technology an effective tool for tourism destination marketing? A flow perspective. *Journal of Hospitality and Tourism Technology*, *13*(3), 427–440. doi:10.1108/JHTT-03-2021-0076

Chang, L., Chen, Y., & Liu, H. (2015). Explaining innovation in tourism-retailing contexts by applying Simon's sciences of the artificial. *Journal of Hospitality and Tourism Technology*, *6*(1), 40–58. doi:10.1108/JHTT-02-2015-0012

Chan, J. H., Iankova, K., Zhang, Y., McDonald, T., & Qi, X. (2016). The role of self-gentrification in sustainable tourism: Indigenous entrepreneurship at Honghe Hani rice terraces world heritage site, china. *Journal of Sustainable Tourism*, *24*(8-9), 1262–1279. doi:10.1080/09669582.2016.1189923

Cheng, H., Jia, R., Li, D., & Li, H. (2019). The rise of robots in China. *The Journal of Economic Perspectives*, *33*(2), 71–88. doi:10.1257/jep.33.2.71

Chen, S. S., Choubey, B., & Singh, V. (2021). A neural network based price sensitive recommender model to predict customer choices based on price effect. *Journal of Retailing and Consumer Services*, *61*, 102573. doi:10.1016/j.jretconser.2021.102573

Chen, Y. F., & Law, R. (2016). A Review of Research on Electronic Word-of-Mouth in Hospitality and Tourism Management. *International Journal of Hospitality & Tourism Administration*, *17*(4), 347–372. doi:10.1080/15256480.2016.1226150

Chen, Y.-Y. K., Jaw, Y.-L., & Wu, B.-L. (2016). Effect of digital transformation on organizational performance of SMEs: Evidence from the Taiwanese textile industry's web portal. *Internet Research*, *26*(1), 186–212. doi:10.1108/IntR-12-2013-0265

Cheong, A., Lau, M. W. S., Foo, E., Hedley, J., & Bo, J. W. (2016). Development of a robotic waiter system. *IFAC-PapersOnLine*, *49*(21), 681–686. doi:10.1016/j.ifacol.2016.10.679

Cheung, C. M. K., & Thadani, D. R. (2012). The impact of electronic word-of-mouth communication: A literature analysis and integrative model. *Decision Support Systems, 54*(1), 461–470. doi:10.1016/j.dss.2012.06.008

Chien, W. C., & Hassenzahl, M. (2020). Technology-mediated relationship maintenance in romantic long-distance relationships: An autoethnographical research through design. *Human-Computer Interaction, 35*(3), 240–287. doi:10.1080/07370024.2017.1401927

Chikuta, O. (2016). Is There Room in the Inn? Towards Incorporating People with Disability in Tourism Planning. *Review of Disability Studies: An International Journal, 11*(3).

Chi, O. H., Denton, G., & Gursoy, D. (2020). Artificially intelligent device use in service delivery: A systematic review, synthesis, and research agenda. *Journal of Hospitality Marketing & Management, 29*(7), 757–786. doi:10.1080/19368623.2020.1721394

Choi, D., R'bigui, H., & Cho, C. (2021). Candidate digital tasks selection methodology for automation with robotic process automation. *Sustainability (Basel), 13*(16), 8980. doi:10.3390u13168980

Choi, D., R'bigui, H., & Cho, C. (2021). Robotic Process Automation Implementation Challenges. *Lecture Notes in Networks and Systems, 149*, 297–304. doi:10.1007/978-981-15-7990-5_29

Choi, J., Ok, C., & Choi, S. (2016). Outcomes of destination marketing organization website navigation: The role of telepresence. *Journal of Travel & Tourism Marketing, 33*(1), 46–62. doi:10.1080/10548408.2015.1024913

Choi, Y., Mehraliyev, F., & Kim, S. S. (2020). Role of virtual avatars in digitalized hotel service. *International Journal of Contemporary Hospitality Management, 32*(3), 977–997. doi:10.1108/IJCHM-03-2019-0265

Choudhary, N., David, A., & Feleen, F. (2021). Employee Engagement and Commitment in Service Sector. *Wesleyan Journal of Research, 13*(4.7), p107-112.

Chou, S., Horng, J., Liu, C., Huang, Y., & Zhang, S. (2020). The critical criteria for innovation entrepreneurship of restaurants: Considering the interrelationship effect of human capital and competitive strategy a case study in Taiwan. *Journal of Hospitality and Tourism Management, 42*, 222–234. doi:10.1016/j.jhtm.2020.01.006

Chu, F. L. (2014). Using a logistic growth regression model to forecast the demand for tourism in Las Vegas. *Tourism Management Perspectives, 12*, 62–67. doi:10.1016/j.tmp.2014.08.003

Chunhua, S., & Guangqing, S. (2020). Application and development of 3D printing in medical field. *Modern Mechanical Engineering, 10*(3), 25–33. doi:10.4236/mme.2020.103003

Chwiłkowska-Kubala, A., Malewska, K., & Mierzejewska, K. (2022). Digital transformation of energy sector enterprises in Poland. *Scientific Papers of Silesian University of Technology – Organization and Management Series, 162*, 101-120. http://dx.doi.org/ doi:10.29119/1641-3466.2022.162.5

Chwiłkowska-Kubala, A., Cyfert, S., Malewska, K., Mierzejewska, K., & Szumowski, W. (2021). The relationships among social, environmental, economic CSR practices and digitalization in Polish energy companies. *Energies, 14*(22), 7666. doi:10.3390/en14227666

Ciani, A., & Vörös, M. L. (2020). Rural Tourism and Agrotourism as Drivers of the Sustainable Rural Development – a Proposal for a Cross-Border Cooperation Strategy, In Burkiewicz, L. & Knap- Stefaniuk, A. (eds.) Management Tourism Culture. Studies and Reflections on Tourism Management. Ignatianum University Press Kraków.

Cichosz, M., Wallenburg, C. M., & Knemeyer, A. M. (2020). Digital transformation at logistics service providers: Barriers, success factors and leading practices. *International Journal of Logistics Management, 31*(2), 209–238. doi:10.1108/IJLM-08-2019-0229

CKH. (2012). *Az "Akadálymentes turizmus mindenkinek" ökonómiai impulzusai* [The economic impulses of "Accessible tourism for all"]. A Gazdasági és Technológiai Szövetségi Minisztérium megbízásából készült vizsgálat. Retrieved February 15, 2023, from http://www.ckh.hu/sites/default/files/fajlok_projekt/2012/2012-05-22- okonomiaiimpulzusoke-gyakadalymentesturizmusertm.pdf

Clerides, S., Nearchou, P., & Pashardes, P. (2006). Intermediaries as Quality Assessors: Tour Operators in the Travel Industry. SSRN *Electronic Journal*. doi:10.2139/ssrn.505282

Cobo, M. J., Martínez, M. A., Gutiérrez-Salcedo, M., Fujita, H., & Herrera-Viedma, E. (2015). 25 years at Knowledge-Based Systems: A bibliometric analysis. *Knowledge-Based Systems, 80*, 3–13. doi:10.1016/j.knosys.2014.12.035

Cocca, P., Marciano, F., Rossi, D., & Alberti, M. (2018). Business software offer for industry 4.0: The SAP case. *IFAC-PapersOnLine, 51*(11), 1200–1205. doi:10.1016/j.ifacol.2018.08.427

Cooper, C. (1993). The tourist destination-Introduction. In C. Cooper, J. Fletcher, D. Gilbert, & S. Wanhill (Eds.), *Tourism:principles and practice* (pp. 77–79). Pitman Publishing.

Covas, A., & Covas, M. M. (2015). A Dieta Mediterrânica: entre a tradição e a inovação – Uma oportunidade para o rural tradicional algarvio. In A. Freitas, J.P. Bernardes, M.P. Mateus & N. Braz (Eds.). *Dimensões da Dieta Mediterrânica* (pp. 277-294). Faro: Universidade do Algarve Dieta Mediterrânica (2018).

Covas, A., & Covas, M. M. (2014). *Os territórios-rede, a inteligência territorial da 2.ª ruralidade.* Editora Colibri.

Creswell, J. W. (2007). *Qualitative inquiry research design choosing among five approaches.* Sage Publications.

Cretu, R. C., Stefan, P., & Alecu, I. I. (2021). Has tourism gone on holiday? Analysis of the effects of the COVID-19 pandemic on tourism and post-pandemic tourism behavior. *Scientific Papers. Series Management, Economic, Engineering in Agriculture and Rural Development, 21*, 191–197.

Cronin, J. J. Jr, & Taylor, S. A. (1994). SERVPERF versus SERVQUAL: Reconciling performance-based and perceptions-minus-expectations measurement of service quality. *Journal of Marketing, 58*(1), 125–131. doi:10.1177/002224299405800110

Crook, J. (2014). *Starwood introduces robotic butlers at aloft hotel in Cupertino.* Tech Crunch. https://techcrunch.com/2014/08/13/starwood-introduces-robotic-butlers-at-aloft-hotel-in-paloalto/#:%7E:text=Starwood%2C%20one%20of%20the%20world's,around%20guests%20and%20use%20elevators

Crossley, É. (2012). Poor but happy: Volunteer tourists' encounters with poverty. *Tourism Geographies, 14*(2), 235–253. doi:10.1080/14616688.2011.611165

Crouch, G. I. (2000). Services research in destination marketing: A retrospective and prospective appraisal. *International Journal of Hospitality & Tourism Administration, 1*(2), 65–86. doi:10.1300/J149v01n02_04

Csizmazia, D. J. (2001). To protect our health. *Manuscript., 28*(October), 1–4.

Çuhadar, M. (2013). Türkiye'ye yönelik dış turizm talebinin MLP, RBF ve TDNN yapay sinir ağı mimarileri ile modellenmesi ve tahmini: Karşılaştırmalı bir analiz. *Journal of Yasar University, 8*(31), 5274–5295.

Çuhadar, M. (2020). Türkiye'nin dış aktif turizm gelirlerinin alternatif yaklaşımlarla modellenmesi ve tahmini. *Ankara Hacı Bayram Veli Üniversitesi Turizm Fakültesi Dergisi, 23*(1), 115–141. doi:10.34189/tfd.23.01.006

Çuhadar, M., & Güngör, İ. ve Göksu, A. (2009). Turizm talebinin yapay sinir ağları ile tahmini ve zaman serisi yöntemleri ile karşılaştırmalı analizi: Antalya iline yönelik bir uygulama. *Süleyman Demirel Üniversitesi İktisadi ve İdari Bilimler Fakültesi Dergisi, 14*(1), 99–114.

Czerwińska, T., Głogowski, A., Gromek, T., & Pisany, P. (2021). Digital transformation in banks of different sizes: Evidence from the Polish banking sector. In I. Boitan & K. Marchewka-Bartkowiak (Eds.), *Fostering Innovation and Competitiveness With FinTech, RegTech, and SupTech* (pp. 161–185). IGI Global. doi:10.4018/978-1-7998-4390-0.ch009

Dabiri, S., & Abbas, M. (2018). Evaluation of the gradient boosting of regression trees method on estimating car-following behavior. *Transportation Research Record: Journal of the Transportation Research Board, 2672*(45), 136–146. doi:10.1177/0361198118772689

Darcy, S. (1998). *Anxiety to access: Tourism patterns and experiences of New South Wales people with a physical disability*. Sydney, Australia: Tourism New South Wales.

Darcy, S. (2002). Marginalised participation: Physical disability, high support needs and tourism. *Journal of Hospitality and Tourism Management, 9*(1), 61–72.

Darcy, S. (2007). A methodology for testing accessible accomodation information provision formats. In I. McDonnell, S. Grabowski, & R. March (Eds.), *CAUTHE 2007: Tourism - Past Achievements, Future Challenges* (pp. 1–18). University of Technology Sydney.

Darcy, S., & Dickson, T. (2009). A whole-of-life approach to tourism: The case for accessible tourism experiences. *Journal of Hospitality and Tourism Management, 16*(1), 32–44. doi:10.1375/jhtm.16.1.32

Datourway. (2011). *The development of sustainable tourism in the Béda-Karapancsa area - Tourism development survey, strategy and action plan - Short version*. Béda-Karapancsa Pilot Project (Hungary-Croatia-Serbia). https://www.eubusiness.com/topics/food/door

Davari, D., Vayghan, S., Jang, S., & Erdem, M. (2022). Hotel experiences during the COVID-19 pandemic: High-touch versus high-tech. *International Journal of Contemporary Hospitality Management, 34*(4), 1312–1330. doi:10.1108/IJCHM-07-2021-0919

Davenport, T., Guha, A., Grewal, D., & Bressgott, T. (2020). How artificial intelligence will change the future of marketing. *Journal of the Academy of Marketing Science, 48*(1), 24–42. doi:10.100711747-019-00696-0

David, A., Nagarjuna, K., Mohammed, M., & Sundar, J. (2019a). Determinant Factors of Environmental Responsibility for the Passenger Car Users. *International Journal of Innovative Technology and Exploring Engineering,* 2278-3075.

David, A., Ravi, S., & Reena, R. A. (2018). The Eco-Driving Behaviour: A Strategic Way to Control Tailpipe Emission. *International Journal of Engineering & Technology, 7*(3.3), 21-25.

David, A. (2020). Consumer purchasing process of organic food product: An empirical analysis. [QAS]. *Journal of Management System-Quality Access to Success, 21*(177), 128–132.

David, A., Ganesh Kumar, C., & Jeganathan, G. S. (2022a). Impact of Food Safety and Standards Regulation on Food Business Operators. In *Au Virtual International Conference* (pp. 355-363). SSRN.

David, A., Kumar, C. G., & Paul, P. V. (2022b). Blockchain technology in the food supply chain: Empirical analysis. [IJISSCM]. *International Journal of Information Systems and Supply Chain Management, 15*(3), 1–12. doi:10.4018/IJISSCM.290014

David, A., Thangavel, Y. D., & Sankriti, R. (2019b). Recover, recycle and reuse: An efficient way to reduce the waste. *Int. J. Mech. Prod. Eng. Res. Dev, 9*, 31–42.

De Bernardi, P., Bertello, A., & Shams, R. (2019). Logics hindering digital transformation in cultural heritage strategic management: An Exploratory Case Study. *Tourism Analysis, 24*(3), 315–327. doi:10.3727/108354219X15511864843876

De Freitas, C. R. (2017). Tourism climatology past and present: A review of the role of the ISB Commission on Climate, Tourism and Recreation. *International Journal of Biometeorology*, *61*(S1), 107–114. doi:10.100700484-017-1389-y PMID:28647761

Debasa, F. (2022). Digital wellbeing tourism in the fourth industrial revolution. *Journal of Tourism Sustainability and Well-being*, *10*(3), 227–237.

Dębski, M., Borkowska-Niszczota, M., & Andrzejczyk, R. (2021). Tourist Accommodation Establishments during the Pandemic – Consequences and Aid Report on a Survey among Polish Micro-enterprises Offering Accommodation Services. *Journal of Intercultural Management*, *13*(1), 1–25. doi:10.2478/joim-2021-0001

Decelle, X. (2004). A conceptual and dynamic approach to innovation in tourism. *Maître de Conférences*. https://www.oecd.org/cfe/tourism/34267921.pdf

Decker, M. (2008). Caregiving robots and ethical reflection: The perspective of interdisciplinary technology assessment. *AI & Society*, *22*(3), 315–330. doi:10.100700146-007-0151-0

Dengler, K., & Matthes, B. (2018). The impacts of digital transformation on the labor market: Substitution potentials of occupations in Germany. *Technological Forecasting and Social Change*, *137*, 304–316. doi:10.1016/j.techfore.2018.09.024

Department of Health. (2010). *Making written information easier to understand for people with learning disabilities: Guidance for people who commission or produce easy read information*. Department of Health.

Deri, M. N., Ari Ragavan, N., Niber, A., Zaazie, P., Akazire, D. A., Anaba, M., & Andaara, D. (2023). COVID-19 shock in the hospitality industry: Its effect on hotel operations within the Bono region of Ghana. *African Journal of Economic and Management Studies*. doi:10.1108/AJEMS-07-2022-0264

Derudder, B., Liu, X., Hong, S., Ruan, S., Wang, Y., & Witlox, F. (2019). The shifting position of the Journal of Transport Geography in 'transport geography research': A bibliometric analysis. *Journal of Transport Geography*, *81*, 1–9. doi:10.1016/j.jtrangeo.2019.102538

Dias, Á. L., Silva, R., Patuleia, M., Estêvão, J., & González-Rodríguez, M. R. (2022). Selecting lifestyle entrepreneurship recovery strategies: A response to the COVID-19 pandemic. *Tourism and Hospitality Research*, *22*(1), 115–121. doi:10.1177/1467358421990724

Dias, Á., Cascais, E., Pereira, L., Lopes da Costa, R., & Gonçalves, R. (2022). Lifestyle entrepreneurship innovation and self-efficacy: Exploring the direct and indirect effects of marshaling. *International Journal of Tourism Research*, *24*(3), 443–455. doi:10.1002/jtr.2513

Dias, Á., & Silva, G. M. (2021). Lifestyle entrepreneurship and innovation in rural areas: The case of tourism entrepreneurs. *Journal of Small Business Strategy*, *31*(4), 40–49. doi:10.53703/001c.29474

Dias, Á., Silva, G. M., Patuleia, M., & González-Rodríguez, M. R. (2020). Developing sustainable business models: Local knowledge acquisition and tourism lifestyle entrepreneurship. *Journal of Sustainable Tourism*. doi:10.1080/09669582.2020.1835931

Dias, Á., Silva, G. M., Patuleia, M., & González-Rodríguez, M. R. (2021). Transforming local knowledge into lifestyle entrepreneur's innovativeness: Exploring the linear and quadratic relationships. *Current Issues in Tourism*, *24*(22), 3222–3238. doi:10.1080/13683500.2020.1865288

Dickey, D. A., & Fuller, W. A. (1979). Distribution of the estimators for autoregressive time series with a unit root. *Journal of the American Statistical Association*, *74*(366), 427–431. doi:10.2307/2286348

Dickey, D. A., & Fuller, W. A. (1981). Likelihood ratio statistics for autoregressive time series with a unit root. *Econometrica*, *49*(4), 1057–1072. doi:10.2307/1912517

Dieck, M. C., & Jung, T. (2018). A theoretical model of mobile augmented reality acceptance in urban heritage tourism. *Current Issues in Tourism*, *21*(2), 154–174. doi:10.1080/13683500.2015.1070801

Dieck, M. C., & Jung, T. H. (2017). Value of augmented reality at cultural heritage sites: A stakeholder approach. *Journal of Destination Marketing & Management*, *6*(2), 110–117. doi:10.1016/j.jdmm.2017.03.002

Dieck, M. C., Jung, T., & Moorhouse, N. (2018). Tourists' virtual reality adoption: An exploratory study from Lake District National Park. *Leisure Studies*, *37*(4), 371–383. doi:10.1080/02614367.2018.1466905

Diener, F., & Špaček, M. (2021). Digital Transformation in Banking: A Managerial Perspective on Barriers to Change. *Sustainability*, *13*(4), 1–26. doi:10.3390u13042032

Ding, D. X., Hu, P. J. H., & Sheng, O. R. L. (2011). e-SELFQUAL: A scale for measuring online self-service quality. *Journal of Business Research*, *64*(5), 508–515. doi:10.1016/j.jbusres.2010.04.007

Dinnie, K. (2015). *Nation branding: Concepts, issues, practice*. Routledge. doi:10.4324/9781315773612

Dogruel, L., Joeckel, S., & Bowman, N. D. (2015). The use and acceptance of new media entertainment technology by elderly users: Development of an expanded technology acceptance model. *Behaviour & Information Technology*, *34*(11), 1052–1063. doi:10.1080/0144929X.2015.1077890

Domínguez Vila, T., Alén González, E., & Darcy, S. (2017). Website accessibility in the tourism industry: An analysis of official national tourism organisation websites around the world. *Disability and Rehabilitation*. doi:10.1080/09638288.2017.1362709 PMID:28793789

Donthu, N., Kumar, S., Mukherjee, D., Pandey, N., & Lim, W. M. (2021). How to conduct a bibliometric analysis: An overview and guidelines. *Journal of Business Research*, *133*, 285–296. doi:10.1016/j.jbusres.2021.04.070

Donthu, N., Kumar, S., Pandey, N., Pandey, N., & Mishra, A. (2021). Mapping the electronic word-of-mouth (eWOM) research: A systematic review and bibliometric analysis. *Journal of Business Research*, *135*, 758–773. doi:10.1016/j.jbusres.2021.07.015

Doukas, H., Patlitzianas, K. D., Iatropoulos, K., & Psarras, J. (2007). Intelligent building energy management system using rule sets. *Building and Environment*, *42*(10), 3562–3569. doi:10.1016/j.buildenv.2006.10.024

Dritsakis, N. (2004). Cointegration analysis of German and British tourism demand for Greece. *Tourism Management*, *25*(1), 111–119. doi:10.1016/S0261-5177(03)00061-X

Du Rand, G. E., & Heath, E. (2006). Towards a Framework for Food Tourism as an Element of Destination Marketing. May 2006. *Current Issues in Tourism*, *9*(3), 206–234. doi:10.2167/cit/226.0

Duhart, F. (2020). Territorial Food Identities Tips for Gastronomy Actors. WGI Global Report 2020. WGI.

Dulházi, F., & Zsarnóczky, M. (2018). Az akadálymentes turizmus, mint rehabilitációs "eszköz" [Accessible tourism as a rehabilitation "tool"]. *Conference Paper. LX. Georgikon napok, 60th Georgikon Scientific Conference*, 56-61.

Dupeyras, A. (2016). *Growth of Sharing Economy in Tourism: Developing a Balanced Policy Response*. OECD.

Durgampudi, K. (2022, November 21). How Robotic Process Automation Can Quietly Change The Way People Work. *Forbes*. https://www.forbes.com/sites/forbestechcouncil/2022/11/21/how-robotic-process-automation-can-quietly-change-the-way-people-work/?sh=3179e31a5fa1

Dwivedi, Y. K., & Wang, Y. (2022). Guest editorial: Artificial intelligence for B2B marketing: Challenges and opportunities. *Industrial Marketing Management, 105,* 109–113. doi:10.1016/j.indmarman.2022.06.001

Dybsand, H. N. H. (2022). 'The next best thing to being there'–participant perceptions of virtual guided tours offered during the COVID-19 pandemic. *Current Issues in Tourism,* 1–14. doi:10.1080/13683500.2022.2122417

EC - EUSDR. (2019). European Commission. European Union Strategy for the Danube Region (EUSDR). Danub Economics. https://www.danubecommission.org/dc/en/danube; https://www.interreg-danube.eu/about-dtp/eu-strategy-for-the-danube-region

Edwards, S. M. (2001). The technology paradox: Efficiency versus creativity. *Creativity Research Journal, 13*(2), 221–228. doi:10.1207/S15326934CRJ1302_9

Eichhorn, V., & Buhalis, D. (2011). Accessibility – A Key Objective for the Tourism Industry. In D. Buhalis & S. Darcy (Eds.), *Accessible Tourism: Concepts and Issues* (pp. 46–61). Channel View Publications.

Eichhorn, V., Miller, G., Michopoulou, E., & Buhalis, D. (2008). Enabling disabled tourists? Accessibility tourism information schemes. *Annals of Tourism Research, 35*(1), 189–210. doi:10.1016/j.annals.2007.07.005

EK. (2004). *Gleichstellung, Vielfalt und Erweiterung. Luxemburg: Amt für amtliche Veröffentlichungen der Europäischen Gemeinschaften.* Europäische Kommission.

El Aouni, F. (2015). *Destination branding and the role of emigrants: The case of morocco.* [PhD Thesis, Universitat Rovira i Virgili].

Elkhwesky, Z., El Manzani, Y., & Elbayoumi Salem, I. (2022). Driving hospitality and tourism to foster sustainable innovation: A systematic review of COVID-19-related studies and practical implications in the digital era. *Tourism and Hospitality Research, 14673584221126792.* doi:10.1177/14673584221126792

Emmendoerfer, M., Chagas de Almeida, T., Richards, G., & Marques, L. (2023). Co-creation of local gastronomy for regional development in a slow city. *Tourism & Management Studies, 19*(2), 51–60. doi:10.18089/tms.2023.190204

ENAT. (2013). *Accessibility review of European national tourist boards' websites 2012.* ENAT. https://www.accessibletourism.org/resources/enat-nto-websites-study-2012_public.pdf

England Athletics (n.a.). *Providing accessible information formats.* England Athletics. https://d192th11qal2xm.cloudfront.net/2018/11/Providing-accessible-information-guidance-PDF-189kB.pdf (accessed:03.08.2020)

ENSZ .(2006). *A Fogyatékossággal élő személyek jogairól szóló egyezmény [Convention on the Rights of Persons with Disabilities].* ENSZ. file:///C:/Users/kk/Downloads/A%20fogyat%C3%A9koss%C3%A1ggal%20%C3%A9l%C5%91k%20 jogair%C3%B3l%20sz%C3%B3l%C3%B3%20ENSZ%20egyezm%C3%A9ny.pdf

Erdei, F. (1971). Ethnographic cuisine (in Hungarian). Osiris, Budapest.

Ert, E., Fleischer, A., & Magen, N. (2016). Trust and reputation in the sharing economy: The role of personal photos in Airbnb. *Tourism Management, 55,* 62–73. doi:10.1016/j.tourman.2016.01.013

Eshghi, B. (2022). *Top 12 Use Cases of RPA in Hotels & Hospitality Industry in 2023.* AI Multiple. https://research.aimultiple.com/rpa-in-hospitality-industry/

Eshghi, B. (2023). *RPA in Food Industry: Top 11 Use Cases in 2023.* AI Multiple. https://research.aimultiple.com/rpa-in-food-industry/

Esping-Andersen, G. (1991). Mi a jóléti állam? In Zs. Ferge & K. Lévai (Eds.), *A jóléti állam [The welfare state].* (pp. 116–134). T-Twins.

Estol, J., Camilleri, M. A., & Font, X. (2018). European Union tourism policy: An institutional theory critical discourse analysis. *Tourism Review*, *73*(3), 421–431. doi:10.1108/TR-11-2017-0167

European Commission (2014). *Economic impact and travel patterns of accessible tourism in Europe – Final report*. EC. file:///C:/Users/kk/Downloads/study%20A%20Economic%20Impact%20and%20travel%20patterns%20of%20accessible%20tourism_final

Facebook supprime près de 400 faux comptes. (2021, March 4). Hespress Français. https://fr.hespress.com/192812-facebook-supprime-pres-de-400-faux-comptes-lies-a-une-propagande-dun-media-marocain.html

Fanack. (2023). Governance & Politics of Morocco. *Fanack.Com*. https://fanack.com/morocco/politics-of-morocco/

Fejes, A. (2019). *Newer data on the domestic spread of Paprika and the usage of Paprika pálinka* (Manuscript in Hungarian).

Fejes, A. (2020). Living tradition of making fish soup in Baja. Nomination data sheet for admission to the national register of intangible cultural heritage in Hungary. (Manuscript in Hungarian) p.10.

Fekete, M. (2006). *Hétköznapi turizmus. A turizmuselmélettől a gyakorlatig [Casual tourism. From tourism theory to practice]* [Unpublished doctoral dissertation, Nyugat-magyarországi Egyetem, Sopron, Hungary].

Feleen, F., & David, A. (2021). A Comparative Study of Work From Home vs Work From Office: Preference of Women Employees in IT Industry. *Design Engineering (London)*, *7*(1), 5763–5775.

Feller, C. (2021). 10 robots automating the restaurant industry. *Fast Casual*. https://www.fastcasual.com/blogs/10-robotsautomating-the-restaurant-industry/

Feng, C. M., Park, A., Pitt, L., Kietzmann, J., & Northey, G. (2021). Artificial intelligence in marketing: A bibliographic perspective. *Australasian Marketing Journal*, *29*(3), 252–263. doi:10.1016/j.ausmj.2020.07.006

Feng, J. T. (2015). Innovation and integration of hotel management under the new situation. *China Business*, *3*(1), 32–34.

Fernandes, S. (2021). Which way to cope with covid-19 challenges? Contributions of the IoT for smart city projects. *Big Data and Cognitive Computing*, *5*(2), 26. doi:10.3390/bdcc5020026

Ferri, M. A., & Aiello, L. (2017). Tourism destination management in sustainability development perspective, the role of entrepreneurship and networking ability: Tourist kit. *World Review of Entrepreneurship, Management and Sustainable Development*, *13*(5-6), 647–664. doi:10.1504/WREMSD.2017.086334

Filieri, R., & McLeay, F. (2014). E-WOM and Accommodation: An Analysis of the Factors That Influence Travelers' Adoption of Information from Online Reviews. *Journal of Travel Research*, *53*(1), 44–57. doi:10.1177/0047287513481274

Fischer, M., Imgrund, F., Janiesch, C., & Winkelmann, A. (2020). Strategy archetypes for digital transformation: Defining meta objectives using business process management. *Information & Management*, *57*(5), 103262. doi:10.1016/j.im.2019.103262

Fleming, W. R., & Toepper, L. (1990). Economic impact studies: Relating the positive and negative impacts to tourism development. *Journal of Travel Research*, *29*(1), 35–42. doi:10.1177/004728759002900108

Fong, S. C. (1998). Conceptualizing consumer experiences in cyberspace. *European Journal of Marketing*, *32*(7/8), 655–663. doi:10.1108/03090569810224056

Font, X., Bonilla-Priego, M. J., & Kantenbacher, J. (2019). Trade associations as corporate social responsibility actors: An institutional theory analysis of animal welfare in tourism. *Journal of Sustainable Tourism*, *27*(1), 118–138. doi:10.1080/09669582.2018.1538231

Forbes, S. L., De Silva, T., & Gilinsky, A. (2019). Social sustainability in the global wine industry: Concepts and cases, pp. 1-204. SCOPUS. www.scopus.com doi:10.1007/978-3-030-30413-3

Fornell, C., & Larcker, D. F. (1981). *Structural equation models with unobservable variables and measurement error: Algebra and statistics.* Sage.

Frank, M., Roehrig, P., & Pring, B. (2017). *What To Do When Machines Do Everything: How to Get Ahead in a world of AI, algorithms, bots, and big data.* John Wiley & Sons, Inc.

Frank, O. L. (2007). Intelligent building concept: The challenges for building practitioners in the 21st century. [AARCHES J]. *J. Assoc. Archit. Educ. Niger.*, *6*(3), 107–113.

Freitas, A., Bernardes, J. P., Mateus, M. P., & Braz, N. (2015). *Dimensões da Dieta Mediterrânica.* Universidade do Algarve.

Freitas, A., Braz, N., Bernardes, J. P., Cruz, A. L., Quintas, C., Gonçalves, A., Romano, A., & Mateus, M. P. (2022). *Mediterranean Diet: a multidisciplinary approach to develop a new territorial strategy.*

Friedlander, A., & Zoellner, C. (2020). Artificial intelligence opportunities to improve food safety at retail. *Food Protection Trends*, *40*(4), 272–278.

Friedman, J. H. (1999). Greedy function approximation: A stochastic boosting machin. Technical Report. Department of Statistics Stanford University.

Fung, H. P. (2013). Criteria, Use Cases and Effects of Information Technology Process Automation (ITPA). *Advances in Robotics & Automation*, *03*(03). doi:10.4172/2168-9695.1000124

Gabryelczyk, R. (2020). Has COVID-19 accelerated digital transformation? Initial lessons learned for public administrations. *Information Systems Management*, *37*(4), 303–309. doi:10.1080/10580530.2020.1820633

Gačnik, A. (2012). Gastronomy heritage as a source of development for gastronomy tourism and as a means of increasing Slovenian's tourism visibility. *Academica Turistica: Tourism & Innovation Journal*, 39-60.

Gačnik, A., & Vörös, M. (2018). Protected Food & Wine Products as a Driving Force for Creativity and Innovation of Gastronomy Tourism Development: Case of Slovenia and Hungary. *Agriculture, 15*(1-2), 19-34. https://www.agricultura-online.com/portal/index.php/issues/issue-21

Gajić, T., Đoković, F., Blešić, I., Petrović, M. D., Radovanović, M. M., Vukolić, D., Mandarić, M., Dašić, G., Syromiatnikova, J. A., & Mićović, A. (2023). Pandemic Boosts Prospects for Recovery of Rural Tourism in Serbia. *Land (Basel)*, *12*(3), 624. doi:10.3390/land12030624

Galdon, J. L., Garrigos, F., & Gil-Pechuan, I. (2013). Leakage, entrepreneurship, and satisfaction in hospitality. *Service Industries Journal*, *33*(7-8), 759–773. doi:10.1080/02642069.2013.740464

Gálvez, J. C. P., Granda, M. J., & Guzmán-López, T. (2017). Local gastronomy, culture and tourism sustainable cities: The behavior of the American tourist. *Sustainable Cities and Society*, *32*, 604–512. doi:10.1016/j.scs.2017.04.021

Gamanyuk, A. (2017). Restaurant table reservation chatbot for Facebook bootmaker.

Ganeshkumar, C., & David, A. (2022, August). Digital Information Management in Agriculture—Empirical Analysis. In *Proceedings of the Third International Conference on Information Management and Machine Intelligence: ICIMMI 2021* (pp. 243-249). Springer Nature Singapore.

Ganeshkumar, C., David, A., & Jebasingh, D. R. (2022). Digital transformation: artificial intelligence based product benefits and problems of Agritech industry. In *Agri-Food 4.0.* Emerald Publishing Limited. doi:10.1108/S1877-361120220000027010

Ganeshkumar, C., David, A., Sankar, J. G., & Saginala, M. (2023a). Application of Drone Technology in Agriculture: A Predictive Forecasting of Pest and Disease Incidence. In *Applying Drone Technologies and Robotics for Agricultural Sustainability* (pp. 50–81). IGI Global.

Ganeshkumar, C., Prabhu, M., Reddy, P. S., & David, A. (2020). Value chain analysis of Indian edible mushrooms. *International Journal of Technology*, *11*(3), 599–607. doi:10.14716/ijtech.v11i3.3979

Ganeshkumar, C., Sankar, J. G., & David, A. (2023b). Adoption of Big Data Analytics: Determinants and Performances Among Food Industries. [IJBIR]. *International Journal of Business Intelligence Research*, *14*(1), 1–17. doi:10.4018/ IJBIR.317419

Gao, J., van Zelst, S. J., Lu, X., & van der Aalst, W. M. P. (2019). Automated Robotic Process Automation: A Self-Learning Approach. *On the Move to Meaningful Internet Systems: OTM 2019 Conferences: Confederated International Conferences: CoopIS, ODBASE, C&TC 2019, Rhodes, Greece, October 21–25, 2019, Proceedings*, (pp. 95–112). Springer. 10.1007/978-3-030-33246-4_6

Garay, L., Font, X., & Corrons, A. (2019). Sustainability-oriented innovation in tourism: An analysis based on the decomposed theory of planned behavior. *Journal of Travel Research*, *58*(4), 622–636. doi:10.1177/0047287518771215

Gartner, Inc. (2016). *Gartner says worldwide wearable devices sales to grow 18.4 percent in 2016*. Gartner, Inc. http:// www.gartner.com/newsroom/id/ 3198018/

Gebayew, C., Hardini, I. R., Panjaitan, G. H. A., & Kurniawan, N. B., & Suhardi, (2018). A systematic literature review on digital transformation. *International Conference on Information Technology Systems and Innovation* (pp. 260-265). IEEE. https://doi.org/10.1109/ICITSI.2018.8695912

Geng, W. (2023). Whether and how free virtual tours can bring back visitors. *Current Issues in Tourism*, *26*(5), 823–834. doi:10.1080/13683500.2022.2043253

Geoffrey, C. (2007). *Modelling destination competitivness – A szrvet abd abakzsis of the impact of competitivness atrib-utes*. CRC for Sustainable Tourism Pty Ltd.

GFK. (2014). *Economic impact and travel patterns of accessible tourism in Europe*. GFK. https://ec.europa.eu/docsroom/ documents/5567/attachments/1/translations/en/renditions/native

Ghavifekr, S., & Ibrahim, M. S. (2015). Effectiveness of ICT integration inMalaysian schools: A quantitative analysis. *International ResearchJournal for Quality in Education*, *2*(8), 1–12. /ijres.net/index.php/ijres/article/view/79/43

Giannopoulos, K., Tsartas, P., & Anagnostelos, K. (2022). A targeted multi-parameter approach of Greek start-ups, related to tourism, culture, and leisure. *Paper presented at the Springer Proceedings in Business and Economics,* (pp. 215-226). IEEE. 10.1007/978-3-030-92491-1_13

Giddens, A. (2008). *Szociológia* [Sociology] (2nd ed.). Osiris.

Gienow-Hecht, J. (2019). Nation branding: A useful category for international history. *Diplomacy and Statecraft*, *30*(4), 755–779. doi:10.1080/09592296.2019.1671000

Giuliani, M., Petrick, R. P. A., Foster, M. E., Gaschler, A., Isard, A., Pateraki, M., & Sigalas, M. (2013). Comparing task-based and socially intelligent behavior in a robot bartender. Paper presented at the *ICMI 2013 – 2013 ACM International Conference on Multimodal Interaction*, (pp. 263–270). ACM.

Goffi, G., & Cucculelli, M. (2019). Explaining tourism competitiveness in small and medium destinations: The Italian case. *Current Issues in Tourism*, *22*(17), 2109–2139. doi:10.1080/13683500.2017.1421620

Gold, J., & Mahrer, N. E. (2018). Is virtual reality ready for prime time in the medical space? A randomized control trial of pediatric virtual reality for acute procedural pain management. *Journal of Pediatric Psychology, 43*(3), 266–275. doi:10.1093/jpepsy/jsx129 PMID:29053848

Goleman, D. (1996). Emotional intelligence. Why it can matter more than IQ. *Learning, 24*(6), 49–50.

Gölgeci, I., Ali, I., Ritala, P., & Arslan, A. (2022). A bibliometric review of service ecosystems research: Current status and future directions. *Journal of Business and Industrial Marketing, 37*(4), 841–858. doi:10.1108/JBIM-07-2020-0335

Gomez V. & Maravall A. (1996). *Programs TRAMO and SEATS: instructions for the user.* Banco de España. Servicio de Estudios.

Gondos, B. (2020). *Speciális igények a turizmusban – A mozgáskorlátozottak helye, szerepe és lehetősége a turisztikai szektorban [Special needs in tourism – Role and opportunity of people with reduced mobility in tourism sector]* [Unpublished doctoral dissertation, Széchenyi István University, Győr, Hungary].

Gössling, S. (2017). Tourism, information technologies and sustainability: An exploratory review. *Journal of Sustainable Tourism, 25*(7), 1024–1041. doi:10.1080/09669582.2015.1122017

Gössling, S. (2021). Tourism, technology and ICT: A critical review of affordances and concessions. *Journal of Sustainable Tourism, 29*(5), 733–750. doi:10.1080/09669582.2021.1873353

Gössling, S., & Michael Hall, C. (2019). Sharing versus collaborative economy: How to align ICT developments and the SDGs in tourism? *Journal of Sustainable Tourism, 27*(1), 74–96. doi:10.1080/09669582.2018.1560455

Goudey, A., & Bonnin, G. (2016). Must smart objects look human? Study of the impact of anthropomorphism on the acceptance of companion robots. [English Edition]. *Recherche et Applications en Marketing, 31*(2), 2–20. doi:10.1177/2051570716643961

GOV.UK. (n.a.). *Why GOV.UK content should be published in HTML and not PDF.* GDS. https://gds.blog.gov.uk/2018/07/16/why-gov-uk-content-should-be-published-in-html-and-not-pdf/ (accessed: 23.04.2022)

Goyal, N., & Singh, H. (2021). A Design of Customer Service Request Desk to Improve the Efficiency using Robotics Process Automation. *2021 6th International Conference on Signal Processing, Computing and Control (ISPCC),* (pp. 21–24). IEEE. 10.1109/ISPCC53510.2021.9609338

Graan, A. (2016). The nation brand regime: Nation branding and the semiotic regimentation of public communication in contemporary Macedonia. *Signs and Society (Chicago, Ill.), 4*(S1), S70–S105. doi:10.1086/684613

Graça, P. (2014a). Breve história do conceito de dieta Mediterrânica numa perspetiva de saúde. *Revista Fatores de Risco, 31,* 20–22.

Graça, P. (2014b). Dieta Mediterrânica: uma realidade multifacetada. In A. Freitas, J. P. Bernardes, M. P. Mateus, & N. Braz (Eds.), *Dimensões da Dieta Mediterrânica* (pp. 19–27). Universidade do Algarve.

Granger, C. W. J. (2001). Spurious regressions in econometrics. In B. H. Baltagı (ed.), A Companion to Theoretical Econometrics. Oxford: Blackwell. doi:10.1017/CCOL052179207X.006

Granger, C. W. J., & Newbold, P. (1974). Spurious regressions in econometrics. *Journal of Econometrics, 2*(2), 111–120. doi:10.1016/0304-4076(74)90034-7

Gretzel, U., & Yoo, K. H. (2008). Use and Impact of Online Travel Reviews. In P. O'Connor, W. Höpken, & U. Gretzel (Eds.), *Information and Communication Technologies in Tourism 2008: Proceedings of the International Conference in Innsbruck,* (pp. 35–46). Springer. 10.1007/978-3-211-77280-5_4

Grey, A. (2016). *This robot chef wants to know how you like your pancakes.* We Forum. https://www.weforum.org/agenda/2016/10/robot-chef-makes-pancakes-japan-hennna/

Griffin, T., Guttentag, D., Lee, S. H., Giberson, J., & Dimanche, F. (2023). Is VR always better for destination marketing? Comparing different media and styles. *Journal of Vacation Marketing, 29*(1), 119–140. doi:10.1177/13567667221078252

Grönroos, C. (1984). A service quality model and its marketing implications. *European Journal of Marketing, 18*(4), 36–44. doi:10.1108/EUM0000000004784

Guo, K., Fan, A., Lehto, X., & Day, J. (2021). Immersive digital tourism: The role of multisensory cues in digital museum experiences. *Journal of Hospitality & Tourism Research (Washington, D.C.),* 10963480211030319. doi:10.1177/10963480211030319

Gupta, S. (2018). *Organizational barriers to digital transformation. KTH Royal Institute of Technology.* School of Industrial Engineering and Management.

Gupta, S., Modgil, S., Lee, C. K., Cho, M., & Park, Y. (2022). Artificial intelligence enabled robots for stay experience in the hospitality industry in a smart city. *Industrial Management & Data Systems, 122*(10), 2331–2350. doi:10.1108/IMDS-10-2021-0621

Gursoy, D., & Chi, C. G. (2020). Effects of COVID-19 pandemic on hospitality industry: Review of the current situations and a research agenda. *Journal of Hospitality Marketing & Management, 29*(5), 527–529. doi:10.1080/19368623.2020.1788231

Gursoy, D., Chi, O. H., Lu, L., & Nunkoo, R. (2019). Consumers acceptance of artificially intelligent (AI) device use in service delivery. *International Journal of Information Management, 49,* 157–169. doi:10.1016/j.ijinfomgt.2019.03.008

Guttentag, D. A. (2010). Virtual reality: Applications and implications for tourism. *Tourism Management, 31*(5), 637–651. doi:10.1016/j.tourman.2009.07.003

Guzmán, I., & Pathania, A. (2016). *Chatbots in customer service.* Accenture. http://bit. ly/Accenture-Chatbots-Customer-Service

Hair Jr, J. F., Sarstedt, M., Ringle, C. M., & Gudergan, S. P. (2017). *Advanced issues in partial least squares structural equation modeling.* Sage publications.

Hair, J. F., Harrison, D., & Risher, J. J. (2018). Marketing research in the 21st century: Opportunities and challenges. *Brazilian Journal of Marketing-BJMkt. Revista Brasileira de Marketing–ReMark,* (Special Issue), 17.

Halili, S. H., & Sulaiman, H. (2018). Factors influencing the rural students' acceptance of using ICT for educational purposes. *Kasetsart Journal of Social Sciences.* doi:10.1016/j.kjss.2017.12.022

Hall, J., Matos, S., Sheehan, L., & Silvestre, B. (2012). Entrepreneurship and innovation at the base of the pyramid: A recipe for inclusive growth or social exclusion? *Journal of Management Studies, 49*(4), 785–812. doi:10.1111/j.1467-6486.2012.01044.x

Hankinson, G. (2015). Rethinking the place branding construct. In *Rethinking place branding* (pp. 13–31). Springer. doi:10.1007/978-3-319-12424-7_2

Han, Y., Niyato, D., Leung, C., Miao, C., & Kim, D. I. (2021). A Dynamic Resource Allocation Framework for Synchronizing Metaverse with IoT Service and Data. *Computer Science and Game Theory, 21*(1), 43–58.

Hao, A. W., Paul, J., Trott, S., Guo, C., & Wu, H.-H. (2021). Two decades of research on nation branding: A review and future research agenda. *International Marketing Review, 38*(1), 46–69. doi:10.1108/IMR-01-2019-0028

Harris, J. G., & Davenport, T. H. (2005). Automated decision making comes of age. *MIT Sloan Management Review*, *46*(4), 2–10.

Hashemi, H., Rajabi, R., & Brashear-Alejandro, T. G. (2022). COVID-19 research in management: An updated bibliometric analysis. *Journal of Business Research*, *149*, 795–810. doi:10.1016/j.jbusres.2022.05.082 PMID:35669095

Hauke, J., Bogacka, E., Tobolska, A., & Weltrowska, J. (2021). Students of public and private universities in Wielkopolska region (Poland) facing the challenges of remote education during the COVID-19 pandemic. *Studies of the Industrial Geography Commission of the Polish Geographical Society*, *35*(4), 205–226. doi:10.24917/20801653.354.13

Heavin, C., & Power, D. J. (2018). Challenges for digital transformation: Towards a conceptual decision support guide for managers. *Journal of Decision Systems*, *27*(1), 38–45. doi:10.1080/12460125.2018.1468697

He, D., Ai, B., Guan, K., García-Loygorri, J. M., Tian, L., Zhong, Z., & Hrovat, A. (2017). Influence of typical railway objects in a mmWave propagation channel. *IEEE Transactions on Vehicular Technology*, *67*(4), 2880–2892. doi:10.1109/TVT.2017.2782268

Heerink, M., Kröse, B., Evers, V., & Wielinga, B. (2010). Assessing acceptance of assistive social agent technology by older adults: The Almere Model. *International Journal of Social Robotics*, *2*(4), 361–375. doi:10.100712369-010-0068-5

Hei, A. (2020). Fish as a Functional Food in Human Health. *Diseases and Well – Being, 107th Indian Science Congress*. Banglore.

Hennig-Thurau, T., Gwinner, K. P., Walsh, G., & Gremler, D. D. (2004). Electronic word-of-mouth via consumer-opinion platforms: What motivates consumers to articulate themselves on the Internet? *Journal of Interactive Marketing*, *18*(1), 38–52. doi:10.1002/dir.10073

Henson, S., & Heasman, M. (1998). Food safety regulation and the firm: Understanding the compliance process1Financial support from the Ministry of Agriculture, Fisheries and Food (MAFF) is acknowledged. However, the views expressed in this paper reflect those of the author and not necessarily those of MAFF.1. *Food Policy*, *23*(1), 9–23. https://doi.org/10.1016/S0306-9192(98)00015-3. doi:10.1016/S0306-9192(98)00015-3

Herrmann, B., & Kritikos, A. S. (2013). Growing out of the crisis: Hidden assets to Greece's transition to an innovation economy. *IZA Journal of European Labor Studies*, *2*(1), 14. doi:10.1186/2193-9012-2-14

He, Z. (2019). Integration and innovation of hotel management under the new trend. *Modern Marketing*, *4*, 112–119.

Hicks, A., & Esping-Andersen, G. (2005). Comparative and Historical Studies of Public Policy and the Welfare State. In T. Janoski, R. R. Alford, A. M. Hicks, & M. A. Schwartz (Eds.), *The Handbook of Political Sociology. States, Civil Societies, and Globalization* (pp. 509–525). Cambridge University Press.

Hillman, W., & Radel, K. (2022). The social, cultural, economic and political strategies extending women's territory by encroaching on patriarchal embeddedness in tourism in Nepal. *Journal of Sustainable Tourism*, *30*(7), 1754–1775. doi:10.1080/09669582.2021.1894159

Hines, K. (2022, January 14). Social Media Usage Statistics For Digital Marketers In 2022. *Search Engine Journal*. https://www.searchenginejournal.com/top-social-media-statistics/418826/

Hjalager, A. (2010). A review of innovation research in tourism. *Tourism Management*, *31*(1), 1–12. doi:10.1016/j.tourman.2009.08.012

Hjalager, A., & Johansen, P. H. (2013). Food tourism in protected areas - sustainability for producers, the environment and tourism? *Journal of Sustainable Tourism*, *21*(3), 417–433. doi:10.1080/09669582.2012.708041

Hlee, S., Lee, H., & Koo, C. (2018). Hospitality and tourism online review research: A systematic analysis and heuristic-systematic model. *Sustainability (Basel)*, *10*(4), 1–27. doi:10.3390u10041141

HMOARD. (2000). *Hungarian Ministry of Agriculture and Rural Development. Collection of Hungary's Traditional and Local Agricultural Products. Traditions-Tastes-Regions (in Hungarian HÍR)*. *2000*. CD-ROM.

Hoeven, C. L., Van Zoonen, W., & Fonner, K. L. (2016). The practical paradox of technology: The influence of communication technology use on employee burnout and engagement. *Communication Monographs*, *83*(2), 239–263. doi:10.1080/03637751.2015.1133920 PMID:27226694

Hogg, M. A. (2016). *Social identity theory*. Springer.

Höpken, W., Eberle, T., Fuchs, M., & Lexhagen, M. (2021). Improving tourist arrival prediction: A big data and artificial neural network approach. *Journal of Travel Research*, *60*(5), 1–20. doi:10.1177/0047287520921244

Hoppstadius, F., & Möller, C. (2018). 'You have to try being a role model' – learning for sustainability among tourism entrepreneurs in a Swedish biosphere reserve. *European Journal of Tourism Research*, *20*, 28–45. www.scopus.com. doi:10.54055/ejtr.v20i.338

Horváth, N. (2011). From Development to Sustainability? The EU Strategy for the Danube Region, In Tarrósy, I.; Milford, S. (eds.) *EU Strategy for the Danube Region Perspectives for the future*. EU.

Hosseini, K., Stefaniec, A., & Hosseini, S. P. (2021). World Heritage Sites in developing countries: Assessing impacts and handling complexities toward sustainable tourism. *Journal of Destination Marketing & Management*, *20*, 100616. doi:10.1016/j.jdmm.2021.100616

Hosseini, S. J. F., Niknami, M., & Chizari, M. (2009). To determine thechallenges in the application of ICTs by the agricultural extensionservice in Iran. Journal of Agricultural Extension and Rural Development,1(1), 292e299. *Journal of Agricultural Extension and Rural Development*, *1*(1), 292–299.

Huang, C. Y., Chou, C. J., & Lin, P. C. (2010). Involvement theory in constructing bloggers' intention to purchase travel products. *Tourism Management*, *31*(4), 513–526. doi:10.1016/j.tourman.2009.06.003

Huang, M. H., & Rust, R. T. (2018). Artificial intelligence in service. *Journal of Service Research*, *21*(2), 155–172. doi:10.1177/1094670517752459

Huang, M. H., & Rust, R. T. (2021). A strategic framework for artificial intelligence in marketing. *Journal of the Academy of Marketing Science*, *49*(1), 30–50. doi:10.100711747-020-00749-9

Huang, M. H., & Rust, R. T. (2021). Engaged to a robot? The role of AI in service. *Journal of Service Research*, *24*(1), 30–41. doi:10.1177/1094670520902266

Huang, X. T., Wei, Z. D., & Leung, X. Y. (2020). What you feel may not be what you experience: A psychophysiological study on flow in VR travel experiences. *Asia Pacific Journal of Tourism Research*, *25*(7), 736–747. doi:10.1080/10941665.2019.1711141

Huang, Y. C. (2023). Integrated concepts of the UTAUT and TPB in virtual reality behavioral intention. *Journal of Retailing and Consumer Services*, *70*, 103127. doi:10.1016/j.jretconser.2022.103127

Huang, Y. C., Backman, K. F., Backman, S. J., & Chang, L. L. (2016). Exploring the implications of virtual reality technology in tourism marketing: An integrated research framework. *International Journal of Tourism Research*, *18*(2), 116–128. doi:10.1002/jtr.2038

Hudson, S., Matson-Barkat, S., Pallamin, N., & Jegou, G. (2019). With or without you? Interaction and immersion in a virtual reality experience. *Journal of Business Research*, *100*, 459–468. doi:10.1016/j.jbusres.2018.10.062

Huerta, E.M.L., García, A.E., & Nava, M.R.Z. (2019). Cordodes: Realidad Aumentada, el futuro del Turismo.

Huete-Alcocer, N. (2017). A literature review of word of mouth and electronic word of mouth: Implications for consumer behavior. *Frontiers in Physiology*, *8*, 1–4. doi:10.3389/fpsyg.2017.01256 PMID:28790950

Hungarikum. (n.d.). *Collection of Hungarikums*. Hungariankum. https://hungarikum.hu/

Hung, K., & Petrick, J. F. (2011). Why do you cruise? Exploring the motivations for taking cruise holidays, and the construction of a cruising motivation scale. *Tourism Management*, *32*(2), 386–393. doi:10.1016/j.tourman.2010.03.008

Hurn, B. J. (2016). The role of cultural diplomacy in nation branding. *Industrial and Commercial Training*, *48*(2), 80–85. doi:10.1108/ICT-06-2015-0043

Hyun, M. Y., & O'Keefe, R. M. (2012). Virtual destination image: Testing a telepresence model. *Journal of Business Research*, *65*(1), 29–35. doi:10.1016/j.jbusres.2011.07.011

Imrie, R., & Hall, P. (2001). *Inclusive Design. Designing and Developing Accessible Environments*. Spoon Press.

INE. (2022). *O que nos dizem os censos sobre as dificuldades sentidas pelas pessoas com incapacidade*. INE. https://www.ine.pt (accessed: 02.01.2023)

Io, H. N., & Lee, C. B. (2019). Understanding the Adoption of Chatbot: A Case Study of Siri. In *Advances in Information and Communication Networks: Proceedings of the 2018 Future of Information and Communication Conference (FICC)*, Vol. 1 (pp. 632-643). Springer International Publishing. 10.1007/978-3-030-03402-3_44

Iqbal, J., Khan, Z. H., & Khalid, A. (2017). Prospects of robotics in food industry. Food Science and Technology (Brazil), 37(2), 159–165. https://doi.org/ doi:10.1590/1678-457X.14616

Irefin, I. A., Abdul-Azeez, I. A., & Tijani, A. A. (2012). An Investigative Study of the factors Affecting the Adoption of Information and Communication Technology in Small and Medium Enterprises In Nigeria. *Australian Journal of Business and Management Research*, *02*(02), 01–09. doi:10.52283/NSWRCA.AJBMR.20120202A01

Is cultural heritage the new battleground ? (2021). The Africa Report. https://www.theafricareport.com/146636/algeria-morocco-is-cultural-heritage-the-new-battleground/

Isik, C., Dogru, T., & Turk, E. S. (2018). A nexus of linear and non-linear relationships between tourism demand, renewable energy consumption, and economic growth: Theory and evidence. *International Journal of Tourism Research*, *20*(1), 38–49. doi:10.1002/jtr.2151

Iskakova, M. S., Abenova, M. K., Dzhanmuldaeva, L. N., Zeinullina, A. Z., Tolysbaeva, M. S., Salzhanova, Z. A., & Zhansagimova, A. (2021). Methods of state support of innovative entrepreneurship. The example of rural tourism. *Journal of Environmental Management and Tourism*, *12*(2), 466–472. doi:10.14505//jemt.12.2(50).14

Ismagilova, E., Slade, E. L., Rana, N. P., & Dwivedi, Y. K. (2020). The effect of electronic word of mouth communications on intention to buy: A meta-analysis. *Information Systems Frontiers*, *22*(5), 1203–1226. doi:10.100710796-019-09924-y

Israeli, A. A. (2002). A preliminary investigation of the importance of site accessibility factor for disabled tourists. *Journal of Travel Research*, *41*(1), 101–104. doi:10.1177/004728750204100114

Ivanov, S., & Webster, C. (2019b). What should robots do? A comparative analysis of industry professionals, educators, and tourists. In *Information and Communication Technologies in Tourism 2019: Proceedings of the International Conference in Nicosia*, (pp. 249-262). Springer International Publishing.

Ivanov, D., Dolgui, A., & Sokolov, B. (2019). The impact of digital technology and Industry 4.0 on the ripple effect and supply chain risk analytics. *International Journal of Production Research*, *57*(3), 829–846. doi:10.1080/00207543.2018.1488086

Ivanov, S., & Webster, C. (2020). Robots in tourism: A research agenda for tourism economics. *Tourism Economics*, *26*(7), 1065–1085. doi:10.1177/1354816619879583

Ivanov, S., Webster, C., & Berezina, K. (2017). Adoption of robots and service automation by tourism and hospitality companies. *Revista Turismo & Desenvolvimento (Aveiro)*, *27*(28), 1501–1517.

Jalilvand, M. R., & Samiei, N. (2012). The impact of electronic word of mouth on a tourism destination choice: Testing the theory of planned behavior (TPB). *Internet Research*, *22*(5), 591–612. doi:10.1108/10662241211271563

Jandala, Cs. (2015). *2. modul: Turizmus-ipar* [Module 2: Tourism industry]. UNI NKE. https://www.uni-nke.hu/document/uni-nke-hu/jandala_2modul.pdf

Jayawardena, C., Ahmad, A., Valeri, M., & Jaharadak, A. A. (2023). Technology acceptance antecedents in digital transformation in hospitality industry. *International Journal of Hospitality Management*, *108*, 103350. doi:10.1016/j.ijhm.2022.103350

Jeganathan, G. S., David, A., & Ganesh Kumar, C. (2022). Adaptation of Blockchain Technology In HRM. *Korea Review of International Studies*, 10-22.

Jeganathan, G. S., & David, A. (2022). Determination of Hospitality Services Quality and Customer Satisfaction-A Holserv Approach. In *Au Virtual International Conference* (pp. 325-334).

Jiang, Y., & Balaji, M. S. (2021). Getting unwired: What drives travellers to take a digital detox holiday? *Tourism Recreation Research*, *47*(5-6), 453–469. doi:10.1080/02508281.2021.1889801

Jóhannesson, G. T. (2012). "To get things done": A relational approach to entrepreneurship. *Scandinavian Journal of Hospitality and Tourism*, *12*(2), 181–196. doi:10.1080/15022250.2012.695463

Jóhannesson, G. T., Huijbens, E. H., & Sharpley, R. (2010). Icelandic tourism: Past directions - future challenges. *Tourism Geographies*, *12*(2), 278–301. doi:10.1080/14616680903493670

Johnson, D. S., Bardhi, F., & Dunn, D. T. (2008). Understanding how technology paradoxes affect customer satisfaction with self-service technology: The role of performance ambiguity and trust in technology. *Psychology and Marketing*, *25*(5), 416–443. doi:10.1002/mar.20218

Jonathan, S. (1992). Defining virtual reality: Dimensions determining telepresence. *Journal of Communication*, *42*(4), 73–93. doi:10.1111/j.1460-2466.1992.tb00812.x

Jong, A., & Varley, P. (2018). Food tourism and events as tools for social sustainability? *Journal of Place Management & Development*, *11*(3), 277–295. doi:10.1108/JPMD-06-2017-0048

Jordan, M. I., & Mitchell, T. M. (2015). Machine learning: Trends, perspectives, and prospects. *Science*, *349*(6245), 255–260. doi:10.1126cience.aaa8415 PMID:26185243

Jung, T. H., Lee, H., Chung, N., & tom Dieck, M. C. (2018). Cross-cultural differences in adopting mobile augmented reality at cultural heritage tourism sites. *International Journal of Contemporary Hospitality Management*, *30*(3), 1621–1645. doi:10.1108/IJCHM-02-2017-0084

Kalburgi, N. K., David, A., & Muralidhar, L. B. (2023). Understanding the Perceptions of Students towards YouTube as a Learning Tool-An Empirical Approach. *Central European Management Journal*, 2336-2693.

Kallmuenzer, A., & Peters, M. (2018). Entrepreneurial behaviour, firm size and financial performance: The case of rural tourism family firms. *Tourism Recreation Research*, *43*(1), 2–14. doi:10.1080/02508281.2017.1357782

Kálmán, Zs., & Könczei, Gy. (2002). *A Taigetosztól az esélyegyenlőségig* [*From Taygetos to equal opportunities*]. Osiris.

Kankhuni, Z., & Ngwira, C. (2022). Overland tourists' natural soundscape perceptions: Influences on experience, satisfaction, and electronic word-of-mouth. *Tourism Recreation Research*, *47*(5-6), 591–607. doi:10.1080/02508281.2021.1878653

Karjaluoto, H. (2017). *Influence of Social Media on Corporate Heritage Tourism Brand*. Springer. doi:10.1007/978-3-319-51168-9_50

Karnani, A. (2007). Romanticizing the poor harms the poor. *Metamorphosis*, *6*(2), 151–162. doi:10.1177/0972622520070206

Katz, E., & Lazarsfeld, P. F. (1966). *Personal Influence: The Part Played by People in the Flow of Mass Communications*. Transaction Publishers.

Kaufman-Scarborough, C. (2001). Accessible advertising for visually-disabled persons: The case of color-deficient consumers. *Journal of Consumer Marketing*, *18*(4), 303–318. doi:10.1108/07363760110392985

Kaur, M. Sandhu, N. Mohan, & P. S. Sandhu. (2011). RFID Technology Principles Advantages Limitations Its Applications, *International Journal of Computer and Electrical Engineering*, *3*(1), 151–157.

Keleş, M. B., & Keleş, A. ve Keleş A. (2020). Makine öğrenmesi yöntemleri ile uçuş fiyatlarının tahmini. *Eurosia Journal of Mathematics*, *7*(11), 72–78.

Kemp, S. (2020). Global internet use accelerates. Retrieved from https://wearesocial.com/digital-2020-global-internet-use-accelerates

Kentish, P. (2019). The Danube Swabians: A story of cultural loss and revival. *Emerging Europe*. https://emerging-europe.com/after-hours/the-danube-swabians/a-story-of-cultural-loss-and-revival

Keskin, H. İ. (2019). Using the seemingly unrelated regression model in the estimation of tourism demand of Türkiye. *Journal of Tourism Theory and Research*, *5*(2), 182–190. doi:10.24288/jttr.526021

Khan, S. (2020). Comparative Analysis Of Rpa Tools-Uipath, Automation Anywhere And Blueprism. *International Journal of Computer Science and Mobile Applications*, *8*(11), 1–6. doi:10.47760/ijcsma.2020.v08i11.001

Khatri, I. (2019). Information technology in tourism & hospitality. *Journal of Tourism & Hospitality Education Industry: A Review of Ten Years'*. *Publications*, *9*, 74–87.

Kılıçhan, R., & Yılmaz, M. (2020). Artificial Intelligence and Robotic Technologies In Tourism And Hospitality Industry. *Erciyes Üniversitesi Sosyal Bilimler Enstitüsü Dergisi*, *3*(1), 353–380. doi:10.48070/erusosbilder.838193

Kilu, R. H., Sanda, M.-A., & Alacovska, A. (2023). Demystifying business models (shifts) among Ghanaian creative entrepreneurs in a COVID-19 era. *African Journal of Economic and Management Studies*, *14*(2), 188–204. doi:10.1108/AJEMS-07-2022-0305

Kim, D. H., Park, G. M., Yoo, Y. H., Ryu, S. J., Jeong, I. B., & Kim, J. H. (2017). Realization of task intelligence for service robots in an unstructured environment. *Annual Reviews in Control*, *44*, 9–18. doi:10.1016/j.arcontrol.2017.09.013

Kim, E. E. K., Mattila, A. S., & Baloglu, S. (2011). Effects of gender and expertise on consumers' motivation to read online hotel reviews. *Cornell Hospitality Quarterly*, *52*(4), 399–406. doi:10.1177/1938965510394357

Kim, E., Tang, L., & Bosselman, R. (2018). Measuring customer perceptions of restaurant innovativeness: Developing and validating a scale. *International Journal of Hospitality Management*, *74*, 85–98. doi:10.1016/j.ijhm.2018.02.018

Kim, H., & So, K. K. F. (2022). Two decades of customer experience research in hospitality and tourism: A bibliometric analysis and thematic content analysis. *International Journal of Hospitality Management*, *100*, 103082. doi:10.1016/j.ijhm.2021.103082

Kim, J. J., & Han, H. (2020). Hotel of the future: Exploring the attributes of a smart hotel adopting a mixed-methods approach. *Journal of Travel & Tourism Marketing*, *37*(7), 804–822. doi:10.1080/10548408.2020.1835788

Kiráľová, A., & Hamarneh, I. (2017). Local gastronomy as a prerequisite of food tourism development in the Czech Republic. *Marketing and Management of Innovations*, *2*(2), 15–25. doi:10.21272/mmi.2017.2-01

Kiráľová, A., & Malec, L. (2021). Local Food as a Tool of Tourism Development in Regions. *International Journal of Tourism and Hospitality Management in the Digital Age*, *5*(1), 54–68. doi:10.4018/IJTHMDA.20210101.oa1

Kiráľová, A., & Pavlíčeka, A. (2015). Development of Social Media Strategies in Tourism Destination. *Procedia: Social and Behavioral Sciences*, *175*, 358–366. doi:10.1016/j.sbspro.2015.01.1211

Kirk, M. (2017). *Thoughtful Machine Learning with Python: A Test-Driven Approach*. O'reilly.

Kivela, J., & Crott, J. C. (2006). Tourism and Gastronomy: Gastronomy's influence on how tourism experience influence destination. *Journal of Hospitality & Tourism Research (Washington, D.C.)*, *30*(3), 354–377. doi:10.1177/1096348006286797

Kline, R. B. (2015). *Principles and practice of structural equation modeling*. Guilford publications.

Koc, E. (2020). Do women make better in tourism and hospitality? A conceptual review from a customer satisfaction and service quality perspective. *Journal of Quality Assurance in Hospitality & Tourism*, *21*(4), 402–429. doi:10.1080/1528008X.2019.1672234

Kolenda, P. (2020). *Raport Strategiczny Internet 2019/2020*. ICAN Institute.

Kong, X., Wu, Y., Wang, H., & Xia, F. (2022). Edge Computing for Internet of Everything: A Survey. *IEEE Internet of Things Journal*, *9*(23), 23472–23485. doi:10.1109/JIOT.2022.3200431

Koo, C., Shin, S., Kim, K., Kim, C., & Chung, N. (2013). *Smart tourism of the Korea: A case study*. CORE.

Kordel, S. (2016). Selling ruralities: How tourist entrepreneurs commodify traditional and alternative ways of conceiving the countryside. *Rural Society*, *25*(3), 204–221. doi:10.1080/10371656.2016.1255475

Korinth, B., & Ranasinghe, R. (2020). Covid-19 pandemic's impact on tourism in Poland in March 2020. *Geo Journal of Tourism and Geosites*, *31*(3), 987–990. doi:10.30892/gtg.31308-531

Koski, H. (2010). Firm growth and profitability: The role of mobile IT and organizational practices. *Discussion Paper No. 1222*. The Research Institute of the Finnish Economy. https://www.etla.fi/wp-content/uploads/2012/09/dp1222.pdf

Kosmatos, E. A., Tselikas, N. D., & Boucouvalas, A. C. (2011). Integrating RFIDs and Smart Objects into a Unified Internet of Things Architecture. *Advances in Internet of Things: Scientific Research*, *1*(1), 5–12. doi:10.4236/ait.2011.11002

Kostin, K. B. (2018). Foresight of the global digital trends. *Strategic Management-International Journal of Strategic Management and Decision Support Systems in Strategic Management*, *23*(1).

Kotler, P., & Armstrong, G. (2020). *Principles of Marketing (18th Globa)*. Pearson Educatio n Limited.

Koutras, A., & Panagopoulos, A. ve Nikas, I. A. (2016). Forecasting tourism demand using linear and nonlinear prediction models. *Academica Turistica-Tourism and Innovation Journal*, *9*(1), 85–98.

Kozarkiewicz, A. (2020). General and specific: The impact of digital transformation on project processes and management methods. *Foundations of Management*, *12*(1), 237–248. doi:10.2478/fman-2020-0018

Kozinets, R. V. (2002). The field behind the screen: Using netnography for marketing research in online communities. *JMR, Journal of Marketing Research*, *39*(1), 61–72. doi:10.1509/jmkr.39.1.61.18935

Krasavac-Chroneos, B., Radosavljević, K., & Bradić-Martinović, A. (2018). SWOT analysis of the rural tourism as a channel of marketing for agricultural products in Serbia. *Ekonomika Poljoprivrede*, *65*(4), 1573–1584. doi:10.5937/ekoPolj1804573K

Krauze-Maślankowska, P. (2021). Open data and smart city initiatives for digital transformation in public sector in Poland. A survey. In: Wrycza, S., Maślankowski, J. (eds) *Digital Transformation*. Springer, Cham. https://doi.org/10.1007/978-3-030-85893-3_5

Krippendorff, K. (2004). *Content analysis an introduction to its methodology*. Sage Publications.

Kryk, B. (2021). Generations on the Polish labor market in the context of competencies needed in the economy based on knowledge and 4.0. [Economics and Law]. *Ekonomia I Prawo*, *20*(1), 121–137. doi:10.12775/EiP.2021.008

Kudyba, S. (2020). COVID-19 and the acceleration of digital transformation and the future of work. *Information Systems Management*, *37*(4), 284–287. doi:10.1080/10580530.2020.1818903

Kumar, S., & Nafi, S. M. (2020). *Impact of COVID-19 pandemic on tourism: Perceptions from Bangladesh*. SSRN 3632798.

Kumar, S., & Nafi, S. M. (2020). *Impact of COVID-19 pandemic on tourism: Recovery proposal for future tourism*. *GeoJournal of Tourism and Geosites*, *33*.

Kumar, B. K., Vijayalakshmi, G., Krishnamoorthy, A., & Basha, S. S. (2010). A single server feedback retrial queue with collisions. *Computers & Operations Research*, *37*(7), 1247–1255. doi:10.1016/j.cor.2009.04.019

Kummitha, H. R. (2020). Eco-entrepreneurs organizational attitude towards sustainable community ecotourism development. *DETUROPE*, *12*(1), 85-101. www.scopus.com

Kusumowidagdo, A., & Rembulan, C. L. (2022). *The sense of place value and the actors involved: Indigenous entrepreneurship in Indonesia*. Springer. doi:10.1007/978-981-16-4795-6_7

Lai, I. K. W., Liu, Y., & Lu, D. (2021). The effects of tourists' destination culinary experience on electronic word-of-mouth generation intention: The experience economy theory. *Asia Pacific Journal of Tourism Research*, *26*(3), 231–244. doi:10.1080/10941665.2020.1851273

Laki, I. (2013). A fogyatékossággal élő emberekről [About people with disabilities]. *Neveléstudomány Tanulmányok*, *1*(3), 79–84.

Lakshman, K., & David, A. (2023). *Senior Citizens' Perceptions on E-banking Services*. Exceller Books.

Lasek, M., & Jessa, S. (2013). Chatbots for customer service on hotels' websites. *Information Systems Management*, 2.

Law, R., Buhalis, D., & Cobanoglu, C. (2014). Progress on information and communication technologies in hospitality and tourism. *International Journal of Contemporary Hospitality Management*, *26*(5), 727–750. doi:10.1108/IJCHM-08-2013-0367

Law, R., Qi, S., & Buhalis, D. (2010). A review of website evaluation in tourism research. *Tourism Management*, *31*(3), 297–313. doi:10.1016/j.tourman.2009.11.007

Law, R., Ye, H., & Chan, I. C. C. (2022). A critical review of smart hospitality and tourism research. *International Journal of Contemporary Hospitality Management*, *34*(2), 623–641. doi:10.1108/IJCHM-08-2021-0986

Laws, E., Scott, N., & Brouder, P. (2020). *Accessibility, disability and inclusive tourism: Concepts, issues and global perspectives*. Channel View Publications.

Le, D., Scott, N., & Lohmann, G. (2019). Applying Experiential Marketing in Selling Tourism Dreams. *Journal of Travel & Tourism Marketing*, *36*(2), 220–235. doi:10.1080/10548408.2018.1526158

Le, D., Scott, N., & Wang, Y. (2021). Impact of prior knowledge and psychological distance on tourist imagination of a promoted tourism event. *Journal of Hospitality and Tourism Management*, *49*, 101–111. doi:10.1016/j.jhtm.2021.09.001

Lee, J. W., & Syah, A. M. (2018). *Economic and Environmental Impacts of Mass Tourism on Regional Tourism Destinations in Indonesia*. SSRN Scholarly Paper No. 3250133. https://papers.ssrn.com/abstract=3250133 doi:10.13106/jafeb.2018.vol5.no3.31

Lee, C., Hallak, R., & Sardeshmukh, S. (2016). Innovation, entrepreneurship, and restaurant performance: A higher-order structural model. *Tourism Management*, *53*, 215–228. doi:10.1016/j.tourman.2015.09.017

Leech, J. (2019). 11 Evidence-Based Health Benefits of Eating Fish. *Healthline*. https://www.healthline.com/nutrition/11-evident-based-health-benefits-of-fish

Lee, W., & Gretzel, U. (2012). Designing persuasive destination websites: A mental imagery processing perspective. *Tourism Management*, *33*(5), 1270–1280. doi:10.1016/j.tourman.2011.10.012

Leiper, N. (1995). *Tourism Management*. RMIT Press.

Lengyel, M. (2004). *A turizmus általános elmélete [General theory of tourism]*. Budapest: Heller Farkas Gazdasági és Turisztikai Szolgáltatások Főiskolája.

Lester, R. K., & Piore, M. J. (2004). *Innovation—The missing dimension*. Harvard University Press. doi:10.4159/9780674040106

Leung, R. (2021). Hospitality technology progress towards intelligent buildings: A perspective article. *Tourism Review*, *76*(1), 69–73. doi:10.1108/TR-05-2019-0173

Leung, W. K., Cheung, M. L., Chang, M. K., Shi, S., Tse, S. Y., & Yusrini, L. (2022). The role of virtual reality interactivity in building tourists' memorable experiences and post-adoption intentions in the COVID-19 era. *Journal of Hospitality and Tourism Technology*, *13*(3), 481–499. doi:10.1108/JHTT-03-2021-0088

Lhuer, X. (2016). The next acronym you need to know about: RPA (robotic process automation). *Digital McKinsey*, *17*, 1–5. https://www.mckinsey.com/business-functions/digital-mckinsey/our-insights/the-next-acronym-you-need-to-know-about-rpa%0Ahttps://usblearn.belpark.sun.ac.za/pluginfile.php/40119/mod_resource/content/0/2018MPhilFS_Robotic_process_automation_McKinsey_Company

Li, Y. (2021). *Hoteles inteligentes y nuevas tecnologías aplicadas en la industria hotelera. Estudio del caso de FlyZoo Hotel*. Zaragoza: Facultad de Empresa y Gestión Pública de la Universidad de Zaragoza. https://zaguan.unizar.es/record/106337/files/TAZ-TFM-2021-324.pdf

Liere-Netheler, K., Packmohr, S., & Vogelsang, K. (2018). Drivers of digital transformation in manufacturing. Proceedings of the 51st Hawaii International Conference on System Sciences, (pp. 3926-3935). Semantic Scholar. https://pdfs.semanticscholar.org/5783/7648a8ca127462f2ef35f2e4a6e3a4f7508e.pdf

Lindvert, M., Laven, D., & Gelbman, A. (2022). Exploring the role of women entrepreneurs in revitalizing historic Nazareth. *Journal of Sustainable Tourism*, 1–19. doi:10.1080/09669582.2022.2145291

Lin, L. Z., & Yeh, H. R. (2022). Using ZMET to explore consumers' cognitive model in virtual reality: Take the tourism experience as an example. *Current Issues in Tourism*, 1–15. doi:10.1080/13683500.2022.2147052

Lippi, G., Massimo, F., & Guidi, G. C. (2010). Red wine and cardiovascular health: The "French Paradox" revisited. *International Journal of Wine Research*, 2, 1–7. doi:10.2147/IJWR.S8159

Li, T., & Chen, Y. (2019). Will virtual reality be a double-edged sword? Exploring the moderation effects of the expected enjoyment of a destination on travel intention. *Journal of Destination Marketing & Management*, 12, 15–26. doi:10.1016/j.jdmm.2019.02.003

Litvin, S. W., Goldsmith, R. E., & Pan, B. (2008). Electronic word-of-mouth in hospitality and tourism management. *Tourism Management*, 29(3), 458–468. doi:10.1016/j.tourman.2007.05.011

Litvin, S. W., Goldsmith, R. E., & Pan, B. (2018). A retrospective view of electronic word-of-mouth in hospitality and tourism management. *International Journal of Contemporary Hospitality Management*, 30(1), 313–325. doi:10.1108/IJCHM-08-2016-0461

Liu, C., & Huang, X. (2023). Does the selection of virtual reality video matter? A laboratory experimental study of the influences of arousal. *Journal of Hospitality and Tourism Management*, 54, 152–165. doi:10.1016/j.jhtm.2022.12.002

Liu, Y., & Hu, H. (2021). Digital-free tourism intention: A technostress perspective. *Current Issues in Tourism*, 24(23), 3271–3274. doi:10.1080/13683500.2021.1883560

Li, W. (2022). Prediction of Tourism Demand in Liuzhou Region Based on Machine Learning. *Mobile Information Systems*, 2022, 1–9. doi:10.1155/2022/9362562

Li, X., Ma, E., & Qu, H. (2017). Knowledge mapping of hospitality research – A visual analysis using CiteSpace. *International Journal of Hospitality Management*, 60, 77–93. doi:10.1016/j.ijhm.2016.10.006

Liya, M. L., Aswathy, M., & Jayakrishnan, V. M. (2022, June). An Overview of Radio Frequency Identification systems. In *2022 7th International Conference on Communication and Electronics Systems (ICCES)* (530-535). IEEE. 10.1109/ICCES54183.2022.9835782

Lonely Planet. (2020a). *Travel for all: Join Lonely Planet's accessible travel project*. Lonely Planet. https://www.lonelyplanet.com/articles/travel-for-all-join-lonely-planets-accessible-travel-project

Lonely Planet. (2020b). *Travel for all: Accessible travel solutions*. Lonely Planet. https://www.lonelyplanet.com/travel-tips-and-articles/travel-for-all-accessible-travel-solutions

Loureiro, S. M. C., Guerreiro, J., & Ali, F. (2020). 20 years of research on virtual reality and augmented reality in tourism context: A text-mining approach. *Tourism Management*, 77, 104028. doi:10.1016/j.tourman.2019.104028

Lukanova, G., & Ilieva, G. (2019). Robots, artificial intelligence, and service automation in hotels, In: Ivanov, S., Webster, C., (eds) Robots, artificial intelligence, and service automation in travel, tourism and hospitality. Emerald Publishing Limited.

Lundberg, C., Fredman, P., & Wall-Reinius, S. (2014). Going for the green? The role of money among nature-based tourism entrepreneurs. *Current Issues in Tourism*, 17(4), 373–380. doi:10.1080/13683500.2012.746292

Luo, J., Wong, I. A., King, B., Liu, M. T., & Huang, G. (2019). Co-creation and co-destruction of service quality through customer-to-customer interactions: Why prior experience matters. *International Journal of Contemporary Hospitality Management*, 31(3), 1309–1329. doi:10.1108/IJCHM-12-2017-0792

Luo, X., & Pan, Y. (2021). A Study on the customer experience design through analyzing smart hotels in China. *Journal of the Korea Convergence Society, 12*(3), 115–124.

Luu, T. T. (2021). Green creative behavior in the tourism industry: The role of green entrepreneurial orientation and a dual-mediation mechanism. *Journal of Sustainable Tourism, 29*(8), 1290–1318. doi:10.1080/09669582.2020.1834565

Lu, V. N., Wirtz, J., Kunz, W. H., Paluch, S., Gruber, T., Martins, A., & Patterson, P. G. (2020). Service robots, customers and service employees: What can we learn from the academic literature and where are the gaps? *Journal of Service Theory and Practice, 30*(3), 361–391. doi:10.1108/JSTP-04-2019-0088

Machado, C. G., Winroth, M., Carlsson, D., Almström, P., Centerholt, V., & Hallin, M. (2019). Industry 4.0 readiness in manufacturing companies: Challenges and enablers towards increased digitalization. *Procedia CIRP, 81*, 1113–1118. doi:10.1016/j.procir.2019.03.262

Mackenzie, R. (2019). Tranforming the terrior into a tourism destination. In S. K. Dixit (Ed.), *The Routledge Handbook of Gastronomic Tourism. Routledge* (pp. 70–78). doi:10.4324/9781315147628-10

Madaan, G., Swapna, H. R., Kumar, A., Singh, A., & David, A. (2021). Enactment of sustainable technovations on healthcare sectors. *Asia Pacific Journal of Health Management, 16*(3), 184–192. doi:10.24083/apjhm.v16i3.989

Madakam, S., Holmukhe, R. M., & Kumar Jaiswal, D. (2019). The Future Digital Work Force: Robotic Process Automation (RPA). *Journal of Information Systems and Technology Management, 16*, 1–17. doi:10.4301/S1807-1775201916001

Madani, A., Boutebal, S. E., Benhamida, H., & Bryant, C. R. (2020). The impact of the COVID-19 outbreak on the tourism needs of the Algerian population. *Sustainability, 12*(21), 1–11. doi:10.3390u12218856

Ma, E., Liu, Y., Li, J., & Chen, S. (2016). Anticipating Chinese tourists arrivals in Australia: A time series analysis. *Tourism Management Perspectives, 17*, 50–58. doi:10.1016/j.tmp.2015.12.004

Makadia, J., Pashchapur, R., & Dhulasawant, T., & PY, D. R. (2020, July). Autonomous Flight Vehicle Incorporating Artificial Intelligence. In *2020 International Conference on Computational Performance Evaluation (ComPE)* (pp. 419-426). IEEE. 10.1109/ComPE49325.2020.9200061

Makuyana, T., & Saayman, M. (2018). The postulate for the systematic mainstreaming of impairments in Tourism Education in South Africa: A literature 204 synthesis. *African Journal of Hospitality, Tourism and Leisure, 7*(5), 1–28.

Malhotra, N. K. (2008). *Marketing research: An applied orientation, 5/e*. Pearson Education India. doi:10.1108/S1548-6435(2008)4

Manczak, I., & Gruszka, I. (2021). Averting the effects of the COVID-19 pandemic in tourism - a semantic field analysis. *Studies of the Industrial Geography Commission of the Polish Geographical Society, 35*(3), 164–176. doi:10.24917/20801653.353.10

Manios, Y., Detopoulou, V., Visioli, F., & Galli, C. (2006). Mediterranean diet as a nutrition education and dietary guide: Misconceptions and the neglected role of locally consumed foods and wild green plants. *Forum of Nutrition, 59*, 154–170. doi:10.1159/000095212 PMID:16917178

Manzoor, F., Wei, L., Hussain, A., Asif, M., & Shah, S. I. A. (2019). Patient satisfaction with health care services; an application of physician's behavior as a moderator. *International Journal of Environmental Research and Public Health, 16*(18), 3318. doi:10.3390/ijerph16183318 PMID:31505840

Marasco, A., Buonincontri, P., Van Niekerk, M., Orlowski, M., & Okumus, F. (2018). Exploring the role of next-generation virtual technologies in destination marketing. *Journal of Destination Marketing & Management, 9*, 138–148. doi:10.1016/j.jdmm.2017.12.002

Marjański, A., & Sułkowski, Ł. (2021). Consolidation strategies of small family firms in Poland during the Covid-19 crisis. *Entrepreneurial Business and Economics Review*, 9(2), 167–182. doi:10.15678/EBER.2021.090211

Marr, B. (2021, June 11). Extended reality in tourism: 4 Ways VR and AR can enhance the travel experience. *Forbes*. https://www.forbes.com/sites/bernardmarr/2021/06/11/extended-reality-in-tourism-4-ways-vr-and-ar-can-enhance-the-travel-experience/?sh=50729fbd82ff

Marsden, P. (2020, July 17). *What is digital wellbeing? A list of definitions*. Digital Wellbeing: https://digitalwellbeing.org/what-is-digital-wellbeing-a-list-of-definitions/

Martínez, J. A., & Martínez, L. (2010). Some insights on conceptualizing and measuring service quality. *Journal of Retailing and Consumer Services*, 17(1), 29–42. doi:10.1016/j.jretconser.2009.09.002

Martin, J. F. (2018). *Unlocking success in digital transformations*. McKinsey & Company.

Marx, S., & Klotz, M. (2021). Entrepreneurship during crisis: Innovation practices of micro and small tour operators. *International Journal of Entrepreneurship and Innovation*. doi:10.1177/14657503211061025

Matsiliza, N. S. (2017). Seeking strategies for sustainability in tourism entrepreneurship in South Africa. *African Journal of Hospitality, Tourism and Leisure*, 6(4), 1–10. www.scopus.com

McCabe, S., & Qiao, G. (2020). A review of research into social tourism: Launching the annals of tourism research curated collection on social tourism. *Annals of Tourism Research*, 85, 103103. doi:10.1016/j.annals.2020.103103

McCarthy, J. (2007). *What is Artificial Intelligence?* Stanford University, Computer Science Department.

McKechnie, S. (2011). Consumer confidence in financial services after the crunch: New theories and insights. *International Journal of Bank Marketing*, 29(2). Advance online publication. doi:10.1108/ijbm.2011.03229baa.001

McKinsey & Company. (2016). *Digital Poland: Capturing the opportunity to join leading global economies*. McKinsey. https://www.mckinsey.com/~/media/McKinsey/Business%20Functions/McKinsey%20Digital/Our%20Insights/Digital%20Poland/Digital%20Poland.ashx

McLean, G., & Barhorst, J. B. (2022). Living the experience before you go... but did it meet expectations? The role of virtual reality during hotel bookings. *Journal of Travel Research*, 61(6), 1233–1251. doi:10.1177/00472875211028313

McLean, G., Osei-Frimpong, K., Wilson, A., & Pitardi, V. (2020). How live chat assistants drive travel consumers' attitudes, trust and purchase intentions: The role of human touch. *International Journal of Contemporary Hospitality Management*, 32(5), 1795–1812. doi:10.1108/IJCHM-07-2019-0605

Medai, N., & Wu, L. (2022). A study of determinants that affect the intention to participate in online tours and the role of constraints under COVID-19 pandemic. *Current Issues in Tourism*, 1–15.

Medarić, Z., Sulyok, J., Kardos, S., & Gabruć, J. (2021). Lake Balaton as an accessible tourism destination – the stakeholders' perspectives. *Hungarian Geographical Bulletin*, 70(3), 233–247. doi:10.15201/hungeobull.70.3.3

Melović, B., Jocović, M., Dabić, M., Vulić, T. B., & Dudic, B. (2020). The impact of digital transformation and digital marketing on the brand promotion, positioning, and electronic business in Montenegro. *Technology in Society*, 63, 101425. doi:10.1016/j.techsoc.2020.101425

Mende, M., Scott, M. L., van Doorn, J., Grewal, D., & Shanks, I. (2019). Service robots rising: How humanoid robots influence service experiences and elicit compensatory consumer responses. *JMR, Journal of Marketing Research*, 56(4), 535–556. doi:10.1177/0022243718822827

Mendes-Filho, L., Mills, A. M., Tan, F. B., & Milne, S. (2018). Empowering the traveler: An examination of the impact of user-generated content on travel planning. *Journal of Travel & Tourism Marketing, 35*(4), 425–436. doi:10.1080/10 548408.2017.1358237

Merriam, S. B. (2009). *Qualitative research a guide to design and implementation.* Jossey-Bass- John Wiley & Sons.

Meuter, M. L., Ostrom, A. L., Roundtree, R. I., & Bitner, M. J. (2000). Self-service technologies: Understanding customer satisfaction with technology-based service encounters. *Journal of Marketing, 64*(3), 50–64. doi:10.1509/jmkg.64.3.50.18024

Meyer-Waarden, L., Pavone, G., Poocharoentou, T., Prayatsup, P., Ratinaud, M., Tison, A., & Torné, S. (2020). How service quality influences customer acceptance and usage of chatbots? *SMR-Journal of Service Management Research, 4*(1), 35–51. doi:10.15358/2511-8676-2020-1-35

Michalkó, G. (2010). *Boldogító utazás – a turizmus és az életminőség kapcsolatának magyarországi vonatkozásai [A happy trip - aspects of the relationship between tourism and quality of life in Hungary].* MTA Földrajztudományi Kutatóintézet.

Mikolajková, M., Ladicka, N., Janusova, M., Ondrova, K., Mikulaskova, H.K. & Dordevic, D. (2021). *Resveratrol content in wine – resveratrol biochemical properties.* MASO INTERNATIONAL. doi:10.2478/mjfst-2022-0005

Mills, A., Wood, R., & Darcy, S. (2008). The Internet and accessible tourism: Representing disability online. *Journal of Vacation Marketing, 14*(3), 207–217.

Mills, J. E., Han, J.-H., & Clay, J. M. (2008). Accessibility of hospitality and tourism websites. *Cornell Hospitality Quarterly, 49*(1), 28–41. doi:10.1177/1938965507311499

Milojević, L. (2004). The Social and Cultural Aspects of Rural Tourism. In Rural Tourism in Europe: Experiences, Development and Perspectives (pp. 115–121). UN WTO.

Ministry of Agriculture and Water management. (2009). *Analysis of the budgetary support to the development of rural tourism in Serbia and diversification of economic activities in the countryside.* MAWM. http://www.minpolj.gov.rs/?script=lat

Mithas, S., Krishnan, M. S., & Fornell, C. (2005). Why do customer relationship management applications affect customer satisfaction? *Journal of Marketing, 69*(4), 201–209. doi:10.1509/jmkg.2005.69.4.201

Mitroulis, D., & Kitsios, F. (2019). MCDA for assessing the impact of digital transformation on hotel performance in Thessaloniki. *Proceedings of the 8th International Symposium & 30th National Conference on Operational Research,* (pp. 53–57). HELORS. http://eeee2019.teiwest.gr/docs/HELORS_2019_proceedings.pdf#page=54 .

Mittal, B., & Lassar, W. M. (1996). The role of personalization in service encounters. *Journal of Retailing, 72*(1), 95–109. doi:10.1016/S0022-4359(96)90007-X

Mizrahi, A. B., & Zoran, A. (2023). Digital gastronomy testcase: A complete pipeline of robotic induced dish variations. *International Journal of Gastronomy and Food Science, 31.*

Mkwizu, K. H. (2020). Digital marketing and tourism: Opportunities for Africa. *International Hospitality Review, 34*(1), 5–12. doi:10.1108/IHR-09-2019-0015

Moffitt, K. C., Rozario, A. M., & Vasarhelyi, M. A. (2018). Robotic Process Automation for Auditing. *Journal of Emerging Technologies in Accounting, 15*(1), 1–10. doi:10.2308/jeta-10589

Moher, D., Liberati, A., Tetzlaff, J., & Altman, D. G. (2009). Preferred Reporting Items for Systematic Reviews and Meta-Analyses: The PRISMA Statement. *PLoS Medicine, 6*(7), e1000097. doi:10.1371/journal.pmed.1000097 PMID:19621072

Moley. (2023). *The future is served at work in the kitchen.* Moley. https://moley.com/

Moore, M., & Westley, F. (2011). Surmountable Chasms: Networks and Social Innovation for Resilient Systems. *Ecology and Society*, *16*(1), 5. doi:10.5751/ES-03812-160105

Moreno, I. (2015). Culturas mediterrânicas e sistemas alimentares: continuidades, imaginários e novos desafios. In A. Freitas, J. P. Bernardes, M. P. Mateus, & N. Braz (Eds.), *Dimensões da Dieta Mediterrânica* (pp. 51–79). Universidade do Algarve.

Moro, S., & Rita, P. (2018). Brand strategies in social media in hospitality and tourism. *International Journal of Contemporary Hospitality Management*, *30*(1), 343–364. doi:10.1108/IJCHM-07-2016-0340

Moroz, M. (2018). Acceleration of digital transformation as a result of launching programs financed from public funds: Assessment of the implementation of the operational program digital Poland. *Foundations of Management*, *10*(1), 59–74. doi:10.2478/fman-2018-0006

Morris, M.C.,Evans, A.D., Tangney, C.C., Bienias, J.L & Wilson, R.S. (2005). Fish consumption and cognitive decline with age in a large community study. *Arch Neurol, 62*. doi:10.1001/archneur.62.12.noc

Morrison, A. (2013). *Marketing and managing tourism destinations*. Routledge. doi:10.4324/9780203081976

Moscardo, G. (2008). Sustainable tourism innovation: Challenging basic assumptions. *Tourism and Hospitality Research*, *8*(1), 4–13. doi:10.1057/thr.2008.7

Mosurović, M., & Kutlača, D. (2011). Organizational design as a driver for firm innovativeness in Serbia. *Innovation (Abingdon)*, *24*(4), 427–447. doi:10.1080/13511610.2011.633432

Mouloudj, K., Bouarar, A. C., & Stojczew, K. (2021). Analyzing the students' intention to use online learning system in the context of COVID-19 pandemic: A theory of planned behavior approach. In W. B. James, C. Cobanoglu, & M. Cavusoglu (Eds.), Advances in global education and research (Vol. 4, pp. 1–17). USF M3 Publishing.

Mouloudj, K., Bouarar, A. C., Asanza, D. M., Saadaoui, L., Mouloudj, S., Njoku, A. U., Evans, M. A., & Bouarar, A. (2023). Factors Influencing the Adoption of Digital Health Apps: An Extended Technology Acceptance Model (TAM). In A. Bouarar, K. Mouloudj, & D. Martínez Asanza (Eds.), *Integrating Digital Health Strategies for Effective Administration* (pp. 116–132). IGI Global. doi:10.4018/978-1-6684-8337-4.ch007

Mouloudj, K., Bouarar, A. C., & Fechit, H. (2020). The impact of COVID-19 pandemic on food security. Les cahiers du CREAD, 36(3), 159-184. https://doi.org/ doi:10.6084/m9.figshare.13991939.v1

Moussawi, S. (2016). *Investigating personal intelligent agents in everyday life through a behavioral lens*. City University of New York.

Moyle, C., Moyle, B., & Burgers, H. (2020). Entrepreneurial strategies and tourism industry growth. *Tourism Management Perspectives*, *35*, 100708. doi:10.1016/j.tmp.2020.100708

MTÜ (2019, October). *Trendek és gyakorlatok a szenior, az akadálymentes és az orvosi turizmusban* [Trends and practices in senior, accessible and medical tourism]. Paper presented Tourism Academy at a series of events. Budapest: Magyar Turisztikai Ügynökség.

Mubarak, M. F., Shaikh, F. A., Mubarik, M., Samo, K. A., & Mastoi, S. (2019). The impact of digital transformation on business performance: A study of Pakistani SMEs. Engineering Technology & *Applied Scientific Research*, *9*(6), 5056–5061.

Mukhopadhyay, S., Pandey, R., & Rishi, B. (2022). Electronic word of mouth (eWOM) research – a comparative bibliometric analysis and future research insight. *Journal of Hospitality and Tourism Insights*. doi:10.1108/JHTI-07-2021-0174

Mukhopadhyay, S., Pandey, R., & Rishi, B. (2022). Electronic word of mouth (eWOM) research–a comparative bibliometric analysis and future research insight. *Journal of Hospitality and Tourism Insights*.

Mulet-Forteza, C., Martorell-Cunill, O., Merigó, J. M., Genovart-Balaguer, J., & Mauleon-Mendez, E. (2018). Twenty five years of the Journal of Travel & Tourism Marketing: A bibliometric ranking. *Journal of Travel & Tourism Marketing*, *35*(9), 1201–1221. doi:10.1080/10548408.2018.1487368

Müller, E., & Hopf, H. (2017). Competence center for the digital transformation in small and medium-sized enterprises. *Procedia Manufacturing*, *11*, 1495–1500. doi:10.1016/j.promfg.2017.07.281

Munar, A. M., & Jacobsen, J. K. S. (2014). Motivations for sharing tourism experiences through social media. *Tourism Management*, *43*, 46–54. doi:10.1016/j.tourman.2014.01.012

Murphy, R. R., Adams, J., & Gandudi, V. B. M. (2020). How robots are on the frontlines in the battle against COVID-19. *Smithsonian Magazine*. https://www.smithsonianmag.com/innovation/how-robots-are-on-front-linesbattle against-covid-19-180974720/

Murphy, A., & Liszewski, B. (2019). Artificial intelligence and the medical radiation profession: How our advocacy must inform future practice. *Journal of Medical Imaging and Radiation Sciences*, *50*(4), S15–S19. doi:10.1016/j.jmir.2019.09.001 PMID:31611013

Murphy, J., Hofacker, C., & Gretzel, U. (2017). Dawning of the age of robots in hospitality and tourism: Challenges for teaching and research. *European Journal of Tourism Research*, *15*, 104–111. doi:10.54055/ejtr.v15i.265

Mustak, M., Salminen, J., Plé, L., & Wirtz, J. (2021). Artificial intelligence in marketing: Topic modeling, scientometric analysis, and research agenda. *Journal of Business Research*, *124*, 389–404. doi:10.1016/j.jbusres.2020.10.044

Nagy, S. A. (2010). Elek Woynarovich 95 years (in Hungarian), In: Pisces Hunagarici 4.

Nakanishi, J., Baba, J., Kuramoto, I., Ogawa, K., Yoshikawa, Y., & Ishiguro, H. (2020, October). Smart speaker vs. social robot in a case of hotel room. In *2020 IEEE/RSJ International Conference on Intelligent Robots and Systems (IROS)* (11391-11396). IEEE. 10.1109/IROS45743.2020.9341537

Nam, K., Baker, J., Ahmad, N., & Goo, J. (2020). Determinants of writing positive and negative electronic word-of-mouth: Empirical evidence for two types of expectation confirmation. *Decision Support Systems*, *129*, 113168. doi:10.1016/j.dss.2019.113168

Napierała, T., Leśniewska-Napierała, K., & Burski, R. (2020). Impact of geographic distribution of COVID-19 cases on hotels' performances: Case of Polish cities. *Sustainability*, *12*(11), 4697. doi:10.3390u12114697

Nárai, M. (2013). *Fogyatékossággal élők szociális* ellátása [Social care for people with disabilities]. MTA KRTK RKI NYUTO, Győr, Hungary.

Nawaz, N. (2019). Article ID: IJARET_10_02_057 Cite this Article Dr Nishad Nawaz, Robotic Process Automation for Recruitment Process. [IJARET]. *International Journal of Advanced Research in Engineering and Technology*, *10*(2), 608–611. http://www.iaeme.com/IJARET/index.asp608http://www.iaeme.com/IJARET/issues.asp?JType=IJARET&VType=10&IType=02http://www.iaeme.com/IJARET/issues.asp?JType=IJARET&VType=10&IType=2. doi:10.34218/IJARET.10.2.2019.057

Ndou, V., Mele, G., & Del Vecchio, P. (2019). Entrepreneurship education in tourism: An investigation among European universities. *Journal of Hospitality, Leisure, Sport and Tourism Education*, *25*, 100175. Advance online publication. doi:10.1016/j.jhlste.2018.10.003

Neuhofer, B., Buhalis, D., & Ladkin, A. (2015). Technology as a catalyst of change: Enablers and barriers of the tourist experience and their consequences. In I. Tussyadiah & A. Inversini (Eds.), *Information and Communication Technologies in Tourism 2015* (pp. 789–802). Springer. doi:10.1007/978-3-319-14343-9_57

Neuhofer, B., Magnus, B., & Celuch, K. (2021). The impact of artificial intelligence on event experiences: A scenario technique approach. *Electronic Markets*, *31*(3), 601–617. doi:10.100712525-020-00433-4

NFÜ. (2012). *Fogyatékos személyek akadálymentes hozzáférésének és megkülönböztetés-mentességének biztosítása* [Ensuring accessible access and non-discrimination of disabled persons]. Nemzeti Fejlesztési Ügynökség. https://www.kozbeszerzes.hu/data/filer_public/fa/fb/fafbe76a-2678-47d1-bc6d9737d614765d/fogyatekos_szemelyek_akadalymentes_hozzaferesenek_es_megkulonbo ztetes-mentessegenek_biztositasa.pdf

Nguyen, T. H., Newby, M., & Macaulay, M. J. (2015). Information Technology Adoption in Small Business: Confirmation of a Proposed Framework. *Journal of Small Business Management*, *53*(1), 207–227. doi:10.1111/jsbm.12058

Nhamo, G., Dube, K., Chikodzi, D., Nhamo, G., Dube, K., & Chikodzi, D. (2020). Impacts and implications of COVID-19 on the global hotel industry and Airbnb. *Counting the Cost of COVID-19 on the global tourism industry,* 183-204. Research Gate.

Niavis, S., & Tsiotas, D. (2019). Assessing the tourism performance of the Mediterranean coastal destinations: A combined efficiency and effectiveness approach. Journal of Destination Marketing & Management, 14, 100379. doi:10.1016/j.jdmm.2019.100379

Nicholls, R. (2010). New directions for customer-to-customer interaction research. *Journal of Services Marketing*, *24*(1), 87–97. doi:10.1108/08876041011017916

Nieto, J., Hernández-Maestro, R. M., & Muñoz-Gallego, P. A. (2014). Marketing decisions, customer reviews, and business performance: The use of the Toprural website by Spanish rural lodging establishments. *Tourism Management*, *45*, 115–123. doi:10.1016/j.tourman.2014.03.009

Nihmathullah, Z., Ramasamy, R., & Raj David, A. (2022). *Event Impact Assessment: A Case of Puducherry*. Book Rivers.

Nikopoulou, M., Kourouthanassis, P., Chasapi, G., Pateli, A., & Mylonas, N. (2023). Determinants of digital transformation in the hospitality industry: Technological, organizational, and environmental drivers. *Sustainability*, *15*(3), 2736. doi:10.3390u15032736

Nilashi, M., Abumalloh, R. A., Alghamdi, A., Minaei-Bidgoli, B., Alsulami, A. A., Thanoon, M., Asadi, S., & Samad, S. (2021). What is the impact of service quality on customers' satisfaction during COVID-19 outbreak? New findings from online reviews analysis. *Telematics and Informatics*, *64*, 101693. doi:10.1016/j.tele.2021.101693 PMID:34887617

Nirala, K. K., Singh, N. K., & Purani, V. S. (2022). A survey on providing customer and public administration based services using AI: Chatbot. *Multimedia Tools and Applications*, *81*(16), 22215–22246. doi:10.100711042-021-11458-y PMID:35002470

Noor, N., Hill, S. R., & Troshani, I. (2022). Developing a service quality scale for artificial intelligence service agents. *European Journal of Marketing*, *56*(5), 1301–1336. doi:10.1108/EJM-09-2020-0672

Noppen, P., Beerepoot, I., van de Weerd, I., Jonker, M., & Reijers, H. A. (2020). How to Keep RPA Maintainable? In D. Fahland, C. Ghidini, J. Becker, & M. Dumas (Eds.), *Business Process Management* (pp. 453–470). Springer International Publishing. doi:10.1007/978-3-030-58666-9_26

NTG. (2020). *Accessibility in Tourism: challenges and opportunities*. Next Tourism Generation. NTG. https://next-tourismgeneration.eu/accessibility-in-tourism-challenges-and-opportunities/?fbclid=IwAR2thEAC_SRvZaXTGO3zX-zlcskcN

Nunberg, G. (2012). *The Advent of the Internet*. Citation Times.

Nunnally, J. C. (1978). An overview of psychological measurement. *Clinical diagnosis of mental disorders: A handbook*, 97-146.

Nwankpa, J. K., & Roumani, Y. (2016). IT capability and digital transformation: A firm performance perspective. In *37th International Conference on Information Systems*, vol.5 (pp. 3839-3854). Dublin, Ireland.

Nye Jr, J. S. (2004). *Soft power: The means to success in world politics*. Public affairs.

Ojo, S. (2020). Interrogating place brand–a case of two cities. *Qualitative Market Research*, 23(4), 907–932. doi:10.1108/QMR-11-2017-0151

Olgunoglu, I. A. (2017). Review on Omega-3 (n-3) Fatty Acids in Fish and Seafood. *Journal of Biology, Agriculture and Healthcare*. 7(12), 37-45.

Ongori, H., & Migiro, S. O. (2010). Information and communication technologies adoption in SMEs: Literature review. *Journal of Chinese Entrepreneurship*, 2(1), 93–104. doi:10.1108/17561391011019041

Ortega, C., Nogueira, C., & Pinto, H. (2013). Sea and Littoral Localities' Economy: Exploring Potentialities for a Maritime Cluster - An Integrated Analysis of Huelva, Spain and Algarve, Portugal. *Journal for Maritime Research*, 10(2), 35–42.

OSSATE/University of Surrey. (2006). *Accessibility market and stakeholder analysis: One-stop-shop for accessible tourism in Europe*. OSSATE.

Ožegović, J. (2019). *Report on digital inclusion in the Republic of Serbia for the period from 2014 to 2018*. Government of the Republic of Serbia.

Öztemel, E. (2003). *Yapay Sinir Ağları*. Papatya Yayıncılık.

Packer, T. L., Mckercher, B., & Yau, M. K. (2009). Understanding the complex interplay between tourism, disability and environmental contexts. *Disability and Rehabilitation*, 29(4), 281–292. doi:10.1080/09638280600756331 PMID:17364778

Pailis, E. A., Fatkhurahman, & Arif, A. (2020). Indigenous community approach through indigenous leaders social entrepreneurship in five Luhak in Rokan Hulu regency Riau Indonesia. *International Journal of Scientific and Technology Research, 9*(3), 5249-5255. www.scopus.com

Palácios, H., de Almeida, M. H., & Sousa, M. J. (2021). A bibliometric analysis of trust in the field of hospitality and tourism. *International Journal of Hospitality Management*, 95, 102944. doi:10.1016/j.ijhm.2021.102944

Panisoara, I. O., Lazar, I., Panisoara, G., Chirca, R., & Ursu, S. A. (2020). Motivation and continuance intention towards online instruction among teachers during the Covid-19 pandemic: The mediating effect of burnout and technostress. *International Journal of Environmental Research and Public Health*, 17(21), 1–28. doi:10.3390/ijerph17218002 PMID:33143180

Paral, T. (2022). Robotics. *Smart Manufacturing: The Lean Six Sigma Way*, 311-329.

Parasotskaya, N., Berezyuk, V., Prasolov, V., Nazarova, V., & Mezentseva, T. (2021). Comparative analysis of small and medium-sized businesses and its impact on the development of tourism. *Journal of Environmental Management and Tourism*, 12(6), 1586–1602. doi:10.14505//jemt.12.6(54).15

Parasuraman, R., Greenwood, P. M., & Alexander, G. E. (1995). Selective impairment of spatial attention during visual search in Alzheimer's disease. *Neuroreport: An International Journal for the Rapid Communication of Research in Neuroscience*.

Parasuraman, A., Berry, L. L., & Zeithaml, V. A. (1993). More on improving service quality measurement. *Journal of Retailing, 69*(1), 140–147. doi:10.1016/S0022-4359(05)80007-7

Parasuraman, A., Zeithaml, V. A., & Berry, L. L. (1985). A conceptual model of service quality and its implications for future research. *Journal of Marketing, 49*(4), 41–50. doi:10.1177/002224298504900403

Parasuraman, A., Zeithaml, V. A., & Berry, L. L. (1994). Alternative scales for measuring service quality: A comparative assessment based on psychometric and diagnostic criteria. *Journal of Retailing, 70*(3), 201–230. doi:10.1016/0022-4359(94)90033-7

Parasuraman, A., Zeithaml, V. A., & Malhotra, A. (2005). ES-QUAL: A multiple-item scale for assessing electronic service quality. *Journal of Service Research, 7*(3), 213–233. doi:10.1177/1094670504271156

Parviainen, P., Tihinen, M., Kääriäinen, J., & Teppola, S. (2017). Tackling the digitalization challenge: How to benefit from digitalization in practice. *International Journal of Information Systems and Project Management, 5*(1), 63–77. doi:10.12821/ijispm050104

Paunović, M., Lazarević-Moravčević, M., & Mosurović Ružičić, M. (2022). Business Process Innovation of Serbian Entrepreneurial Firms. *Economic Analysis*. doi:10.28934/ea.22.55.2.pp66-78

Pearce, D. G., & Schott, C. (2005). Tourism Distribution Channels: The Visitors' Perspective. *Journal of Travel Research, 44*(1), 50–63. doi:10.1177/0047287505276591

Pedersen, J. S., & Wilkinson, A. (2018). The digital society and provision of welfare services. *The International Journal of Sociology and Social Policy, 38*(3/4), 194–209. doi:10.1108/IJSSP-05-2017-0062

Peeters, P., Gossling, S., & Becken, S. (2006). Innovation towards tourism sustainability: Climate change and aviation. *International Journal of Innovation and Sustainable Development, 1*(3), 184–200. doi:10.1504/IJISD.2006.012421

Pekmezci, A., & Bozkurt, K. (2016). Döviz kuru ve ekonomik büyüme: Türk turizm sektörü için bir analiz. *SDÜ Sosyal Bilimler Enstitüsü Dergisi, 2*(24), 98–110.

Pentland, A. (2001). Perceptual intelligence. In *Handheld and Ubiquitous Computing: First International Symposium, HUC'99 Karlsruhe,* (pp. 74-88). Springer Berlin Heidelberg.

Pentland, A. (2000). Perceptual user interfaces: Perceptual intelligence. *Communications of the ACM, 43*(3), 35–44. doi:10.1145/330534.330536

Pérez, D., & Velasco, D. (2003). *Turismo accesible: Hacia un turismo para todos*. Madrid: CERMI.

Perron, P. (1989). The great crash, the oil price shock and the unit root hypothesis. *Econometrica, 57*(6), 1361–1401. doi:10.2307/1913712

Petrevska, B., Cingoski, V., & Gelev, S. (2016). From smart rooms to smart hotels. *Zbornik radova sa XXI međunarodnog naučno-stručnog skupa Informacione tehnologije-sadašnjost i budućnost, Žabljak*.

Pfau, J., Porzel, R., Pomarlan, M., Cangalovic, V. S., Grudpan, S., Höffner, S., & Malaka, R. (2019). Give MEANinGS to robots with kitchen clash: a VR human computation serious game for world knowledge accumulation. In *Entertainment Computing and Serious Games: First IFIP TC 14 Joint International Conference*. Springer International Publishing.

Phua, J., & Kim, J. J. (2018). Starring in your own Snapchat advertisement: Influence of self-brand congruity, self-referencing and perceived humor on brand attitude and purchase intention of advertised brands. *Telematics and Informatics*, *35*(5), 1524–1533. doi:10.1016/j.tele.2018.03.020

Pike, A. (2015). *Origination: The geographies of brands and branding.* John Wiley & Sons. doi:10.1002/9781118556313

Pilipczuk, O. (2021). Transformation of the business process manager profession in Poland: The impact of digital technologies. *Sustainability*, *13*(24), 13690. doi:10.3390u132413690

Pillai, R., & Sivathanu, B. (2020). Adoption of AI-based chatbots for hospitality and tourism. *International Journal of Contemporary Hospitality Management*, *32*(10), 3199–3226. doi:10.1108/IJCHM-04-2020-0259

Pimentel de Oliveira, D., & Pitarch-Garrido, M. D. (2022). Measuring the sustainability of tourist destinations based on the SDGs: the case of Algarve in Portugal: tourism agenda-2030. *Tourism Review*. Advance online publication. doi:10.1108/TR-05-2022-0233

Pinto, A. S., Abreu, A., Costa, E., & Paiva, J. (2022). Augmented reality for a new reality: Using UTAUT-3 to assess the adoption of mobile augmented reality in tourism (MART). *Journal of Information Systems Engineering & Management*, *7*(2), 14550.

Poland Statistics. (2021a). *Enterprises having access to the Internet.* SWAID. http://swaid.stat.gov.pl/en/NaukaTechnika_dashboards/Raporty_predefiniowane/RAP_DBD_NTSI_10.aspx .

Poland Statistics. (2021b). *Occupancy of tourist accommodation establishments in Poland in April and May 2021.* Poland Statistics. https://stat.gov.pl/en/topics/culture-tourism-sport/tourism/occupancy-of-tourist-accommodation-establishments-in-poland-in-april-and-may-2021,5,31.html

Poland, S. (2020). *Tourism in Poland in the face of COVID-19 pandemic.* Statistics Poland. https://stat.gov.pl/en/topics/culture-tourism-sport/tourism/tourism-in-poland-in-the-face-of-covid-19-pandemic,6,1.html

Polat, N., & Hermans, E. (2016). A model proposed for sustainable accessible tourism (SAT). *TÉKHNE - Review of Applied Management Studies, 14*(2), 125-133.

Polgár, J. (n.d.). *Szakmai jegyzetek az akadálymentességről, akadálymentes turizmusról [Professional notes on accessibility and accessible tourism].* Unpublished professional notes.

Pollice, F. (2003). *The role of territorial identity in rural development processes.* Research Gate. https://www.researchgate.net/publication/242122046.

Poole, D. L., & Mackworth, A. K. (2010). *Artificial Intelligence: foundations of computational agents.* Cambridge University Press. doi:10.1017/CBO9780511794797

Popović-Pantić, S., Semenčenko, D., & Vasilić, N. (2019). The influence of digital transformation on business performance: Evidence of the women-owned companies. *Ekonomika Preduzeća*, *67*(7-8), 397–414. doi:10.5937/EKOPRE1908397P

Popović-Pantić, S., Semenčenko, D., & Vasilić, N. (2020). Digital technologies and the financial performance of female SMEs in Serbia: The mediating role of innovation. *Economic Annals*, *65*(224), 53–82. doi:10.2298/EKA2024053P

Pourfakhimi, S., Duncan, T., & Coetzee, W. J. L. (2020). Electronic word of mouth in tourism and hospitality consumer behaviour: State of the art. *Tourism Review*, *75*(4), 637–661. doi:10.1108/TR-01-2019-0019

Preiser, W., & Ostroff, E. (2001). *Universal design handbook.* McGraw-Hill.

Preko, A. (2020). Tourism development: National policies and tourism priorities in Ghana. *International Journal of Tourism Policy*, *10*(4), 380–391. doi:10.1504/IJTP.2020.112644

Premkumar, G., & Roberts, M. (1999). Adoption of new information technologies in rural small businesses. *Omega*, *27*(4), 467–484. doi:10.1016/S0305-0483(98)00071-1

Prentice, C., Dominique Lopes, S., & Wang, X. (2020). The impact of artificial intelligence and employee service quality on customer satisfaction and loyalty. *Journal of Hospitality Marketing & Management*, *29*(7), 739–756. doi:10.1080/19368623.2020.1722304

Prideaux, S. (2019). Robot host welcomes guests at new Dubai Mall restaurant. *The National News.* https://www.thenationalnews.com/lifestyle/food/robot-host-welcomes-guests-at-new-dubai-mall-restaurant-1.813177

Protiviti. (2019). *Taking RPA to the Next Level - How companies are using robotic process automation to beat the competition.* Protiviti.

Puah, C. H., Jong, M. C., Ayob, N., & Ismail, S. (2018). The impact of tourism on the local economy in Malaysia. *International Journal of Business and Management*, *13*(12), 151–157. doi:10.5539/ijbm.v13n12p151

Purgel, I., & Szabó, I. (2019). Hey fishermen, anglers, and their activities in Fejér County (in Hungarian). Ministry of Agriculture.

Pusztai, B. (2007). Identity, Canonisation and Branding at the Baja Festival, In: *Tourism, Festivals and Local Identity: Fish Soup Cooking in Baja,* 83-101. Research Gate. https://www.researchgate.net/publication/329686693

Queiroz, J. (2015). A Dieta Mediterrânica e a UNESCO: memória breve de um reconhecimento mundial. In A. Freitas, J. P. Bernardes, M. P. Mateus, & N. Braz (Eds.), *Dimensões da Dieta Mediterrânica* (pp. 29–47). Universidade do Algarve.

Quivy, R., & Campenhoudt, L. V. (2008). *Manual de investigação em ciências sociais.* Gradiva.

Rajan, K., & Saffiotti, A. (2017). Towards a science of integrated AI and Robotics. *Artificial Intelligence*, *247*, 1–9. doi:10.1016/j.artint.2017.03.003

Raji, C.A., Erickson, K.I., Lopez, O.L.,Kuller, L.H., Gach. M. H. Thompson, P.M., Riverol,, M. &Becker, J.T. (2014). Regular fish consumption and age-related brain gray matter loss. *Am J Prev Med 47*(4).444-51. . doi:10.1016/j.amepre.2014.05.037

Ramachandran, K. K., Mary, A. A. S., Hawladar, S., Asokk, D., Bhaskar, B., & Pitroda, J. R. (2022). Machine learning and role of artificial intelligence in optimizing work performance and employee behavior. *Materials Today: Proceedings*, *51*, 2327–2331. doi:10.1016/j.matpr.2021.11.544

Ran, L., Zhenpeng, L., Bilgihan, A., & Okumus, F. (2021). Marketing China to U.S. travelers through electronic word-of-mouth and destination image: Taking Beijing as an example. *Journal of Vacation Marketing*, *27*(3), 267–286. doi:10.1177/1356766720987869

Rasulovich, K. A. (2021). The role of agro-tourism in the development of socio-economic infrastructure in rural areas. *Наука и образование сегодня, 3*(62), 13-14.

Ravi, S., David, A., & Imaduddin, M. (2018). Controlling & calibrating vehicle-related issues using RFID technology. *International Journal of Mechanical and Production Engineering Research and Development*, *8*(2), 1125–1132. doi:10.24247/ijmperdapr2018130

Ray, N. M., & Ryder, M. E. (2003). "Ebilities" tourism: An exploratory discussion of the travel needs and motivations of the mobility-disabled. *Tourism Management*, *24*(1), 57–72. doi:10.1016/S0261-5177(02)00037-7

Raza, S. A., Ashrafi, R., & Akgunduz, A. (2020). A bibliometric analysis of revenue management in airline industry. *Journal of Revenue and Pricing Management*, *19*(6), 436–465. doi:10.105741272-020-00247-1

Real, H. & Graça, P. (2019a). Marcos da história da Dieta Mediterrânica, desde Ancel Keys. *Acta Portuguesa de Nutrição, 17*, 06-14. doi:10.21011/apn.2019.1702

Real, H. & Graça, P. (2019b). Perceções de utilização do conceito de Dieta Mediterrânica, potencial utilização indevida e perspetivas a explorar. *Revista española de comunicación en salud, 10*(2), 147-159. doi:10.20318/recs.2019.4824

Reid, S. (2019). Wonderment in tourism land: Three tales of innovation. *Journal of Teaching in Travel & Tourism, 19*(1), 79–92. doi:10.1080/15313220.2018.1560533

Ren, X. X. (2014). *Research on the Construction and Application of the Smart Hotel Construction Evaluation Index System [D]* [Doctoral dissertation, Thesis for Master Degree in Hebei Normal University]

Rennecker, J., & Godwin, L. (2005). Delays and interruptions: A self-perpetuating paradox of communication. *Information and Organization, 15*(3), 247–266. doi:10.1016/j.infoandorg.2005.02.004

Ribeiro, J., Lima, R., Eckhardt, T., & Paiva, S. (2021). Robotic Process Automation and Artificial Intelligence in Industry 4.0 – A Literature review. *Procedia Computer Science, 181*, 51–58. doi:10.1016/j.procs.2021.01.104

Richards, G. (2012). Food and the tourism experience: major findings and policy orientations. In D. Dodd (Ed.), *Food and the Tourism Experience* (pp. 13–46). OECD. doi:10.1787/9789264171923-3-en

Riedmiller, M., & Braun, H. (1993). A direct adaptive method for faster backpropagation learning: The RPROP algorithm. *International Conference On Neural Networks*. IEEE. 10.1109/ICNN.1993.298623

Ristić, L., Despotović, D., & Dimitrijević, M. (2020). Multifunctionality of Agriculture as a Significant Factor for Sustainable Rural Development of the Republic of Serbia. *Economic Themes, 58*(1), 17–32. doi:10.2478/ethemes-2020-0002

Rita, P., Oliveira, T., & Farisa, A. (2019). The impact of e-service quality and customer satisfaction on customer behavior in online shopping. *Heliyon, 5*(10), e02690. doi:10.1016/j.heliyon.2019.e02690 PMID:31720459

Ritzer, G. (2021). The McDonaldization of society. In *In the Mind's Eye* (pp. 143–152). Routledge. doi:10.4324/9781003235750-15

Robbins, P., & Devitt, F. (2017). Collaboration, creativity and entrepreneurship in tourism: A case study of how design thinking created a cultural cluster in Dublin. *International Journal of Entrepreneurship and Innovation Management, 21*(3), 185–211. doi:10.1504/IJEIM.2017.083454

Roblek, V., Petrović, N. N., Gagnidze, I., & Khokhobaia, M. (2021). Role of a Digital Transformation in Development of a Rural Tourism Destinations. *VI International Scientific Conference Challenges of Globalization in Economics and Business,* (pp. 297–305). Research Gate.

Rodgers, W. (2020). *Artificial intelligence in a throughput model: Some major algorithms.* CRC Press.

Rodríguez-Antón, J. M., & Alonso-Almeida, M. D. M. (2020). COVID-19 impacts and recovery strategies: The case of the hospitality industry in Spain. *Sustainability (Basel), 12*(20), 8599. doi:10.3390u12208599

Romagnoli, S., & Tarabu', C., MalekiVishkaei, B., & De Giovanni, P. (2023). The impact of digital technologies and sustainable practices on circular supply chain management. *Logistics, 7*(1), 1. doi:10.3390/logistics7010001

Romão, J., Machino, K., & Nijkamp, P. (2018). Integrative diversification of wellness tourism services in rural areas– an operational framework model applied to east Hokkaido (Japan). *Asia Pacific Journal of Tourism Research, 23*(7), 734–746. doi:10.1080/10941665.2018.1488752

Rosário, A., Vilaça, F., Raimundo, R., & Cruz, R. (2021). Literature review on Health Knowledge Management in the last 10 years (2009-2019). *The Electronic Journal of Knowledge Management, 18*(3), 338-355. doi:10.34190/ejkm.18.3.2120

Rosário, A. T. (2021). The Background of articial intelligence applied to marketing. *Academy of Strategic Management Journal, 20*(6), 1–19. 1939-6104-20-S6-118

Rosário, A. T., & Dias, J. C. (2022). Sustainability and the Digital Transition: A Literature Review. *Sustainability (Basel), 14*(7), 4072. doi:10.3390u14074072

Rosenbaum, M. S., & Russell-Bennett, R. (2020). Service research in the new (post-COVID) marketplace. *Journal of Services Marketing, 34*(5), I–V. doi:10.1108/JSM-06-2020-0220

Rossi, A., & Fusco, F. (2019). Wine Index of Salubrity and Health (WISH). An evidence-based instrument to evaluate the impact of good wine on well-being. *International Journal of Wine Research., 2019*(11), 23–37. doi:10.2147/IJWR.S177394

Rowley, T. D., & Shirley, L. P. (2019). Removing rural development barriers through telecommunications: Illusion or reality. In *In Economic Adaptation* (pp. 247–264). Routledge. doi:10.4324/9780429041082-14

Ruiz-Ortega, M. J., Parra-Requena, G., & García-Villaverde, P. M. (2021). From entrepreneurial orientation to sustainability orientation: The role of cognitive proximity in companies in tourist destinations. *Tourism Management, 84*, 104265. doi:10.1016/j.tourman.2020.104265

Rumelhart, D. E., & McClelland, J. L. (1986). *Parallel Distributed Processing. Explorations in the Microstructure of Cognition* (Vol. 1). MIT Press. Cambridge. doi:10.7551/mitpress/5236.001.0001

Russell, S., & Norvig, P. (2021). Artificial intelligence: a modern approach. Global Foundations, 19, 23.

Rust, R. T., & Oliver, R. L. (Eds.). (1993). *Service quality: New directions in theory and practice*. Sage Publications.

Ružičić, M. M., Miletić, M., & Dobrota, M. (2021). Does a national innovation system encourage sustainability? Lessons from the construction industry in Serbia. *Sustainability (Basel), 13*(7), 3591. doi:10.3390u13073591

Saiidi, U. (2019). *Facial recognition is coming to hotels*. CNBC. https://www.cnbc.com/2019/10/04/alibaba-china-flyzoo-hotel-uses-facial-recognition-tech-and-robots.html

Salemink, K., Strijker, D., & Bosworth, G. (2017). Rural development in the digital age: A systematic literature review on unequal ICT availability, adoption, and use in rural areas. *Journal of Rural Studies, 54*, 360–371. doi:10.1016/j.jrurstud.2015.09.001

Salinas Fernández, J. A., Serdeira Azevedo, P., Martín Martín, J. M., & Rodríguez Martín, J. A. (2020). Determinants of tourism destination competitiveness in the countries most visited by international tourists: Proposal of a synthetic index. *Tourism Management Perspectives, 33*, 100582. doi:10.1016/j.tmp.2019.100582

Salome, L. R., van Bottenburg, M., & van den Heuvel, M. (2013). 'We are as green as possible': Environmental responsibility in commercial artificial settings for lifestyle sports. *Leisure Studies, 32*(2), 173–190. doi:10.1080/02614367.2011.645247

Sancho-Esper, F., Ostrovskaya, L., Rodriguez-Sanchez, C., & Campayo-Sanchez, F. (2022). Virtual reality in retirement communities: Technology acceptance and tourist destination recommendation. *Journal of Vacation Marketing*, 13567667221080567.

Sanjeev, G. M., & Birdie, A. K. (2019). The tourism and hospitality industry in India: emerging issues for the next decade. In Worldwide Hospitality and Tourism Themes (Vol. 11, pp. 355–361). Emerald Group Publishing Ltd. doi:10.1108/WHATT-05-2019-0030

Santos, J. A. C., Santos, M. C., Pereira, L. N., Richards, G., & Caiado, L. (2020). Local food and changes in tourist eating habits in a sun-and-sea destination: A segmentation approach. *International Journal of Contemporary Hospitality Management*, *35*(3), 3501–3521. doi:10.1108/IJCHM-04-2020-0302

Sarabdeen, M., & Alofaysan, H. (2023). Investigating the impact of digital transformation on the labor market in the era of changing digital transformation dynamics in Saudi Arabia. *Economies*, *11*(1), 12. doi:10.3390/economies11010012

Sari, T., Güleş, H. K., & Yiğitol, B. (2020). Awareness and readiness of Industry 4.0: The case of the Turkish manufacturing industry. *Advances in Production Engineering & Management*, *15*(1), 57–68. doi:10.14743/apem2020.1.349

Saulle, R., & la Torre, G. (2010). The Mediterranean Diet, recognized by UNESCO as a cultural heritage of humanity. *Italian Journal of Public Health*, *7*(4), 414–415. doi:10.2427/5700

Saviera, T. M., Kusumastuti, R., & Hidayanto, A. N. (2022). Importance-performance analysis towards sustainable indigenous tourism (a lesson learned from Indonesia). *International Journal of Innovation and Learning*, *31*(1), 91–116. doi:10.1504/IJIL.2022.119638

Sawy, O. A. E., Amsinck, H., Kræmmergaard, P., & Vinther, A. L. (2016). How LEGO built the foundations and enterprise capabilities for digital leadership. *MIS Quarterly Executive*, *15*(2), 141–166.

Scandic Press Release. (2019). *Scandic Hotels introduces standard for allergy-friendly rooms*. Scandic.

Scheyvens, R., & Biddulph, R. (2018). Inclusive tourism development. *Tourism Geographies*, *20*(4), 589–609. doi:10. 1080/14616688.2017.1381985

Schilirò, D. (2021). Digital transformation, COVID-19, and the future of work. *International Journal of Business Management and Economic Research*, *12*(3), 1945–1952.

Sedra, D., & El Bayed, H. (2022). Branding the city: The case of Casablanca-Morocco. *Place Branding and Public Diplomacy*, *18*(2), 181–189. doi:10.105741254-020-00195-y

Serra Cantallops, A., & Salvi, F. (2014). New consumer behavior: A review of research on eWOM and hotels. *International Journal of Hospitality Management*, *36*, 41–51. doi:10.1016/j.ijhm.2013.08.007

Serra, M. (2016). Algarve - relação enogastronómica. Tese de mestrado. Faro: ESGHT-Universidade do Algarve.

Serra-Majem, L., Bach-Faig, A., & Raidó-Quintana, B. (2012). Nutritional and cultural aspects of the Mediterranean diet. *International Journal for Vitamin and Nutrition Research*, *82*(3), 157–162. doi:10.1024/0300-9831/a000106 PMID:23258395

Seth, N., Deshmukh, S. G., & Vrat, P. (2005). Service quality models: A review. *International Journal of Quality & Reliability Management*, *22*(9), 913–949. doi:10.1108/02656710510625211

Sevin, E. (2021). Computational approaches to place branding: A call for a theory-driven research agenda. In *A Research Agenda for Place Branding* (pp. 33–45). Edward Elgar Publishing. doi:10.4337/9781839102851.00010

SHA, Singapore Hotel Association (2019). Smart Hotel Technology Guide. *Using Technology to Transform the "Heart of House"*. SHA.

SHA, Singapore Hotel Association. (2018). Smart Hotel Technology Guide. Using Technology to Navigate the Guest Experience Journey.

Shabani, N., Munir, A., & Hassan, A. (2018). E-Marketing via augmented reality: A case study in the tourism and hospitality industry. *IEEE Potentials*, *38*(1), 43–47. doi:10.1109/MPOT.2018.2850598

Shareef, M. A., Mukerji, B., Dwivedi, Y. K., Rana, N. P., & Islam, R. (2019). Social media marketing: Comparative effect of advertisement sources. *Journal of Retailing and Consumer Services*, *46*, 58–69. doi:10.1016/j.jretconser.2017.11.001

Sharma, A., & Guleria, K. (2021). A Framework for Hotel Inventory Control System for Online Travel Agency using Robotic Process Automation. *2021 International Conference on Advance Computing and Innovative Technologies in Engineering, ICACITE 2021*, (pp. 764–768). IEEE. 10.1109/ICACITE51222.2021.9404613

Sheehan, K. B. (2018). Crowdsourcing research: Data collection with Amazon's Mechanical Turk. *Communication Monographs*, *85*(1), 140–156. doi:10.1080/03637751.2017.1342043

Shelton, E., & Tucker, H. (2005). Tourism and disability: Issues beyond access. *Tourism Review International*, *8*(3), 211–219. doi:10.3727/154427205774791528

Shen, Z., Chen, J., Bai, K., Li, Y., Cui, Y., & Song, M. (2023). The digital impact on environmental performance: Evidence from Chinese publishing. *Emerging Markets Finance & Trade*. doi:10.1080/1540496X.2022.2164188

Shin, H. H., & Jeong, M. (2022). Does a virtual trip evoke travelers' nostalgia and derive intentions to visit the destination, a similar destination, and share?: Nostalgia-motivated tourism. *Journal of Travel & Tourism Marketing*, *39*(1), 1–17. doi:10.1080/10548408.2022.2044972

Shi, Y., Prentice, C., & He, W. (2014). Linking service quality, customer satisfaction and loyalty in casinos, does membership matter? *International Journal of Hospitality Management*, *40*, 81–91. doi:10.1016/j.ijhm.2014.03.013

SHTG (Smart Hotel Technology Guide). (2018). *Using technology to navigate the guest experience journey*. SHA. https://sha.org.sg/userfiles/ckeditor/Files/Smart%20Hotel%20Technology%20Guide%202018.pdf

SHTG (Smart Hotel Technology Guide). (2019). *Using technology to navigate the guest experience journey*. SHA. https://sha.org.sg/userfiles/ckeditor/Files/Smart%20Hotel%20Technology%20Guide%202019.pdf

Shu, Q., Tu, Q., & Wang, K. (2011). The impact of computer self-efficacy and technology dependence on computer-related technostress: A social cognitive theory perspective. *International Journal of Human-Computer Interaction*, *27*(10), 923–939. doi:10.1080/10447318.2011.555313

Siddaway, A. P., Wood, A. M., & Hedges, L. V. (2019). How to do a systematic review: A best practice guide for conducting and reporting narrative reviews, meta-analyses, and meta-syntheses. *Annual Review of Psychology*, *70*(1), 747–770. doi:10.1146/annurev-psych-010418-102803 PMID:30089228

Siderska, J. (2020). Robotic Process Automation-a driver of digital transformation? *Engineering Management in Production and Services*, *12*(2), 21–31. doi:10.2478/emj-2020-0009

Sigala, M. (2020). Tourism and COVID-19: Impacts and implications for advancing and resetting industry and research. *Journal of Business Research*, *117*, 312–321. doi:10.1016/j.jbusres.2020.06.015 PMID:32546875

Sigala, M., & Gkritzali, A. (2020). Accessible tourism at the crossroads: Lessons learnt, challenges and future directions. *Inaugural International Conference on Smart Tourism, Smart Cities and Enabling Technologies (STATE 2020)* (pp. 56-65). Springer.

Singh, A. (2018). Facebook, WhatsApp, and Twitter: Journey towards Education. *SOSHUM: Jurnal Sosial Dan Humaniora*, *8*(2), 139–149. doi:10.31940oshum.v8i2.987

Singh, S., & Wagner, R. (2022). Indian wine tourism: New landscape of international spillovers. *Journal of Asia Business Studies*. doi:10.1108/JABS-01-2022-0004

Sinno, N. (2019). *The effect of digital transformation on innovation and entrepreneurship in the tourism sector: The case of Lebanese tourism services providers.* Springer. doi:10.1007/978-3-030-30874-2_3

Siuta-Tokarska, B., Kruk, S., Krzemiński, P., Thier, A., & Żmija, K. (2022). Digitalisation of enterprises in the energy sector: Drivers—business models—prospective directions of changes. *Energies, 15*(23), 8962. doi:10.3390/en15238962

Slevitch, L., Chandrasekera, T., & Sealy, M. D. (2022). Comparison of virtual reality visualizations with traditional visualizations in hotel settings. *Journal of Hospitality & Tourism Research (Washington, D.C.), 46*(1), 212–237. doi:10.1177/1096348020957067

Smartvel. (2020). *Smart Hotels: what to expect from the future of hospitality.* Smartvel. https://blog.smartvel.com/blog/smart-hotels-what-to-expect-from-the-future-of-hospitality

Smith, C. (2021). Barney' the Swiss robot bartender can mix dozens of cocktails, tell jokes and sanitise itself. *The Drinks Business.* https://www.thedrinksbusiness.com/2021/04/barney-the-swiss-robot-bartender-can-mix-dozens-of-cocktails-tell-jokes-and-sanitise-itself/

Snyder, H. (2019). Literature review as a research methodology: An overview and guidelines. *Journal of Business Research, 104*, 333–339. doi:10.1016/j.jbusres.2019.07.039

Sobczak, A. (2022). Robotic Process Automation as a Digital Transformation Tool for Increasing Organizational Resilience in Polish Enterprises. *Sustainability (Basel), 14*(3), 1333. doi:10.3390u14031333

Sofield, T., & Lia, S. (2011). Tourism governance and sustainable national development in China: A macro-level synthesis. *Journal of Sustainable Tourism, 19*(4-5), 501–534. doi:10.1080/09669582.2011.571693

Solvoll, S., Alsos, G. A., & Bulanova, O. (2015). Tourism entrepreneurship – review and future directions. *Scandinavian Journal of Hospitality and Tourism, 15*(sup1), 120–137. doi:10.1080/15022250.2015.1065592

Solymos, E. (1997). Nationalities in Baja in the 18th-19th century. In J. Bárth (Ed.), *Mirror-images on the Sugovica* (pp. 101–103). (in Hungarian)

Song, H., & Turner, L. (2006). Tourism demand forecasting. *International handbook on the economics of tourism*, 89-114.

Song, H., Qiu, R. T. R., & Park, J. (2019). A review of research on tourism demand forecasting: Launching the Annals of Tourism Research Curated Collection on tourism demand forecasting. *Annals of Tourism Research, 75*, 338–362. doi:10.1016/j.annals.2018.12.001

Sørensen, F., & Grindsted, T. S. (2021). Sustainability approaches and nature tourism development. *Annals of Tourism Research, 91*, 103307. doi:10.1016/j.annals.2021.103307

Sotiriadis, M. (2018). Entrepreneurship and Entrepreneurs in Tourism. In *The Emerald Handbook of Entrepreneurship in Tourism, Travel and Hospitality* (pp. 3–17). Emerald Publishing Limited. doi:10.1108/978-1-78743-529-220181001

Sotiriadis, M. D. (2017). Sharing tourism experiences in social media: A literature review and a set of suggested business strategies. *International Journal of Contemporary Hospitality Management, 29*(1), 179–225. doi:10.1108/IJCHM-05-2016-0300

Sotiriadis, M. D., & van Zyl, C. (2013). Electronic word-of-mouth and online reviews in tourism services: The use of twitter by tourists. *Electronic Commerce Research, 13*(1), 103–124. doi:10.100710660-013-9108-1

Soto-Acosta, P. (2020). COVID-19 pandemic: Shifting digital transformation to a high-speed gear. *Information Systems Management, 37*(4), 260–266. doi:10.1080/10580530.2020.1814461

Souca, M. L. (2010). Accessible Tourism – The Ignored Opportunity. *Annals of Faculty of Economics, 1*(2), 1154–1157.

Sousa-Zomer, T. T., Neely, A., & Martinez, V. (2020). Digital transforming capability and performance: A micro foundational perspective. *International Journal of Operations & Production Management, 40*(7/8), 1095–1128. doi:10.1108/IJOPM-06-2019-0444

Sparks, B. A., & Browning, V. (2011). The impact of online reviews on hotel booking intentions and perception of trust. *Tourism Management, 32*(6), 1310–1323. doi:10.1016/j.tourman.2010.12.011

Stankić, R. (2014). Key Success Factors for Implementation of Business Information System University in East Sarajevo, Faculty of Economic. *Bčko*, (November), 18–25.

Stankov, U., & Gretzel, U. (2021). Digital wellbeing in the tourism domain: Mapping new roles and responsibilities. *Information Technology & Tourism, 23*(5), 5–17. doi:10.100740558-021-00197-3

Statista. (2021). *Tourism in Morocco.* Statistia. https://www.statista.com/topics/8256/tourism-industry-in-morocco/

Stephens Balakrishnan, M. (2009). Strategic branding of destinations: A framework. *European Journal of Marketing, 43*(5/6), 611–629. doi:10.1108/03090560910946954

Sternberg, S. R. (1981). Parallel architectures for image processing. *Real-Time Parallel Computing: Imaging Analysis*, 347-359. Book Authority.

Sternberg, S. R. (2013, March). Environmental Research Institute of Michigan. In *Biomedical Images and Computers: Selected Papers Presented at the United States-France Seminar on Biomedical Image Processing, St. Pierre de Chartreuse, France, May 27–31, 1980 (Vol. 17*, p. 294). Springer Science & Business Media.

Sternberg, R. J. (2005). Creativity or creativities? *International Journal of Human-Computer Studies, 63*(4-5), 370–382. doi:10.1016/j.ijhcs.2005.04.003

Sternberg, S. R. (1984). Parallel processing in machine vision. *Robotica, 2*(1), 33–40. doi:10.1017/S026357470000881X

Stoffelen, A., Adiyia, B., Vanneste, D., & Kotze, N. (2020). Post-apartheid local sustainable development through tourism: An analysis of policy perceptions among 'responsible' tourism stakeholders around Pilanesberg national park, South Africa. *Journal of Sustainable Tourism, 28*(3), 1–20. doi:10.1080/09669582.2019.1679821

Stojczew, K. (2021). Ocena wpływu pandemii koronawirusa na branżę turystyczną w Polsce. *Prace Naukowe Uniwersytetu Ekonomicznego we Wrocławiu, 65*(1), 157-172. https://doi.org/ doi:10.15611/pn.2021.1.09

Stroink, M. L., & Nelson, C. H. (2013). Complexity and food hubs: Five case studies from Northern Ontario. *Local Environment, 18*(5), 620–635. doi:10.1080/13549839.2013.798635

Stumbo, N. J., & Pegg, S. (2005). Travelers and Tourists with Disabilities: A Matter of Priorities and Loyalties. *Tourism Review International, 8*(3), 195–209. doi:10.3727/154427205774791537

Suárez-Henríquez, C., Ricoy-Cano, A. J., Hernández-Galan, J., & de la Fuente-Robles, Y. M. (2022). The past, present, and future of accessible tourism research: A bibliometric analysis using the Scopus database. *Journal of Accessibility and Design for All, 7*(2), 26–60.

Subawa, N. S., Widhiasthini, N. W., Astawa, I. P., Dwiatmadja, C., & Permatasari, N. P. I. (2021). The practices of virtual reality marketing in the tourism sector, a case study of Bali, Indonesia. *Current Issues in Tourism, 24*(23), 3284–3295. doi:10.1080/13683500.2020.1870940

Sudhakar, B. D., Kattepogu, N., & David, A. (2017). Marketing assistance and digital branding-an insight for technology up-gradation for MSME's. *International Journal of Management Studies & Research, 5*(1), 2455–1562.

Suh, K. S., & Chang, S. (2006). User interfaces and consumer perceptions of online stores: The role of telepresence. *Behaviour & Information Technology*, *25*(2), 99–113. doi:10.1080/01449290500330398

Sümegi, J. (2019). The foundation and early history of the Benedictine abbey in Báta (in Hungarian: A bátai bencés apátság alapítása és korai története). [Szekszárd, Hungary.]. *Year Book of the Wosinsky Mor County Museum.*, *XLI*, 2019.

Sundbo, J., Orfila-Sintes, F., & Sørensen, F. (2007). The innovative behavior of tourism firms-comparative studies of Denmark and Spain. *Research Policy*, *36*(1), 88–106. doi:10.1016/j.respol.2006.08.004

Su, Y., Cherian, J., Sial, M. S., Badulescu, A., Thu, P. A., Badulescu, D., & Samad, S. (2021). Does tourism affect economic growth of China? A panel granger causality approach. *Sustainability (Basel)*, *13*(3), 1349. doi:10.3390u13031349

Swanson, K. K., & DeVereaux, C. (2017). A theoretical framework for sustaining culture: Culturally sustainable entrepreneurship. *Annals of Tourism Research*, *62*, 78–88. doi:10.1016/j.annals.2016.12.003

Sweezey, M. (2020). *The Context Marketing Revolution: How to Motivate Buyers in the Age of Infinite Media*. Harvard Business Press.

Syed, R., Suriadi, S., Adams, M., Bandara, W., Leemans, S. J. J., Ouyang, C., ter Hofstede, A. H. M., van de Weerd, I., Wynn, M. T., & Reijers, H. A. (2020). Robotic Process Automation: Contemporary themes and challenges. *Computers in Industry*, *115*, 103162. doi:10.1016/j.compind.2019.103162

Szilágyi, M. (1997). Fish foods as man foods. In *Variations of the food culture in the 18th-20th centuries (in Hungarian: A halételek, mint férfiételek, In: A táplálkozáskultúra változatai a 18-20. században)* (pp. 117–132). Kalocsa.

Tabachnick, B. G., & Fidell, L. S. (2013). *Using Multivariate Statistics*. Pearson.

Takeda, K., & Card, J. A. (2002). US tour operators and travel agencies: Barriers encountered when providing package tours to people who have difficulty walking. *Journal of Travel & Tourism Marketing*, *12*(1), 47–61. doi:10.1300/J073v12n01_03

Talwar, S., Kaur, P., Nunkoo, R., & Dhir, A. (2022). Digitalization and sustainability: Virtual reality tourism in a post pandemic world. *Journal of Sustainable Tourism*, 1–28. doi:10.1080/09669582.2022.2029870

Tang, G. N., Ren, F., & Zhou, J. (2022). Does the digital economy promote "innovation and entrepreneurship" in rural tourism in China? *Frontiers in Psychology*, *13*, 979027. doi:10.3389/fpsyg.2022.979027 PMID:36312131

Tanrıverdi, G., & Durak, M. Ş. (2022). A visualized bibliometric analysis of mapping research trends of airline business models (ABMs) from 1985 to 2021. *Journal of Aviation*, *6*(3), 387–403. doi:10.30518/jav.1172121

Tarafdar, M., Pullins, E. B., & Ragu-Nathan, T. S. (2015). Technostress: Negative effect on performance and possible mitigations. *Information Systems Journal*, *25*(2), 102–132. doi:10.1111/isj.12042

Taste Atlas. (2023a). The 3 Most Popular Fish-dishes in Hungary (What to eat in Hungary). *Taste Atlas*. https://www.tasteatlas.com/most-popular-fish-dishes

Taste Atlas. (2023b). Where to Eat the Best Fiš Paprikaš in the World? *Taste Atlas*. https://www.tasteatlas.com/fish-paprikas/wheretoeat

Teixeira, S. J., & Ferreira, J. J. M. (2019). Entrepreneurial artisan products as regional tourism competitiveness. *International Journal of Entrepreneurial Behaviour & Research*, *25*(4), 652–673. doi:10.1108/IJEBR-01-2018-0023

Temelkov, Z. (2022). Financial performance of selected hotel groups and resorts during COVID-19 pandemic: 2019/2020 comparison. *Менаџмент у Хотелијерству и Туризму*, *10*(1), 41–51.

Tengilimoğlu, E. ve Kuzucu, S. C. (2019). Döviz kuru oynaklıkları turist başına düşen ortalama harcamayı etkiler mi? 2003-2018 yılları arasında bir nedensellik analizi. *International Human and Civilization Congress, From Past to Future, Ines*, (pp. 148-156). IEEE.

Terdpaopong, K. (2020). Digital Transformation in the Hospitality Industry in an Emerging Country. In K. Sandhu (Ed.), *Leadership, Management, and Adoption Techniques for Digital Service Innovation* (pp. 223–243). IGI Global. doi:10.4018/978-1-7998-2799-3.ch012

Teruel-Sánchez, R., Briones-Peñalver, A. J., Bernal-Conesa, J. A., & de Nieves-Nieto, C. (2021). Influence of the entrepreneur's capacity in business performance. *Business Strategy and the Environment*, 30(5), 2453–2467. doi:10.1002/bse.2757

The Economist (2021). Thanks to the pandemic, diplomats have a bigger, better toolkit. *The Economist*.

Thellefsen, T., Sørensen, B., Vetner, M., & Andersen, C. (2006). *Negotiating the meaning of artefacts: Branding in a semeiotic perspective*. De Gruyter.

Theodoropoulou, I. (2020). The case of Mykonos, Greece on Facebook. *Research Companion to Language and Country Branding*, 313.

Thompson, B. S., Gillen, J., & Friess, D. A. (2018). Challenging the principles of ecotourism: Insights from entrepreneurs on environmental and economic sustainability in Langkawi, Malaysia. *Journal of Sustainable Tourism*, 26(2), 257–276. doi:10.1080/09669582.2017.1343338

Thummula, E., Yadav, R. K., & David, A. (2019). A cost-effective technique to avoid communication and computation overhead in vehicle insurance database for online record monitoring. [IJMPERD]. *International Journal of Mechanical and Production Engineering Research and Development*, 9(2), 711–722.

Tian, W., Liu, Y. J., Liu, B., He, D., Wu, J. Y., Han, X. G., Zhao, J., & Fan, M. (2019). Guideline for thoracolumbar pedicle screw placement assisted by orthopaedic surgical robot. *Orthopaedic Surgery*, 11(2), 153–159. doi:10.1111/os.12453 PMID:31025807

Tiganis, A., & Tsakiridou, E. (2022). Local food consumption by foreign tourists in Greece. *International Journal of Tourism Policy*, 12(2), 70–83. doi:10.1504/IJTP.2022.121898

Todorivić, M., & Bjelac, Ž. (2009). Rural tourism in Serbia as a Concept of Development in Undeveloped Regions, journal. *Acta Geographica Slovenica*, 49(2), 453–473. doi:10.3986/AGS49208

Török, Á. (2019). The recognitions and the embeddedness of the TTR trademark among the Hungarian consumers. Nutrition Marketing, 6, 81-97.

Torraco, R. J. (2016). Writing integrative literature reviews: Using the past and present to explore the future. *Human Resource Development Review*, 15(4), 404–428.

Torres, E. N., Adler, H., & Behnke, C. (2014). Stars, diamonds, and other shiny things: The use of expert and consumer feedback in the hotel industry. *Journal of Hospitality and Tourism Management*, 21, 34–43. doi:10.1016/j.jhtm.2014.04.001

Tovar, J. (2020). Reimaging and branding a post-reunification Germany. *Research Companion to Language and Country Branding*, 183.

Trauer, B. (2006). Conceptualizing special interest tourism-frameworks for analyses. *Tourism Management*, 27(2), 183–200. doi:10.1016/j.tourman.2004.10.004

Trejos, N. (2014). *Ready for the Hotel Industry's First Robotic Butler?* https://www.usatoday.com/story/travel/hotels/2014/08/12/aloft-hotels-starwood-robotic-bultler/13954231/

Triantafillidou, E., & Tsiaras, S. (2018). Exploring entrepreneurship, innovation and tourism development from a sustainable perspective: Evidence from greece. *Journal for International Business and Entrepreneurship Development, 11*(1), 53–64. doi:10.1504/JIBED.2018.090020

Trichopoulou, A. (2021). Mediterranean diet as intangible heritage of humanity: 10 years on. *Nutrition, Metabolism, and Cardiovascular Diseases, 31*(7), 1943–1948. doi:10.1016/j.numecd.2021.04.011 PMID:34059382

Trienekens, J., & Zuurbier, P. (2008). Quality and safety standards in the food industry, developments and challenges. *International Journal of Production Economics, 113*(1), 107–122. https://doi.org/https://doi.org/10.1016/j.ijpe.2007.02.050. doi:10.1016/j.ijpe.2007.02.050

Trisic, I., Privitera, D., Stetic, S., Petrovic, M. D., Radovanovic, M. M., Maksin, M., Simicevic, D., Jovanovic, S. S., & Lukic, D. (2022). Sustainable Tourism to the Part of Transboundary UNESCO Biosphere Reserve "Mura-Drava-Danube". A Case of Serbia, Croatia and Hungary. *Sustainability (Basel), 2022*(14), 6006. doi:10.3390u14106006

Tsai, T.-H., Lin, W.-Y., Chang, Y.-S., Chang, P.-C., & Lee, M.-Y. (2020). Technology anxiety and resistance to change behavioral study of a wearable cardiac warming system using an extended TAM for older adults. *PLoS One, 15*(1), 1–24. doi:10.1371/journal.pone.0227270 PMID:31929560

Tsinghua-Chinese Academy of Engineering Joint Research Center for Knowledge and Intelligence, Centre for Intelligent Research, Institute of Artificial Intelligence, Tsinghua University, Chinese Association for Artificial Intelligence. (2020). *Report on Artificial Intelligence Development 2011-2020*. Aminer. https://static.aminer.cn/misc/pdf/zpAIreport2020.pdf

Tu, Q., & Liu, A. (2014). Framework of smart tourism research and related progress in China. In *International conference on management and engineering (CME 2014)* (140-146). DEStech Publications, Inc.

Tuan, L. T. (2022a). Employee mindfulness and proactive coping for technostress in the COVID-19 outbreak: The roles of regulatory foci, technostress, and job insecurity. *Computers in Human Behavior, 129*, 1–9. doi:10.1016/j.chb.2021.107148 PMID:34975214

Tuan, L. T. (2022b). How and when does hospitality employees' core beliefs challenge foster their proactive coping for technostress?: Examining the roles of promotion focus, job insecurity, and technostress. *Journal of Hospitality and Tourism Management, 22*, 86–99. doi:10.1016/j.jhtm.2022.05.017

Tuffield, P., & Elias, H. (2003). The shadow robot mimics human actions. *The Industrial Robot, 30*(1), 56–60. doi:10.1108/01439910310457715

Tung, V. W. S., & Law, R. (2017). The potential for tourism and hospitality experience research in human-robot interactions. *International Journal of Contemporary Hospitality Management, 29*(10), 2498–2513. doi:10.1108/IJCHM-09-2016-0520

Turco, D. M., Stumbo, N. J., & Garncarz, J. (1998). Tourism Constraints for People with Disabilities. *Parks & Recreation, 33*(9), 78–84.

Turismo de Portugal. (2017). *Accessible tourism destination handbook*. Turismo de Portugal.

Turismo de Portugal. (2020) *Portugal é o destino turíststico acessível*. Turismo de Portugal. https://travelbi.turismodeportugal.pt/en-us/Pages/Portugal-%C3%A9-o-Destino-Tur%C3%ADstico-Acess%C3%ADvel-2019.aspx

TURKSTAT. (2020). *İstatiksel Tablolardan ve MEDAS Veritabanından Derlenmiş bilgiler*. TURKSTAT. https://www.tuik.gov.tr/

Tussyadiah, I. (2020). A review of research into automation in tourism: Launching the Annals of Tourism Research Curated Collection on Artificial Intelligence and Robotics in Tourism. *Annals of Tourism Research, 81*, 102883. doi:10.1016/j.annals.2020.102883

Tussyadiah, I. P., Wang, D., Jung, T. H., & tom Dieck, M. C. (2018). Virtual reality, presence, and attitude change: Empirical evidence from tourism. *Tourism Management*, *66*, 140–154. doi:10.1016/j.tourman.2017.12.003

Tussyadiah, I. P., Zach, F. J., & Wang, J. (2020). Do travelers trust intelligent service robots? *Annals of Tourism Research*, *81*, 102886. doi:10.1016/j.annals.2020.102886

Tüzünkan, D. (2017). The Relationship between Innovation and Tourism: The Case of Smart Tourism. *International Journal of Applied Engineering Research: IJAER*, *12*(23), 13861–13867. http://www.ripublication.com

UiPath Academy. (n.d.). *About*. UiPath Academy. https://academy.uipath.com/

Ukpabi, D., Karjaluoto, H., Olaleye, S. A., & Mogaji, E. (2018). Dual perspectives on the role of artificially intelligent robotic virtual agents in the tourism, travel and hospitality industries. In *EuroMed Academy of Business Conference Book of Proceedings*. EuroMed Press.

UNFAO. (2021). *Cultivating Our Futures - Issues Paper: The Multifunctional Character of Agriculture and Land, webl*. UNFAO. https://www.fao.org/3/x2777e/x2777e00.htm

United Nations Environmental Development Programme/World Tourism Organization. (2005). Making tourism more sustainable. A guide for policymakers. United Nations Environment Programme, Division of Technology, Industry and Economics and World Tourism Organization. Paris, Madrid: UNEP, WTO.

United Nations Report. (2021, April 12). *UN Tourism News #23*. UNWTO. https://www.unwto.org/un-tourism-news-23

United Nations. (2023). *International Day of Argania*. UN. https://www.un.org/en/observances/argania-day

UNWTO - BCC. (2019). *Guidelines for the Development of Gastronomy Tourism*. International Tourism Organization UNWTO and Basque Culinary Center., doi:10.18111/9789284420957

UNWTO (UN World Tourism Organization). (2020). *A compilation of data on inbound tourism by country*. UNWTO. https://www.unwto.org/country-profile-inbound-tourism

UNWTO. (2007). *A Practical Guide to Tourism Destination Management*. UNWTO.

UNWTO. (2015). *Manual on Accessible Tourism for All Public-private Partnerships and Good Practices*. UNWTO. https://cf.cdn.unwto.org/sites/all/files/docpdf/aamanualturismoaccesibleomt-facseng.pdf

UNWTO. (2016). *Manual on accessible tourism for all: Principles, tools and best practices – Module I: Accessible Tourism – Definition and context*. UNWTO. https://cf.cdn.unwto.org/sites/all/files/docpdf/moduleieng13022017.pdf

UNWTO. (2016a). *Accessible Tourism for All: An Opportunity within Our Reach*. UNWTO. https://cf.cdn.unwto.org/sites/all/files/docpdf/turismoaccesiblewebenok.pdf

UNWTO. (2016b). Manual on Accessible Tourism for All: Principles, Tools and Best Practices. *Module I: Accessible Tourism – Definition and Context. World Tourism Organization*. https://www.e-unwto.org/doi/pdf/10.18111/9789284418077

UNWTO. (2019). *Accessible tourism for all: An opportunity within our reach*. UNWTO. https://www.e-unwto.org/doi/abs/10.18111/9789284421147

UNWTO. (2020). *Accessible tourism*. UNTWO. https://www.unwto.org/fr/accessibility

UNWTO. (2021). *Portugal takes first step to becoming a more accessible destination*. UNWTO. https://www.unwto.org/news/portugal-takes-first-step-to-becoming-a-more-accessible-destination

Ursache, M. (2015). Tourism–significant driver shaping a destinations heritage. *Procedia: Social and Behavioral Sciences*, *188*, 130–137. doi:10.1016/j.sbspro.2015.03.348

Uysal, M., & El Roubi, M. S. (1999). Artificial neural networks versus multiple regression in tourism demand analysis. *Journal of Travel Research*, *38*(2), 111–118. doi:10.1177/004728759903800203

Valagão, M. M. (2015). Identidade alimentar mediterrânica de Portugal e do Algarve. In A. Freitas, J. P. Bernardes, M. P. Mateus, & N. Braz (Eds.), *Dimensões da Dieta Mediterrânica* (pp. 155–179). Universidade do Algarve.

Valeri, M., & Baiocco, S. (2012). The integration of a Swedish minority in the hotel business culture: The case of Riva del sole. *Tourism Review*, *67*(1), 51–60. doi:10.1108/16605371211216378

Valeri, M., & Fadlon, L. (2016). Sustainability in tourism: An originality and hospitality business in Italy. *Tourismos*, *11*(1), 1–18. www.scopus.com

Van der Meulen, H. S. (2007, March). A normative deðnition method for origin food products. *Anthropology of Food*, *S2*(S2). doi:10.4000/aof.406

Van Doorn, J., Mende, M., Noble, S. M., Hulland, J., Ostrom, A. L., Grewal, D., & Petersen, J. A. (2017). Domo arigato Mr. Roboto: Emergence of automated social presence in organizational frontlines and customers' service experiences. *Journal of Service Research*, *20*(1), 43–58. doi:10.1177/1094670516679272

Van Wijk, J., Van der Duim, R., Lamers, M., & Sumba, D. (2015). The emergence of institutional innovations in tourism: The evolution of the African wildlife foundation's tourism conservation enterprises. *Journal of Sustainable Tourism*, *23*(1), 104–125. doi:10.1080/09669582.2014.927878

Vanneschi, L., Horn, D. M., Castelli, M., & Popovič, A. (2018). An artificial intelligence system for predicting customer default in e-commerce. *Expert Systems with Applications*, *104*, 1–21. doi:10.1016/j.eswa.2018.03.025

Vărzaru, A. A., Bocean, C. G., & Cazacu, M. (2021). Rethinking tourism industry in pandemic COVID-19 period. *Sustainability*, *13*(12), 6956. doi:10.3390u13126956

Veenhoven, R. (2003). Hedonism and happiness. *Journal of Happiness Studies*, *4*(4), 437–457. doi:10.1023/B:JOHS.0000005719.56211.fd

VermaK.ChahalA.KumarV.NayakP.SinghP. (n.d.). *3 RD INTERNATIONAL CONFERENCE ON INNOVATIVE COMPUTING AND COMMUNICATION (ICICC-2020) Automated Order Management using Robotic Process Automation.* SSRN. https://ssrn.com/abstract=3563094

Vermeulen, I. E., & Seegers, D. (2009). Tried and tested: The impact of online hotel reviews on consumer consideration. *Tourism Management*, *30*(1), 123–127. doi:10.1016/j.tourman.2008.04.008

Vidas-Bubanja, M., & Bubanja, I. (2017). The importance of ICT for the competitiveness of tourism companies. *Tourism in Function of Development of the Republic of Serbia*, (pp. 470–489). TISC.

Vid, G. (Ed.). (2017). *Ne parázz! Információk az egyenlő esélyű hozzáférésről* [*Do not be afraid! Information on equal access*]. Camelot Mozgássérült Fiatalok Győri Egyesülete.

Vila, T. D., González, E. A., & Darcy, S. (2019). Accessible tourism online resources: A Northern European perspective. *Scandinavian Journal of Hospitality and Tourism*, *19*(2), 140–156. doi:10.1080/15022250.2018.1478325

Virtanen, J. K.,Mozaffarian, D..Chiuve, S. E. & Rimm, E.B. (2007). Fish consumption and risk of major chronic disease in men. *Am J Clin Nutr.* *88*(6), 1618–1625. . doi:10.3945/ajcn.2007.25816

Vörös, M., & Gačnik, A. (2018). Gastronomy tourism enterprises in the digital economy: Case study on fish soup as a tourism attraction – Comparing fish-dish cuisine of 'Gold Carp' Fishing Inn in Rétimajor and brewing tradition in South-Danube region. Edutus College, Budapest.

Vörös, M., & Gemma, M. (2011). Current Status, Future Trends & Real-life Cases from Japan. In M. Bourlakis, V. Zeimpekis, I. Vlachos, (eds.) Intelligent Agrifood Chains and Networks. Wiley-Blackwell Publishing.

Voros, M., & Gemma, M. (2015). Promotion of Local Agricultural and Food Products by Using Geographical Indications and Traditional Specialties Guaranteed Schemes in Hungary, In: *The Proceedings for the International Farm Management.* International Farm Management Association (IFMA), Laval University. http://ifmaonline.org/proceedings/20th-vol1/

Vučetić, Š. (2017). The importance of using ICT in the rural tourism of the Zadar county. *DIEM, 3*(1), 176–187. hrcak. srce.hr/187378

Vuković, P. (2017). Character and dynamics of development rural tourism in the Republic of Serbia, *Ekonomika, 63.* http://www.ekonomika.org.rs/sr/PDF/ekonomika/2017/Ekonomika-2017-4.pdf

Vuković, P., Simonović, Z., & Kljajić, N. (2016). Complementarity of Multifunctional Agriculture and Rural Development with Rural Tourism and Possibilities for their Implementation in the Republic of Serbia, *International Journal of Scientific and Technology Research,* 195-212.

Vuković, P., & Subić, J. (2016). Sustainable Tourism Development of Rural Areas in Serbia as a Precondition to Competitiveness. In *Global Perspectives on Trade Integration and Economies in Transition* (pp. 342–361). IGI Global. doi:10.4018/978-1-5225-0451-1.ch017

Wagner, Zs. (2010). *Fogyatékos fogadókészség [Impaired reception skills].* Turizmus. https://turizmus.com/migracio/fogyatekos-fogadokeszseg-1092077

Wahyuni, N. M., & Sara, I. M. (2020). The effect of entrepreneurial orientation variables on business performance in the SME industry context. *Journal of Workplace Learning, 32*(1), 35–62. doi:10.1108/JWL-03-2019-0033

Wang, F., Huang, S., Morrison, A. M., & Wu, B. (2022). The effects of virtual reality tourism involvement on place attachment and behavioral intentions: Virtual reality tourism of the Yellow Crane Tower in Wuhan. *Asia Pacific Journal of Tourism Research, 27*(3), 274–289. doi:10.1080/10941665.2022.2061363

Wang, W., Kumar, N., Chen, J., Gong, Z., Kong, X., Wei, W., & Gao, H. (2020). Realizing the potential of the internet of things for smart tourism with 5G and AI. *IEEE Network, 34*(6), 295–301. doi:10.1109/MNET.011.2000250

Webster, C. (2021). Demography as a driver of robotics. *ROBONOMICS: The Journal of the Automated Economy, 1,* 12–12.

Wei, W. (2019). Research progress on virtual reality (VR) and augmented reality (AR) in tourism and hospitality: A critical review of publications from 2000 to 2018. *Journal of Hospitality and Tourism Technology.*

Weithmann, M. (2011). The Danube: A European River as Street, Bridge and Frontier. An Introductory Essay. In: Tarrósy, I. – Milford, S. (eds.) EU Strategy for the Danube Region Perspectives for the future. Pécs, Hungary.

Westbrook, R. A. (1987). Product/Consumption-Based Affective Responses and Postpurchase Processes. *JMR, Journal of Marketing Research, 24*(3), 258–270. doi:10.1177/002224378702400302

WHO. (1980). *International classification of impairments, disabilities and handicap.* World Health Organization. https://apps.who.int/iris/bitstream/handle/10665/41003/9241541261_eng.pdf?sequence=1

WHO. (2001). *International classification of functioning, disability and health (ICF).* WHO. https://apps.who.int/iris/bitstream/handle/10665/42407/9241545429.pdf?sequence=1

WHO. (2007). *Global age-friendly cities guide.* World Health Organization.

WHO. (2011). *World report on disability.* World Health Organization.

WHO. (2021). *Disability*. WHO. https://www.who.int/news-room/fact-sheets/detail/disability

Widiastuti, I. A. M. S., Astawa, I. N. D., Mantra, I. B. N., & Susanti, P. H. (2021). The Roles of English in the Development of Tourism and Economy in Indonesia. *SOSHUM: Jurnal Sosial Dan Humaniora, 11*(3), 305–313. doi:10.31940oshum.v11i3.305-313

Williams, A. M. (2014). Tourism innovation products, processes, and people. The Wiley Blackwell companion to tourism (pp. 168-178). Wiley. doi:10.1002/9781118474648.ch13

Williams, A. M., & Shaw, G. (2011). Internationalization and innovation in tourism. *Annals of Tourism Research, 38*(1), 27–51. doi:10.1016/j.annals.2010.09.006

Winarsih, I. M., & Fuad, K. (2021). Impact of COVID-19 on digital transformation and sustainability in small and medium enterprises (SMEs): A conceptual framework. In: Barolli L., Poniszewska-Maranda A., Enokido T. (Eds), Complex, Intelligent and Software Intensive Systems. CISIS 2020. Advances in Intelligent Systems and Computing, (vol. 1194 pp. 471-476). Cham: Springer. https://doi.org/ doi:10.1007/978-3-030-50454-0_48

Winston, P. H. (1993). *Artificial Intelligence* (III. Edition). Addison-Wesley Publishing Company the computer intelligence. USA: Massachusetts.

Wirtz, J., Patterson, P. G., Kunz, W. H., Gruber, T., Lu, V. N., Paluch, S., & Martins, A. (2018). Brave new world: Service robots in the frontline. *Journal of Service Management, 29*(5), 907–931. doi:10.1108/JOSM-04-2018-0119

World Bank. (2022). *Number of arrivals—Morocco*. World Bank. https://data.worldbank.org/indicator/ST.INT.ARVL?locations=MA

WTO. (2007). *A practical guide to tourism destination management*. World Tourism Organization.

WTTC (World Travel and Tourism Council). (2019). *Economic Impact Reports*. WTTC. https://wttc.org/Research/Economic-Impact

Wuenderlich, N. V., & Paluch, S. (2017). *A nice and friendly chat with a bot: User perceptions of AI-based service agents*. Semantic Scholar.

Wu, H.-C., & Cheng, C.-C. (2018). Relationships between technology attachment, experiential relationship quality, experiential risk and experiential sharing intentions in a smart hotel. *Journal of Hospitality and Tourism Management, 37*, 42–58. doi:10.1016/j.jhtm.2018.09.003

Wüst, A., & Nicolai, K. (2022). Cultural diplomacy and the reconfiguration of soft power: Evidence from Morocco. *Mediterranean Politics*, 1–26. doi:10.1080/13629395.2022.2033513

Xiao, L., & Kumar, V. (2021). Robotics for customer service: A useful complement or an ultimate substitute? *Journal of Service Research, 24*(1), 9–29. doi:10.1177/1094670519878881

Xiao, Y., & Watson, M. (2019). Guidance on conducting a systematic literature review. *Journal of Planning Education and Research, 39*(1), 93–112. doi:10.1177/0739456X17723971

Xu, X. (2018). Research on the construction and development of smart hotels from the perspective of serving customers. In *2018 2nd International Conference on Education Science and Economic Management (ICESEM 2018)* (975-978). Atlantis Press. 10.2991/icesem-18.2018.228

Xue, Y., Fang, C., & Dong, Y. (2021). The impact of new relationship learning on artificial intelligence technology innovation. *International Journal of Innovation Studies, 5*(1), 2–8. doi:10.1016/j.ijis.2020.11.001

Xu, Y., Shieh, C. H., van Esch, P., & Ling, I. L. (2020). AI customer service: Task complexity, problem-solving ability, and usage intention. *Australasian Marketing Journal, 28*(4), 189–199. doi:10.1016/j.ausmj.2020.03.005

Yachin, J. M. (2019). The entrepreneur–opportunity nexus: Discovering the forces that promote product innovations in rural micro-tourism firms. *Scandinavian Journal of Hospitality and Tourism, 19*(1), 47–65. doi:10.1080/15022250.2017.1383936

Yasin, E. S., Abdelmaboud, A. E., Saad, H. E., & Qoura, O. E. (2021). Side and the light side of technostress related to hotel innovations: Transforming the hospitality industry or threat to human touch. *International Journal of Tourism. Archaeology and Hospitality, 1*(1), 44–59.

Yau, M. K., McKercher, B., & Packer, T. L. (2004). Traveling with a Disability: More than an Acces Issue. *Annals of Tourism Research, 31*(4), 946–960. doi:10.1016/j.annals.2004.03.007

Ye, Q., Law, R., & Gu, B. (2009). The impact of online user reviews on hotel room sales. *International Journal of Hospitality Management, 28*(1), 180–182. doi:10.1016/j.ijhm.2008.06.011

Ye, Q., Law, R., Gu, B., & Chen, W. (2011). The influence of user-generated content on traveler behavior: An empirical investigation on the effects of e-word-of-mouth to hotel online bookings. *Computers in Human Behavior, 27*(2), 634–639. doi:10.1016/j.chb.2010.04.014

Yin, R. K. (2009). Case study research. Design and methods (4. ed.). Thousand Oaks.

Ying, T., Tang, J., Ye, S., Tan, X., & Wei, W. (2022). Virtual reality in destination marketing: Telepresence, social presence, and tourists' visit intentions. *Journal of Travel Research, 61*(8), 1738–1756. doi:10.1177/00472875211047273

Yin, R. K. (2009). *Case study research: Design and methods.* Sage Publications.

Yoo, Y. J., Anthony, J., Saliba, A. J., & Prenzler, P. D. (2010). *Should Red Wine Be Considered a Functional Food? Comprehensive Reviews in FoodScience and FoodSafety* (Vol. 9). Institute of Food Technologists. doi:10.1111/j.1541-4337.2010.00125

Yu, C. E. (2020). Humanlike robots as employees in the hotel industry: Thematic content analysis of online reviews. *Journal of Hospitality Marketing & Management, 29*(1), 22–38. doi:10.1080/19368623.2019.1592733

Yu, C.-E., & Zhang, X. (2020). The embedded feelings in local gastronomy: A sentiment analysis of online reviews. *Journal of Hospitality and Tourism Technology, 11*(3), 461–478. doi:10.1108/JHTT-02-2019-0028

Żabińska, J. (2016). Cyfryzacja jako determinanta zmian w strukturze europejskiego sektora bankowego. *Zeszyty Naukowe Wydziału Zamiejscowego w Chorzowie Wyższej Szkoły Bankowej w Poznaniu.* Yadda. http://yadda.icm.edu.pl/yadda/element/bwmeta1.element.ekon-element-000171481860

Zaibaf, M., Taherikia, F., & Fakharian, M. (2013). Effect of perceived service quality on customer satisfaction in hospitality industry: Gronroos' service quality model development. *Journal of Hospitality Marketing & Management, 22*(5), 490–504. doi:10.1080/19368623.2012.670893

Zajadacz, A., & Lubarska, A. (2019). Development of a catalogue of criteria for assessing the accessibility of cultural heritage sites. *Studia Periegetica, 2*(26). doi:10.26349t.per.0026.06

Zapalska, A. M., & Brozik, D. (2015). The life-cycle growth and development model and leadership model to analyzing tourism female businesses in poland. *Problems and Perspectives in Management, 13*(2), 82–90. www.scopus.com

Zapalska, A. M., Brozik, D., & Zieser, N. (2015). Factors affecting success of small business enterprises in the polish tourism industry. *Tourism (Zagreb), 63*(3), 365–381.

Zdorov, A. B. (2009). Comprehensive development of tourism in the countryside. *Studies on Russian Economic Development, 20*(4), 453–455. doi:10.1134/S107570070904011X

Zenker, S., Braun, E., & Petersen, S. (2017). Branding the destination versus the place: The effects of brand complexity and identification for residents and visitors. *Tourism Management, 58*, 15–27. doi:10.1016/j.tourman.2016.10.008

Zhang, J. (2016). Weighing and realizing the environmental, economic and social goals of tourism development using an analytic network process-goal programming approach. *Journal of Cleaner Production, 127*, 262–273. doi:10.1016/j.jclepro.2016.03.131

Zhang, L., Li, N., & Liu, M. (2012). On the Basic Concept of Smarter Tourism and Its Theoretical System. *Luyou Xuekan, 27*(5), 66–73.

Zhang, T., Chen, J., & Hu, B. (2019). Authenticity, Quality, and Loyalty: Local Food andSustainable Tourism Experience. *Sustainability (Basel), 11*(2), 3437. doi:10.3390u11123437

Zhang, Z., Ye, Q., Law, R., & Li, Y. (2010). The impact of e-word-of-mouth on the online popularity of restaurants: A comparison of consumer reviews and editor reviews. *International Journal of Hospitality Management, 29*(4), 694–700. doi:10.1016/j.ijhm.2010.02.002

Zhao, F., Meng, T., Wang, W., Alam, F., & Zhang, B. (2023). Digital transformation and firm performance: Benefit from letting users participate. [JGIM]. *Journal of Global Information Management, 31*(1), 1–23. doi:10.4018/JGIM.322104

Zhao, X., Ke, Y., Zuo, J., Xiong, W., & Wu, P. (2020). Evaluation of sustainable transport research in 2000–2019. *Journal of Cleaner Production, 256*, 120404. doi:10.1016/j.jclepro.2020.120404

Zhao, X., Xia, Q., & Huang Wayne, W. (2020). Impact of technostress on productivity from the theoretical perspective of appraisal and coping processes. *Information & Management, 57*(8), 1–11. doi:10.1016/j.im.2020.103265

Zheng, C., Chen, Z., Zhang, Y., & Guo, Y. (2022). Does vivid imagination deter visitation? The role of mental imagery processing in virtual tourism on tourists' behavior. *Journal of Travel Research, 61*(7), 1528–1541. doi:10.1177/00472875211042671

Zhou, C., Yu, H., Ding, Y., Guo, F., & Gong, X.-J. (2017). Multi-scale encoding of amino acid sequences for predicting protein interactions using gradient boosting decision tree. *PLoS One, 12*(8), 1–18. doi:10.1371/journal.pone.0181426 PMID:28792503

Ziółkowska, M. (2020). Managers' decisions and strategic actions of enterprises in Poland in the face of digital transformation. [Economics and Law]. *Ekonomia I Prawo, 19*(4), 817–825. doi:10.12775/EiP.2020.053

Ziółkowska, M. J. (2021). Digital transformation and marketing activities in small and medium-sized enterprises. *Sustainability, 13*(5), 2512. doi:10.3390u13052512

Zivot, E., & Andrews, D. W. K. (1992). Further evidence of the great crash, the oil price shock and the unit root hypothesis. *Journal of Business & Economic Statistics, 10*, 251–270.

Zsarnóczky, M. (2017a). Accessible Accessible tourism in the European Union. In K. Borsekova, A. Vanova, & K. Vitálisowa (Eds.), *6th Central European Conference in Regional Science "Engines of Urban and Regional Development": Conference Proceedings.* (pp. 30-39). Banská Bystrica: Faculty of Economics, Matej Bel University.

Zsarnóczky, M. (2017b). Developing Senior Tourism in Europe. *Pannon Management Review, 6*(3-4), 201–214.

About the Contributors

Gonçalo José Poeta Fernandes is a Professor at Polytechnic Institute of Guarda – School of Tourism and Hospitality Management/Dept. Tourism and Hotel Industry. PhD in Geography and Regional Planning, specializing in Tourism and Territorial Management, Faculty of Social Sciences and Humanities, New University of Lisbon. Research Areas: Tourism and Development, Planning and Territorial Management, Public Policy and Border Areas, Rural Entrepreneurship, Territory and Tourist Destinations, Mountains and geotourism. Researcher at Centre for Tourism Research, Development and Innovation – CITUR & Interdisciplinary Centre of Social Sciences - CICS:NOVA. Member of the Scientific Council of the Douro International Natural Park and Strategic Council of the Serra da Estrela Natural Park. Member of the Scientific Committee Estrela UNESCO Global Geopark, Member of the Executive Board of the National Mountain Research Network of FCT and Scientific Council of the Iberian Mountain Research Network (RIIM). Publisher of books in the area of Tourism and Heritage with publications in several international books and journals in the area of tourism, destination and heritage management.

António Melo, Professor at the School of Hospitality and Tourism, Polytechnic Institute of Porto, member of CiTUR - Centre for Tourism Research, Development and Innovation and collaborating member of CIDTFF -Research Centre on Didactics and Technology in the Education of Trainers. His educational background includes a graduate degree in Hotel Management, a master's degree in Management and PhD in Education. Antonio has more than 30 years of experience in practice of hospitality management and almost 20 years in academia. Currently, he is coordinating the Master Degree in Hospitality Management and in parallel, he also carries out research in the area of tourism, hospitality, gastronomy and education practices.

* * *

Nesenur Altinigne is an Assistant Professor at the Marketing Department of Istanbul Bilgi University. She is a Ph.D. graduate of Istanbul Technical University. She has visited SKEMA Business School (France) as a visiting scholar supported by the French Embassy Research Fellowship. Her research interests mainly focus on consumer-technology interactions, consumer ethics, brand relationships, and consumers' reactions to product harm/firm crises. Her work has been published in the Journal of Business Research, Journal of Retailing and Consumer Services, Journal of Product & Brand Management and Journal of Vacation Marketing, and presented at several leading international conferences.

Mahmut Bakır graduated from the Department of Civil Air Transport Management, Mustafa Kemal University, Turkey. He obtained his master's and PhD degrees in Civil Aviation Management Programme from the Graduate School of Social Sciences, Anadolu University. He is currently working as a faculty member at the Department of Aviation Management, Samsun University. His primary research interests include airline business, service marketing, consumer behavior, multi-attribute decision analysis, and multivariate statistics.

Hüseyin Baran is an academic staff at Duzce University, Arts and Architecture Faculty, visual communication. Additionally, he holds the position of Art Director in the field of visual communication at Duzce University. Conducts research on a wide range of topics; visual designs, 3D drawing techniques, VR, and new-generation technologies.

Züleyhan Baran is an academic staff at Duzce University, Akcakoca Campus, Tourism and Hotel Management Department. Additionally, Erasmus Coordinator and Head of the Department. Conducts research on a wide range of topics; gastronomy, tourism, QFD, halal food, and next-gen technologies (VR, AR, MR, XR, etc.).

Zélia Breda holds a PhD in Tourism, a Master's degree in Chinese Studies (Business and International Relations) and a degree in Tourism Management and Planning from the University of Aveiro, where she is an Assistant Professor, at the Department of Economics, Management, Industrial Engineering, and Tourism, holding the position of Vice-Director of the Master in Tourism Management and Planning. She is a full member of the Research Unit' Governance, Competitiveness and Public Policies' of the University of Aveiro; a founding member and vice-president of the Observatory of China (), and a senior researcher of the spin-off company Idtour (). Furthermore, Zélia Breda actively contributes to the academic community by serving on the editorial and scientific boards of several national and international journals. She also participates in organizing and scientific committees for various international tourism conferences. She has authored and co-authored several national and international papers and communications on tourism, and has been taking part in several research projects in the tourism field, both as a member of the team and as a consultant.

Onur Çelen graduated from tourism management with a bachelor's degree and master's degree in tourism management. He has PhD degree in tourism management at Ankara Hacı Bayram Veli University (Ankara, Turkey) in 2022. He has been working as a lecturer at Bursa Uludağ University (Bursa, Turkey) since 2017. His research areas include tourism management, tourism marketing and sustainable tourism. He continues to work on articles and book chapters and conference presentations in national and international fields.

Arokiaraj David serves as an Associate Professor & Academic Coordinator of SFIMAR-PGDM in Mumbai. He has successfully cleared the UGC-National Eligibility Test for Lectureship (NET) and has been honored with the Junior and Senior Research Fellow (JRF & SRF) titles. With over a decade of teaching and research experience, his areas of expertise include Green Marketing, Consumer Behavior, Product & Brand Management, Global Marketing, Retail Management, Strategic Management, Data Analytics, Sustainable Practices, Resources Management, Product Development, Environmental Responsibility, and General Management. Dr. David has authored more than 60 research articles, including 11 published in

Scopus, 08 in ABDC, and 05 in Web of Science. He has also contributed to 07 book chapters, published 04 books, registered 07 patents, and received 01 copyright and 01 consultant project. He is proficient in various research tools and techniques such as SPSS, AMOS, E-Views, STATA, and PLS-SEM and has hands-on experience in both primary and secondary data analysis. Recently, Dr. David has been recognized with the Best Research Award, Young Academician Award, and Young Scientist Award.

Nilgün Demirel İli graduated from tourist guiding with a bachelor's degree and has master's degree in tourism management at Nevsehir Haci Bektas Veli University. She has PhD degree in tourism management in the same university in 2019. Her research areas include tourism marketing, sustainable tourism, customer relations and complaint management as well as tourist guiding. She continues to work on articles and book chapters and conference presentations in national and international fields.

Maria Gorete Dinis holds a Ph.D. in Tourism, an MA in Innovation, Planning and Development, and a BSc in Tourism Management and Planning from the University of Aveiro. She is coordinator and professor of the BSc in Tourism at the Polytechnic Institute of Portalegre (PIP) / School of Education and Social Sciences. She is a full member of the Research Unit 'Governance, Competitiveness, and Public Policies' and collaborating member of Centre for Tourism Research, Development and Innovation (CITUR).She is also a member of the editorial and scientific boards of international journals, as well as a member of the scientific committees of international tourism conferences. During recent years, she has published several scientific articles and presented communications in areas such as management of tourist destinations, tourism indicators, ICT applied to tourism and Big data in tourism.

Yassine El Bouchikhi, as an Assistant Professor in Marketing within the Al Akhawayn University in Ifrane, Morocco, has been involved in Marketing practice for over 15 years both as a professional and as an academic, in the USA, Europe, and North Africa. Yassine received his Ph.D. in Marketing from the Paris Dauphine University in 2019. Before joining Paris Dauphine's Ph.D. program, he held positions in the Marketing and information technology departments where he conducted several internal research and development projects. In parallel with his PhD thesis, he was a post-doctoral fellow at ESSEC Business School, where he taught advanced Marketing engineering methods in the Master of Management Science program (Ranked 2nd in the world by Financial Times 2022). He has also lectured in the United States as a Visiting Assistant Professor of Marketing for the Maine Business School at the University of Maine. Yassine intervened regularly in various institutions in France like Neoma Business School, ISG, Christian university of Lyon in the undergraduate and graduate programs.

Aleš Gačnik, PhD, is an ethnologist and sociologist of culture. He is an Assistant Professor and a researcher at the Faculty of Tourism Studies – Turistica, University of Primorska. From 2017 to 2017 he was a head of the Department of Cultural Tourism and Cultural Heritage and from 2017 onwards he is a deputy head of the Department of Cultural Tourism, joined in the UNESCO UNITWIN Network "Culture, Tourism, Development" (Sorbona, Paris).He is the founder and a head of the Centre of gastronomy and wine culture of the University of Primorska. In his research he deals with issues in heritage / cultural tourism and cultural heritage, wine / culinary / gastronomic tourism, tourism development in rural areas. museums / museology and tourism, culture of masks and tourism… He was the editor or co-editor of several scientific, professional and artistic monographs and the author or co-author of numerous scientific articles and articles published in domestic and foreign scientific journals / publishing houses.

He was also a head or member of the steering committees of several domestic and international scientific conferences, summer schools, research camps. He is a professional member of the International Institute of Gastronomy, Culture, Art in Tourism (Barcelona, Spain) and member of the Expert Group of the The Slovenia Restaurant Award. Receiver of several national and international awards and recognitions.

Jeganathan Gomathi Sankar is a dedicated educationist who specializes in the field of marketing. He is a member of the BSSS Institute of Advanced Studies, where he teaches courses on various marketing subjects. Dr. Sankar has previously served as a full-time research scholar in the Department of International Business at Pondicherry University, as well as a faculty member at the Saveetha School of Management. Dr. Sankar's research interests are varied and include areas such as service quality, information diffusion, ICT, marketing technology, and AI. He has published his work in well-regarded journals that are indexed in Scopus and ABDC. Through his teaching and research, Dr. Sankar strives to contribute to the field of marketing and provide his students with a comprehensive understanding of the subject matter.

Alexandra Gonçalves is an Assistant Professor at the University of Algarve, School of Management, Hospitality and Tourism. Integrated Researcher of the Research Centre for Tourism, Sustainability and Well-being (CinTurs) for the areas of Tourism, Sustainability, Territory, Heritage, Museums, Cultural Management and Creative Industries. PhD Degree in Tourism. Master's Degree in Cultural Heritage Management. Post-graduate degree in Cultural Heritage Law. Graduate in Marketing. Regional Director of Culture of the Algarve from December 16, 2013 to December 15, 2018; councilor of the Municipality of Faro (Oct. 2009-Oct. 2013). Since October 30th, 2019, is the Director of the School of Management, Hospitality and Tourism (University of Algarve). Has published research in the areas of tourism, cultural experiences, cultural and creative tourism, cultural heritage management and museums, events evaluation. Was the regional researcher responsible for the project funded by the Portuguese Science and Technology Foundation -CREATOUR-Creative Tourism Destination Development in Small Cities and Rural Areas (2016-2020). Currently is the UALG coordinator for the iHERITAGE project, ENI CBC MED Program (2020-2023).

Borbála Gondos (1982, Budapest) is a sociologist-economist, she works as an associate professor and Head of Tourism Department at Edutus University, Budapest (Hungary). She doctorated (PhD) in Regional and management sciences in 2020, she wrote her PhD work about „Special needs in tourism – Role and opportunities of people with reduced mobility in tourism sector". Her research fields are accessible tourism, people with disabilities, tourism and quality of life, volunteers at sport events.

Ozgur Guler is a statistical analysis expert, business analyst and also enterprise resource planning system consultant in a company. He holds MSc degree in Industrial Engineering in the Department of Industrial Engineering at Pamukkale University. His research interests include machine learning, simulation, mathematical modelling, enterprise resource planning programs, optimization and information systems.

Sukran Karaca is an academic staff at Sivas Cumhuriyet University, Faculty of Tourism, Tourism Management. She received her associate professorship in Marketing in 2019. Conducts research on a wide range of topics; Marketing, Consumer Behavior, Sustainable Marketing, and Digital Marketing.

Sameera Khan is B.TECH (CSE), M.TECH (CSE). She has 12+ years of teaching and research experience, currently working at Vardhaman College of Engineering, Hyderabad as Assistant Professor in IT. She is currently pursuing PhD. Her research interests include biometric technology, Machine learning, artificial intelligence, digital image processing and Robotic Process automation. She has published 30+ research papers in various international journals and conference proceedings. She has also published three book chapters published under Taylor and Francis publication and IGI-global publication. She is trained Robotics Process Automation(RPA) professional licensed from Automation Anywhere Inc. She is a certified Trainer on RPA licensed from Automation Anywhere Inc. and UiPath. She has conducted various RPA training sessions for students and professionals. She has delivered Key Note Addresses, Guest Talks and Chaired the Sessions in International/National Conferences, Workshops, and Technical Events. She has conducted various training sessions on developing computer programming skills in various languages like C, Java (Core & advanced), Python, MatLab etc. She has also worked for various Industrial Collaborated activities, training programmes under "Industry Institute Interaction" in association with MNCs like Texas Instruments, Automation Anywhere Inc. and UiPath, etc.

Marija Mosurović Ružičić, PhD, started career at the Ministry of Science, then at Mihailo Pupin Institute, and now at the Institute of Economic Sciences in Innovation Economics department as Research Associate. She obtained Ph.D in Economics science and Technical science as well. She has carried out research in the field of strategic management, R&D management, organizational theory, management and business, industrial policy development, EU integration, and science and technology policy research. She published the research results in more than 30 scientific and professional papers including journals, monographs, and conference proceedings, both national and international. Since 2005. she has been involved in the projects funded by the national Ministry of Science but also, in different EU-funded projects. Participated in more than 10 research projects funded by European Commission.

Kamel Mouloudj has examined the relationship between constructs of theory of planned behavior and intentions to implement green practices and tourist's intention to stay in green hotels; green food purchase intentions; intention to use online food delivery services; and students' intention to use online learning system. More broadly, the concepts of word-of-mouth, online learning, green hotel and green food. More recently, he has focused on the Impacts Of COVID-19 Pandemic.

Hande Mutlu Öztürk was born in Ordu on 14.06.1982. She completed her primary education in France and secondary and high school education in Istanbul, Turkey. In 2005, she graduated from Pamukkale University Food Engineering Department. In 2007, she completed her master's degree at Pamukkale University, Institute of Science. In 2011, she completed her PhD at Ege University, Institute of Science and Technology. Between 2009-2013, she worked as a lecturer in Pamukkale University, School of Tourism and Hotel Management, Food and Beverage Management Department. In 2013, she started to work as Assistant Professor at Pamukkale University, Faculty of Tourism, Department of Gastronomy and Culinary Arts. She speaks fluent French and good English.

Márta Nárai (1970, Sárvár) is a sociologist, she works as an associate professor at the Széchenyi István University, Győr (Hungary). She is program leader of the Community and Civil Development Studies MA at the University. She doctorated (PhD) in Sociology in 2009, she wrote her PhD work about the position and roles of the non-profit organisations. She has been the chief editor of the Journal

Civil Rewiev since 2014. Her research fields are the non-profit organisations, social engagement and responsibility, social inequalities, youth sociology, family sociology, settlement sociology.

Olcay Polat received his Industrial Engineering PhD degree from the Technical University of Berlin (Germany) with the scholarship of German Academic Exchange Service (DAAD) in 2013. He has been working as an Associate Professor in the Department of Industrial Engineering at Pamukkale University (Turkey). He has a special interest in supply chain management, logistics and transportation optimization, production management and risk management. He has been published more that 100 papers in respected journals, conference proceedings and books.

Célia M. Q. Ramos is graduated in Computer Engineering from the University of Coimbra, obtained her Master in Electrical and Computers Engineering from the Higher Technical Institute, Lisbon University, and the PhD in Econometrics in the University of the Algarve (UALG), Faculty of Economics, Portugal. Holds a Professional Diploma in Digital Marketing from Digital Marketing Institute in Dublin, Ireland. She is Coordinator Professor at School for Management, Hospitality and Tourism, also in the UALG, where she lectures computer science, mainly Information Systems, Business Intelligence, Information Systems for Management Support Decision. Currently, Celia was President of the Pedagogical Council, since 2017 until 2021, Director of the degree in Marketing and member of the coordination of the master's degree in Hospitality Management, from September 2015 to October 2019, of the School of Management, Hospitality and Tourism. Areas of research and special interest include conception and development of information systems, tourism information systems, big data, business intelligence, electronic tourism, websites evaluation, marketing digital, econometric modeling and panel-data models. Célia Ramos has published in the fields of information systems, marketing and tourism, namely, she has authored a book, book chapters, conference papers and journal articles. She is reviewer of several scientific journals. At the level of applied research, she has participated in several funded projects, such as: Mobile Five Senses Augmented Reality System for Museums (M5SAR) project funded by Portugal2020, CRESC Algarve 2020 I&DT (n° 3322); SRM: Smart Revenue Management, QREN I&DT, n.° 38962; FootData: Integrated information system for football, QREN I&DT, n.° 23119; and IMPACTUR: Indicators for the Monitoring and Forecast of Tourism Activity, protocol with the Portuguese Tourism Institute.

Albérico Travassos Rosário is a Ph.D. Marketing and Strategy of the Universities of Aveiro (UA), Minho (UM) and Beira Interior (UBI). With affiliation to the GOVCOPP research center of the University of Aveiro. Master in Marketing and Degree in Marketing, Advertising and Public Relations, degree from ISLA Campus Lisbon-European University | Laureate International Universities. Has the title of Marketing Specialist and teaches with the category of Assistant Professor at IADE-Faculty of Design, Technology and Communication of the European University and as a visiting Associate Professor at the Santarém Higher School of Management and Technology (ESGTS) of the Polytechnic Institute of Santarém. He taught at IPAM-School of Marketing | Laureate International Universities, ISLA- Higher Institute of Management and Administration of Santarém (ISLA-Santarém), was Director of the Commercial Management Course, Director of the Professional Technical Course (TeSP) of Sales and Commercial Management, Chairman of the Pedagogical Council and Member of the Technical Council and ISLA-Santarém Scientific Researcher. He is also a marketing and strategy consultant for SMEs.

Ali Emre Sarilgan is a distinguished academician and researcher in the field of Aviation Management. He completed his undergraduate studies at the School of Civil Aviation at Anadolu University in 1998, followed by a master's degree in Civil Aviation Management from the same institution in 2001. He later earned his Ph.D. from Anadolu University in 2007, with a research focus on regional air transport. Dr. Sarilgan is currently a faculty member at Eskişehir Technical University, where he serves as a member of the Department of Aviation Management. He has made significant contributions to the field of aviation management and marketing through his research, writing, and teaching. His scholarly work has been widely published in many academic journals and has been presented at national and international conferences.

Dileep Singh is Ph.D., UGC NET, MBA, M.COM, PGDIBO, BCS, BA (ELT), and BEd. He has significant years of academic and Industry experience. He is Currently Working as an Assistant Professor at SVKM's NMIMS, HYDERABAD. He is a certified soft skill trainer and has special expertise in 50 topics of life skills. He has conducted many guest lectures in Nagpur, Amravati and Pune. He has written exceptional Articles in Hitavada. He has exposure of Future Group, Landmark Pvt Ltd, Volga International, IMRB and Hyundai Motors. He has presented many Papers in Scopus, UGC Journals, National and International Conferences. He has authored books Information Technology for Managers and Industrial Relation and labour law.

Carina Viegas holds a Master's degree in Marketing Management (UAlg- University of the Algarve, 2022) a Post-Graduate degree in Travel Journalism (UAB- Autonomous University of Barcelona, 2012) and a Graduation in Tourism (UAlg- University of the Algarve, 2011). Researcher in the emergent topic of Sensory Marketing, with a dissertation about the issue and, currently working on research about innovation in the Mediterranean diet, component of the HostLab´s Project. Specialist in market research for consumer behaviour, having contributed to international consultancy companies such as Nielsen and Euromonitor, developing strategic solutions for the main FMCG big players in the market.

Mihály Vörös obtained his PhD and Candidate of Science (C.Sc.) degree from Hungarian Academy of Sciences (HAS) in 1994 as well as Dr. Habil degree from Debrecen University in 2000. Former Council Member of the International Farm Management Association (IFMA) for Eastern Europe, Reading, UK (1988 – 2013). Research scholar and visiting professor in Japan (2008-2009) supported by the Japan Foundation. Implemented ERASMUS Teaching Exchange Mobility Grant Programs in Perugia University Italy between 2012 – 2015 years and in University of Primorska, Portoroz Slovenia in 2017. Elected as professor and research supervisor of Hungarian University of Agriculture and Life Sciences (MATE), Economic and Regional Science Doctorate School, Gödöllő, Hungary in 2016 . Currently he is professor emeritus in the Department of Tourism of Edutus University and head of research for the Local Food Chains Innovation HÉLIA Working Group, a "spin-off" university organization established in 2011 operating also a Scientific Student College. Author and co-author of books, book chapters, several journal articles and scientific conference proceedings.

Index

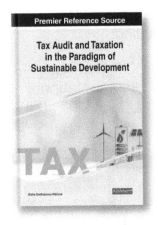

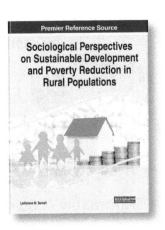

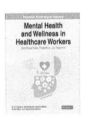

Printed in the USA
CPSIA information can be obtained
at www.ICGtesting.com
LVHW011722121023
760831LV00030B/344

9 781668 469859